Plasma Medicine

Plasma Medicine

Alexander Fridman and Gary Friedman
Drexel University, Philadelphia, PA, USA

A John Wiley & Sons, Ltd., Publication

Library of Congress Cataloging-in-Publication Data

Fridman, Alexander A., 1953–
    Plasma medicine / Alexander Fridman, Gary Friedman.
        p. ; cm.
    Includes index.
    ISBN 978-0-470-68970-7 (cloth) – ISBN 978-0-470-68969-1 (paper)
    I. Friedman, Gary (Gary G.) II. Title.
    [DNLM: 1. Biomedical Engineering.   2. Plasma Gases–therapeutic use.   3. Plasma Gases–pharmacology.   4. Wound Healing.   QT 36]
    610.28–dc23

                                                                                              2012025754

A catalogue record for this book is available from the British Library.

Print ISBN: Cloth: 9780470689707 Paper: 9780470689691

Typeset in 10/12pt Times by Aptara Inc., New Delhi, India
Printed and bound in Singapore by Markono Print Media Pte Ltd

Contents

Preface   xv

Acknowledgements   xvii

1   Introduction to Fundamental and Applied Aspects of Plasma Medicine   **1**
   1.1   Plasma medicine as a novel branch of medical technology   1
   1.2   Why plasma can be a useful tool in medicine   4
   1.3   Natural and man-made, completely and weakly ionized plasmas   5
   1.4   Plasma as a non-equilibrium multi-temperature system   7
   1.5   Gas discharges as plasma sources for biology and medicine   9
   1.6   Plasma chemistry as the fundamental basis of plasma medicine   13
   1.7   Non-thermal plasma interaction with cells and living tissues   14
   1.8   Applied plasma medicine   15

**2   Fundamentals of Plasma Physics and Plasma Chemistry for Biological
and Medical Applications**   **19**
   2.1   Elementary plasma generation processes   19
      2.1.1   Classification of ionization processes   19
      2.1.2   Direct and stepwise ionization by electron impact   21
      2.1.3   Other ionization mechanisms in plasma-medical discharges   23
      2.1.4   Mechanisms of electron-ion recombination in plasma   24
      2.1.5   Elementary processes of negative ions in plasma-medical systems: Electron
attachment and detachment   26
      2.1.6   Heterogeneous ionization processes, electron emission mechanisms   29
   2.2   Excited species in plasma medicine: Excitation, relaxation and dissociation of neutral
particles in plasma   31
      2.2.1   Vibrational excitation of molecules by electron impact   31
      2.2.2   Rotational and electronic excitation of molecules by electron impact   33
      2.2.3   Plasma generation of atoms and radicals: Dissociation of molecules
by electron impact   34
      2.2.4   Distribution of non-thermal plasma discharge energy between different
channels of excitation of atoms and molecules   35
      2.2.5   Relaxation of plasma-excited species: Vibrational-translational (VT) relaxation   36
      2.2.6   Vibrational energy transfer between molecules: VV-relaxation   37
      2.2.7   Relaxation of rotational and electronic excitation in plasma-medical systems   38
   2.3   Elementary plasma-chemical reactions of excited neutrals and ions   39
      2.3.1   Rate coefficient of reactions of excited molecules in plasma   39
      2.3.2   Efficiency α of excitation energy in overcoming activation energy of chemical
reactions of plasma-generated active species   40

		2.3.3	Fridman-Macheret α-model for chemical reactions of plasma-generated active species	40
		2.3.4	Ion-molecular polarization collisions and reactions in plasma	43
		2.3.5	Ion-molecular chemical reactions of positive and negative ions	45
	2.4		Plasma statistics, thermodynamics, and transfer processes	46
		2.4.1	Statistical distributions in plasma: Boltzmann distribution function, Saha equation, and dissociation equilibrium	46
		2.4.2	Complete and local thermodynamic equilibrium in plasma	48
		2.4.3	Thermodynamic functions of quasi-equilibrium thermal plasma systems	49
		2.4.4	Non-equilibrium statistics of thermal and non-thermal plasmas	50
		2.4.5	Non-equilibrium statistics of vibrationally excited molecules in plasma and Treanor distribution	51
		2.4.6	Transfer processes in plasma: Electron and ion drift	52
		2.4.7	Transfer processes in plasma: Diffusion of electrons and ions	54
		2.4.8	Transfer processes in plasma: Thermal conductivity	55
		2.4.9	Radiation energy transfer in plasma	56
	2.5		Plasma kinetics: Energy distribution functions of electrons and excited atoms and molecules	58
		2.5.1	Electron energy distribution functions (EEDF) in plasma: Fokker-Planck kinetic equation	58
		2.5.2	Relation between electron temperature and the reduced electric field	60
		2.5.3	Non-equilibrium vibrational distribution functions of plasma-excited molecules, Fokker-Plank kinetic equation	61
		2.5.4	Vibrational distribution functions of excited molecules in plasma controlled by VV-exchange and VT-relaxation processes	64
		2.5.5	Kinetics of population of electronically excited states in plasma	65
		2.5.6	Macrokinetics of chemical reactions of vibrationally excited molecules	66
	2.6		Plasma electrodynamics	68
		2.6.1	Ideal and non-ideal plasmas	68
		2.6.2	Plasma polarization, Debye shielding of electric field in plasma	69
		2.6.3	Plasmas and sheaths	69
		2.6.4	Electrostatic plasma oscillations, Langmuir plasma frequency	72
		2.6.5	Penetration of slow-changing fields into plasma: Skin-effect in plasma	72
		2.6.6	Plasma magneto-hydrodynamics: Generalized Ohm's law, Alfven velocity and magnetic Reynolds' number	73
		2.6.7	High-frequency conductivity and dielectric permittivity of plasma	75
		2.6.8	Propagation of electromagnetic waves in plasma	77
		2.6.9	Plasma absorption and reflection of electromagnetic waves: Bouguer law and critical electron density	78
3	**Selected Concepts in Biology and Medicine for Physical Scientists**			**81**
	3.1		Molecular basis of life: Organic molecules primer	81
		3.1.1	Essential primer on bonds and organic molecules	81
		3.1.2	Main classes of organic molecules in living systems	84
	3.2		Function and classification of living forms	96
		3.2.1	What is life: Functionality of living forms	96
		3.2.2	Major classification of living forms	96

3.3		Cells: Organization and functions	97
	3.3.1	Primary cell components	98
	3.3.2	Transport across cell membranes	105
	3.3.3	Cell cycle and cell division	109
	3.3.4	Cellular metabolism	110
	3.3.5	Reactive species in cells and living organisms	115
3.4		Overview of anatomy and physiology	124
	3.4.1	Tissues	124
	3.4.2	The body covering: The integumentary system	130
	3.4.3	Circulatory system	132
	3.4.4	Immune system	139
	3.4.5	Digestive system	144
	3.4.6	The nervous system	149
	3.4.7	The endocrine system	155
	3.4.8	The muscular and skeletal systems	157
	3.4.9	The respiratory system	160
	3.4.10	The excretory system	163

4 Major Plasma Disharges and their Applicability for Plasma Medicine — **165**

4.1		Electric breakdown and steady-state regimes of non-equilibrium plasma discharges	165
	4.1.1	Townsend mechanism of electric breakdown, Paschen curves	165
	4.1.2	Streamer or spark breakdown mechanism	168
	4.1.3	Meek criterion of streamer formation and streamer propagation models	171
	4.1.4	Streamers and microdischarges	172
	4.1.5	Interaction of streamers and microdischarges	173
	4.1.6	Steady-state regimes of non-equilibrium electric discharges	174
	4.1.7	Discharge regime controlled by electron-ion recombination	175
	4.1.8	Discharge regime controlled by electron attachment	175
	4.1.9	Non-thermal discharge regime controlled by diffusion of charged particles to the walls	176
4.2		Glow discharge and its application to biology and medicine	177
	4.2.1	Glow discharge structure	177
	4.2.2	Current-voltage characteristics of DC discharges	180
	4.2.3	Townsend dark discharge	180
	4.2.4	Current-voltage characteristics of the cathode layer	182
	4.2.5	Abnormal, subnormal and obstructed regimes of glow discharges	184
	4.2.6	Positive column of glow discharge	186
	4.2.7	Atmospheric pressure glow discharges and applications in plasma medicine	186
	4.2.8	Resistive barrier discharge (RBD) as modification of APG discharges	188
	4.2.9	Atmospheric pressure micro glow-discharges	188
	4.2.10	Hollow-cathode glow discharge and hollow-cathode APG microplasma	191
4.3		Arc discharge and its medical applications	191
	4.3.1	Major types of arc discharges	191
	4.3.2	Cathode and anode layers of arc discharges and spots	193
	4.3.3	Positive column of high-pressure arcs	196
	4.3.4	Steenbeck – Raizer 'channel' model of positive column of arc discharges	198

	4.3.5	Configurations of arc discharges and their applicability to plasma medicine	199
	4.3.6	Gliding arc discharge as a powerful source of non-equilibrium plasma	204
4.4	Radio-frequency and microwave discharges in plasma medicine	208
	4.4.1	Generation of thermal plasma in radio-frequency discharges	208
	4.4.2	Atmospheric-pressure microwave discharges and their biomedical applications	210
	4.4.3	Non-thermal RF discharges: CCP and ICP coupling	213
	4.4.4	Non-thermal RF-CCP discharges at moderate pressure regime	215
	4.4.5	Low-pressure CCP RF discharges	217
	4.4.6	Low-pressure RF magnetron discharges for surface treatment	219
	4.4.7	Low-pressure non-thermal ICP RF discharges in cylindrical coil	219
	4.4.8	Planar-coil and other configurations of low-pressure non-thermal RF-ICP discharges	222
	4.4.9	Non-thermal RF atmospheric pressure plasma jets as surface-treatment device	224
	4.4.10	Atmospheric-pressure non-thermal RF plasma microdischarges: Plasma needle	226
	4.4.11	Non-thermal low-pressure microwave and other wave-heated discharges	227
	4.4.12	Non-equilibrium microwave discharges of moderate and elevated pressures: Energy-efficient plasma source of chemically active species	229
4.5	Coronas, DBDs, plasma jets, sparks and other non-thermal atmospheric-pressure streamer discharges	232
	4.5.1	Corona and pulsed corona discharges	232
	4.5.2	Dielectric-barrier discharges (DBDs)	234
	4.5.3	Special modifications of DBD: Surface, asymmetric, packed bed and ferroelectric discharges	235
	4.5.4	OAUGDP as quasi-homogeneous DBD (APG) modification	237
	4.5.5	Electronically stabilized DBD in APG discharge mode	238
	4.5.6	Arrays of DBD-based microdischarges and kilohertz-frequency microdischarges	239
	4.5.7	Floating-electrode dielectric barrier discharge (FE-DBD)	241
	4.5.8	Micro- and nanosecond pulsed uniform FE-DBD plasma	243
	4.5.9	Spark discharges	243
	4.5.10	Pin-to-hole spark discharge (PHD), thermal microplasma source-generating ROS and NO for medical applications	247
	4.5.11	Atmospheric-pressure cold helium microplasma jets and plasma bullets	250
	4.5.12	Propagation of plasma bullets in long dielectric tubes and splitting and mixing of plasma bullets	252
4.6	Discharges in liquids	253
	4.6.1	General features of electrical discharges in liquids in relation to their biomedical applications	253
	4.6.2	Mechanisms and characteristics of plasma discharges in water	254
	4.6.3	Physical kinetics of water breakdown: Thermal breakdown mechanism	254
	4.6.4	Non-thermal short pulse electrostatic (electrostriction) water breakdown	256
	4.6.5	Nanosecond pulse breakdown and plasma generation in liquid water without bubbles	261
	4.6.6	Nanosecond-pulse uniform cold plasma in liquid water without bubbles: Analysis and perspectives for biomedical applications	263

5 Mechanisms of Plasma Interactions with Cells   **269**
  5.1   Main interaction stages and key players   269
  5.2   Role of plasma electrons and ions   272
        5.2.1   Selection of biological targets and plasma generation methods   272
        5.2.2   Comparison of direct DBD plasma treatment to indirect treatment with and
                without ion flux   272
        5.2.3   Effect of gas composition on antibacterial efficacy of direct DBD   274
        5.2.4   Effect of positive and negative ions in nitrogen corona discharge   276
  5.3   Role of UV, hydrogen peroxide, ozone and water   276
        5.3.1   Effect of UV in DBD treatment   276
        5.3.2   Effect of hydrogen peroxide   277
        5.3.3   Effect of ozone   279
        5.3.4   Effects of water and its amount   280
  5.4   Biological mechanisms of plasma interaction for mammalian cells   281
        5.4.1   Intracellular ROS as key mediators of plasma interaction
                with mammalian cells   281
        5.4.2   DNA damage and repair as a consequence of DBD plasma treatment   283
        5.4.3   Effect of the cell medium in plasma interaction with mammalian cells   286
        5.4.4   Crossing the cell membrane   289

6 Plasma Sterilization of Different Surfaces and Living Tissues   **293**
  6.1   Non-thermal plasma surface sterilization at low pressures   293
        6.1.1   Direct application of low-pressure plasma for biological sterilization   293
        6.1.2   Effect of low-pressure plasma afterglow on bacteria deactivation   294
  6.2   Surface microorganism inactivation by non-equilibrium high-pressure plasma   295
        6.2.1   Features of atmospheric-pressure air plasma sterilization   295
        6.2.2   Kinetics of atmospheric-pressure plasma sterilization   295
        6.2.3   Cold plasma inactivation of spores: Bacillus cereus and Bacillus
                anthracis (anthrax)   297
        6.2.4   Atmospheric-pressure air DBD plasma inactivation of Bacillus cereus and
                Bacillus anthracis spores   298
        6.2.5   Decontamination of surfaces from extremophile organisms using non-thermal
                atmospheric-pressure plasma   301
        6.2.6   Plasma sterilization of contaminated surgical instruments: Prion proteins 303
  6.3   Plasma species and factors active for sterilization   304
        6.3.1   Direct effect of charged particles in plasma sterilization   304
        6.3.2   Biochemical effect of plasma-generated electrons in plasma sterilization   304
        6.3.3   Bio-chemical effect of plasma-generated negative and positive ions   306
        6.3.4   Sterilization effect of ion bombardment   307
        6.3.5   Sterilization effect of electric fields related to charged plasma particles 307
        6.3.6   Effect of plasma-generated active neutrals: ROS and RNS   308
        6.3.7   Effect of plasma-generated active neutrals: OH-radicals and ozone   309
        6.3.8   Effects of plasma-generated active neutrals: Hydrogen peroxide (H_2O_2) 311
        6.3.9   Contribution of plasma-generated heat and temperature to plasma sterilization 311
        6.3.10 Effect of UV radiation   312

	6.4	Physical and biochemical effects of atmospheric-pressure air plasma on microorganisms	313
		6.4.1 Direct and indirect effects of non-thermal plasma on bacteria	313
		6.4.2 FE-DBD experiments demonstrating higher effectiveness of direct plasma treatment	315
		6.4.3 Surface versus penetrative plasma sterilization	317
		6.4.4 Apoptosis versus necrosis	320
	6.5	Animal and human living tissue sterilization	320
		6.5.1 Direct FE-DBD for living tissue treatment	320
		6.5.2 Direct FE-DBD plasma source for living tissue sterilization	321
		6.5.3 Toxicity (non-damaging) analysis of direct plasma treatment of living tissue	322
	6.6	Generated active species and plasma sterilization of living tissues	324
		6.6.1 Physico-chemical in vitro tissue model: Production and delivery in tissue of active species generated in plasma	324
		6.6.2 FE-DBD plasma system for analysis of deep tissue penetration of plasma-generated active species	325
		6.6.3 Deep tissue penetration of plasma-generated active species	327
	6.7	Deactivation/destruction of microorganisms due to plasma sterilization: Are they dead or just scared to death?	329
		6.7.1 Biological responses of *Bacillus stratosphericus* to FE-DBD plasma treatment	329
		6.7.2 FE-DBD plasma treatment of *Bacillus stratosphericus*	331
		6.7.3 Analysis of deactivation/destruction of *Bacillus stratosphericus* due to non-thermal plasma sterilization	336
		6.7.4 Bottom line for plasma physicists: Plasma sterilization can lead to VBNC state of microorganisms	338
7	**Plasma Decontamination of Water and Air Streams**		**339**
	7.1	Non-thermal plasma sterilization of air streams	339
		7.1.1 Direct sterilization versus application of filters	339
		7.1.2 Pathogen detection and remediation facility	340
		7.1.3 The dielectric barrier grating discharge (DBGD) applied in the PDRF	343
		7.1.4 Rapid and direct plasma deactivation of airborne bacteria in the PDRF	343
		7.1.5 Phenomenological kinetic model of non-thermal plasma sterilization of air streams	345
		7.1.6 Kinetics and mechanisms of rapid plasma deactivation of airborne bacteria at the PDRF	346
	7.2	Direct and indirect effects in non-thermal plasma deactivation of airborne bacteria	347
		7.2.1 Major sterilization factors	347
		7.2.2 PDRF: experimental procedure	348
		7.2.3 PDRF: experimental results	351
	7.3	Non-thermal plasma in air-decontamination: Air cleaning from SO_2 and NO_x	353
		7.3.1 Plasma cleaning of industrial SO_2 emissions	353
		7.3.2 SO_2 oxidation to SO_3 using relativistic electron beams	354
		7.3.3 SO_2 oxidation to SO_3 using continuous and pulsed corona discharges	354
		7.3.4 Plasma-stimulated liquid-phase chain oxidation of SO_2 in droplets	355
		7.3.5 Plasma-catalytic chain oxidation of SO_2 in clusters	358

	7.3.6	Simplified mechanism and energy balance of the plasma-catalytic chain oxidation of SO$_2$ in clusters	359
	7.3.7	Plasma-stimulated combined oxidation of NO$_x$ and SO$_2$ in air; simultaneous industrial exhaust gas cleaning from nitrogen and sulfur oxides	360
7.4		Non-thermal plasma decontamination of air from volatile organic compound (VOC) emissions	361
	7.4.1	General features of non-thermal plasma treatment of VOC emissions in air	361
	7.4.2	Mechanisms and energy balance of treatment of VOC exhaust gases from paper mills and wood processing plants	362
	7.4.3	Removal of acetone and methanol from air using pulsed corona discharge	364
	7.4.4	Removal of dimethyl sulfide from air using pulsed corona discharge	364
	7.4.5	Removal of α-pinene from air using pulsed corona discharge	368
	7.4.6	Treatment of paper mill exhaust gases using wet pulsed corona discharge	369
	7.4.7	Non-thermal plasma decontamination of diluted large-volume emissions of chlorine-containing VOC	370
	7.4.8	Non-thermal plasma removal of elemental mercury from coal-fired power plants and other industrial air-based off-gases	376
	7.4.9	Mechanism of non-thermal plasma removal of elemental mercury from air streams	377
7.5		Plasma desinfection and sterilization of water	378
	7.5.1	Plasma water disinfection using UV-radiation, ozone and pulsed electric fields	378
	7.5.2	Applications of pulsed plasma discharges for water treatment	379
	7.5.3	Energy-effective water treatment using pulsed spark discharges	380
	7.5.4	Characterization of the pulsed spark discharge system applied for energy-effective water sterilization	381
	7.5.5	Analysis of D-value and role of UV radiation in inactivation of microorganisms in water	385
8		**Plasma Treatment of Blood**	**389**
8.1		Plasma-assisted blood coagulation	389
	8.1.1	General features of plasma-assisted blood coagulation	389
	8.1.2	Experiments with non-thermal atmospheric-pressure plasma-assisted in vitro blood coagulation	389
	8.1.3	In-vivo blood coagulation using FE-DBD plasma	391
	8.1.4	Mechanisms of non-thermal plasma-assisted blood coagulation	391
	8.1.5	Influence of protein activity	393
8.2		Effect of non-thermal plasma on improvement of rheological properties of blood	395
	8.2.1	Control of low-density-lipoprotein (LDL) cholesterol and blood viscosity	395
	8.2.2	Plasma-medical system for DBD plasma control of blood properties	395
	8.2.3	DBD plasma control of whole blood viscosity (WBV)	397
	8.2.4	DBD plasma effect on improvement of rheological properties of blood	400
9		**Plasma-assisted Healing and Treatment of Diseases**	**403**
9.1		Wound healing and plasma treatment of wounds	403
	9.1.1	Wounds and healing processes	403
	9.1.2	Treatment of wounds using thermal and nitric-oxide-producing plasmas	407
	9.1.3	Experience with other thermal discharges	416

9.2   Treatment of inflammatory dysfunctions   418
    9.2.1   Examples of anti-inflammatory treatment by Plazon   418
    9.2.2   Pin-to-hole microdischarge for ulcerative colitis treatment   419
9.3   Plasma treatment of cancer   422
    9.3.1   Observations of cultured malignant cells   424
    9.3.2   Non-thermal plasma treatment of explanted tumors in animal models 426
9.4   Plasma applications in dentistry   428
    9.4.1   Brief overview of structure of teeth and dental health   429
    9.4.2   Recent promising results of plasma applications in dentistry   431
9.5   Plasma surgery   433

10   Plasma Pharmacology   **435**
10.1   Non-thermal plasma treatment of water   435
10.2   Deionized water treatment with DBD in different gases: Experimental setup 436
    10.2.1   Changing the working gas and sample degassing   436
10.3   Deionized water treatment with DBD in different gases: Results and discussion 438
10.4   Enhanced antimicrobial effect due to organic components dissolved in water 442
    10.4.1   Setup and sample preparation   442
    10.4.2   Comparison of DBD treated water, PBS and NAC solutions   444
10.5   Summary   446

11   Plasma-assisted Tissue Engineering and Plasma Processing of Polymers **447**
11.1   Regulation of biological properties of medical polymer materials   447
    11.1.1   Tissue engineering and plasma control of biological properties of medical
             polymers   447
    11.1.2   Wettability or hydrophilicity of medical polymer surfaces for biocompatibility 448
11.2   Plasma-assisted cell attachment and proliferation on polymer scaffolds 448
    11.2.1   Attachment and proliferation of bone cells on polymer scaffolds   448
    11.2.2   DBD plasma effect on attachment and proliferation of osteoblasts cultured over
             PCL scaffolds   449
11.3   Plasma-assisted tissue engineering in control of stem cells and tissue regeneration 451
    11.3.1   About plasma-assisted tissue regeneration   451
    11.3.2   Control of stem cell behavior on non-thermal plasma-modified polymer surfaces 452
    11.3.3   Plasma-assisted bioactive liquid micro-xerography   453
11.4   Plasma-chemical polymerization of hydrocarbons and formation of thin polymer films 453
    11.4.1   Biological and medical applications of plasma polymerization   453
    11.4.2   Mechanisms and kinetics of plasma polymerization   454
    11.4.3   Initiation of polymerization by dissociation of hydrocarbons in plasma volume 455
    11.4.4   Heterogeneous mechanisms of plasma-chemical polymerization of C_1/C_2
             hydrocarbons   456
    11.4.5   Plasma-initiated chain polymerization: Mechanisms of plasma polymerization
             of MMA   457
    11.4.6   Plasma-initiated graft polymerization   458
    11.4.7   Formation of polymer macro-particles in non-thermal plasma in hydrocarbons 459
    11.4.8   Specific properties of plasma-polymerized films   460
    11.4.9   Electric properties of plasma-polymerized films   460

11.5   Interaction of non-thermal plasma with polymer surfaces   461
    11.5.1   Plasma treatment of polymer surfaces   461
    11.5.2   Major initial chemical products created on polymer surfaces during
          non-thermal plasma interaction   463
    11.5.3   Formation kinetics of main chemical products in pulsed RF treatment of PE   464
    11.5.4   Kinetics of PE treatment in continuous RF discharge   466
    11.5.5   Non-thermal plasma etching of polymer materials   466
    11.5.6   Contribution of electrons and UV radiation in plasma treatment of polymer
          materials   467
    11.5.7   Interaction of chemically active heavy particles generated in non-thermal
          plasma with polymer materials   469
    11.5.8   Synergetic effect of plasma-generated active particles and UV radiation
          with polymers   471
    11.5.9   Aging effect in plasma-treated polymers   471
    11.5.10 Plasma modification of wettability of polymer surfaces   472
    11.5.11 Plasma enhancement of adhesion of polymer surfaces   474
    11.5.12 Plasma modification of polymer fibers and polymer membranes   475

References   **483**
Index   **497**

Preface

Plasma medicine has inspired the last decade of the authors' professional activities at Drexel Plasma Institute of the Drexel University. Plasma medicine is a very exciting and new multidisciplinary branch of modern science and technology. Even the term 'plasma medicine' has only been in existence since the start of the 21st century. Plasma medicine embraces physics required to develop novel plasma discharges relevant for medical applications, medicine to apply the technology not only in vitro but also in vivo testing and, last but not least, biology to understand the complicated biochemical processes involved in plasma interaction with living tissues.

While an understanding of the mechanisms by which non-thermal plasma interacts with living systems has begun to emerge only recently, a significant number of journal publications and even reviews focused on plasma medicine have appeared during the last 5–10 years. Several prestigious journals have published special issues dedicated to the topic, the new *Plasma Medicine* journal has been recently inaugurated, multiple world symposiums have created special sessions in this new field and plasma medical workshops have been organized in the USA, Germany, France, Korea, Japan, China and other countries. Four successful International Conferences on Plasma Medicine (ICPM) took place during the last 7 years in the USA, Germany and France; the 5th ICPM is planned to be held in Japan. Finally, the International Society on Plasma Medicine has been organized to coordinate the efforts of physicist, chemists, biologists, engineers, medical doctors and representatives of the industry working in this new field.

Despite the tremendous interest in plasma medicine, no single monograph has published in this field. There is no book where recent developments in plasma medicine, both technological and scientific, are described in a fashion accessible to the highly interdisciplinary audience of doctors, physicists, biologists, chemists and other scientists, university students and professors, engineers and medical practitioners. This is exactly the goal of the present *Plasma Medicine* book. The book is written for numerous scientists and medical practitioners, students, professors and industrial professionals who are involved today in plasma medicine.

When writing the book, we kept in mind the multidisciplinary nature of the field of plasma medicine. Physicists, chemists and engineers should be able to learn the different terminology of their biologist and medical practitioner partners, and vice versa. The book is beneficial to sides and should promote more effective development of the field of plasma medicine. The subject of plasma medicine has recently been included in the academic curriculum of universities, and we hope that the book will be helpful in this regard to students (as well as professors) involved in plasma-medical education.

Plasma Medicine consists of 11 chapters; Chapters 1–5 are focused on the **fundamentals of plasma medicine** and Chapter 6–11 are focused on **applied plasma medicine**.

Chapter 1 introduces the subject of plasma medicine. Chapter 2 describes the fundamental physical and chemical processes in plasmas relevant to its interaction with living tissues, providing a basic introduction to plasma medicine. Chapter 3 describes fundamental biology relevant to an understanding of the major principles of plasma interaction with living tissues. This topic covers the basic biological and medical introduction to plasma medicine, and will help physicists and engineers understand that even simple living organisms are much more complicated than electric devices. Chapter 4 describes plasma physics and engineering of the systems and devices relevant for medical applications. This chapter covers physical, chemical and engineering aspects of major electric discharges used for plasma–medical applications. In chapter 5, a description of the biophysical and biochemical mechanisms of plasma interaction with living tissues is provided. This chapter enables an understanding of the kinetics of plasma interaction with eukaryotic and prokaryotic cells, starting from gas phase and surface processes stimulated by active plasma species and including the consequent biochemical processes inside the cells.

Chapters 6 and 7 describe plasma sterilization of inanimate surfaces, as well as plasma sterilization of water and air. These chapters cover multiple applications of different low-pressure and atmospheric-pressure non-thermal discharges for disinfection and sterilization of different surfaces (e.g. medical instruments, food, space-crafts etc.); natural, drinking and industrial water; and large-volume air flows. Chapter 8 is focused on plasma-induced cauterization and blood coagulation. Plasma control of blood composition and relevant plasma treatment of blood diseases is also discussed in this chapter. Chapter 9 describes plasma treatment and healing of different wounds and diseases, in particular, plasma abatement of skin, dental and internal infections, treatment and healing of wounds and plasma treatment of oncological (cancerous), gastrointestinal, cardiovascular and other diseases. Chapter 10 describes plasma pharmacology, which suggests preliminary plasma treatment of water or special organic or inorganic solutions. These plasma-treated solutions can then be utilized for sterilization or healing purposes. The last chapter is focused on basic aspects of plasma-medical tissue engineering. This topic covers major modern aspects of plasma treatment of biomaterials and plasma-supported tissue engineering. This very important topic of applied plasma medicine is not directly related to plasma interaction with living tissue. This part of modern plasma medicine is very interesting, significant and relatively better developed (in particular, by our colleagues from University of Bari) than other branches; only a concise review of the subject is provided in this book.

Instructors can access PowerPoint files of the illustrations presented within this text, for teaching, at http://booksupport.wiley.com.

Alexander Fridman and Gary Friedman
Philadelphia

Acknowledgements

The authors gratefully acknowledge the support of their families, support of Drexel Plasma Institute (DPI) by John and Chris Nyheim and Kaplans family, support from leaders of Drexel University (President John Fry, Provost Mark Greenberg, Dean Joe Hughes, and senior vice-provost Deborah Crawford). The authors greatly appreciate support of their plasma-medical research by the National Institute of Health, National Science Foundation, US Department of Defense (especially DARPA and TATRC), W.M. Keck Foundation, Coulter Foundation, NASA and USDA; as well as the support of our industrial sponsors, especially Johnson & Johnson and GoJo.

For stimulating discussions on the topic of plasma medicine and assistance in development of the book, the authors gratefully acknowledge their colleagues and friends from Drexel University, especially Professors Danil Dobrynin, Gregory Fridman, Young Cho, Ari Brooks, Jane Azizkhan-Clifford, Richard Hamilton, Donna Murasko and Banu Onoral as well as Dr Terry Freeman from Jefferson University, Dr Steve Davis from University of Miami, Dr Alexander Gutsol from Chevron, Professor Ken Blank from Temple University, Professor Richard Satava from the University of Washington (formerly with DARPA) and many wonderful graduate students. Special thanks are addressed to Kirill Gutsol for assistance with illustrations.

1

Introduction to Fundamental and Applied Aspects of Plasma Medicine

1.1 Plasma medicine as a novel branch of medical technology

New ideas bring new hopes: plasma medicine is definitely one of those. Recent developments in physics and engineering have resulted in many important medical advances. The various medical technologies that have been widely described in the existing literature include applications of ionizing radiation, lasers, ultrasound, magnetism, and others. Plasma technology is a relative newcomer to the field of medicine. Very recent exponential developments in physical electronics and pulsed power engineering have promoted consequent significant developments in non-thermal atmospheric-pressure plasma science and engineering. Space-uniform and well-controlled cold atmospheric-pressure plasma sources have become a reality, creating the opportunity to safely and controllably apply plasma to animal and human bodies. This has instigated the creation of a novel and exciting area of medical technolgy: plasma medicine.

Experimental work conducted at several major universities, research centers, and hospitals around the world over the last decade demonstrates that non-thermal plasma can provide breakthrough solutions to challenging medical problems. It is effective in sterilization of different surfaces including living tissues, disinfects large-scale air and water streams, deactivates dangerous pathogens including those in food and drinks, and is able to stop serious bleeding without damaging healthy tissue. Non-thermal plasma can be directly used to promote wound healing and to treat multiple diseases including skin, gastrointestinal, cardiovascular, and dental diseases, as well as different forms of cancer. It has also proven effective in the treatment of blood, controlling its properties. Non-thermal discharges have also proven to be very useful in the treatment of different biomaterials and in tissue engineering, tissue analysis and diagnostics of diseases. Research indicates that non-thermal plasma may prove to be useful in pharmacology by changing properties of existing drugs and creating new medicines. Non-thermal plasma, developed recently due to the rapid progress in electronics, is clearly a promising new tool which should be provided to medical doctors to resolve medical problems. Plasma medicine, the subject of this book, is a source of great interest today.

When talking about the novel plasma sources which it is possible to apply to human and animal bodies, as well as for the treatment of cells and tissues in detailed biomedical experiments, we have to stress the *safety* and *controllability* of these novel plasma devices. As an example, the floating-electrode dielectric barrier discharge (FE-DBD) plasma source widely used for medical applications, in particular in Drexel University,

Plasma Medicine, First Edition. Alexander Fridman and Gary Friedman.
© 2013 John Wiley & Sons, Ltd. Published 2013 by John Wiley & Sons, Ltd.

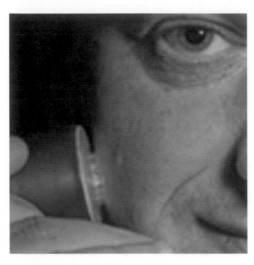

Figure 1.1   *Non-thermal short-pulsed 40 kV FE-DBD plasma sustained directly between a dielectric-coated electrode and a human body (see color plate).*

applies c. 30–40 kV directly to the human body (see one of the authors of this book in Figure 1.1). Obviously, safety is the main issue in this case. Of no less importance is the controllability of the plasma parameters. The uniform cold atmospheric-pressure plasmas as well as some other plasma-medical devices developed recently can be effectively controlled; this is important not only for prescribing specific doses of medical treatment, but also for investigation of the mechanisms of plasma-medical treatment. Without detailed understanding of physical, chemical, and biomedical mechanisms, plasma tools have little chance of successful application in medicine.

Non-thermal plasma is very far from thermodynamic equilibrium, which is discussed below. Such strongly non-equilibrium medium can be very 'creative' in its interactions with biomolecules. As first demonstrated in the 1950s by Stanley Miller (see Figure 1.2) and his colleagues from the University of Chicago, plasma is even able to generate amino acids from methane and inorganics. It is very much possible that plasma, being a strongly non-equilibrium and multi-parametric medium, can even be responsible for the creation of life itself. Recent experiments prove that controllable changes of DNA after non-thermal plasma treatment are very sensitive to plasma parameters. This explains the great importance of the controllability of plasma parameters and a deep understanding of mechanisms for successful progress of plasma-medical science. The success of plasma medicine requires a detailed understanding of physical, chemical, and biomedical mechanisms of the strongly non-equilibrium plasma interaction with cells and living tissues. Without a fundamental understanding, plasma medicine is at risk of become a modernized medieval magic (see Figure 1.3).

Plasma medicine is a multidisciplinary branch of modern science and technology. It embraces physics (required to develop novel plasma discharges relevant for medical applications), medicine (to apply the technology for not only in vitro but also in vivo testing), and last but not least biology (to understand the complicated biochemical processes involved in plasma interaction with living tissues). While an understanding of the mechanisms by which non-thermal plasma can interact with living systems has begun to emerge only recently, a significant number of original journal publications and even reviews have appeared since the mid-2000s. Several prestigious journals have published special issues dedicated to the plasma medicine, the new *Plasma Medicine* journal has been recently launched, multiple world symposiums have created special sessions in this new field, and plasma-medical workshops have been organized in the USA, Germany, France,

Figure 1.2 *In the 1950s, Stanley Miller of the University of Chicago synthesized amino acids in plasma from methane and inorganic compounds (see color plate).*

Figure 1.3 *International Society for Plasma Medicine (ISPM) signifies crucial importance of deep and detailed research focused on fundamental understanding of physical, chemical and biological bases of plasma medicine.*

Korea, Slovakia, and other countries. The most important world forum of plasma-medical research is the International Conferences on Plasma Medicine (ICPM). Four of these biannual conferences have already been successfully organized: ICPM-1 in Corpus Christi, Texas, USA; ICPM-2 in San-Antonia, Texas, USA; ICPM-3 in Greifswald, Germany; and finally ICPM-4 in Orleans, France in 2012. Finally, the International Society on Plasma Medicine was launched this year (2012) to coordinate the efforts of physicist, chemists, biologists, engineers, medical doctors and representatives of the industry in the new field of plasma medicine.

Hopefully, this book will be helpful to this entire and very multidisciplinary group of researchers and industry representatives. Plasma scientists and medical doctors speak different languages; they even have two different meanings for the word 'plasma' itself. Plasma scientists, medical doctors and biologists often have very different approaches to fundamental knowledge as well as practical applicability, but this book recognizes that they are united by a mutual interest in this new field of plasma medicine and by the common idea that development of plasma medicine brings new opportunities for treating human conditions.

1.2   Why plasma can be a useful tool in medicine

While the term 'medicine' in the title of the book does not require a special introduction, the term 'plasma' may require some elucidation (especially for medical practitioners). Plasma is an ionized gas and a distinct fourth state of matter. 'Ionized' means that at least one electron is not bound to an atom or molecule, converting them into positively charged ions. As temperature increases, atoms and molecules become more energetic and the state of matter transforms in the sequence: solid to liquid, liquid to gas and finally gas to plasma, which justifies the label of 'fourth state of matter'.

The free electric charges, electrons and ions make plasma electrically conductive (with magnitudes of conductivity sometimes exceeding that of gold and copper), internally interactive, and strongly responsive to electromagnetic fields. Ionized gas is defined as plasma when it is electrically neutral (electron density is balanced by that of positive ions) and contains a significant number of electrically charged particles, sufficient to affect its electrical properties and behavior. In addition to being important in many aspects of our daily lives, plasmas are estimated to constitute more than 99% of the known universe.

The term 'plasma' was first introduced by Irving Langmuir in 1928 when the multi-component, strongly interacting ionized gas reminded him of blood plasma; the term 'plasma' itself therefore has a strong relation to medicine. This can however be confusing: for example, read the discussions regarding plasma treatment of blood plasma in Chapter 8 of this book. Defining the term plasma, Irving Langmuir wrote: "Except near the electrodes, where there are sheaths containing very few electrons, the ionized gas contains ions and electrons in about equal numbers so that the resultant space charge is very small. We shall use the name **plasma** to describe this region containing balanced charges of ions and electrons". Plasmas occur naturally, but can also be effectively produced in laboratory settings and in industrial or hospital operations, providing opportunities for numerous applications including thermonuclear synthesis, electronics, lasers, fluorescent lamps, cauterization and tissue ablation during surgeries, and many others. We remind the reader that most computer and cell-phone hardware is based on plasma technology, not to forget about the plasma TV. In this book, we will focus on the fundamental and practical aspects of plasma applications to medicine, biology, and related disciplines, which represent today probably the most novel and exciting component of plasma science and engineering. Plasma is widely used in practice today. Generally, plasma offers three major features which are attractive for major practical applications.

1. *Temperatures* and energy densities of some plasma components can significantly exceed those in conventional technologies. These temperatures can easily exceed the level of c. 10 000 K. For example, if melted ceramics are needed to make relevant coatings, requiring temperatures above 3000 K, there is no

choice but to use plasma. In medical settings, high temperatures and energy densities can be useful for cauterization and tissue ablation during surgery, for example.

2. Plasmas are able to produce a *very high concentration of energetic and chemically active species* (e.g., electrons, positive and negative ions, atoms and radicals, excited atoms and molecules, as well as photons that span wide spectral ranges). A high concentration of active species is crucial for important plasma applications such as plasma-assisted ignition and combustion (probably the oldest plasma application) and plasma generation of ozone for water cleaning. In medical settings, generation of the high concentration of excited and reactive species can be useful for sterilization of surfaces, air, and water streams, as well as for tissue engineering.

3. Plasma systems can be very far from thermodynamic equilibrium, providing an *extremely high concentration of the chemically active species while maintaining bulk temperatures as low as room temperature.* This feature determines exclusiveness of plasma use in microelectronics and semiconductor industries: most elements of modern computers, cell phones, television equipment, cold lighting, and other electronic devices are manufactured using cold plasma technology. This important feature also determines the wide application of cold plasma in treatment of polymers: most textiles for our clothes, photographic paper, wrapping materials and so on are today plasma treated. In medical settings, the generation of an extremely high concentration of the chemically active species, while maintaining bulk temperatures as low as room temperature, can be useful for: non-thermal blood coagulation; corrections of blood composition and properties; sterilization of skin and other living tissues; healing wounds; and treating diseases not effectively treated before.

The three specific plasma features described above permit significant intensification of traditional chemical and biochemical processes, improvements in their efficiency, and often successful stimulation of chemical and biochemical reactions that are not possible using conventional techniques.

1.3 Natural and man-made, completely and weakly ionized plasmas

Plasma comprises the majority of the mass in the known universe: the solar corona, solar wind, nebula, and the Earth's ionosphere are all plasmas. The most readily recognized form of natural plasma phenomenon in the Earth's atmosphere is lightning. The breakthrough experiments with this natural form of plasma were performed long ago by Benjamin Franklin (Figure 1.4), which explains the special interest in plasma research in the Philadelphia area where the authors of this book are based (Drexel Plasma Institute, Drexel University).

At altitudes of approximately 100 km, the atmosphere no longer remains non-conducting due to significant ionization and formation of plasma by solar radiation. As one progresses further into near-space altitudes, the Earth's magnetic field interacts with charged particles streaming from the sun. These particles are diverted and often become trapped by the Earth's magnetic field. The trapped particles are most dense near the poles, creating the beautifully rendered Aurora Borealis (Figure 1.5). Lightning and the Aurora Borealis are the most common forms of natural plasmas observed on earth.

Natural and man-made or manufactured plasmas (generated in gas discharges) occur over a wide range of pressures, electron temperatures, and electron densities (see Figure 1.6). Temperatures of manufactured plasmas range from slightly above room temperature to temperatures comparable to the interior of stars, with electron densities that span over 15 orders of magnitude. Most plasmas of practical significance, however, have electron temperatures of 1–20 eV with electron densities in the range 10^6–10^{18} cm^{-3} (high temperatures are conventionally expressed in electron-volts, with 1 eV *c.* 11 600 K).

Not all particles need to be ionized in plasma; a common condition in plasma chemistry is for the gases to be only partially ionized. The ionization degree (ratio of density of major charged species to that of neutral gas)

Figure 1.4   *Benjamin Franklin successfully performed the first experiments with the atmospheric plasma phenomenon of lightning.*

in conventional plasma-chemical systems is in the range 10^{-7}–10^{-4}. When the ionization degree is close to unity, such plasma is referred to as *completely ionized plasma*. Completely ionized plasmas are conventional for thermonuclear plasma systems (tokomaks, stellarators, plasma pinches, focuses, etc.). When ionization degree is low, the plasma is called *weakly ionized plasma*. Weakly ionized plasmas and the important chemical and biochemical processes stimulated in such plasmas is the focus of this book.

Figure 1.5   *Aurora borealis.*

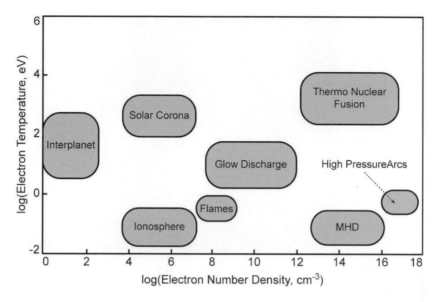

Figure 1.6   *General chart of plasma temperatures and densities.*

Both natural and manufactured or man-made laboratory plasmas are quasi-neutral, which means that concentrations of positively charged particles (positive ions) and negatively charged particles (electrons and negative ions) are well balanced. Langmuir was one of the pioneers who studied gas discharges, and defined plasma to be a region not influcnced by its boundaries. The transition zone between the plasma and its boundaries was termed the plasma *sheath*. The properties of the sheath differ from those of the plasma and these boundaries influence the motion of the charge particles in this sheath. They form an electrical screen for the plasma from influences of the boundary. Very important concepts group plasma physics, plasma chemistry, and plasma medicine into two major classes – those of thermal and non-thermal plasmas – which are discussed in the following section.

1.4   Plasma as a non-equilibrium multi-temperature system

Temperature in plasma is determined by the average energies of the plasma particles (neutral and charged) and their relevant degrees of freedom (translational, rotational, vibrational, and those related to electronic excitation). As multi-component systems, plasmas are therefore able to exhibit multiple temperatures. In electric discharges common for plasmas generated in the laboratory, energy from the electric field is first accumulated by the electrons through collisions; it is subsequently transferred from the electrons to the heavy particles. Electrons receive energy from the electric field during their mean free path. During the following collision with a heavy particle, they only lose a small portion of that energy (because electrons are much lighter than the heavy particles). That is why electron temperature in plasma is initially higher than that of heavy particles. Subsequently, collisions of electrons with heavy particles (Joule heating) can equilibrate their temperatures unless time or energy are not sufficient for the equilibration (such as the situation in coronas and pulsed discharges), or there is an intensive cooling mechanism preventing heating of the entire gas (as for wall-cooled low-pressure discharges).

Figure 1.7   *Solar plasma.*

The temperature difference between electrons and heavy neutral particles due to Joule heating in the collisional weakly ionized plasma is conventionally proportional to the square of the ratio of the electric field (E) to the pressure (p). Only in the case of small values of E/p, the temperatures of electrons and heavy particles approach each other. This is a basic requirement for the so-called local thermodynamic equilibrium (LTE) in plasma. Additionally, LTE conditions require chemical equilibrium as well as restrictions on the gradients. The LTE plasma follows major laws of the equilibrium thermodynamics and can be characterized by a single temperature at each point of space. Ionization and chemical processes in such plasmas are determined by temperature (and only indirectly by the electric fields through Joule heating). The quasi-equilibrium plasma of this kind is usually called *thermal plasma*. In nature, thermal plasmas can be represented by solar plasma (see Figure 1.7).

Numerous plasmas are sustained very far from the thermodynamic equilibrium and are characterized by multiple temperatures related to different plasma particles and different degrees of freedom. The electron temperature often significantly exceeds those of heavy particles ($T_e \gg T_0$). Ionization and chemical processes in such non-equilibrium plasmas are directly determined by electron temperature, and are therefore not very sensitive to thermal processes and temperature of the gas. The non-equilibrium plasma of this kind is usually referred to as *non-thermal plasma*. Non-thermal plasmas in nature are represented by the Aurora Borealis (see Figure 1.5) as opposed to thermal plasmas which are represented by lightening.

Although the relation between different plasma temperatures in non-thermal plasmas can be complex, it can be conventionally presented in the collisional weakly ionized plasmas as: $T_e > T_v > T_r \approx T_i \approx T_0$. Electron temperature T_e is the highest in the system, followed by the temperature of vibrational excitation of molecules T_v. The lowest temperature is usually shared in plasma by heavy neutrals (T_0, temperature of translational degrees of freedom or simply bulk gas temperature), ions (T_i), and rotational degrees of freedom of molecules (T_r). In many non-thermal plasma systems, electron temperature is $c.$ 1 eV ($c.$ 10 000 K), while gas temperature is close to room temperature.

Non-thermal plasmas are usually generated either at low pressures, at lower power levels, or in a different kind of pulsed discharge systems. The engineering aspects and application realms are quite different for thermal and non-thermal plasmas. Thermal plasmas are usually more powerful (up to 30 MW and above), while non-thermal plasmas are more selective and can be used in delicate applications without degrading the surrounding environment. However, these diverse forms of ionized gases share many common characteristics.

It is interesting to note that both thermal and non-thermal plasmas usually have the highest temperature (T_e in one case and T_0 in the other) of the order of magnitude of 1 eV, which is c. 10% of the total energy required for ionization (c. 10 eV). This reflects the general axiom formulated by Zeldovich and Frank-Kamenetsky for atoms and small molecules in chemical kinetics: the temperature required for a chemical process is typically c. 10% of the total required energy, which is the Arrhenius activation energy. Plasma temperatures can be somewhat identified as the 'down payment' for the ionization process (since a similar rule i.e. that of 10% is usually applied to calculate a down payment for a mortgage).

Thermal and non-thermal plasmas have their own specific niches for biological and medical applications. High temperatures and high energy densities typical of thermal plasmas determine their applications for cauterization and tissue ablation during surgeries. Such devices are widely used today in medical practice; some of them even combine the above-mentioned features with tissue sterilization. Thermal plasma in air is also very productive in the generation of NO, which determines its application in the so-called plasma-induced NO-therapy effective in plasma treatment of wounds and different diseases. Non-thermal plasma permits the generation of an extremely high concentration of the chemically active species, while maintaining bulk temperatures as low as room temperature. It determines the specific application niche of the non-thermal plasma, which is usually: non-thermal blood coagulation and corrections of blood composition and properties; sterilization of skin and other living tissues; sterilization of medical instruments and other fragile materials and devices; processing of biopolymers; tissue engineering; and finally non-thermal plasma healing of wounds and different diseases not effectively treated before.

1.5   Gas discharges as plasma sources for biology and medicine

Plasma medicine is based on a sequence of plasma-chemical and biochemical processes involving ionized gases. A plasma source, which in most laboratory conditions is a gas discharge, therefore represents a physical and engineering basis of the plasma medicine. For simplicity, an electric discharge as a plasma source in general can be viewed as two electrodes inserted into a glass tube and connected to a power supply. The tube can be filled with various gases or evacuated. As the voltage applied across the two electrodes increases, the current suddenly increases sharply at a threshold voltage required for sufficiently intensive electron avalanches. If the pressure is low (of the order a few Torr) and the external circuit has a large resistance to prohibit a large current, a glow discharge develops. This is the low-current high-voltage discharge widely used to generate non-thermal plasmas. A similar discharge is the basis of operation for fluorescent lamps, a common plasma discharge device. The glow discharge can be considered as a major example of the low-pressure non-thermal plasma sources (see Figure 1.8). Low-pressure plasma discharges can be effective as UV sources or sources of some active species for sterilization. They can be also effective for the treatment of biopolymers and in tissue engineering. It should be mentioned that, to keep treatment areas clean from the products of erosion of electrodes, low-pressure plasma-medical technologies are often based on electrode-less low-pressure plasma sources such as low-pressure radio-frequency plasma sources.

Historically, some important developments in the area of plasma medicine probably started with the work on surface treatment and subsequent surface interactions with cells (e.g. the work carried out by the groups of Riccardo D'Agostino, Pietro Favia and Michael Wertheimer) and sterilization using low-pressure non-thermal plasma (e.g. the work carried out by the group led by Michel Moisan).

Most plasma-medical applications require operation at atmospheric pressure, therefore use of atmospheric-pressure plasma discharges. Igor Alexeff and Mounir Laroussi were some of the first researchers to employ atmospheric-pressure plasma for sterilization, while Eva Stoffels was probably one of the first to apply such discharges directly to cells. Dr. Richard Satava, who managed various projects at the Defense Advanced Research Projects Agency (DARPA), helped develop initial applications of non-thermal plasma in medicine in the United States.

Figure 1.8   *Glow discharge.*

Probably the simplest example of such discharges is the corona discharge (see Figure 1.9). A non-thermal corona discharge occurs at high pressures (including atmospheric pressure) in regions of sharply non-uniform electric fields. The field near one or both electrodes must be stronger than in the rest of the gas. This occurs near sharp points, edges or small diameter wires. These tend to be low-power plasma sources, limited by the onset of electrical breakdown of the gas. However, it is possible to circumvent this restriction through

Figure 1.9   *Corona discharge.*

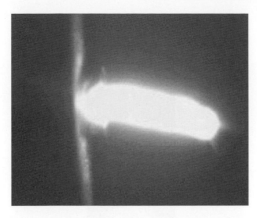

Figure 1.10   *Arc discharge.*

the use of pulsating power supplies. Electron temperature in the corona discharges can exceed 1 eV, while gas temperatures remain at the level of room temperature. The corona discharges are widely applied in the treatment of polymer materials: most synthetic fabrics used for clothing are treated using corona-like discharges to provide sufficient adhesion prior to dye applications.

The corona discharge can be considered as a major example of the atmospheric-pressure non-thermal plasma sources (see Figure 1.9). Atmospheric-pressure non-thermal plasma discharges can be effective in non-thermal blood coagulation and corrections of blood composition and properties; sterilization of skin and other living tissues; sterilization of medical instruments and other fragile materials and devices; processing of biopolymers; tissue engineering; and healing of wounds and different diseases. It should be mentioned that to keep treatment areas free from the products of erosion of electrodes and to provide uniform (or focused) treatment, other more complicated atmospheric pressure discharges such as dielectric barrier discharge (DBD) or atmospheric pressure glow (APG) plasma jet are normally used in medical applications.

For conditions that involve high pressures (i.e. of the order of an atmosphere) and when the external circuit resistance is low, a thermal arc discharge can be organized between two electrodes. Thermal arcs usually carry large currents, greater than 1 A at voltages of the order of tens of volts. Furthermore, they release large amounts of thermal energy at very high temperatures, often exceeding 10 000 K. The arcs are often coupled with a gas flow to form high-temperature plasma jets. The arc discharges are well known not only to scientists and engineers, but also to the general public because of their wide application in welding devices. The arc discharge can be considered as a major example of a thermal plasma source (see Figure 1.10). As mentioned above, arc discharges are widely used in medicine as cauterization and tissue ablation devices.

Thermal discharges can also be used to generate active species and UV for medical applications. This approach has been pursued, for example, by the groups headed by Gregor Morfill at the Max Plank Institute in Munich, Germany and Anatoly Shekhter in Moscow, Russia.

Other electric discharges widely applied in plasma engineering include non-equilibrium low-pressure radio-frequency (RF) discharges that play key roles in sophisticated etching and deposition processes of modern microelectronics, as well as in the treatment of polymer materials. More and more chemical processes are organized in gliding arc discharges (powerful generators of non-equilibrium atmospheric-pressure plasma), especially with plasma stabilization in reverse vortex 'tornado' flow (see Figure 1.11). The gliding arc 'tornado' discharges provide the unique combination of high power typical of arc discharges with a relatively high level of non-equilibrium typical of non-thermal atmospheric-pressure discharges. Regarding biomedical applications, the gliding arc 'tornado' discharges have been effectively used for sterilization of liquids and for plasma treatment of water and water solutions for sterilization of skin and wounds.

Figure 1.11   *Gliding arc discharge stabilized in the 'tornado' gas flow.*

Some electric discharges which are not conventional plasma sources attract significant interest for biological and medical applications. Of such non-traditional but practically interesting biomedical plasma discharges, we can point out the non-thermal high-voltage atmospheric-pressure FE-DBD, which is able to use the human body as a second electrode without damaging the living tissue. Such a discharge, developed independently as a plasma-medical tool by Wolfgang Viol and the authors of this book, obviously provides important opportunities for direct plasma applications in biology and medicine (Figure 1.12). Similar discharges have been developed by a number of different groups including that of Klaus-Dieter Weltmann in Greifwald, Germany, Jean-Michel Pouvesle in Orleans, France and David Graves at the University of California, Berkley.

Another example of this kind is related to atmospheric-pressure non-thermal plasma helium jets (see Figure 1.13), developed by a number of different groups including Mounir Laroussi in the US, Michael Kong

Figure 1.12   *Floating-electrode dielectric barrier discharge (FE-DBD) with a finger as a second electrode.*

Figure 1.13   *Professor Mounir Laroussi of Old Dominion University working with the non-thermal atmospheric-pressure plasma jet, a convenient tool for topical focused plasma treatment of living tissues (see color plate).*

in UK, Jean-Michel Pouvesle in France, Jae Koo Lee in South Korea. All these plasma sources, and especially those widely applied in plasma medicine, are discussed in detail in Chapter 4.

1.6   Plasma chemistry as the fundamental basis of plasma medicine

Chemically active plasma is a multi-component system that is highly reactive due to large concentrations of charged particles (electrons, negative and positive ions); excited atoms and molecules (electronic and vibrational excitations make a major contribution); active atoms and radicals; and UV-photons. Each component of the chemically active plasma plays its own specific role in plasma-chemical kinetics. For example, electrons are usually first to receive the energy from the electric field and then distribute it between other plasma components and specific degrees of freedom of the system. Changing the parameters of the electron gas (density, temperature, electron energy distribution function or EEDF) often allows the plasma-chemical processes to be controlled and optimized.

Ions are charged heavy particles that are able to make a significant contribution to plasma-chemical kinetics either due to their high energy (as in the case of sputtering and reactive ion etching) or to their ability to suppress activation barriers of chemical reactions. This second feature of the plasma ions results in the so-called ion- or plasma-catalysis, which is particularly essential in plasma-assisted ignition and flame stabilization; fuel conversion; hydrogen production; exhaust gas cleaning; and even in the direct plasma treatment of living tissue.

Vibrational excitation of molecules often makes a major contribution to the plasma-chemical kinetics because the plasma electrons with energies around 1 eV primarily transfer most of the energy in gases such as N_2, CO, CO_2, and H_2 into vibrational excitation. Stimulation of plasma-chemical processes through vibrational excitation allows the highest values of energy efficiency to be reached. Electronic excitation of atoms and molecules can also play a significant role, especially when the lifetime of the excited particles is quite long (as in the case of the metastable electronically excited atoms and molecules). As an example, we can mention plasma-generated metastable electronically excited oxygen molecules $O_2(^1\Delta_g)$ (singlet oxygen), which effectively participate in the plasma-stimulated oxidation process in polymer processing and biological and medical applications.

The contribution of atoms and radicals is obviously significant. For example, O-atoms and OH radicals effectively generated in atmospheric air discharges can play key roles in numerous plasma-stimulated oxidation processes. Plasma-generated photons also play key roles in a wide range of applications from plasma light sources to UV sterilization of water.

Plasma is not only a multi-component and multi-parametric system, but it can also be very far from thermodynamic equilibrium. Non-thermal plasmas are often strongly non-equilibrium systems. Concentrations of the active species described above can exceed many orders of magnitude of the concentrations achieved in quasi-equilibrium systems at the same gas temperature. Successful control of plasma permits the chemical or biochemical process to be steered in a particular direction and through an optimal mechanism. Control of a plasma-chemical system requires detailed understanding of elementary processes and kinetics of the chemically active plasma. Major fundamentals of plasma physics, elementary processes in plasma, and plasma kinetics are discussed in Chapter 2.

1.7   Non-thermal plasma interaction with cells and living tissues

As mentioned above, the mechanisms of chemical processes in plasma are complex. The complexity of mechanisms of plasma interaction with cells and living tissues is however much higher. Probably the easiest way to approach this challenging topic is to compare the biological effect of non-thermal plasma, which has been researched for less than a decade, with the biological effect of ionizing radiation, which has been intensively investigated for more than half a century.

Initially, there is a fundamental similarity between the biomolecular action of ionizing radiation (IR) and of non-equilibrium plasma. To a significant extent, both actions influence biological molecules and living organisms through generation of reactive oxygen species (ROS). Since methods of studying IR have been extensively developed, they can be conveniently adapted to investigate the effects of non-equilibrium plasma. Results of such biochemical investigations will be discussed in detail later in Chapter 5.

However, there are also important differences in the mechanisms by which IR and non-equilibrium plasma affect biological systems. For one thing, non-equilibrium plasma can generate many important species that IR usually does not. This includes reactive nitrogen species (RNS), large amount of freely moving charges (mostly positive and negative ions), and strong electric fields. Studying the mechanism of the non-equilibrium plasma treatment therefore has to involve characterization of the effects of these additional species. Even more importantly, IR is a penetrating radiation that generates ROS directly inside cells and, to some extent, causes direct damage to biological molecules within cells.

Non-equilibrium plasma, on the other hand, does not directly generate reactive species inside cells. Rather, it acts by initiating chains of reactions that begin in the plasma generation region. The 'plasma effect' then proceeds through the extracellular region (biological medium), modifying biomolecules which subsequently initiate signaling, modify cell membranes, or diffuse across them to create observable effects on living organisms (see Figure 1.14). For simplicity, the plasma-medical effect can be compared to a three-layer sandwich (illustrated by the Figure 1.14). Primary active species are generated inside the plasma (first layer) and are then are transported from the gas-phase into the intermediate biological medium (second liquid layer), from where the active species reach cells and trigger sophisticated intracellular biochemistry through multiple signaling mechanisms (third layer). Only the first steps toward an understanding of complex mechanism at

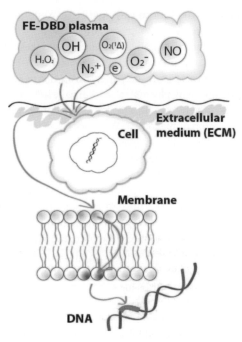

Figure 1.14 *Overall sequence of the non-equilibrium plasma-induced biomolecular processes: charges and active species (OH, H_2O_2, N_2^+, O_2^-, NO, $O_2(^1\Delta)$, etc.) from non-thermal plasma act on the cell membrane through the extracellular medium (ECM) leading to intracellular effects (e.g., protein modification, changes in transcription rates, DNA-related effects, etc.).*

each of these three levels of plasma interaction with cells and living tissues have been made; these will be reviewed in Chapter 5.

1.8 Applied plasma medicine

Plasma technologies successfully compete with conventional approaches in many practical applications, for example: thermal plasma deposition of protective coatings; plasma stabilization of flames; plasma conversion of fuels; plasma light sources; plasma cleaning of exhaust gases; and plasma sterilization of water. All these plasma technologies are practically interesting, commercially viable, and have made an important contribution to the development of our society.

The most exciting applications of plasma, however, have no conventional analogies and no (or almost no) competition. A good relevant example is plasma applications in microelectronics, such as etching of deep trenches (0.2 μm wide and 4 μm deep) in single-crystal silicon (very important in the fabrication of integrated circuits). The capabilities of plasma processing in microelectronics are extraordinary and unique; without plasma processing, we would not have achieved such powerful and compact computers and cell phones. When all alternatives fail, plasma remains as a viable and valuable tool. Selective modulation of regimes, power, density, and temperatures enable plasma to meet challenges and solve problems which cannot be solved by alternative technologies. There are other applications where plasma processes are not only highly efficient, but actually unique. For example, there are no other technologies which are able to compete with plasma for production of ozone (for more than 100 years); we should also keep in mind thermonuclear plasma as a unique major source of energy in the future.

In a similar way, plasma medicine attracts significant interest today because of the opportunities to tackle unresolved medical problems and to treat previously untreatable diseases. As a strongly non-equilibrium multi-parametric medium, plasma is able to shift the paradigm in therapeutics, wound-healing, and disease control. Some of the earlier applications of plasma in medicine relied mainly on its thermal effects. Heat and high temperature have been exploited in medicine for a long time for the purposes of tissue removal, sterilization, and cauterization (cessation of bleeding). Warriors have cauterized wounds by bringing them in contact with red hot metal objects and even flame (plasma) since ancient times.

Electrocautery is a more modern technique which applies controlled heat to surface layers of tissue by passing a sufficiently high current through it. Contact of tissue with the metal surface of a cautery device often results in adhesion of charred tissue to the metal, however. Subsequent removal of the metal can peel the charred tissue away, re-starting bleeding. Some of the earlier applications of plasma in medicine provided an alternative to metal-contact electrocautery. In argon plasma coagulation (APC, also sometimes called argon beam coagulation), highly conductive plasma replaced the metal contacts in order to pass a current through tissue, hence avoiding the difficulty with tissue adhesion. Hot plasma is also employed to cut tissue, although the exact mechanism by which this cutting occurs remains unclear. Heat delivered by plasma has also been employed recently for cosmetic re-structuring of tissue.

What differentiates more recent research on applications of plasma in medicine is the exploitation of the non-thermal effects. Why are non-thermal effects of plasma so interesting and promising? The main reason is that non-thermal plasma effects can be tuned for various sub-lethal purposes such as genetic transfection, cell detachment, wound healing, and others. Moreover, non-thermal effects can be selective in achieving a desired result for some living matter, while having little effect on the surrounding tissue. This is the case, for example, with recent plasma blood coagulation and bacteria deactivation, which does not cause toxicity in the surrounding living tissue.

Many examples demonstrating the effectiveness of plasma wound healing and treatments of different diseases are discussed in Chapter 9. Here, we quote an example of when a human life has been saved as a result of plasma treatment: the plasma treatment of corneal infections. In this regard, a special microplasma system has been developed for local medical treatment of skin diseases, and especially for the treatment of corneal infections (Gostev and Dobrynin, 2006). Details regarding this discharge are provided in Chapter 4 (see pin-to-hole spark discharge or PHD).

A series of in vitro experiments on bacterial cultures and in vivo experiments on rabbit eyes using this plasma discharge were conducted by Misyn *et al.* (2000). The experiments affirm the strong bactericidal effect of this microdischarge with minimal and reversible changes (if any) in biological tissues, even in such delicate tissues as cornea. During the investigation of plasma treatment of ulcerous dermatitis of rabbit cornea, two important observations were made: (1) plasma treatment has a pronounced and immediate bactericidal effect, and (2) the treatment has an effect on wound pathology and the rate of tissue regeneration and wound healing process.

These results provided a strong grounding for the successful application of the medical microplasma system (Gostev and Dobrynin, 2006) for treatment of a human patient with complicated ulcerous eyelid wounds, depicted in Figure 1.15. Necrotic phlegm on the surface of the upper eyelid was treated by an air plasma plume of 3 mm diameter for 5 seconds once every few days. By the fifth day of treatment (two 5-second plasma treatment sessions), the eyelid edema and inflammation were reduced. By the sixth day (third session), the treated area was free of edema and inflammation and a rose granular tissue appeared. Three more plasma treatments were administered (six in total), and the patient was discharged from the hospital six days after the final treatment (Figure 1.15).

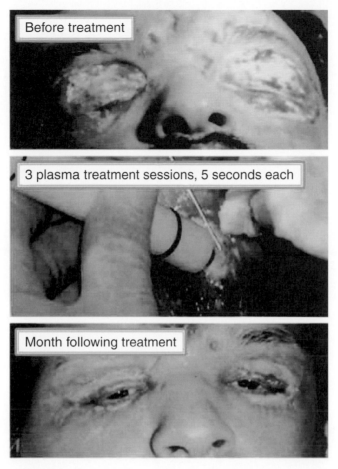

Figure 1.15 *Result of plasma treatment sessions (top: before; lower: after) of plasma treatment (middle) of a complicated ulcerous eyelid wound (see color plate).*

From these initial thoughts on the exciting and motivating results of the application of plasma medicine, the following chapters will discuss in detail the fundamentals of plasma medicine (first the fundamentals related to plasma physics and chemistry, and then basic medicine fundamentals).

2

Fundamentals of Plasma Physics and Plasma Chemistry for Biological and Medical Applications

Plasma-medical procedures are always based on very complex sequences of physical, chemical, biochemical, and biomedical pathways. Although the final medical steps in these sequences are very diverse (because of the enormous diversity of the biomedical pathways), the initial plasma-chemical steps of the sequences are more general and can therefore be discussed in more detail. The plasma-chemical effect on living tissues is due to the synergistic contributions of numerous different elementary reactions taking place simultaneously in a discharge plasma system. The sequence of the plasma-induced transformations of air or other initial chemical substances and electric energy into chemical products (including radicals, ions, excited and active species), radiation, and thermal energy is usually referred to as the *plasma-chemical mechanism* of a plasma-medical process. The plasma-chemical mechanisms can be quite complicated, especially in non-equilibrium discharges, and include a wide variety of elementary reactions of charged, excited, and chemically active neutral particles. A detailed consideration of physical kinetics of elementary plasma processes can be found in our previous books (Fridman and Kennedy, 2004, 2011; Fridman, 2008). In this chapter we focus on the basics of plasma-chemical kinetics, statistics, transfer processes, fluid- and electrodynamics relevant for plasma medicine.

2.1 Elementary plasma generation processes

2.1.1 Classification of ionization processes

Plasma is an ionized gas. The key process in plasma is ionization, which means conversion of neutral atoms or molecules into *electrons* and *positive ions*. Usually the number densities of electrons and positive ions are equal or close in *quasi-neutral plasmas*, but in 'electronegative' gases (e.g. O_2, air) with high electron affinity, *negative ions* are also effectively formed. Plasma-generated charged particles, in particular O_2^- (superoxide), play a significant role in the biochemistry of plasma interaction with living tissues.

Plasma Medicine, First Edition. Alexander Fridman and Gary Friedman.
© 2013 John Wiley & Sons, Ltd. Published 2013 by John Wiley & Sons, Ltd.

Electrons are first to obtain energy from electric fields, because of their low mass and high mobility. Electrons then transmit the energy to all other plasma components, providing energy for ionization, excitation, dissociation, and other chemical processes. The rates of such processes depend on how many electrons have enough energy to do the job. This can be described by means of the electron energy distribution function (EEDF) $f(\varepsilon)$, which is the probability of an electron to have energy density ε. The EEDF strongly depends on electric field and gas composition in plasma (especially in non-thermal discharges), and can often be very far from the equilibrium distribution. Sometimes however (even in non-equilibrium plasmas), the EEDF is determined mostly by the electron temperature T_e and can therefore be described by the quasi-equilibrium Maxwell-Boltzmann distribution function:

$$f(\varepsilon) = 2\sqrt{\varepsilon/\pi(kT_e)^3}\exp(-\varepsilon/kT_e), \qquad (2.1)$$

where k is the Boltzmann constant (when temperature is given in energy units, then $k = 1$ and can be omitted). The *mean electron energy* $\langle\varepsilon\rangle$, which is the first moment of the distribution function, is proportional to temperature in the conventional way:

$$\langle\varepsilon\rangle = \int_0^\infty \varepsilon f(\varepsilon)\,dE = \frac{3}{2}T_e. \qquad (2.2)$$

In most plasmas under consideration, the mean electron energy has a value within the range 1–5 eV. Atoms or molecules lose their electrons in ionization processes and form the positive ions. In hot thermonuclear plasmas, the ions are multi-charged; in relatively cold plasmas of technological interest, their charge is usually equal to $+1e$ (1.6×10^{-19} C). Electron attachment leads to formation of negative ions with charge $-1e$ (1.6×10^{-19} C). Attachment of another electron and formation of multi-charged negative ions is actually impossible in the gas phase because of electric repulsion. Ions are heavy particles, so they cannot receive high energy directly from electric fields because of intensive collisional energy exchange. Ion temperature T_i at elevated pressures is usually close to that of neutral gas T_0. Mechanisms of ionization can be very different, and are generally divided into the following five groups.

1. *Direct ionization by electron impact* is ionization of neutral and preliminary not excited atoms, radicals or molecules by an electron, whose energy is high enough to provide the ionization act in one collision. These processes are the most important in cold discharges, where electric fields and therefore electron energies are quite high but the excitation level of neutral species is relatively moderate.
2. *Stepwise ionization by electron impact* is ionization of preliminary excited neutral species. These processes are mostly important in thermal or energy-intense discharges, when the ionization degree n_e/n_0 is high as well as the concentration of excited neutral species.
3. *Ionization by collision of heavy particles* takes place during ion–molecular or ion–atomic collisions and in collisions of electronically or vibrationally excited species, when the total energy of the collision partners exceeds the ionization potential. The chemical energy of colliding neutral species can also contribute to ionization in the so-called associative ionization processes.
4. *Photo-ionization* takes place in collisions of neutrals with photons, which result in formation of an electron–ion pair. It can be important in thermal plasmas and in propagation of non-thermal discharges.
5. *Surface ionization (electron emission)* is provided by electron, ion and photon collisions with different surfaces, or simply by surface heating.

2.1.2 Direct and stepwise ionization by electron impact

Direct ionization is a result of interaction of an incident electron with a high energy ε and a valence electron of a preliminary neutral atom or molecule. Ionization occurs when the energy $\Delta\varepsilon$ transferred to the valence electron exceeds the ionization potential I. Analysis of the elementary process obviously requires a quantum mechanical consideration, but a clear physical understanding can be obtained from the classical *Thomson model*. The valence electron is assumed to be at rest in this model, and interaction of the two colliding electrons with the rest of the atom is neglected. The differential cross-section of the incident electron scattering with energy transfer $\Delta\varepsilon$ to the valence electron can be defined by the *Rutherford formula*:

$$d\sigma_i = \frac{1}{(4\pi\varepsilon_0)^2} \frac{\pi e^4}{\varepsilon(\Delta\varepsilon)^2} d(\Delta\varepsilon). \tag{2.3}$$

When the transferred energy exceeds the ionization potential, that is $\Delta\varepsilon \geq I$, direct ionization takes place. Integration of Equation (2.3) over $\Delta\varepsilon \geq I$ gives an expression for the ionization cross-section by direct electron impact, known as the *Thomson formula*:

$$\sigma_i = \frac{1}{(4\pi\varepsilon_0)^2} \frac{\pi e^4}{\varepsilon} \left(\frac{1}{I} - \frac{1}{\varepsilon} \right). \tag{2.4}$$

Equation (2.4) should generally be multiplied by the number of valence electrons Z_v. At high electron energies, that is $\varepsilon \gg I$, the Thomson cross-section (2.4) is falling as $\sigma_i \sim 1/\varepsilon$. Quantum mechanical treatment gives a more accurate but close asymptotic approximation $\sigma_i \sim (\ln \varepsilon)/\varepsilon$. When $\varepsilon = 2I$, the Thomson cross-section reaches the maximum value:

$$\sigma_i^{max} = \frac{1}{(4\pi\varepsilon_0)^2} \frac{\pi e^4}{4I^2}. \tag{2.5}$$

Assuming the Maxwellian EEDF, the direct ionization rate coefficient can be presented as:

$$k_i(T_e) = \sqrt{\left(\frac{8T_e}{\pi m} \right)} \sigma_0 \exp\left(-\frac{I}{T_e} \right). \tag{2.6}$$

In this relation, the cross-section $\sigma_0 = Z_v \pi e^4 / I^2 (4\pi\varepsilon_0)^2$ is approximately the geometrical atomic cross-section (for molecular nitrogen 10^{-16} cm^2, and for argon 3×10^{-16} cm^2). When the plasma density and therefore the concentration of excited neutrals are high enough, the energy I necessary for ionization can be provided in two different ways: (1) in the case of direct ionization, it could be provided by the energy of plasma electrons; and (2) the high energy of preliminary electronic excitation of neutrals can be converted in the ionization act referred to as *stepwise ionization*. If the level of electronic excitation is high enough, stepwise ionization is much faster than direct ionization because the statistical weight of electronically excited neutrals is greater than that of free plasma electrons. In other words, when $T_e \ll I$, the probability of obtaining the high ionization energy I is much lower for free plasma electrons (direct ionization) than for excited atoms and molecules (stepwise ionization). In contrast to direct ionization, the stepwise process includes several steps to provide the ionization event.

At first, electron–neutral collisions prepare highly excited species, and then a final collision with a relatively low-energy electron provides the actual ionization event. The stepwise ionization rate coefficient k_i^s can be found by the summation of partial rate coefficients $k_i^{s,n}$, corresponding to the nth electronically excited state, over all states of excitation while taking into account their concentrations:

$$k_i^s = \sum_n k_i^{s,n} \frac{N_n(\varepsilon_n)}{N_0}. \tag{2.7}$$

To calculate the maximum stepwise ionization rate, we can assume that electronically excited neutrals are in quasi-equilibrium with plasma electrons and that the electronically excited states are characterized by the Boltzmann distribution with electron temperature T_e:

$$N_n = \left(\frac{g_n}{g_0}\right) N_0 \exp\left(-\frac{\varepsilon_n}{T_e}\right). \tag{2.8}$$

In this relation, N_n, g_n and ε_n are number densities, statistical weights and energies of the electronically excited atoms, radicals or molecules; the index n is the principal quantum number. From statistical thermodynamics, the statistical weight of an excited particle $g_n = 2\, g_i\, n^2$, where g_i is the statistical weight of an ion. N_0 and g_0 are concentration and statistical weights of ground-state particles. The typical energy transfer from a plasma electron to an electron, sitting in an excited atomic level, is $c. \, T_e$. This means that excited particles with energy of about $\varepsilon_n = I - T_e$ make the major contributions in Equation (2.7). Taking into account that $I_n \sim 1/n^2$, the number of states with energy of about $\varepsilon_n = I - T_e$ and ionization potential of about $I_n = T_e$ has an order of n. From Equations (2.7) and (2.8), we can therefore derive:

$$k_i^s \approx \frac{g_i}{g_0} n^3 \langle \sigma v \rangle \exp\left(-\frac{I}{T_e}\right). \tag{2.9}$$

The cross-section σ in Equation (2.9), corresponding to an energy transfer of about T_e between electrons, can be estimated as $e^4 / T_e^2 (4\pi \varepsilon_0)^2$, with velocity v:

$$v = \sqrt{\frac{T_e}{m}}.$$

The principal quantum number can be taken from:

$$I_n \approx \frac{1}{(4\pi \varepsilon_0)^2} m e^4 / \hbar^2 n^2 \approx T_e. \tag{2.10}$$

As a result, the stepwise ionization rate coefficient can be expressed:

$$k_i^s \approx \frac{g_i}{g_0} \frac{1}{(4\pi \varepsilon_0)^5} \frac{m e^{10}}{\hbar^3 T_e^3} \exp\left(-\frac{I}{T_e}\right). \tag{2.11}$$

Comparing direct ionization (Equation (2.6)) with stepwise ionization (Equation (2.11)), we see that the latter can be much faster because of the high statistical weight of excited species involved in the stepwise ionization. The ratio of rate coefficients for these two competing mechanisms of ionization is:

$$\frac{k_i^s(T_e)}{k_i(T_e)} \approx \frac{g_i a_0^2}{g_0 \sigma_0} \left(\frac{1}{(4\pi \varepsilon_0)^2} \frac{m e^4}{\hbar^2 T_e}\right)^{7/2} \approx \left(\frac{I}{T_e}\right)^{7/2}. \tag{2.12}$$

In this relation, a_0 is the atomic unit of length, $\sigma_0 \sim a_0{}^2$ (taking into account the estimations for geometric collisional cross-section), and for the ionization potential:

$$I \approx \frac{1}{(4\pi\varepsilon_0)^2} \frac{me^4}{\hbar^2}.$$

For typical discharges with $I/T_e \sim 10$, stepwise ionization can be 10^3–10^4 times faster than direct ionization.

2.1.3 Other ionization mechanisms in plasma-medical discharges

First, we consider ionization by *high-energy electrons and electron beams*. The electron energy in electron beams usually varies over the range 50 keV to 1–2 MeV. Typical energy losses of the beams in atmospheric-pressure air are about 1 MeV per 1 m ($\approx$ 1 keV mm^{-1}). For this reason, the generation of large plasma volumes requires the high-energy electrons in the beams. The beams with electron energies exceeding 500 keV are referred to as *relativistic electron beams*. To describe ionization induced by the electrons with velocities significantly exceeding the velocities of atomic electrons, the Born approximation can be applied. Electron energy losses per unit length dE/dx can be evaluated in the non-relativistic case by the *Bethe-Bloch formula*:

$$-\frac{dE}{dx} = \frac{2\pi Z e^4}{(4\pi\varepsilon_0)^2 m v^2} n_0 \ln \frac{2m E v^2}{I^2}. \tag{2.13}$$

In this relation, Z is the atomic number of neutral particles providing the beam stopping; n_0 is their number density; and v is the stopping electron velocity. In the case of relativistic electron beams with electron energies 0.5–1 MeV, the energy losses can be numerically calculated by:

$$-\frac{dE}{dx} = 2 \times 10^{-22} n_0 Z \ln \frac{183}{Z^{1/3}}, \tag{2.14}$$

where dE/dx is expressed in MeV cm^{-1} and n_0 is the concentration of neutral particles expressed in cm^{-3}. Equation (2.14) can be rewritten in terms of effective ionization rate coefficient k_i^{eff} for relativistic electrons:

$$k_i^{\text{eff}} \approx 3 \times 10^{-10} Z \ln \frac{183}{Z^{1/3}}. \tag{2.15}$$

Numerically, this ionization rate coefficient is about 10^{-8}–10^{-7} cm^3 s^{-1}. Another important ionization mechanism is *photoionization*. Photoionization of a neutral particle A with ionization potential I (in eV) by a photon $\hbar\omega$ with wavelength λ (in Angstroms or Å) can be illustrated as:

$$\hbar\omega + A \rightarrow A^+ + e, \qquad \lambda < \frac{12\,400}{I}. \tag{2.16}$$

To provide the photoionization, the photon wavelength should usually be less than 1000 Å, which is ultraviolet radiation. The photoionization cross-section increases sharply from zero at the threshold energy (Equation (2.16)) to much higher values up to the geometrical cross-section. The contribution of the photoionization process is usually not very significant because of low concentrations of high-energy photons in most plasma-medical discharge systems.

Finally, we consider ionization in *collisions of heavy particles*. An electron with a kinetic energy only slightly exceeding the ionization potential is often quite effective in producing the ionization event. That is not true for ionization by collisions of heavy particles such as ions and neutrals. Even when they have enough kinetic energy, they cannot provide ionization because their velocities are much less than those of electrons in atoms. Even with enough energy, a heavy particle is very often unable to transfer this energy to an electron inside an atom because the process is far from resonance.

Such slow motion may be termed 'adiabatic', that is, unable to transfer energy to a fast moving particle. The *adiabatic principle* can be explained in terms of the relationship between low interaction frequency $\omega_{int} = \alpha v$ and high frequency of electron transfers in atoms $\omega_{tr} = \Delta E/\hbar$. Here $1/\alpha$ is the characteristic size of the interacting neutral particles, v is their velocity, and ΔE is the change of electron energy in the atom during the interaction. Only fast Fourier components of the slow interaction potential between particles with frequencies of about $\omega_{tr} = \Delta E/\hbar$ provide the energy transfer between the interacting particles. The relative weight or probability of these fast Fourier components is very low if $\omega_{tr} > \omega_{int}$; numerically it is about $\exp(-\omega_{tr}/\omega_{int})$. As a result, the probability P_{EnTr} of energy transfer processes (including the ionization process under consideration) are usually proportional to the so-called Massey parameter:

$$P_{EnTr} \propto \exp\left(-\frac{\omega_{tr}}{\omega_{int}}\right) \propto \exp\left(-\frac{\Delta E}{\hbar \alpha v}\right) = \exp(-P_{Ma}). \qquad (2.17)$$

where $P_{Ma} = \Delta E/\hbar \alpha v$ is the *adiabatic Massey parameter*. If $P_{Ma} \gg 1$ the process of energy transfer is adiabatic, and its probability is exponentially low. It takes place in collisions of heavy neutrals and ions. To get the Massey parameter close to 1 for ionization, the kinetic energy of the colliding heavy particle has to be about 10–100 keV, which is about three orders of magnitude greater than the ionization potential.

If the electronic excitation energy of a metastable atom A* exceeds the ionization potential of another atom B, their collision can however lead to the so-called *Penning ionization*. The Penning ionization usually proceeds through intermediate formation of an unstable excited quasi-molecule in the state of auto-ionization; cross-sections of the process can be very high. Cross-sections for the Penning ionization of N_2, CO_2, Xe and Ar by metastable helium atoms $He(2^3S)$ with an excitation energy 19.8 eV reach gas-kinetic values of about 10^{-15} cm^2. Similar cross-sections can be attained in collisions of metastable neon atoms (excitation energy 16.6 eV) with argon atoms (ionization potential 15.8 eV). Exceptionally high cross-sections of 1.4×10^{-14} cm^2 can be achieved in the Penning ionization of mercury atoms (ionization potential 10.4 eV) by collisions with metastable helium atoms He $(2^3S, 19.8$ eV). If the total electronic excitation energy of colliding particles is not sufficient, ionization is still possible when heavy species stick to each other, forming a molecular ion. Such processes are referred to as *associative ionization*, for example:

$$Hg(6\,^3P_1, E = 4.9) + Hg(6\,^3P_0, E = 4.7) \rightarrow Hg_2^+ + e. \qquad (2.18)$$

Total electronic excitation energy here (9.6 eV) is less than the ionization potential of mercury atoms (10.4 eV), but higher than the ionization potential for a Hg_2- molecule. Cross-sections of the associative ionization can be quite high and close to the gas-kinetic value (about 10^{-15} cm^2).

2.1.4   Mechanisms of electron-ion recombination in plasma

Electron-ion recombination is a highly exothermic process. The process should therefore have a specific channel of accumulation of the energy released during the neutralization of a positive ion and an electron. Most of these channels of recombination energy consumption are related to either dissociation of molecules, three-body collisions, or radiation, defining the three major groups of mechanisms of electron-ion recombination as follows.

2.1.4.1   *Dissociative electron-ion recombination*

The fastest electron neutralization mechanism in molecular gases or, in the presence of molecular ions, is *dissociative electron-ion recombination*:

$$e + AB^+ \rightarrow (AB)^* \rightarrow A + B^*. \tag{2.19}$$

Recombination energy in these processes goes into dissociation of the intermediately formed molecule ion and to excitation of the dissociation products. These processes are common for molecular gases, but they can also be important in atomic gases because of the formation of molecular ions in the *ion conversion processes*:

$$A^+ + A + A \rightarrow A_2^+ + A. \tag{2.20}$$

The recombination mechanism described by Equation (2.19) is quite fast and plays the major role in molecular gases. Reaction rate coefficients for most of the diatomic and three-atomic ions are of the order 10^{-7} cm^3 s^{-1}. The process has no activation energy, so its dependencies on both electron T_e and gas T_0 temperatures are not strong:

$$k_r^{ei}(T_e, T_0) \propto \frac{1}{T_0\sqrt{T_e}}. \tag{2.21}$$

If the pressure is high enough, the recombination of atomic ions such as Xe$^+$ proceeds through the preliminary formation of molecular ions as described by Equation (2.20). The molecular ion can then be fast neutralized in the rapid process of dissociative recombination. The ion conversion reaction rate coefficients are quite high. When the pressure exceeds 10 Torr the ion conversion is usually faster than the following process of dissociative recombination, which becomes a limiting stage in the overall recombination kinetics.

2.1.4.2   *Three-body electron-ion recombination*

The electron-ion neutralization can be provided by a *three-body electron-ion recombination* in atomic gases in the absence of molecular ions:

$$e + e + A^+ \rightarrow A^* + e. \tag{2.22}$$

Energy excess in this case goes into kinetic energy of a free electron, which participates in the recombination act as a third body partner. Note that heavy particles (ion and neutrals) are unable to accumulate electron recombination energy fast enough in their kinetic energy, and are ineffective as the third body partner. The three-body recombination process described by Equation (2.22) is the most important one in high-density quasi-equilibrium plasmas. Concentrations of molecular ions are very low in this case (because of thermal dissociation) for the fast mechanism of dissociative recombination, and the three-body reaction dominates. The three-body electron-ion recombination process described by Equation (2.22) is a reverse process with respect to the stepwise ionization. For this reason, the rate coefficient of the recombination can be derived from stepwise ionization rate coefficient k_i^s (Equation (2.11)) and the Saha thermodynamic equation for ionization/recombination balance:

$$k_r^{eei} = k_i^s \frac{n_0}{n_e n_i} = k_i^s \frac{g_0}{g_e g_i} \left(\frac{2\pi\hbar}{mT_e}\right)^{3/2} \exp\left(\frac{I}{T_e}\right) \approx \frac{e^{10}}{(4\pi\varepsilon_0)^5\sqrt{mT_e^9}}. \tag{2.23}$$

In this relation, n_e, n_i, n_0 are number densities of electrons, ions and neutrals; g_e, g_i, g_0 are their statistical weights; e and m are electron charge and mass; and I is an ionization potential. For practical calculations, Equation (2.23) can be presented in the numerical form:

$$k_r^{eei} = \frac{\sigma_0}{I} 10^{-14} \left(\frac{I}{T_e}\right)^{4.5}, \tag{2.24}$$

where k_r^{eei} is measured in cm^6 s^{-1}, σ_0 (cm^2) is the gas-kinetic cross section and I and T_e are the ionization potential and electron temperature (eV). Typical values of k_r^{eei} at room temperature are about 10^{-20} cm^6 s^{-1}; at $T_e = 1$ eV, this rate coefficient is about 10^{-27} cm^6 s^{-1}. At room temperature, the three-body recombination is able to compete with the dissociative recombination when the electron concentration is quite high and exceeds 10^{13} cm^{-3}. If the electron temperature is about 1 eV, the three-body recombination can compete with the dissociative recombination only in the case of exotically high electron densities exceeding 10^{20} cm^{-3}.

2.1.4.3   *Radiative electron-ion recombination*

Finally, the recombination energy can be converted into radiation in the process of *radiative electron-ion recombination*:

$$e + A^+ \rightarrow A^* \rightarrow A + \hbar\omega. \tag{2.25}$$

The cross-section of this process is relatively low, and can compete with the three-body recombination only when the plasma density is not high. It is relatively slow, because it requires a photon emission during a short interval of the electron-ion interaction. This type of recombination can play a major role only in the absence of molecular ions and when plasma density is quite low when three-body mechanisms are suppressed. Cross-sections of radiative recombination are usually about 10^{-21} cm^2. Rate coefficients can be estimated as a function of electron temperature:

$$k_{rad.rec.}^{ei} \simeq 3 \times 10^{-13} (T_e)^{-3/4} \tag{2.26}$$

(cm^3 s^{-1}) and T_e is measured in eV.

The radiative recombination is faster than three-body recombination when the electron concentration (cm^{-3}) is relatively low:

$$n_e < 3 \times 10^{13} (T_e)^{3.75}. \tag{2.27}$$

2.1.5   Elementary processes of negative ions in plasma-medical systems: Electron attachment and detachment

Dissociative electron attachment to molecules is a major mechanism of negative ion formation in electronegative molecular gases. It is effective when the products have positive electron affinities:

$$e + AB \rightarrow (AB^-)^* \rightarrow A + B^-. \tag{2.28}$$

The dissociative attachment proceeds by intermediate formation of an auto-ionization state $(AB^-)^*$. This excited state is unstable and decays leading either to the reverse reaction of auto-detachment $(AB + e)$, or to the dissociation $(A + B^-)$. The maximum cross-section for dissociative attachment,

$$\sigma_{d.a.}^{max} \approx \sigma_0 \sqrt{\frac{m(M_A + M_B)}{M_A M_B}}, \tag{2.29}$$

is two orders of magnitude less than the gas-kinetic cross-section σ_0, and is about 10^{-18} cm^2. The dissociative attachment cross-section as a function of electron energy $\sigma_a(\varepsilon)$ has a resonant structure. It allows the dissociative attachment rate coefficient k_a to be estimated as a function of electron temperature T_e by integration of $\sigma_a(\varepsilon)$ over the Maxwellian distribution:

$$k_a(T_e) \approx \sigma_{d.a.}^{max}(\varepsilon_{max}) \sqrt{\frac{2\varepsilon_{max}}{m}} \frac{\Delta\varepsilon}{T_e} \exp\left(-\frac{\varepsilon_{max}}{T_e}\right). \tag{2.30}$$

In this relation, a single $\sigma_a(\varepsilon)$ resonance is taken into account, ε_{max} and $\sigma^{max}{}_{d.a.}$ are the electron energy and maximum cross-section corresponding to the resonance, and $\Delta\varepsilon$ is its energy width. *Three-body electron attachment* also can result in the formation of negative ions:

$$e + A + B \rightarrow A^- + B. \tag{2.31}$$

This process can be a principal channel for electron losses when electron energies are not high enough for the dissociative attachment, and when pressure is elevated (usually more than 0.1 atm) and the third kinetic order processes are preferable. In contrast to the dissociative attachment, the three-body process is exothermic and its rate coefficient does not depend strongly on electron temperature. Electrons are usually kinetically less effective as a third body B because of a low degree of ionization. Atmospheric pressure non-thermal discharges in air are probably the most important plasma-medical systems, where the three-body attachment plays a key role in the balance of charged particles:

$$e + O_2 + M \rightarrow O_2^- + M. \tag{2.32}$$

This process proceeds by the two-stage *Bloch-Bradbury mechanism* starting with the formation of a negative ion (rate coefficient k_{att}) in an unstable auto-ionization state (τ is time of collisionless detachment):

$$e + O_2 \xleftrightarrow{k_{att}, \tau} (O_2^-)^*. \tag{2.33}$$

The second stage of the Bloch-Bradbury mechanism includes collision with the third body particle M (density n_0) leading to relaxation and stabilization of O_2^- or collisional decay of the unstable ion:

$$(O_2^-)^* + M \xrightarrow{k_{at}} O_2^- + M, \tag{2.34}$$

$$(O_2^-)^* + M \xrightarrow{k_{dec}} O_2 + e + M. \tag{2.35}$$

Taking into account the steady-state conditions for number density of the intermediate excited ions $(O_2^-)^*$, the rate coefficient for the total attachment process can be expressed:

$$k_{3M} = \frac{k_{att}k_{st}}{\frac{1}{\tau} + (k_{st} + k_{dec})\,n_0}. \tag{2.36}$$

Usually, when the pressure is not too high, $(k_{st} + k_{dec})\,n_0 \ll \tau^{-1}$, and Equation (2.36) can be simplified as:

$$k_{3M} \approx k_{att}\,k_{st}\,\tau. \tag{2.37}$$

The three-body attachment has a third kinetic order and depends equally on the rate coefficients of formation and stabilization of negative ions. The latter strongly depends on the type of the third particle: the more complex the molecule of a third body (M), the easier it stabilizes the $(O_2^-)^*$ and the higher the rate coefficient k_{3M}. For simple estimations, when $T_e = 1$ eV and $T_0 = 300$ K, k_{3M} can be estimated as 10^{-30} cm^6 s^{-1}. The rate of the three-body process is greater than dissociative attachment (k_a) when gas number density exceeds a critical value $n_0 > k_a\,(T_e)/k_{3M}$. Numerically in oxygen with $T_e = 1$ eV and $T_0 = 300$ K, it requires $n_0 > 10^{18}$ cm^{-3} or, in pressure units, $p > 30$ Torr.

We now consider the three *detachment mechanisms* most important in plasma- medical systems. The first mechanism of electron detachment from a negative ion, important in non-thermal discharges in particular, is *associative detachment*:

$$A^- + B \rightarrow (AB^-)^* \rightarrow AB + e. \tag{2.38}$$

This is a reverse process with respect to the dissociative attachment. It is a non-adiabatic process, which occurs by intersection of electronic terms of a complex negative ion $A^- - B$ and corresponding molecule AB. Rate coefficients of the non-adiabatic reactions are quite high; typically, $k_d = 10^{-10}$–10^{-9} cm^3 s^{-1}. Another detachment mechanism, *electron impact detachment*, is important for high degrees of ionization:

$$e + A^- \rightarrow A + e + e. \tag{2.39}$$

This process is somewhat similar to direct ionization by electron impact (Thomson mechanism). For electron energies of about 10 eV, the cross-section of the detachment process can be fairly high, about 10^{-14} cm^2. The cross-section dependence on incident electron velocity v_e can be illustrated by electron detachment from a hydrogen ion: $e + H^- \rightarrow H + 2e$:

$$\sigma(v_e) \approx \frac{\sigma_0 e^4}{(4\pi\varepsilon_0)^2 \hbar^2 v_e^2}\left(-7.5\ln\frac{e^2}{4\pi\varepsilon_0 \hbar v_e} + 25\right). \tag{2.40}$$

In this relation, σ_0 is the geometrical atomic cross-section. Maximum values of the cross sections are about 10^{-15}–10^{-14} cm^2 and correspond to electron energies of 10–50 eV. The third mechanism is *detachment in collisions with excited particles*:

$$A^- + B^* \rightarrow A + B + e. \tag{2.41}$$

When particle B is *electronically excited,* the process is similar to Penning ionization. If the electronic excitation energy of a collision partner B exceeds the electron affinity of particle $-A$, the detachment process can proceed effectively as an electronically non-adiabatic reaction (without essential energy exchange with translational degrees of freedom of the heavy particles). Exothermic detachment of an electron from an oxygen ion in collision with a metastable electronically excited oxygen molecule (excitation energy 0.98 eV) is an example of such processes:

$$O_2^- + O_2 \left({}^1\Delta_g \right) \rightarrow O_2 + O_2 + e, \, \Delta H = -0.6 \tag{2.42}$$

where enthalpy change is measured in units of eV. The rate coefficient of the detachment is very high: 2×10^{-10} cm^3 s^{-1} at room temperature. Electron detachment can also be effective in collisions with *vibrationally excited molecules*, for example:

$$O_2^- + O_2 \times (v > 3) \rightarrow O_2 + O_2 + e. \tag{2.43}$$

Kinetics of the detachment process can be described in this case in the conventional manner for all reactions stimulated by vibrational excitation of molecules. The traditional Arrhenius formula: $k_d \propto \exp(-E_a/T_v)$ is applicable here. The activation energy of the detachment process can in this case be taken to be equal to the electron affinity to oxygen molecules ($E_a \approx 0.44$ eV).

Actual losses of charged particles in electronegative gases are mostly due to the *ion-ion recombination*, which is the neutralization of positive and negative ions in binary or three-body collisions. The ion-ion recombination can proceed by a variety of different mechanisms which dominate at different pressure ranges, all usually characterized by very high rate coefficients. At high pressures (usually above 30 Torr), a three-body mechanism dominates the recombination. The recombination rate coefficient in this case reaches a maximum of about 1–3 $\times$ 10^{-6} cm^3 s^{-1} at near-atmospheric pressures for room temperature.

2.1.6   Heterogeneous ionization processes, electron emission mechanisms

The ionization processes considered in Sections 2.1.1–2.1.3 take place in the discharge volume. Here we discuss surface-related ionization processes, particularly important in supporting electric currents in cathode layers. The most important surface ionization process in thermal plasmas is *thermionic emission*, which is electron emission from a high temperature metal surface due to the high thermal energy of electrons located in the metal. The emitted electrons can stay in the surface vicinity, creating a negative space charge and preventing further emission. The electric field in the cathode vicinity is sufficient to push the negative space charge out of the electrode and reach the saturation current density, which is the main characteristic of the cathode thermionic emission and can be quantified by the *Sommerfeld formula*:

$$j = \frac{4\pi me}{(2\pi \hbar)^3} T^2 (1 - R) \exp\left(-\frac{W}{T}\right). \tag{2.44}$$

where W is the *work function*, which is the minimum energy necessary to extract an electron from a metal surface; $R = 0 - 0.8$ is a quantum mechanical coefficient describing the reflection of electrons from a potential barrier at the metal surface; the Sommerfeld constant: $4\pi me/(2\pi \hbar)^3 = 120$ (A cm^{-2} K^{-2}); T is the surface temperature; e and m are electron charge and mass; and $\hbar$ is Planck's constant. The thermionic current grows with electric field until the negative space charge near the cathode is eliminated and saturation is achieved. This saturation, however, is rather relative. A further increase of electric field leads gradually to an increase of the saturation current level, which is related to a reduction of work function due to an electric field known as the *Schottky effect*:

$$W = W_0 - 3.8 \times 10^{-4} \sqrt{E}. \tag{2.45}$$

This decrease of work function $W(\text{eV})$ is relatively small at reasonable values of electric field $E(\text{V cm}^{-1})$. The Schottky effect can however result in essential change of the thermionic current, because of its strong exponential dependence on the work function in accordance with the Sommerfeld formula. High electric fields (of about 1 to 3×10^6 V cm^{-1}) are not only able to decrease the work function but also directly extract electrons from cold metal surface due to the quantum-mechanical effect of tunneling. Electrons are able to escape from the metal across the barrier due to tunneling, which is called the *field emission* effect. This can be described by the *Fowler-Nordheim formula*:

$$j = \frac{e^2}{4\pi^2\hbar} \frac{1}{(W_0 + \varepsilon_F)} \sqrt{\frac{\varepsilon_F}{W_0}} \exp\left(-\frac{4\sqrt{2m}\,W_0^{3/2}}{3e\hbar E}\right) \qquad (2.46)$$

where ε_F is the Fermi energy of the metal and W_0 is the metal's work function not perturbed by an external electric field.

The field emission is very sensitive to small changes of electric field and work function, including those related to the Schottky effect. When the cathode temperature is high in addition to an external electric field, both thermionic and field emission make an essential contribution, usually referred to as *thermionic field emission*. To compare the emission mechanisms we can subdivide electrons escaping surfaces into four groups. The first group has energies below the Fermi level, so the electrons are able to escape the metal only through tunneling (i.e. by the field emission mechanism). Electrons of the fourth group leave the metal by the thermionic emission mechanism without any support from the electric field. Electrons of the third group overcome the potential energy barrier because of its reduction by the external electric field. This Schottky effect of the applied electric field is obviously a purely classical effect. The second group of electrons is able to escape the metal only quantum-mechanically by tunneling, similarly to the first group. The potential barrier of tunneling is not so great in this case however, because of the relatively high thermal energy of the second-group electrons. These electrons escape the cathode by the mechanism of thermionic field emission. Because thermionic emission is based on the synergistic effect of temperature and electric field, these two key parameters of electron emission can be reasonably high enough to provide the significant emission current. The thermionic field emission dominates over other mechanisms at $T = 3000$ K and $E > 8 \times 10^6$ V cm^{-1}. Note that at high temperatures but lower electric fields ($E < 5 \times 10^6$), electrons of the third group usually dominate the emission. The Sommerfeld relation in this case includes the work function diminished by the Schottky effect.

Electron emission from solids related to surface bombardment by different particles is called *secondary electron emission*. The most important mechanism from this group is *secondary ion-electron emission*. According to the adiabatic principle, heavy ions are not efficient for transferring energy to light electrons to provide ionization. This general statement can be referred to the direct electron emission from solid surfaces induced by ion impact. The secondary electron emission coefficient γ (electron yield per one ion) becomes relatively high only at very high ion energies exceeding 1 keV, when the Massey parameter becomes low. Although the secondary electron emission coefficient γ is low at lower ion energies, it is not negligible and remains almost constant at ion energies below the kilovolt range. It can be explained by the *Penning mechanism of secondary ion-electron emission*, also called the *potential mechanism*. An ion approaching a surface extracts an electron from there because the ionization potential I exceeds the work function W. The defect of energy $I - W$ is usually large enough ($I - W > W$) to enable the escape of more than one electron from the surface. Such a process is non-adiabatic and its probability is not negligible. The secondary ion-electron emission coefficient γ can be estimated using the empirical formula:

$$\gamma \approx 0.016(I - 2W). \qquad (2.47)$$

Another secondary electron emission mechanism is related to surface bombardment by excited meta-stable atoms with an excitation energy exceeding the surface work function. This so-called *potential electron emission induced by meta-stable atoms* can have quite a high secondary emission coefficient γ. Secondary electron emission can also be provided by the photo-effect. *Photo-electron emission* is usually characterized by the *quantum yield* $\gamma_{\hbar\omega}$, which gives the number of emitted electrons per single quantum $\hbar\omega$ of radiation. Visual light and low-energy UV radiation give the quantum yield $\gamma_{\hbar\omega} \approx 10^{-3}$, which is fairly sensitive to the quality of the surface. High-energy UV radiation provides emission with a quantum yield about 0.01–0.1, which is in this case is less sensitive to the surface characteristics.

2.2 Excited species in plasma medicine: Excitation, relaxation and dissociation of neutral particles in plasma

The extremely high chemical activity of plasmas, in plasma interaction with living tissues in particular, is due to high and often a super-equilibrium concentration of active species. The active species generated in plasma include: chemically aggressive atoms and radicals; charged particles; and excited atoms and molecules. Excited species can be subdivided into three groups: (1) electronically excited atoms and molecules; (2) vibrationally excited molecules; and (3) rotationally excited molecules.

2.2.1 Vibrational excitation of molecules by electron impact

Vibrational excitation is probably the most important elementary process in non-thermal molecular plasmas. It is responsible for the major part of energy exchange between electrons and molecules, and makes a significant contribution to the kinetics of non-equilibrium plasma-assisted chemical and biochemical processes. Elastic collisions of electrons and molecules are not effective in the process of vibrational excitation because of the significant difference in their masses ($m/M \ll 1$). Typical energy transfer from an electron with kinetic energy ε to a molecule in an elastic collision is about $\varepsilon \, (m/M)$; a vibrational quantum can be estimated as $\hbar\omega \approx I\sqrt{m/M}$, which is a much higher energy. For this reason, the classical cross-section of the vibrational excitation in an elastic electron- molecular collision is low:

$$\sigma_{\text{vib}}^{\text{elastic}} \approx \sigma_0 \frac{\varepsilon}{I}\sqrt{\frac{m}{M}}, \tag{2.48}$$

where $\sigma_0 \sim 10^{-16}$ cm^2 is the gas-kinetic cross-section and I is an ionization potential. Equation (2.48) yields numerical values of the cross-section of about 10^{-19} cm^2 for electron energies of about 1 eV. It is much lower than relevant experimental cross-sections, which are about the same as atomic values (10^{-16} cm^2). Experimental cross-sections are non-monotonic functions of electron energy and the probability of multi-quantum excitation is not very low. Vibrational excitation of a molecule AB from vibrational level v_1 to v_2 is not a direct elastic process, but a resonant process proceeding through formation of an intermediate non-stable negative ion:

$$AB(v_1) + e \xleftrightarrow{\Gamma_{1i},\Gamma_{i1}} AB^-(v_i) \xrightarrow{\Gamma_{i2}} AB(v_2) + e, \tag{2.49}$$

where v_i is the vibrational quantum number of a non-stable negative ion and $\Gamma_{\alpha\beta}$ (s^{-1}) are probabilities of transitions between vibrational states. The cross-section of the resonant vibrational excitation process described by Equation (2.49) can be found in the quasi-steady-state approximation using the *Breit-Wigner formula*:

$$\sigma_{12}(v_i, \varepsilon) = \frac{\pi\hbar^2}{2m\varepsilon} \frac{g_{AB^-}}{g_{AB}g_e} \frac{\Gamma_{1i}\Gamma_{i2}}{\frac{(\varepsilon-\Delta E_{1i})^2}{\hbar^2} + \Gamma_i^2}, \tag{2.50}$$

where ε is electron energy; ΔE_{1i} is the energy of transition to the intermediate state $AB(v_1) \rightarrow AB^-(v_i)$; g_{AB^-}, g_{AB} and g_e are statistical weights; and Γ_i is the probability of $AB^-(v_i)$ decay through all channels. Equation (2.50) illustrates the resonance structure of the cross-sectional dependence on electron energy. The energy width of the resonance pikes is about $\hbar\Gamma_i$, which is related to the lifetime of the non-stable intermediate negative ion $AB^-(v_i)$. The maximum value of the cross-section (Equation (2.50)) is about the same as the atomic value (10^{-16} cm^2).

The energy dependence of vibrational excitation cross-section depends on the lifetime of the intermediate ionic states (the so-called 'resonances'). First let us consider the so-called *short-lifetime resonances* (e.g., H_2, N_2O, H_2O, etc.), where the lifetime of the auto-ionization states $AB^-(v_i)$ is much shorter than the period of oscillation ($\tau \ll 10^{-14}$ s). The energy width of the auto-ionization level $\sim\hbar\Gamma_i$ is very large for the short-lifetime resonances, in accordance with the Uncertainty Principle. It results in wide maximum peaks (several eV) and no fine energy structure for $\sigma_{12}(\varepsilon)$. Because of the short lifetime of the auto-ionization state $AB^-(v_i)$, displacement of nuclei during the lifetime period is small which leads mostly to excitation of low vibrational levels.

If an intermediate-ion lifetime is similar to a molecular oscillation period ($\sim10^{-14}$ s), such a resonance is referred to as a *boomerang resonance* (e.g., low energy resonances in N_2, CO, CO_2, etc.). The boomerang model treats the formation and decay of the negative ion during one oscillation as interference of incoming and reflected waves. The interference of the nuclear wave packages results in an oscillating dependence of excitation cross-section on electron energy, with typical peaks in period of about 0.3 eV. Boomerang resonances require higher electron energies for excitation of higher vibrational levels. For example, the excitation threshold of $N_2(v = 1)$ is 1.9 eV and that of $N_2(v = 10)$ from the ground state is about 3 eV. Excitation of CO($v = 1$) requires a minimal electron energy of 1.6 eV; the threshold for CO($v = 10$) excitation from the ground state is about 2.5 eV. The maximum value of vibrational excitation cross-section decreases with vibrational quantum number.

Long-lifetime resonances (e.g., low-energy resonances in O_2, NO, C_6H_6, etc.) correspond to auto-ionization states $AB^-(v_i)$ with a much longer lifetime ($\tau = 10^{-14}$–10^{-10} s). The long-lifetime resonances result in quite narrow isolated peaks (about 0.1 eV) in cross-section, depending on electron energy. In contrast to boomerang resonances, the maximum value of the vibrational excitation cross-section is the same for different vibrational quantum numbers. Electron energies most effective in vibrational excitation are 1–3 eV, which usually correspond to a maximum in the electron energy distribution function. The vibrational excitation rate coefficients, which are the results of integration of the cross-sections over the electron energy distribution function, are obviously very high in this case and reach 10^{-7} cm^3 s^{-1}. For molecules such as N_2, CO, CO_2, almost each electron-molecular collision leads to a vibrational excitation at $T_e = 2$ eV.

This explains why a significant fraction of electron energy in non-thermal discharges goes into vibrational excitation at $T_e = 1$–3 eV. Vibrational excitation by electron impact is preferably a one-quantum process. Nevertheless, multi-quantum vibrational excitation is also important. Rate coefficients $k_{eV}(v_1, v_2)$ for excitation of molecules from an initial vibrational level v_1 to a final level v_2 can be found using the semi-empirical *Fridman approximation* for multi-quantum vibrational excitation. This approach allows the excitation rate

Table 2.1   *Parameters of the multi-quantum vibrational excitation by electron impact.*

Molecule	α	β	Molecule	α	β
N_2	0.7	0.05	H_2	3	—
CO	0.6	—	O_2	0.7	—
$CO_2(\nu 3)$	0.5	—	NO	0.7	—

coefficients $k_{eV}(v_1, v_2)$ to be determined from the much better-known vibrational excitation rate coefficient $k_{eV}(0,1)$, corresponding to excitation from the ground state to the first vibrational level. According to the Fridman approximation:

$$k_{eV}(v_1, v_2) = k_{eV}(0, 1)\frac{\exp\left[-\alpha\left(v_2 - v_1\right)\right]}{1 + \beta v_1}. \tag{2.51}$$

Parameters α and β of the Fridman approximation are listed in Table 2.1.

2.2.2   Rotational and electronic excitation of molecules by electron impact

If electron energies exceed ~ 1 eV, rotational excitation can proceed resonantly through the auto-ionization states similarly to the case of vibrational excitation. However, the relative contribution of this multi-stage rotational excitation is small, taking into account the low value of a rotational energy quantum with respect to vibrational quanta. Non-resonant rotational excitation by electron impact can be illustrated using the classical approach. The energy transfer from an electron (kinetic energy ε, mass m) to a molecule (mass M) in an elastic collision, inducing rotational excitation, is $\sim \varepsilon\,(m/M)$. The typical spacing between rotational levels (rotational quantum) is $\sim I\,(m/M)$, where I is the ionization potential. Cross-sections of the non-resonant rotational excitation can therefore be related to $\sigma_0 \sim 10^{-16}$ cm^2 (gas-kinetic collisional cross-section) as:

$$\sigma_{\text{rotational}}^{\text{elastic}} \approx \sigma_0 \frac{\varepsilon}{I}. \tag{2.52}$$

The cross-section of the resonant rotational excitation can exceed the non-resonant value by a factor of 100. Numerically, this cross-section is about $1–3 \times 10^{-16}$ cm^2 at $\varepsilon = 0.1$ eV. To evaluate 'quasi-elastic' energy transfer from an electron gas to neutral molecules, rotational excitation can be combined with the elastic collisions. The process is then characterized by a gas-kinetic rate coefficient $k_{e0} \simeq \sigma_0 \langle v_e \rangle \simeq 3 \times 10^{-8}$ (cm^2 s^{-1}), where $\langle v_e \rangle$ is the average thermal velocity of electrons, and each collision is considered as a loss of $\sim \varepsilon\,(m/M)$ of electron energy.

Electronic excitation by electron impact requires higher electron energies ($\varepsilon > 10$ eV) than vibrational and rotational excitation. The *Born approximation* can be applied to calculate the cross-section of these processes when the electron energies are sufficiently high. For excitation of optically permitted transitions from an atomic state i to another state k, the Born approximation gives the following cross-section:

$$\sigma_{ik}(\varepsilon) = 4\pi a_0^2 f_{ik} \left(\frac{\text{Ry}}{\Delta E_{ik}}\right)^2 \frac{\Delta E_{ik}}{\varepsilon} \ln\left(\frac{\varepsilon}{\Delta E_{ik}}\right) \tag{2.53}$$

where Ry is the Rydberg constant; a_0 is the Bohr radius; f_{ik} is the force of the oscillator for transition $i \rightarrow k$; and ΔE_{ik} is the energy of the transition. The formula is valid for high electron energies, that is, for $\varepsilon \gg \Delta E_{ik}$. Semi-empirical formulas can be used for such calculations:

$$\sigma_{ik}(\varepsilon) = 4\pi a_0^2 f_{ik} \left(\frac{\text{Ry}}{\Delta E_{ik}}\right)^2 \frac{x-1}{x^2} \ln(2.5x) \tag{2.54}$$

where $x = \varepsilon / \Delta E_{ik}$; the semi-empirical formulas are valid for $x \gg 1$. Maximum cross-sections for the excitation of optically permitted transitions are about the same as the gas-kinetic cross-section, that is, $\sigma_0 \sim 10^{-16} \text{ cm}^2$. To reach this cross-section, the electron energy should be 2–3 times greater than the transition energy ΔE_{ik}. The equation for the cross-section $\sigma_{ik}(\varepsilon)$ is different for excitation of electronic terms, from which optical transitions (radiation) are forbidden. The maximum cross-section, which here is about the same as atomic value $\sigma_0 \sim 10^{-16} \text{ cm}^2$, can be reached at much lower electron energies $\varepsilon / \Delta E_{ik} \approx 1.2 - 1.6$. It leads to an interesting effect of predominant excitation of the optically forbidden and metastable states by electron impact in non-thermal discharges, where the electron temperature T_e is usually much less than the transition energy ΔE_{ik}. Rate coefficients of electronic excitation are calculated by integration of the cross-sections $\sigma_{ik}(\varepsilon)$ over the electron energy distribution functions (EEDF). In the simplest case of a Maxwellian EEDF ($T_e \ll \Delta E_{ik}$):

$$k_{\text{el.excit.}} \propto \exp\left(-\frac{\Delta E_{ik}}{T_e}\right). \tag{2.55}$$

A semi-empirical relation for electronic excitation and ionization rate coefficients as a function of reduced electric field E/n_0 can be expressed as:

$$\log k_{\text{el.excit.}} = -C_1 - \frac{C_2}{E/n_0}. \tag{2.56}$$

The rate coefficient $k_{\text{el.excit.}}$ is given here in $\text{cm}^3 \text{ s}^{-1}$; E is the electric field strength in V cm^{-1}; and n_0 is gas density in cm^{-3}. Numerical values of the parameters C_1 and C_2 for different electronically excited states (and ionization) of CO_2 and N_2 are listed in Table 2.2.

2.2.3   Plasma generation of atoms and radicals: Dissociation of molecules by electron impact

Dissociation of molecules through electronic excitation can occur in a single collision; this is referred to as stimulated by direct electron impact. The elementary process can proceed through different intermediate steps of intramolecular transitions as follows.

- *Mechanism A* starts with direct electronic excitation from the ground state to a repulsive state with resultant dissociation. In this case, the required electron energy can significantly (by a few eV) exceed the dissociation energy. This mechanism therefore generates hot (high-energy) neutral fragments, which could for example significantly affect surface chemistry in low-pressure non-thermal discharges.
- *Mechanism B* includes direct electronic excitation of a molecule from the ground state to an attractive state with energy exceeding the dissociation threshold. Excitation of the state then results in dissociation. Energies of the dissociation fragments are lower in this case.
- *Mechanism C* consists of direct electronic excitation from the ground state to an attractive state corresponding to electronically excited dissociation products. Excitation of this state can lead to radiative

Table 2.2   *Parameters for semi-empirical approximation of rate coefficients of electronic excitation and ionization of CO_2 and N_2 by electron impact.*

Molecule	Excitation level or ionization	C_1	$C_2 \times 10^{16}$ (V cm²)	Molecule	Excitation level or ionization	C_1	$C_2 \times 10^{16}$ (V cm²)
N_2	$A^3\Sigma_u^+$	8.04	16.87	N_2	$c^1\Pi_u$	8.85	34.0
N_2	$B^3\Pi_g$	8.00	17.35	N_2	$a^1\Pi_u$	9.65	35.2
N_2	$W^3\Delta_u$	8.21	19.2	N_2	$b'^1\Sigma_u^+$	8.44	33.4
N_2	$B'^3\Sigma_u^-$	8.69	20.1	N_2	$c^3\Pi_u$	8.60	35.4
N_2	$a'^1\Sigma_u^-$	8.65	20.87	N_2	$F^3\Pi_u$	9.30	32.9
N_2	$a^1\Pi_g$	8.29	21.2	N_2	Ionization	8.12	40.6
N_2	$W^1\Delta_u$	8.67	20.85	CO_2	$^3\Sigma_u^+$	8.50	10.7
N_2	$C^3\Pi_u$	8.09	25.5	CO_2	$^1\Delta_u$	8.68	13.2
N_2	$E^3\Sigma_g^+$	9.65	23.53	CO_2	$^1\Pi_g$	8.84	14.8
N_2	$a''^1\Sigma_g^+$	8.88	26.5	CO_2	$^1\Sigma_g^+$	8.23	18.9
N_2	$b^1\Pi_u$	8.50	31.88	CO_2	Other levels	8.34	20.9
N_2	$c'^1\Sigma_u^+$	8.56	35.6	CO_2	ionization	8.38	25.5

transition to a low-energy repulsive state with subsequent dissociation. The energy of the dissociation fragments in this case is similar to that of mechanism *A*.

- *Mechanism D* (similarly to mechanism *C*) starts with direct electronic excitation from the ground state to an attractive state corresponding to electronically excited dissociation products. Excitation of this state leads to radiation-less transfer to a highly excited repulsive state, with subsequent dissociation into electronically excited fragments. This mechanism is usually referred to as *predissociation*.
- *Mechanism E* is similar to mechanism *A* and consists of direct electronic excitation from the ground state to a repulsive state, but with the following dissociation into electronically excited fragments. This mechanism requires the highest values of electron energies.

Kinetic yield of atoms and radicals in plasma is often characterized in plasma-medical systems by means of the *G-factors*, showing number of the active species generated per 100 eV. In the case of atmospheric pressure discharges, typical values of the G-factors for relatively strong nitrogen molecules are $\sim$0.3–0.6; for relatively weak oxygen molecules, the G-factors are $\sim$3–5.

2.2.4   Distribution of non-thermal plasma discharge energy between different channels of excitation of atoms and molecules

Electron gas energy received from an electric field in non-thermal plasma is distributed between elastic energy losses and different channels of excitation and ionization. Such a distribution for typical plasma-medical discharge in atmospheric air is shown in Figure 2.1 as a function of reduced electric field E/n_0. Such energy distributions have quite similar general features for different molecular gases. For example, the contribution of rotational excitation of molecules and elastic energy losses are essential only at low values of E/n_0. It is natural, because these processes are non-resonant and take place at low electron energies ($\ll$ 1 eV). At electron temperatures of about 1 eV (conventional for non-thermal discharges), most of the electron energy and therefore most of the discharge power can be localized within vibrational excitation of molecules. It makes the process of vibrational excitation very important and special in the non-equilibrium plasma chemistry and therefore plasma biochemistry of molecular gases. Obviously, the contribution of

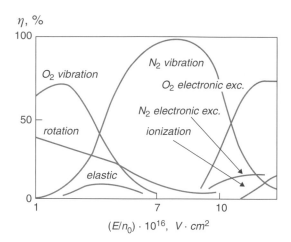

Figure 2.1   *Electron energy distribution between excitation channels in air.*

electron attachment processes (including dissociative attachment) can effectively compete with vibrational excitation at similar electron temperatures, but only in strongly electronegative gases. Finally, the contribution of electronic excitation and ionization becomes significant at higher electron temperatures, because of the high-energy thresholds of these processes.

2.2.5   Relaxation of plasma-excited species: Vibrational-translational (VT) relaxation

We have discussed the mechanisms of formation of excited atoms and molecules in plasma by electron impact. We now consider elementary mechanisms of deactivation and conversion of the excited species, usually referred to as relaxation processes. The elementary process of energy transfer between vibrational and translational degrees of freedom is usually called vibrational relaxation or vibrational-translation or VT-relaxation. Qualitative features of vibrational relaxation can be demonstrated by consideration of collision of a classical harmonic oscillator with an atom or molecule. The oscillator with frequency ω and reduced mass μ_0 is considered in this case under the influence of an external force $F(t)$, which represents the intermolecular collision. Energy transfer during the VT-relaxation process can be then defined as:

$$\Delta E_v = \frac{1}{2\mu_0} \left| \int_{-\infty}^{+\infty} F(t) \exp(-i\omega t)\, dt \right|^2 . \tag{2.57}$$

Only the Fourier component of the perturbation force on the oscillator frequency is effective in collisional excitation (or deactivation) of vibrational degrees of freedom of molecules. Considering $\tau_{\mathrm{col}} = 1/\alpha v$ as the collisional time (where α is the reverse radius of interaction between molecules and v is a relative velocity of the colliding particles), integration of Equation (2.57) gives:

$$\Delta E_v \propto \exp(-2\omega\tau_{\mathrm{col}}). \tag{2.58}$$

This relation demonstrates the adiabatic behavior of vibrational relaxation. Usually the *Massey parameter* (Equation (2.17)) at low gas temperatures is high for molecular vibration $\omega\tau_{\mathrm{col}} \gg 1$, which explains the adiabatic behavior and results in the exponentially slow vibrational energy transfer during the VT-relaxation. During the adiabatic collision, a molecule has enough time for many vibrations and the oscillator can actually be considered as 'structureless', which explains such a low level of energy transfer. An exponentially slow adiabatic VT-relaxation and intensive vibrational excitation by electron impact result in a unique role of vibrational excitation in plasma chemistry. Molecular vibrations for some gases such as N_2, CO_2, H_2 and CO are able to 'trap' energy of non-thermal discharges; it is therefore easy to activate them and difficult to deactivate them. Quantitatively, classical VT-energy transfer between atom A and molecule BC can be defined:

$$\Delta E_v = \frac{8\pi \omega^2 \mu^2 \lambda^2}{\alpha^2 \mu_0^2} \exp\left(-\frac{2\pi \omega}{\alpha v}\right) \tag{2.59}$$

where μ is the reduced mass of A and BC; μ_0 is the reduced mass of the molecule BC; and $\lambda = m_C/(m_B + m_C)$. The probability of a single quantum $\hbar\omega$ transfer can be expressed, according to Equation (2.59), as:

$$P_{01}^{VT}(v) = \frac{\Delta E_v}{\hbar \omega} = \frac{8\pi^2 \omega \mu^2 \lambda^2}{\hbar \alpha^2 \mu_0^2} \exp\left(-\frac{2\pi \omega}{\alpha v}\right). \tag{2.60}$$

To calculate the rate coefficient for VT-relaxation of a harmonic oscillator, the relaxation probability as a function of relative velocity v of colliding particles should be integrated over the Maxwellian distribution function. The result is known as the *Landau-Teller formula*:

$$k_{VT}^{10} \propto \exp\left(-\frac{B}{T_0^{1/3}}\right), \quad \text{where } B = \sqrt[3]{\frac{27\hbar^2 \mu \omega^2}{2\alpha^2 T_0}} \tag{2.61}$$

where temperature and vibrational quantum are measured in units of K and reduced mass is expressed in atomic mass units (amu).

A semi-empirical Milliken-White formula for VT-relaxation time (s) as a function of pressure (atm) can be applied for numerical calculations:

$$\ln(p\tau_{VT}) = 1.16 \times 10^{-3} \mu^{1/2} (\hbar\omega)^{4/3} \left(T_0^{-1/3} - 0.015\mu^{1/4}\right) - 18.42. \tag{2.62}$$

Some VT-relaxation rate coefficients at room temperature are listed in Table 2.3; most of them follow the Landau-Teller functionality.

Vibrational relaxation is therefore slow in adiabatic collisions when there is no chemical interaction between colliding partners. For example, the probability of deactivation of a vibrationally excited N_2 in collision with another N_2 molecule can be as low as 10^{-9} at room temperature. The vibrational relaxation process can be much faster in non-adiabatic collisions, when the colliding partners interact chemically.

2.2.6 Vibrational energy transfer between molecules: VV-relaxation

Kinetics of reactions of vibrationally excited molecules in plasma-medical systems is determined not only by their concentration but even more by the fraction of highly excited molecules able to dissociate or participate in endothermic chemical reactions. The formation of the highly vibrationally excited molecules at elevated pressures is due not to direct electron impact, but to collisional energy exchange referred to

Table 2.3   *Vibrational VT-relaxation rate coefficients $k_{VT}^{10}(T_0 = 300\,\text{K})$ for one-component gases at room temperature.*

Molecule	$k_{VT}^{10}(T_0 = 300\,\text{K})$ $(\text{cm}^3\ \text{s}^{-1})$	Molecule	$k_{VT}^{10}(T_0 = 300\,\text{K})$ $(\text{cm}^3\ \text{s}^{-1})$
O_2	5×10^{-18}	F_2	2×10^{-15}
Cl_2	3×10^{-15}	D_2	3×10^{-17}
Br_2	10^{-14}	$CO_2(01^10)$	5×10^{-15}
J_2	3×10^{-14}	$H_2O(010)$	3×10^{-12}
N_2	3×10^{-19}	N_2O	10^{-14}
CO	10^{-18}	COS	3×10^{-14}
H_2	10^{-16}	CS_2	5×10^{-14}
HF	2×10^{-12}	SO_2	5×10^{-14}
DF	5×10^{-13}	C_2H_2	10^{-12}
HCl	10^{-14}	CH_2Cl_2	10^{-12}
DCl	5×10^{-15}	CH_4	10^{-14}
HBr	2×10^{-14}	CH_3Cl	10^{-13}
DBr	5×10^{-15}	$CHCl_3$	5×10^{-13}
HJ	10^{-13}	CCl_4	5×10^{-13}
HD	10^{-16}	NO	10^{-13}

as vibrational-vibrational (VV)-relaxation. Most conventional resonant VV-processes usually imply vibrational energy exchange between molecules of the same kind, for example: $N_2^*\ (v = 1) + N_2\ (v = 0) \rightarrow N_2\ (v = 0) + N_2^*\ (v = 1)$, and are characterized by a probability $q_{mn}^{sl}(v)$ of a transition, when one oscillator changes its vibrational quantum number from s to l and other from m to n. VV-relaxation probability of the one-quantum exchange as a function of translational temperature T_0 can be found by averaging the probability $q_{mn}^{sl}(v)$ over Maxwellian distribution:

$$Q_{n+1,n}^{m,m+1}(T_0) = (m + 1)(n + 1)Q_{10}^{01}(T_0). \tag{2.63}$$

Resonant VV-exchange is usually much faster at room temperature than VT-relaxation, which promotes a population of highly vibrationally excited and reactive molecules.

2.2.7   Relaxation of rotational and electronic excitation in plasma-medical systems

Relaxation of electronically excited atoms and molecules is due to different mechanisms. Super-elastic collisions (energy transfer back to plasma electrons) and radiation are essential mostly in thermal plasma. Relaxation in collision with other heavy particles dominates in non-thermal discharges. *Relaxation of electronic excitation into translational degrees of freedom* (several eV) is a strongly adiabatic process with very high Massey parameters ($\omega\tau_{col} \sim 100$–$1000$). The adiabatic relaxation is very slow. For example,

$$Na(3^2P) + Ar \rightarrow Na + Ar \tag{2.64}$$

is characterized by a cross-section $<10^{-19}\ \text{cm}^2$, which corresponds to probability 10^{-9} or less. We should note that some specific relaxation processes of this kind can be fast, for example: $O(^1D)$ relaxation on atoms of noble gases can proceed through an intermediate complex and only require several collisions to occur. Electronically excited atoms and molecules transfer energy not only into translational, but also into *vibrational and rotational degrees of freedom*, which is less adiabatic and faster. Fast relaxation takes place by the formation of intermediate ionic complexes. For example, electronic energy transfer from excited metal atoms Me* to vibrational excitation of nitrogen takes place almost for each collision:

$$\text{Me}^* + \text{N}_2(v = 0) \rightarrow \text{Me}^+\text{N}_2^- \rightarrow \text{Me} + \text{N}_2(v > 0). \tag{2.65}$$

Initial and final Me N_2 energy terms cross an ionic term, which leads to fast non-adiabatic relaxation transition. *Electronic excitation energy transfer processes* are only effective very close to a resonance (0.1 eV or less), which limits them to some specific collision partners. For example, electronic excitation transfer from Hg atoms to sodium has cross-section reaching 10^{-14} cm^2. Rotational-rotational *(RR)* and rotational-translational *(RT)* energy transfer processes are usually non-adiabatic and fast because rotational quanta (and therefore the Massey parameter) are small. As a result, the collision of a rotator with an atom or another rotator can be considered as a classical collision accompanied by essential energy transfer. Rotational degrees of freedom are therefore often considered in quasi-equilibrium with translational degrees of freedom, and characterized by the same temperature T_0 even in strongly non-equilibrium systems.

2.3   Elementary plasma-chemical reactions of excited neutrals and ions

2.3.1   Rate coefficient of reactions of excited molecules in plasma

High rates of plasma-stimulated chemical and biochemical reactions are often due to a high concentration of excited atoms and molecules in electric discharges. Vibrational and electronic excitation plays the most important role in stimulation of such processes, especially if they are endothermic.

In this section we focus on reactions of vibrationally excited molecules, which are easier to analyze. The kinetic relations can also be applied, to some extent, to reactions of electronically excited particles. A convenient formula for calculation of rate coefficients of elementary reactions of an excited molecule with vibrational energy E_v at translational gas temperature T_0 can be expressed in the framework of the *theoretical-informational approach*:

$$k_R(E_v, T_0) = k_{R0} \exp\left(-\frac{E_a - \alpha E_v}{T_0}\right) \theta\left(E_a - \alpha E_v\right) \tag{2.66}$$

where E_a is the Arrhenius activation energy of elementary chemical reaction; the coefficient α is the efficiency of excitation energy in overcoming the activation barrier; k_{R0} is the pre-exponential factor; and $\theta(x - x_0)$ is the Heaviside function ($\theta(x - x_0) = 1$ when $x > 0$; $\theta(x - x_0) = 0$ when $x < 0$).

According to Equation (2.66), the reaction rate coefficients of vibrationally excited molecules follow the traditional Arrhenius law with an activation energy reduced on the value of vibrational energy, taken with efficiency α. If the vibrational temperature exceeds the translational temperature, that is $T_v \gg T_0$, and the chemical reaction is mostly determined by the vibrationally excited molecules, then Equation (2.66) can be simplified:

$$k_R(E_v) = k_{R0}\theta(\alpha E_v - E_a). \tag{2.67}$$

The effective activation energy in this case is E_a/α. Probabilities of reactions of excited molecules without averaging over the Maxwellian distribution can be calculated by using the *Le Roy formula*, which gives the reaction probability $P_v(E_v, E_t)$ as a function of vibrational and translational energies, E_v and E_t:

$$P_v(E_v, E_t) = 0, \quad \text{if} \quad E_t < E_a - \alpha E_v \text{ and } E_a \geq \alpha E_v, \tag{2.68}$$

$$P_v(E_v, E_t) = 1 - \frac{E_a - \alpha E_v}{E_t}, \quad \text{if} \quad E_t > E_a - \alpha E_v \text{ and } E_a \geq \alpha E_v, \tag{2.69}$$

$$P_v(E_v, E_t) = 1, \quad \text{if} \quad E_a < \alpha E_v, \quad \text{any} \quad E_t. \tag{2.70}$$

Averaging $P_v(E_v, E_t)$ over the Maxwellian distribution yields Equation (2.67), which is actually the most important relation for kinetic calculations of elementary reactions of excited particles.

2.3.2   Efficiency α of excitation energy in overcoming activation energy of chemical reactions of plasma-generated active species

The coefficient α is a key parameter describing the influence of plasma excitation of molecules on their chemical reaction rates (Equation (2.66)). Numeric values of the α coefficients for some specific chemical reactions are listed in Table 2.4 (Levitsky, Macheret and Fridman, 1983; Levitsky *et al.*, 1983; Rusanov and Fridman, 1984).

Table 2.4 allows chemical reactions to be classified according to specific probable values of the α coefficients. Such a classification is presented in Table 2.5. Reactions are divided in this table into endothermic, exothermic and thermoneutral categories, and into simple- and double-exchange elementary processes. The classification also separates reactions with breaking bonds in excited or non-excited molecules. This classification table approach is useful for determination of the efficiency of vibrational energy α in elementary chemical and biochemical reactions, if not known experimentally.

2.3.3   Fridman-Macheret α-model for chemical reactions of plasma-generated active species

The Fridman-Macheret α-model allows the α-coefficient for the efficiency of excitation energy in elementary chemical and biochemical processes to be calculated, mostly from information about the activation energies of the correspondent direct and reverse reactions. The model describes the exchange reaction $A + BC \rightarrow AB + C$ with the energy profile depicted by Figure 2.2. Vibration of the molecule BC can be taken into account using the approximation of *vibronic terms*, which are shown by dash-dotted lines. The energy profile, corresponding to the reaction of a vibrationally excited molecule with energy E_v (vibronic term), can be obtained in this approach by a parallel shifting of the initial profile $A + BC$ upwards by the value of E_v. Part of the reaction path profile corresponding to products $AB + C$ remains the same if the products are not excited. Simple geometry (Figure 2.2) shows that the effective decrease of activation energy related to vibrational excitation E_v is equal to:

$$\Delta E_a = E_v \frac{F_{A+BC}}{F_{A+BC} + F_{AB+C}}. \tag{2.71}$$

where F_{A+BC} and F_{AB+C} are the characteristic slopes of the terms $A+BC$ and $AB+C$ (Figure 2.2b). If these energy terms depend exponentially on reaction coordinates with decreasing parameters γ_1 and γ_2 (reverse radii of corresponding exchange forces), then:

$$\frac{F_{A+BC}}{F_{AB+C}} = \frac{\gamma_1 E_a^{(1)}}{\gamma_2 E_a^{(2)}} \tag{2.72}$$

Table 2.4 *Efficiency α of excitation energy of molecules in overcoming activation energy barriers.*

Reaction	α_{exp}	α_{MF}	Reaction	α_{exp}	α_{MF}
$F + HF^* \rightarrow F_2 + H$	0.98	0.98	$F + DF^* \rightarrow F_2 + D$	0.99	0.98
$Cl + HCl^* \rightarrow H + Cl_2$	0.95	0.96	$Cl + DCl^* \rightarrow D + Cl_2$	0.99	0.96
$Br + HBr^* \rightarrow H + Br_2$	1.0	0.98	$F + HCl^* \rightarrow H + ClF$	0.99	0.96
$Cl + HF^* \rightarrow ClF + H$	1.0	0.98	$Br + HF^* \rightarrow BrF + H$	1.0	0.98
$J + HF^* \rightarrow JF + H$	1.0	0.98	$Br + HCl^* \rightarrow BrCl + H$	0.98	0.98
$Cl + HBr^* \rightarrow BrCl + H$	1.0	0.97	$J + HCl^* \rightarrow JCl + H$	1.0	0.98
$SCl + HCl^* \rightarrow SCl_2 + H$	0.96	0.98	$S_2Cl + HCl^* \rightarrow S_2Cl_2 + H$	0.98	0.97
$SOCl + HCl^* \rightarrow SOCl_2 + H$	0.95	0.96	$SO_2Cl + HCl^* \rightarrow SO_2Cl_2 + H$	1.0	0.96
$NO + HCl^* \rightarrow NOCl + H$	1.0	0.98	$FO + HF^* \rightarrow F_2O + H$	1.0	0.98
$O_2 + OH^* \rightarrow O_3 + H$	1.0	1.0	$NO + OH^* \rightarrow NO_2 + H$	1.0	1.0
$ClO + OH^* \rightarrow ClO_2 + H$	1.0	1.0	$CrO_2Cl + HCl^* \rightarrow CrO_2Cl_2 + H$	0.94	0.9
$PBr_2 + HBr^* \rightarrow PBr_3 + H$	1.0	0.97	$SF_5 + HBr^* \rightarrow SF_5Br + H$	1.0	0.98
$SF_3 + HF^* \rightarrow SF_4 + H$	0.89	0.98	$SF_4 + HF^* \rightarrow SF_5 + H$	0.97	0.99
$H + HF^* \rightarrow H_2 + F$	1.0	0.95	$D + HF^* \rightarrow HD + F$	1.0	0.95
$Cl + HF^* \rightarrow HCl + F$	0.96	0.97	$Br + HF^* \rightarrow HBr + F$	1.0	0.98
$J + HF^* \rightarrow HJ + F$	0.99	0.99	$Br + HCl^* \rightarrow HBr + Cl$	1.0	0.95
$J + HCl^* \rightarrow HJ + Cl$	1.0	0.98	$J + HBr^* \rightarrow HJ + Br$	1.0	0.96
$OH + HF^* \rightarrow H_2O + F$	1.0	0.95	$HS + HF^* \rightarrow H_2S + F$	1.0	0.98
$HS + HCl^* \rightarrow H_2S + Cl$	1.0	1.0	$HO_2 + HF^* \rightarrow H_2O_2 + F$	1.0	0.98
$HO_2 + HF^* \rightarrow H_2O_2 + F$	1.0	0.98	$NH_2 + HF^* \rightarrow NH_3 + F$	1.0	0.97
$SiH_3 + HF^* \rightarrow SiH_4 + F$	1.0	1.0	$GeH_3 + HF^* \rightarrow GeH_4 + F$	1.0	1.0
$N_2H_3 + HF^* \rightarrow N_2H_4 + F$	0.97	0.98	$CH_3 + HF^* \rightarrow CH_4 + F$	0.98	0.97
$CH_2F + HF^* \rightarrow CH_3F + F$	1.0	0.97	$CH_2Cl + HF^* \rightarrow CH_3Cl + F$	1.0	0.97
$CCl_3 + HF^* \rightarrow CHCl_3 + F$	0.98	0.98	$CH_2Br + HF \rightarrow CH_3Br + F$	1.0	0.97
$C_2H_5 + HF^* \rightarrow C_2H_6 + F$	1.0	0.99	$CH_2CF_3 + HF^* \rightarrow CH_3CF_3 + F$	1.0	0.98
$C_2H_5O + HF^* \rightarrow (CH_3)_2O + F$	1.0	1.0	$C_2H_5Hg + HF^* \rightarrow (CH_3)_2Hg \mid F$	0.99	0.97
$HCO + HF^* \rightarrow H2CO + F$	1.0	0.99	$FCO + HF^* \rightarrow HFCO + F$	1.0	0.99
$C_2H_3 + HF^* \rightarrow C_2H_4 + F$	0.99	0.97	$C_3H_5 + HF^* \rightarrow C_3H_6 + F$	1.0	0.98
$C_6H_5 + HF^* \rightarrow C_6H_6 + F$	1.0	0.97	$CH_3 + JF^* \rightarrow CH_3J + F$	0.81	0.9
$S + CO^* \rightarrow CS + O$	1.0	0.99	$F + CO^* \rightarrow CF + O$	1.0	1.0
$CS + CO^* \rightarrow CS_2 + O$	0.83	0.9	$CS + SO^* \rightarrow CS_2 + O$	0.90	0.95
$SO + CO^* \rightarrow COS + O$	0.96	0.92	$F_2 + CO^* \rightarrow CF_2 + O$	1.0	1.0
$C_2H_4 + CO^* \rightarrow (CH_2)_2 C + O$	0.94	0.98	$CH_2 + CO^* \rightarrow C_2H_2 + O$	0.90	0.94
$H_2 + CO^* \rightarrow CH_2 + O$	0.96	—	$OH + OH^* \rightarrow H_2O + O(^1D_2)$	1.0	0.97
$NO + NO^* \rightarrow N_2O + O(^1D_2)$	1.0	—	$CH_3 + OH^* \rightarrow CH_4 + O(^1D_2)$	1.0	1.0
$O + NO^* \rightarrow O2 + N$	0.94	0.86	$H + BaF^* \rightarrow HF + Ba$	0.99	1.0
$O + AlO^* \rightarrow Al + O_2$	0.67	0.7	$O + BaO^* \rightarrow Ba + O_2$	1.0	0.99
$CO + HF^* \rightarrow CHF + O$	1.0	1.0	$CO_2 + HF^* \rightarrow O_2 + CHF$	1.0	—
$CO^* + HF \rightarrow CHF + O$	1.0	1.0	$CF_2O + HF^* \rightarrow CHF_3 + O(^1D_2)$	1.0	1.0
$H_2CO + HF^* \rightarrow CH_3F + O(^1D_2)$	1.0	1.0	$J + HJ^* \rightarrow J_2 + H$	1.0	1.0
$O + N_2 \rightarrow NO + N$ (non-adiabatic)	1.0	1.0	$O + N_2 \rightarrow NO + N$ (adiabatic)	0.6	1.0
$F + ClF^* \rightarrow F_2 + Cl$	1.0	0.93	$O + (CO^+)^* \rightarrow O_2 + C^+$	0.9	1.0
$H + HCl^* \rightarrow H_2 + Cl$	0.3	0.4	$NO + O_2^* \rightarrow NO_2 + O$	0.9	1.0
$O + H_2^* \rightarrow OH + H$	0.3	0.5	$O + HCl^* \rightarrow OH + Cl$	0.60	0.54
$H + HCl^* \rightarrow HCl + H$	0.3	0.5	$H + H_2^* \rightarrow H_2 + H$	0.4	0.5
$NO + O_3^* \rightarrow NO_2(^3B_2) + O_2$	0.5	0.3	$SO + O_3^* \rightarrow SO_2(^1B_1) + O_2$	0.25	0.15
$O^+ + N_2^* \rightarrow NO^+ + N$	0.1	—	$N + O_2^* \rightarrow NO + O$	0.24	0.19
$OH + H_2^* \rightarrow H_2O + H$	0.24	0.22	$F + HCl^* \rightarrow HF + Cl$	0.4	0.1
$H + N_2O^* \rightarrow OH + N_2$	0.4	0.2	$O_3 + OH^* \rightarrow O_2 + O + OH$	0.02	0
$H_2 + OH^* \rightarrow H_2O + H$	0.03	0	$Cl_2 + NO^* \rightarrow ClNO + Cl$	0	0
$O_3 + NO^* \rightarrow NO_2(^2A_1) + O_2$	0	0	$O_3 + OH^* \rightarrow HO_2 + O_2$	0.02	0

Table 2.5 *Classification of chemical and biochemical reactions for determination of the energy efficiency coefficient α.*

Reaction type	Simple exchange	Simple exchange	Simple exchange	Double exchange
	Bond break in excited molecule	Bond break in not-excited molecule Through complex	Bond break in not-excited molecule Direct	
Endothermic	0.9–1.0	0.8	<0.04	0.5–0.9
Exothermic	0.2–0.4	0.2	0	0.1–0.3
Thermo-neutral	0.3–0.6	0.3	0	0.3–0.5

where the subscripts (1) and (2) represent direct ($A+BC$) and reverse ($AB + C$) reactions, respectively. The value of the coefficient α (which is actually equal to $\Delta E_a / E_v$) is then determined via:

$$\alpha = \frac{\gamma_1 E_a^{(1)}}{\gamma_1 E_a^{(1)} + \gamma_2 E_a^{(2)}} = \frac{E_a^{(1)}}{E_a^{(1)} + \frac{\gamma_1}{\gamma_2} E_a^{(2)}}. \tag{2.73}$$

Usually the exchange force parameters for direct and reverse reactions γ_1 and γ_2 are similar, that is $\gamma_1/\gamma_2 \approx 1$. This leads to the approximate but very convenient *Fridman-Macheret formula* for calculations of reactions of molecules excited in plasma:

$$\alpha \approx \frac{E_a^{(1)}}{E_a^{(1)} + E_a^{(2)}}. \tag{2.74}$$

Equation (2.74) is in a good agreement with experimental data (see Table 2.4) and reflects the three most important tendencies of the α-coefficients:

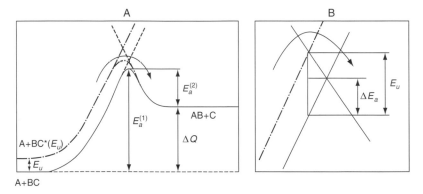

Figure 2.2 *Efficiency of vibrational energy in a simple exchange reaction $A + BC \rightarrow AB + C$: (a) solid curve: reaction profile; dashed line: a vibronic term, corresponding to A interaction with a vibration excited molecule BC^* (E_v); and (b) part of the reaction profile near the top of the reaction energy barrier.*

- the efficiency α of vibrational energy is the highest (i.e., close to 100%) for strongly endothermic reactions with activation energies close to the reaction enthalpy;
- the efficiency α of vibrational energy is the lowest (i.e., close to zero) for exothermic reactions without activation energies;
- the sum of α-coefficients for direct and reverse reactions is equal to unity, that is, $\alpha^{(1)} + \alpha^{(2)} = 1$.

Equation (2.74) does not include any detailed information on dynamics of the elementary chemical reaction and type of excitation. For this reason, we can expect good results when applying the formula to the wide range of chemical reactions of different excited species.

2.3.4 Ion-molecular polarization collisions and reactions in plasma

Ion-molecular reactions play a significant role in plasma-medical processes, especially taking into account that exothermic ion-molecular reactions have no activation barrier and therefore provide the so-called ionic catalysis of chemical and biochemical reactions. The ion-molecular processes start with scattering in polarization potential, leading to the so-called *Langevin capture* of a charged particle and formation of an intermediate ion-molecular complex. If a neutral particle itself has no permanent dipole moment, the ion-neutral charge-dipole interaction is due to dipole moment p_m induced in the neutral particle by the electric field E of an ion:

$$p_m = \alpha \varepsilon_0 E = \alpha \frac{e}{4\pi r^2} \tag{2.75}$$

where r is the distance between the interacting particles and α is polarizability of a neutral atom or molecule, which is numerically about their volume. Typical orbits of relative ion and neutral motion during the polarization scattering are shown in Figure 2.3. When the impact parameter is high, the orbit is hyperbolic; when

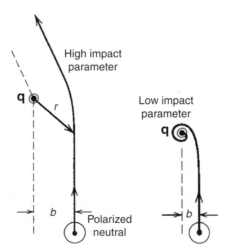

Figure 2.3 *The Langevin scattering in the polarization potential.*

the impact parameter is low, the scattering leads to the *Langevin polarization capture* and formation of an ion-molecular complex. The capture occurs when the charge-dipole interaction

$$p_m E = \alpha \frac{e}{4\pi r^2} \frac{e}{4\pi \varepsilon_0 r^2}$$

becomes of the order of kinetic energy $Mv^2/2$, where M is the reduced mass and v the relative velocity. The *Langevin cross-section* of the ion-molecular polarization capture can then be found as $\sigma_L \approx \pi r^2$, that is:

$$\sigma_L = \sqrt{\frac{\pi \alpha e^2}{\varepsilon_0 M v^2}}. \tag{2.76}$$

The Langevin cross-section is inversely proportional to velocity. The Langevin capture rate coefficient $k_L \approx \sigma_L v$ in this case does not depend on velocity and therefore does not depend on temperature:

$$k_L = \sqrt{\frac{\pi \alpha e^2}{\varepsilon_0 M}}. \tag{2.77}$$

Numerically, the Langevin capture rate coefficient can be defined:

$$k_L^{\text{ion/neutral}} = 2.3 \times 10^{-9} \times \sqrt{\frac{\alpha}{M}}. \tag{2.78}$$

where the Langevin capture rate coefficient is expressed in $cm^3\ s^{-1}$; polarizability α is in $10^{-24}\ cm^3$ units; and reduced mass M is in atomic mass units (amu). The typical value of the rate coefficient for ion-molecular reactions is $10^{-9}\ cm^3\ s^{-1}$, which is 10 times higher than the gas-kinetic rate for neutral particles ($k_0 \approx 10^{-10}\ cm^3\ s^{-1}$). The above relations described the interaction of a charge with an induced dipole. If an ion interacts with a molecule having permanent dipole moment μ_D, the Langevin capture cross-section is increased as follows:

$$k_L = \sqrt{\frac{\pi e^2}{\varepsilon_0 M}} \left(\sqrt{\alpha} + c\mu_D \sqrt{\frac{2}{\pi T_0}} \right) \tag{2.79}$$

where T_0 is gas temperature and the parameter c ($0 < c < 1$) describes the effectiveness of the dipole orientation in the electric field of an ion. For molecules with high permanent dipole moment, the ratio of the second to the first terms is about $\sqrt{I/T_0}$ where I is an ionization potential. Note that Langevin rate coefficient for electron-neutral collision is defined:

$$k_L^{\text{electron/neutral}} = 10^{-7} \times \sqrt{\alpha}. \tag{2.80}$$

where the Langevin capture rate coefficient is expressed in $cm^3\ s^{-1}$ and polarizability α is in $10^{-24}\ cm^3$ units. The numerical value of the Langevin rate coefficient for the electron-neutral collision is about $10^{-7}\ cm^3\ s^{-1}$ and does not depend on temperature.

Consider now the *ion–atomic charge transfer processes*. The charge exchange processes refer to an electron transfer from a neutral particle to a positive ion or from a negative ion to a neutral particle. If the reaction has no defect of electronic energy ΔE it is called *resonant charge transfer*; otherwise, it is referred to as *non-resonant*. The resonant charge transfer is non-adiabatic and characterized by a large cross-section. Let us consider charge exchange between a neutral particle B and a positive ion A^+, assuming particle B/B^+ is at rest:

$$A^+ + B \rightarrow A + B^+. \tag{2.81}$$

The cross-section of the process, which is related to electron tunneling from B to A^+, leading to the resonant or not-energy-limited charge exchange is:

$$\sigma_{\text{ch.tr}}^{\text{tunn}} \approx \frac{1}{I_B} \left(\frac{\pi \hbar^2}{8me} \right) \left(\ln \frac{I_B d}{\hbar} - \ln v \right)^2. \tag{2.82}$$

where I_B is the ionization potential of atom B.

Numerically, Equation (2.82) can be written:

$$\sqrt{\sigma_{\text{ch.tr}}^{\text{tunn}}} \approx \frac{1}{\sqrt{I_B}} \left(6.5 \times 10^{-7} - 3 \times 10^{-8} \ln v \right) \tag{2.83}$$

(where cross-section is measured in cm^2, ionization potential in eV and velocity in cm s^{-1}), which can be applied in the velocity range v of 10^5–10^8 cm s^{-1}; the cross-section reaches 10^{-14} cm^2 at 10^5 cm s^{-1}.

2.3.5   Ion-molecular chemical reactions of positive and negative ions

Probably the simplest reaction to qualify as an ion-molecular chemical reaction is the non-resonant charge transfer. The O-to-N electron transfer is an example of non-resonant charge exchange with a 1 eV energy defect:

$$N^+ + O \rightarrow N + O^+, \Delta E = -0.9. \tag{2.84}$$

The ionization potential of oxygen ($I = 13.6$) is lower than that of nitrogen ($I = 14.5$), which is why the electron transfer from oxygen to nitrogen is an exothermic process and the separated $N + O^+$ energy level is located 0.9 eV lower than that for $O + N^+$. The reaction starts with N^+ approaching O by the attractive NO^+ term. This term is crossed with the repulsive NO^+ term and the system undergoes a non-adiabatic transfer, which results in the formation of $O^+ + N$. The cross-section of such exothermic charge exchange reactions at low energies is of the order of resonant cross-sections with tunneling or Langevin capture. The endothermic reactions of charge exchange, such as a reverse process $N + O^+ \rightarrow N^+ + O$, are usually very slow at low energies.

The contribution of the non-resonant charge exchange can be illustrated by the *acidic behavior of non-thermal air plasma*, which plays a significant role in plasma medicine. Ionization of air in non-thermal discharges leads primarily to a very large number of N_2^+ ions (with respect to other positive ions) because of high molar fraction of nitrogen in air. The low ionization potential and high dipole moment of water molecules enable the following charge exchange:

$$N_2^+ + H_2O \rightarrow N_2 + H_2O^+, \quad k(T = 300) = 2.2 \times 10^{-9} \tag{2.85}$$

where T is temperature (K) and k is reaction rate coefficient (cm^3 s^{-1}). As a result of this reaction, the whole ionization process in air can be significantly focused on the formation of water ions H_2O^+, even though the molar fraction of water is low. The generated water ions can then react with neutral water molecules in the relatively fast ion-molecular reaction:

$$H_2O^+ + H_2O \rightarrow H_3O^+ + OH, \quad \Delta H = -12, \quad k(T = 350) = 0.5 \times 10^{-9} \qquad (2.86)$$

where ΔH is enthalpy (kcal mole^{-1}). The production of H_3O^+ ions and OH radicals determines the acidic behavior of air plasma, which is a fundamental basis for plasma application in air purification of different pollutants.

The ion-molecular chemical reactions (e.g., Equation (2.86)) include *rearrangement of chemical bonds*. These elementary processes can be subdivided into many groups as follows:

- $(A)B^+ + C \rightarrow A + (C)B^+$, reactions with an ion transfer (and for a negative ion),
- $A(B^+) + C \rightarrow (B^+) + AC$, reactions with a neutral transfer (and for a negative ion),
- $(A)B^+ + (C)D \rightarrow (A)D + (C)B^+$, double exchange reactions (and for a negative ion),
- $(A)B^+ + (C)D \rightarrow AC^+ + BD$, reconstruction processes (and for a negative ion).

A significant contribution of ion-molecular reactions in different plasma-medical processes is due to the interesting and important fact that most exothermic ion-molecular reactions have no activation energy. Quantum mechanical repulsion between molecules, which provides the activation barrier even in the exothermic reactions of neutrals, can be suppressed by the charge-dipole attraction in the case of ion-molecular reactions. Rate coefficients of the reactions are therefore very high and often correspond to the Langevin relations. The effect obviously applies to both positive and negative ions.

An absence of activation energies in exothermic ion-molecular reactions facilitates organization of chain reactions in ionized media. This concept is known as *plasma catalysis*, which can be responsible for peroxidation and activation of the lipid layer of cell membranes during their plasma treatment.

2.4　Plasma statistics, thermodynamics, and transfer processes

2.4.1　Statistical distributions in plasma: Boltzmann distribution function, Saha equation, and dissociation equilibrium

Plasma reaction rates depend on the probability of relevant elementary processes from a fixed quantum mechanical state with fixed energy, and the number density of particles with this energy and in this particular quantum mechanical state. A straightforward determination of the distribution of the particles in plasma over different energies and different quantum mechanical states is related to detailed physical kinetics (see Fridman and Kennedy, 2004, 2011). Wherever possible, application of quasi-equilibrium statistical distributions is the easiest and the clearest way to describe kinetics and thermodynamics of plasma-chemical systems. Assume an isolated system with energy E consists of N of particles in different states i:

$$N = \sum_i n_i, E = \sum_i E_i n_i. \qquad (2.87)$$

The objective of a statistical approach is to find a distribution function of particles over the different states i, taking into account that the probability to find n_i particles in these states is proportional to the number of ways in which the distribution can be arranged. *Thermodynamic probability* $W(n_1, n_2, \ldots, n_i, \ldots)$ is the probability of having n_1 particles in the 1st state, n_2 particles in the 2nd state, and so on:

$$W(n_1, n_2, \ldots, n_i, \ldots) = A \frac{N!}{N_1! N_2! \ldots N_i! \ldots} = A \frac{N!}{\prod_i n_i!} \tag{2.88}$$

where A is a normalizing factor. Let us find the most probable number of particles $\overline{n_i}$ in a state i, where the probability (Equation (2.88)) and its logarithm:

$$\ln W(n_1, n_2, \ldots, n_i, \ldots) = \ln(AN!) - \sum_i \ln n_i! \approx \ln(AN!) - \sum_i \int_0^{n_i} \ln x \, dx \tag{2.89}$$

have a maximum. Maximizing the function $\ln W$ over many variables requires:

$$0 = \sum_i \left(\frac{\partial \ln W}{\partial n_i} \right)_{n_i = \overline{n_i}} dn_i = \sum_i \ln \overline{n_i} \, dn_i. \tag{2.90}$$

Differentiation of the conservation laws (2.87) results in:

$$\sum_i dn_i = 0, \quad \sum_i E_i dn_i = 0. \tag{2.91}$$

Multiplying Equations (2.91) by parameters $-\ln C$ and $1/T$ respectively, and summing (2.90) gives:

$$\sum_i \left(\ln \overline{n_i} - \ln C + \frac{E_i}{T} \right) dn_i = 0. \tag{2.92}$$

Equation (2.92) equates zero at any independent value of dn_i. This is possible only in the case of the *Boltzmann distribution function*:

$$\overline{n_i} = C \exp \left(-\frac{E_i}{T} \right) \tag{2.93}$$

where C is the normalizing factor related to the total number of particles and T is the temperature related to the average particle energy (same units are assumed for energy and temperature). If level i is degenerated, we should add statistical weight g, showing the number of states with the given quantum number:

$$\overline{n_j} = C g_j \exp \left(-\frac{E_j}{T} \right). \tag{2.94}$$

The Boltzmann distribution can be expressed in terms of number densities N_j and N_0 of particles in j states with statistical weights g_j and the ground state (0) with statistical weight g_0:

$$N_j = N_0 \frac{g_j}{g_0} \exp \left(-\frac{E_j}{T} \right). \tag{2.95}$$

Assume the vibrational energy of diatomic molecules with respect to ground state ($v = 0$) is $E_v = \hbar\omega v$, rotational energy is $E_r = BJ(J + 1)$ and rotational statistical weight is $2J + 1$. As an example, according to Equation (2.95) the *Boltzmann vibrational-rotational distribution* for diatomic molecules is:

$$N_{vJ} = N\frac{B}{T}(2J + 1)\left[1 - \exp\left(-\frac{\hbar\omega}{T}\right)\right]\exp\left[-\frac{\hbar\omega v + BJ(J + 1)}{T}\right]. \tag{2.96}$$

The Boltzmann distribution (2.95) can also be applied to describe the ionization equilibrium $A^+ + e \Leftrightarrow A$ in plasma:

$$\frac{N_e N_i}{N_a} = \frac{g_e g_i}{g_a}\left(\frac{mT}{2\pi\hbar^2}\right)^{3/2}\exp\left(-\frac{I}{T}\right) \tag{2.97}$$

where I is the ionization potential; g_a, g_i and g_e are the statistical weights of atoms, ions and electrons; N_a, N_i and N_e are their number densities; and m is the electron mass.

Equation (2.97) is known as the *Saha equation for ionization equilibrium in plasma*. The Saha equation (2.97) presented above for ionization equilibrium $A^+ + e \Leftrightarrow A$ can be generalized to describe the *dissociation equilibrium* $X + Y \Leftrightarrow XY$, which is especially important in thermal plasma systems. From Equation (2.97), the relation between densities N_X of atoms X, N_Y of atoms Y and N_{XY} of the molecules XY in the ground state can be written:

$$\frac{N_X N_Y}{N_{XY}(v = 0, J = 0)} = \frac{g_X g_Y}{g_{XY}}\left(\frac{\mu T}{2\pi\hbar^2}\right)^{3/2}\exp\left(-\frac{D}{T}\right) \tag{2.98}$$

where g_X, g_Y and g_{XY} are relevant statistical weights; μ is the reduced mass of atoms X and Y; and D is the dissociation energy of the molecule XY.

2.4.2   Complete and local thermodynamic equilibrium in plasma

Complete thermodynamic equilibrium (CTE) is related to uniform plasma, in which chemical equilibrium and all plasma properties are unambiguous functions of temperature. This temperature is assumed to be homogeneous and the same for all degrees of freedom, all components and all possible reactions. In particular, the following five equilibrium statistical distributions should take place for the same temperature T:

1. Maxwell-Boltzmann translational energy distribution (2.1) is valid for all plasma components.
2. Boltzmann distribution (2.95) describes the population of excited states for all plasma components.
3. Saha equation (2.97) applies for ionization equilibrium.
4. Dissociation balance (2.98) and other thermodynamic relations describe chemical equilibrium.
5. The Plank distribution and other equilibrium relations are relevant for spectral density of electromagnetic radiation.

Plasma in CTE conditions cannot be practically realized in the laboratory. Nevertheless, thermal plasmas are sometimes modeled this way for simplicity. To imagine CTE plasma, consider such a large volume of plasma that its central part is homogeneous and not sensitive to boundaries. Electromagnetic plasma radiation can be considered in this case as that of a black body with the plasma temperature. Actually, even thermal plasmas are quite far from these ideal conditions. Most plasmas are optically thin over a wide range of wavelengths, which results in radiation much less intensive that of a black body. Plasma non-uniformity leads to irreversible losses related to conduction, convection and diffusion, which also disturb CTE.

A more realistic approximation is the so-called *local thermodynamic equilibrium* (LTE). Thermal plasma is considered in this case optically thin, and radiation is not required to be in equilibrium. Collisional (not radiative) processes are required to be locally in equilibrium similar to that described above for CTE with a single temperature T, which can differ from point to point in space and time.

2.4.3   Thermodynamic functions of quasi-equilibrium thermal plasma systems

The partition function Q of an equilibrium particle system at temperature T can be expressed as a statistical sum over states s of the particle with energies E_s and statistical weights g_s:

$$Q = \sum_s g_s \exp\left(-\frac{E_s}{T}\right). \tag{2.99}$$

Translational and internal degrees of freedom of the particles of a chemical component i can be considered independently. Their energy can therefore be expressed as the sum $E_s = E^{\mathrm{tr}} + E^{\mathrm{int}}$ and the partition function for plasma volume v as the product:

$$Q_i = Q_i^{\mathrm{tr}} Q_i^{\mathrm{int}} = \left(\frac{m_i T}{\hbar^2}\right)^{3/2} V Q^{\mathrm{int}}. \tag{2.100}$$

Translational partition function corresponds to continuous spectrum statistical weight; the partition functions of internal degrees of freedom depend on the system characteristics more specifically. Plasma thermodynamic functions can be calculated based on total partition function Q_{tot} of all the particles in an equilibrium system. For example, the Helmholtz free energy F as a function of reference energy F_0 is:

$$F = F_0 - T \ln Q_{\mathrm{tot}}. \tag{2.101}$$

Assuming weak interaction between particles, the total partition function can be expressed as a product of partition functions Q_i of a single particle of a chemical component i:

$$Q_{\mathrm{tot}} = \frac{\prod\limits_i Q_i^{N_i}}{\prod\limits_i N_i!}, \tag{2.102}$$

where N_i is the total number of particles of the species i in the system. Taking into account that $\ln N! = N \ln N - N$, the Helmholtz free energy of non-interacting particles is:

$$F = F_0 - \sum_i N_i T \ln\left(\frac{Q_i e}{N_i}\right). \tag{2.103}$$

Internal energy U and pressure can be expressed in this case as:

$$U = U_0 + T^2 \left(\frac{\partial(F/T)}{\partial T}\right)_{V,N_i} = \sum_i N_i \left[\frac{3}{2}T + T^2 \left(\frac{\partial \ln Q_i^{\mathrm{int}}}{\partial V}\right)_{T,N_i}\right] \tag{2.104}$$

and

$$p = -\left(\frac{\partial(F/T)}{\partial V}\right)_{T,N} = \sum_i N_i T \left(\frac{\partial \ln Q_i}{\partial V}\right)_{T,N_i}. \tag{2.105}$$

2.4.4 Non-equilibrium statistics of thermal and non-thermal plasmas

A detailed description of non-equilibrium plasma systems and processes generally requires the application of kinetic models. In some specific systems, however, statistical approaches can be not only simple but also quite successful in describing non-equilibrium plasma, which is discussed below.

The first example is related to thermal discharges with electron temperature deviating from the temperature of heavy particles, which can take place especially in boundary layers separating plasma from electrodes and walls. In this case, the two-temperature statistics and thermodynamics can be developed. These models assume that partition functions depend on two temperatures. Electron temperature determines the partition functions related to ionization processes, whereas chemical processes are determined by the temperature of heavy particles. The partition functions can then be applied to calculate thermodynamic functions, composition and properties. A second example is related to strongly non-equilibrium process of plasma treatment of solid (condensed) phase:

$$A(\text{solid}) + bB^*(\text{gas}) \rightarrow cC(\text{gas}), \tag{2.106}$$

stimulated by selective excitation of particles B. Deviation from the conventional equation for the equilibrium constant:

$$K = \frac{(Q_C)^c}{Q_A(Q_B)^b}. \tag{2.107}$$

can be introduced here by generalization of the partition functions, taking into account separately translational T_t, rotational T_r, and vibrational T_v temperatures, as well as a non-equilibrium population $f_e^X(\varepsilon_e^X)$ of electronically excited states for each reactants and products $X = A, B, C$:

$$Q_X = \sum_e g_e^X f_e^X\left(\varepsilon_e^X\right) \sum_{(t,r,v)} \prod_{k=t,r,v} g_{ke}^X \exp\left(-\frac{\varepsilon_{ke}^X}{T_k^X}\right) \tag{2.108}$$

where g and ε are statistical weights and energies of corresponding states. A statistical approach can be applied in a consistent way for the reactions described by Equation (2.106) if the non-thermal plasma stimulation of the process is limited to electronic excitation of a single state (energy E_b). Population of the excited state can be expressed by the δ-function and Boltzmann factor with an effective electronic temperature T^*:

$$f_e^B(\varepsilon_e^B) = \delta(\varepsilon_e^B - E_B) \exp\left(-\frac{E_B}{T^*}\right). \tag{2.109}$$

Other degrees of freedom are in quasi-equilibrium with a single gas temperature T_0:

$$f_e^{A,C}\left(\varepsilon_e^{A,C}\right) = 1, \quad T_{k=t,r,v}^X = T_0. \tag{2.110}$$

The single excited state approach leads to the quasi-equilibrium constant of the non-equilibrium process:

$$K \approx \exp\left\{-\frac{1}{T_0}\left[(\Delta H - \Delta F^A) - bE_B\right]\right\} = K_0 \exp\left(\frac{bE_B}{T_0}\right) \tag{2.111}$$

where ΔH is the reaction enthalpy; ΔF_A is the free energy change corresponding to heating to T_0; and K_0 is equilibrium constant at temperature T_0. The equilibrium constant can be significantly increased due to the electronic excitation, and equilibrium significantly shifted to the reaction products.

2.4.5 Non-equilibrium statistics of vibrationally excited molecules in plasma and Treanor distribution

The equilibrium distribution of diatomic molecules over vibrationally excited states N_v (v is the vibrational quantum number) follows the Boltzmann formula, Equation (2.96). Vibrational excitation in non-thermal plasma can be much faster than vibrational VT-relaxation; the vibrational temperature T_v can therefore significantly exceed the translational temperature T_0. The vibrational temperature in this case is usually defined as:

$$T_v = \frac{\hbar\omega}{\ln\left(\frac{N_0}{N_1}\right)}. \tag{2.112}$$

If diatomic molecules are considered as harmonic oscillators, the vibrational distribution function follows the same Boltzmann formula (Equation (2.96)) even when $T_v > T_0$. Interesting non-equilibrium statistical phenomena occur however, taking into account anharmonicity. Then VV-exchange is not resonant and translational degrees of freedom become involved in vibrational distribution, which results in strong deviation from the Boltzmann distribution. Considering vibrational quanta as quasi-particles, and using the Gibbs distribution with variable number of quasi-particles v, the relative population of vibrational levels can be expressed:

$$N_v = N_0 \exp\left(-\frac{\mu v - E_v}{T_0}\right) \tag{2.113}$$

where the parameter μ is the chemical potential and E_v is energy of the vibrational level v taken with respect to the zero-level. Comparing Equation (2.113) with the Boltzmann distribution gives:

$$\mu = \hbar\omega\left(1 - \frac{T_0}{T_v}\right). \tag{2.114}$$

The Gibbs distribution (2.113) together with Equations (2.112) and (2.114) lead to a non-equilibrium vibrational distribution of diatomic molecules known as the *Treanor distribution*:

$$f(v, T_v, T_0) = B \exp\left(-\frac{\hbar\omega v}{T_v} + \frac{x_e \hbar\omega v^2}{T_0}\right). \tag{2.115}$$

where x_e is the coefficient of anharmonicity and B is the normalizing factor. Comparison of the parabolic-exponential Treanor distribution with the linear-exponential Boltzmann distribution is depicted in

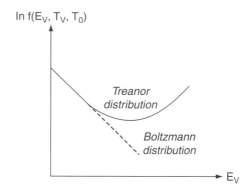

Figure 2.4   *Comparison of the Treanor and Boltzmann distribution functions.*

Figure 2.4. The population of highly vibrationally excited levels at $T_v > T_0$ can be many orders of magnitude higher than predicted by the Boltzmann distribution, even at vibrational temperature. The Treanor distribution results in very high rates of chemical and biochemical reactions stimulated by vibrational excitation in plasma.

2.4.6   Transfer processes in plasma: Electron and ion drift

We begin the discussion of plasma transfer processes with the well-known formula for *electron conductivity*:

$$\sigma = \frac{n_e e^2}{m \nu_{en}}. \tag{2.116}$$

where σ is measured in units of $(\Omega \text{ cm})^{-1}$, n_e is electron number density (cm^{-3}), e and m are charge and mass of an electron and ν_{en} is frequency of electron-neutral collisions (s^{-1}). This formula can be presented in a convenient numerical form:

$$\sigma = 2.82 \times 10^{-4} \frac{n_e}{\nu_{en}}. \tag{2.117}$$

Plasma conductivity is determined by electron density n_e (contribution of ions will be discussed below) and frequency of electron-neutral collisions ν_{en}. Power transferred from electric field to plasma electrons is usually referred to as *Joule heating*:

$$P = \sigma E^2 = \frac{n_e e^2 E^2}{m \nu_{en}}. \tag{2.118}$$

Electron mobility μ_e $(\text{cm}^2 \text{ V}^{-1} \text{ s}^{-1})$ is the coefficient of proportionality between the electron drift velocity ν_d and electric field:

$$\nu_d = \mu_e E, \quad \text{where} \quad \mu_e = \frac{\sigma_e}{e n_e} = \frac{e}{m \nu_{en}}. \tag{2.119}$$

To calculate the electron mobility, the following numerical relation can be used:

$$\mu_e = \frac{1.76 \times 10^{15}}{\nu_{en}}.$$ (2.120)

where electron mobility is expressed in cm^2 $(V\,s)^{-1}$ and electron-neutral collision frequency is expressed in s^{-1}. The relationship between the average energy of the ions $\langle \varepsilon_i \rangle$, gas temperature T_0 and electric field E can be found from the balance of ion-molecular collisions:

$$\langle \varepsilon_i \rangle = \frac{3}{2} T_0 + \frac{M}{2M_i} \left(1 + \frac{M_i}{M} \right)^3 \frac{e^2 E^2}{M_i \nu_{in}^2}$$ (2.121)

where M_i and M are ion and neutral mass and ν_{in} is the frequency of ion-neutral collisions. If the reduced electric field in the plasma is not too high $(E/p < 10\,V\,(cm\,Torr)^{-1})$, the ion energy only slightly exceeds that of neutrals. At high electric fields $(E/p \gg 10)$, ion velocity, collision frequency ν_{in} and ion energy increase with electric field (λ is the ion mean-free-path):

$$\langle \varepsilon_i \rangle \approx \frac{1}{2} \sqrt{\frac{M}{M_i}} \left(1 + \frac{M_i}{M} \right)^{3/2} e E \lambda.$$ (2.122)

At weak and moderate electric fields $(E/p < 10)$, the ion drift velocity is proportional to the electric field and the *ion mobility* μ_i is constant:

$$\vec{v}_{id} = \frac{e\vec{E}}{\nu_{in}\frac{MM_1}{(M+M_1)}}, \qquad \mu_i = \frac{e}{\nu_{in}\frac{MM_1}{(M+M_1)}}.$$ (2.123)

A convenient numerical relation can be used for calculations of ion mobility:

$$\mu_i = \frac{2.7 \times 10^4 \sqrt{1 + M/M_i}}{p\sqrt{A(\alpha/a_0^3)}}$$ (2.124)

where α is polarizability of a neutral particle; A is its molecular mass; and a_0 is the Bohr radius. At high electric fields $(E/p \gg 10)$, the drift velocity is:

$$v_{id} \approx \sqrt[4]{\frac{M_i}{M} \left(1 + \frac{M_i}{M} \right)} \sqrt{\frac{eE\lambda}{M_i}}.$$ (2.125)

Table 2.6   *Coefficients of free diffusion of electrons and ions at room temperature.*

Diffusion	Dp (cm² s⁻¹ Torr)	Diffusion	Dp (cm² s⁻¹ Torr)	Diffusion	Dp (cm² s⁻¹ Torr)
e in He	2.1×10^5	e in Ne	2.1×10^6	e in Ar	6.3×10^5
e in Kr	4.4×10^4	e in Xe	1.2×10^4	e in H$_2$	1.3×10^5
e in N$_2$	2.9×10^5	e in O$_2$	1.2×10^6	N$_2^+$ in N$_2$	40

2.4.7   Transfer processes in plasma: Diffusion of electrons and ions

If electron and ion densities are low, their diffusion in plasma can be considered as independent. Diffusion coefficients can be estimated as a function of thermal velocity, mean free path and collisional frequency ($\langle v_{e,i} \rangle$, $\lambda_{e,i}$, $\nu_{en,in}$):

$$D_{e,i} = \frac{\langle v_{e,i}^2 \rangle}{3\nu_{en,in}} = \frac{\lambda_{e,i}\langle v_{e,i} \rangle}{3}. \tag{2.126}$$

Diffusion coefficients for electrons and ions as defined by Equation (3.126) are inversely proportional to pressure; they are listed for some gases in Table 2.6.

The ratio of free electron diffusion coefficient to electron mobility is proportional to average electron energy, which is known as the *Einstein relation*:

$$\frac{D_e}{\mu_e} = \frac{\frac{2}{3}\langle \varepsilon_e \rangle}{e}. \tag{2.127}$$

Electron and ion diffusion cannot be considered 'free' and independent at high ionization degrees. Electrons move faster than ions and form the charge separation zone with a strong polarization field equalizing the electron and ion fluxes $\vec{\Phi}_e = \vec{\Phi}_i$. This generalized diffusion of electrons and ions $\vec{\Phi}_e = \vec{\Phi}_i$ is called *ambipolar diffusion*:

$$\vec{\Phi}_{e,i} = -D_a \frac{\partial n_{e,i}}{\partial \vec{r}}$$

and is characterized by the diffusion coefficient:

$$D_a = \frac{D_i \mu_e + D_e \mu_i}{\mu_e + \mu_i}. \tag{2.128}$$

In non-equilibrium plasma ($T_e \gg T_i$), the ambipolar diffusion $D_a = (\mu_i/e)\,T_e$ corresponds to temperature of the fast electrons and mobility of the slow ions. The *Debye radius* r_D (cm) is a plasma parameter characterizing the quasi-neutrality. It represents the characteristic size of charge separation and plasma polarization:

$$r_D = \sqrt{\frac{T_e}{e^2 n_e}}. \tag{2.129}$$

If the electron density is high and the Debye radius is small with respect to the size of the discharge system, that is, $r_D \ll R$, then the deviation from quasi-neutrality is small, electrons and ions move 'together', and diffusion is ambipolar. Vice versa, if electron density is relatively low and the Debye radius is large, that is, $r_D \geq R$, then the plasma is not quasi-neutral, electrons and ions move separately, and diffusion is free. Calculations of the Debye radius can be simplified to the numerical formula:

$$r_D = 742\sqrt{\frac{T_e}{n_e}} \tag{2.130}$$

where T_e is in units of eV and n_e in cm^{-3}.

2.4.8 Transfer processes in plasma: Thermal conductivity

The *thermal conductivity coefficient* in a one-component gas without dissociation, ionization and chemical reactions can be estimated as:

$$\kappa \approx \frac{1}{3}\lambda\langle v\rangle n_0 c_v \propto \frac{C_v}{\sigma}\sqrt{\frac{T_0}{M}}. \tag{2.131}$$

where σ is a typical cross-section for neutral collisions, M is the molecular mass and C_v is specific heat at constant volume. Thermal conductivity growth with temperature in plasma at high temperatures can however be much faster than that described by Equation (2.131) because of the influence of dissociation, ionization and chemical reactions. Consider the effect of dissociation and recombination $2A \Leftrightarrow A_2$ on acceleration of thermal conductivity. Molecules are mostly dissociated into atoms in a zone of higher temperature and much less dissociated in a lower-temperature zone. The quasi-equilibrium diffusion of the molecules (D_m) to the higher-temperature zone leads to their intensive dissociation, consumption of dissociation energy E_D and to the related large heat flux:

$$\vec{q}_D = -E_D D_m \nabla n_m = -\left(E_D D_m \frac{\partial n_m}{\partial T_0}\right)\nabla T_0, \tag{2.132}$$

which can be interpreted as acceleration of thermal conductivity related to the dissociation of molecules:

$$\kappa_D = D_m \left(\frac{E_D}{T_0}\right)^2 n_m. \tag{2.133}$$

Thermal conductivity can be also significantly affected by non-equilibrium effects in non-thermal plasmas. Cold gas flowing around a high-temperature thermal plasma zone provides the vibrational-translational non-equilibrium in the area of their contact. This effect in particular is due to the higher rate of vibrational energy transfer from the quasi-equilibrium high-temperature zone with respect to the rate of translational energy transfer. An average value of a vibrational quantum is lower because of anharmonicity at higher vibrational temperatures. The fast VV-exchange during the transfer of the vibrational quanta makes more preferential vibrational quanta transfer from high T_v to lower T_v than in the opposite direction. This Treanor effect in vibrational energy transfer is somewhat similar to the general Treanor effect in plasma statistics, and results in the relative domination of the vibrational energy transfer coefficient over that of the translational energy transfer:

$$\frac{\Delta D_v}{D_0} \approx 4 Q_{01}^{10} q \frac{1 + 30q + 72q^2}{(1 + 2q)^2}, \quad q = \frac{x_e T_v^2}{T_0 \hbar \omega}. \tag{2.134}$$

The effect is significant for molecules such as CO_2, CO and N_2O with VV-exchange provided by long-distance forces, which results in one-quantum transfer probability $Q_{01}^{10} \approx 1$.

2.4.9 Radiation energy transfer in plasma

Radiation is one of the most commonly known plasma properties because of its application in different lighting devices. Radiation also plays an important role in plasma diagnostics, including plasma spectroscopy, in propagation of some electric discharges and even in plasma energy balance.

The total spectral density of quasi-equilibrium plasma emission in continuous spectrum consists of the *bremsstrahlung* and *recombination components*, which can be combined in one general expression:

$$J_\omega d\omega = C \frac{n_i n_e}{T^{1/2}} \Psi \left(\frac{\hbar \omega}{T} \right) d\omega \tag{2.135}$$

where n_i and n_e are electron and ion densities; $C = 1.08 \times 10^{-45}$ (W cm^3 K$^{1/2}$) is a numerical parameter; and $\Psi(x)$ is the dimensionless function defined:

$$\Psi(x) = 1, \quad \text{if} \quad x = \frac{\hbar \omega}{T} < x_g = \frac{|E_g|}{T}, \tag{2.136}$$

$$\Psi(x) = \exp \left[-\left(x - x_g \right) \right], \quad \text{if} \quad x_g < x < x_1, \tag{2.137}$$

$$\Psi(x) = \exp \left[-\left(x - x_g \right) \right] + 2x_1 \exp \left[-\left(x - x_1 \right) \right], \quad \text{if} \quad x = \frac{\hbar \omega}{T} > x_1 = \frac{I}{T} \tag{2.138}$$

where $|E_g|$ is the energy of the first excited state with respect to transition to continuum and I is the ionization potential. Total radiation losses can be found by integration of the spectral density defined by Equation (2.135) as:

$$J = 1.42 \times 10^{-37} \sqrt{T} n_e n_i \left(1 + \frac{|E_g|}{T} \right) \tag{2.139}$$

where J is measured in kW cm^{-3}, temperature is measured in K and electron and ion number densities are measured in cm^{-3}.

The total plasma absorption coefficient in the continuum κ_ω, which corresponds to inverse length of absorption, is measured in units of cm^{-1} and can be expressed as a function of dimensionless parameter $x = \hbar \omega / T$ as:

$$\kappa_\omega = 4.05 \times 10^{-23} \frac{n_e n_i}{(T)^{7/2}} \frac{e^x \Psi(x)}{x^3}. \tag{2.140}$$

The product $n_e n_i$ can be replaced by the gas density n_0 using the Saha equation:

$$\kappa_\omega = \frac{16\pi}{3\sqrt{3}} \frac{e^6 T n_0}{\hbar^4 c \omega^3 (4\pi\varepsilon_0)^3} \frac{g_i}{g_a} \exp\left(\frac{\hbar\omega - I}{T}\right) = 1.95 \times 10^{-7} \frac{n_0}{(T)^2} \frac{g_i}{g_a} \frac{e^{-(x_1-x)}}{x^3}. \qquad (2.141)$$

This relation is usually referred to as the *Unsold-Kramers formula*, where g_a, g_i are statistical weights of an atom and an ion, $x = \hbar\omega/T$, $x_1 = I/T$ and I is the ionization potential. The intensity of radiation I_ω decreases along its path s due to absorption (scattering in plasma can be neglected) and increases because of spontaneous and stimulated emission. The radiation transfer equation in quasi-equilibrium plasma can be then defined:

$$\frac{dI_\omega}{ds} = \kappa'_\omega (I_{\omega e} - I_\omega), \quad \kappa'_\omega = \kappa_\omega \left[1 - \exp\left(-\frac{\hbar\omega}{T}\right)\right] \qquad (2.142)$$

where $I_{\omega e}$ is the quasi-equilibrium Planck radiation intensity, defined:

$$I_{\omega e} = \frac{\hbar\omega^3}{4\pi^3 c^2} \frac{1}{\exp(\hbar\omega/T) - 1}. \qquad (2.143)$$

The optical coordinate ξ is calculated from plasma surface $x = 0$ into the plasma body:

$$\xi = \int_0^x \kappa'_\omega(x) dx, \quad d\xi = \kappa'_\omega(x) dx, \qquad (2.144)$$

and allows the radiation transfer equation to be written in terms of the optical coordinate as:

$$\frac{dI_\omega(\xi)}{d\xi} - I_\omega(\xi) = -I_{\omega e}. \qquad (2.145)$$

If the plasma thickness is d, the radiation intensity on its surface $I_{\omega 0}$ according to Equation (2.145) is:

$$I_{\omega 0} = \int_0^{\tau_\omega} I_{\omega e}[T(\xi)] \exp(-\xi) d\xi, \quad \tau_\omega = \int_0^d \kappa'_\omega dx. \qquad (2.146)$$

The quasi-equilibrium radiation intensity $I_{\omega e}[T(\xi)]$ is shown here as a function of temperature and therefore an indirect function of the optical coordinate. In Equation (2.146), τ_ω is the *optical thickness of plasma* referred to a specific ray and specific spectral range. If optical thickness is small, that is, $\tau_\omega \ll 1$, it is usually referred to as the *transparent* or *optically thin plasma*. In this case the radiation intensity on the plasma surface $I_{\omega 0}$ can be written:

$$I_{\omega 0} = \int_0^{\tau_\omega} I_{\omega e}[T(\xi)] d\xi = \int_0^{\tau_\omega} I_{\omega e} \kappa'_\omega dx = \int_0^d j_\omega dx \qquad (2.147)$$

where the emissivity term $j_\omega = I_{\omega e} \kappa'_\omega$ corresponds to the spontaneous emission. Radiation of the optically thin plasma is a result of summation of independent emission from different intervals dx along the ray. Radiation generated within the plasma volume is able in this case to escape it. Assuming uniformity of plasma parameters, the radiation intensity at the plasma surface can be expressed:

$$I_{\omega 0} = j_\omega d = I_{\omega e} \kappa'_\omega d = I_{\omega e} \tau_\omega \ll I_{\omega e}. \tag{2.148}$$

Radiation intensity of the optically thin plasma is therefore much less ($\tau_\omega \ll 1$) than the equilibrium. The opposite is true when the optical thickness is high ($\tau_\omega \gg 1$), which is referred to as the non-transparent or optically thick system. If temperature is constant, Equation (2.146) yields the quasi-equilibrium Planck intensity for optically thick systems: $I_{\omega 0} = I_{\omega e}(T)$, leading to the Stefan–Boltzmann law of blackbody emission $J = \sigma T^4$. Plasma, however, is usually optically thin for radiation in the continuous spectrum. As a result, the Stefan–Boltzmann law cannot be applied without an emissivity coefficient ε ($J = \varepsilon \sigma T^4$), which coincides in optically thin plasma with the optical thickness, that is, $\varepsilon = \tau_\omega$.

2.5 Plasma kinetics: Energy distribution functions of electrons and excited atoms and molecules

2.5.1 Electron energy distribution functions (EEDF) in plasma: Fokker-Planck kinetic equation

Electron energy distribution functions (EEDFs) in non-thermal discharges can be very sophisticated and quite different from the quasi-equilibrium statistical Boltzmann distribution more relevant for thermal plasma conditions. EEDFs are usually strongly exponential and significantly influence chemical and biochemical reaction rates in plasma. EEDF can obviously be determined from the Boltzmann kinetic equation. Better physical interpretation of EEDF evolution in plasma can however be obtained by using the Fokker-Planck approach. EEDF evolution is considered in this case as an electron diffusion and drift in the space of electron energy. To derive the Fokker-Planck kinetic equation for EEDF $f(\varepsilon)$, we consider the dynamics of energy transfer from an electric field to electrons. An electron has velocity $\vec{v}$ after collision with a neutral and then receives additional velocity during free motion between collisions, corresponding to its drift in electric field $\vec{E}$:

$$\vec{u} = -\frac{e\vec{E}}{m \nu_{\text{en}}} \tag{2.149}$$

where ν_{en} is the frequency of the electron-neutral collisions and e and m are electron charge and mass, respectively. The corresponding change of the electron kinetic energy between two collisions can be expressed:

$$\Delta\varepsilon = \frac{1}{2}m\,(\vec{v} + \vec{u})^2 - m\vec{v}^2 = m\vec{v}\vec{u} + \frac{1}{2}m\vec{u}^2. \tag{2.150}$$

The average contribution of the first term $m\vec{v}\vec{u}$ to $\Delta\varepsilon$ is equal to zero, and the average electron energy increase between two collisions is only related to the square of the drift velocity, that is:

$$\langle\Delta\varepsilon\rangle = \frac{1}{2}m\vec{u}^2. \tag{2.151}$$

Thus electron motion along the energy spectrum (or in energy space) can be considered as a diffusion process. An electron with energy $\varepsilon = mv^2/2$ receives or loses energy of about mvu during one collision, depending on the direction of its motion (along or opposite to the electric field). The energy portion mvu can be considered as the electron 'mean free path along the energy spectrum' (or in energy space). Taking into account the possibility of electron motion across the electric field, we can also introduce a corresponding coefficient of electron diffusion along the energy spectrum as:

$$D_\varepsilon = \frac{1}{3}(mvu)^2 \nu_{en} = \frac{2}{3}mu^2\varepsilon\nu_n. \tag{2.152}$$

As well as the diffusion along the energy spectrum there is also drift in the energy space, related to a permanent average energy gain (2.51) or loss. The average energy loss per collision is mostly due to elastic scattering $(2m/M) \times \varepsilon$ and vibrational excitation $P_{eV}(\varepsilon)\hbar\omega$ in the case of molecular gases, where M and m are a neutral particle and electron mass and $P_{eV}(\varepsilon)$ is the probability of vibrational excitation by electron impact. The drift velocity in energy space can then be expressed:

$$u_\varepsilon = \left(\frac{mu^2}{2} - \frac{2m}{M}\varepsilon - P_{eV}(\varepsilon)\hbar\omega\right)\nu_{en}. \tag{2.153}$$

The EEDF $f(\varepsilon)$ can be considered in this approach as the number density of electrons in energy space, and can be determined from the continuity equation in the energy space referred to as the *Fokker-Planck kinetic equation*:

$$\frac{\partial f(\varepsilon)}{\partial t} = \frac{\partial}{\partial\varepsilon}\left[D_\varepsilon\frac{\partial f(\varepsilon)}{\partial\varepsilon} - f(\varepsilon)u_\varepsilon\right]. \tag{2.154}$$

The steady-state solution of the Fokker-Planck Equation (2.154) for EEDF in non-equilibrium plasma, corresponding to $f(\varepsilon \to \infty) = 0$ and $df/d\varepsilon(\varepsilon \to \infty) = 0$, can be written:

$$f(\varepsilon) = B\exp\left(\int_0^\varepsilon \frac{u_\varepsilon}{D_\varepsilon}d\varepsilon'\right), \tag{2.155}$$

where B is the pre-exponential normalization factor. Using relations for the diffusion coefficient and drift velocity in the energy space, EEDF can be expressed in the integral form:

$$f(\varepsilon) = B\exp\left[-\int_0^\varepsilon \frac{3m^2}{Me^2E^2}\nu_{en}^2\left(1 + \frac{M}{2m}\frac{\hbar\omega}{\varepsilon'}P_{eV}(\varepsilon)\right)d\varepsilon'\right]. \tag{2.156}$$

2.5.1.1   *Maxwellian distribution*

If elastic collisions dominate electron energy losses, that is,

$$P_{eV} \ll \frac{2m}{M}\frac{\varepsilon}{\hbar\omega},$$

and electron-neutral collision frequency $\nu_{en}(\varepsilon)$ is approximately constant, integration of Equation (2.156) gives the Maxwellian distribution (Equation (2.1)) with electron temperature:

$$T_e = \frac{e^2 E^2 M}{3m^2 \nu_{en}^2}. \tag{2.157}$$

2.5.1.2 Druyvesteyn distribution

In contrast to the above, if the mean free path of the electrons λ is constant, then $\nu_{en} = \nu/\lambda$ and the integration of Equation (2.156) gives the exponential-parabolic Druyvesteyn EEDF:

$$f(\varepsilon) = B \exp\left[-\frac{3m}{M}\frac{\varepsilon^2}{(eE\lambda)^2}\right]. \tag{2.158}$$

The Druyvesteyn distribution decreases with energy much faster than the Maxwellian distribution for the same mean energy.

2.5.2 Relation between electron temperature and the reduced electric field

The above-described EEDF allows the reduced electric field to be correlated to average electron energy, which is related to electron temperature:

$$\langle \varepsilon \rangle = \frac{3}{2} T_e$$

even for non-Maxwellian distributions. Such a relation can be derived, for example, from averaging Equation (2.153) (the electron drift velocity in energy space):

$$\frac{e^2 E^2}{m \nu_{en}^2} = \delta \frac{3}{2} T_e \tag{2.159}$$

where the factor δ represents the fraction of electron energy lost in a collision with a neutral particle:

$$\delta \approx \frac{2m}{M} + \langle P_{eV} \rangle \frac{\hbar \omega}{\langle \varepsilon \rangle}. \tag{2.160}$$

Equation (2.160) allows the exact value of the factor $\delta = 2m/M$ to be determined. In molecular gases, however, this factor is usually considered as semi-empirical; for example, $\delta \approx 3 \times 10^{-3}$ in typical conditions of non-equilibrium discharges in nitrogen. Taking into account that $\nu_{en} = n_0 \langle \sigma_{en} \nu \rangle$, $\lambda = 1/n_0 \sigma_{en}$ and $\langle \nu \rangle = \sqrt{8T_e/\pi m}$, Equation (2.159) can be written:

$$T_e = \frac{eE\lambda}{\sqrt{\delta}}\sqrt{\frac{\pi}{12}}. \tag{2.161}$$

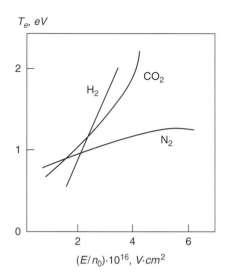

Figure 2.5   *Electron temperature as a function of reduced electric field.*

This relation is in good agreement with both the Maxwell and Druyvesteyn EEDF. It is convenient to rewrite Equation (2.161) as a relation between electron temperature and the reduced electric field E/n_0, that is:

$$T_e = \left(\frac{E}{n_0}\right) \frac{e}{\langle \sigma_{en}\rangle} \sqrt{\frac{\pi}{12\delta}}. \tag{2.162}$$

This linear relation between electron temperature and the reduced electric field is a qualitative one. As one can see from Figure 2.5, the relations $T_e(E/n_0)$ are more complicated.

2.5.3   Non-equilibrium vibrational distribution functions of plasma-excited molecules, Fokker-Plank kinetic equation

Electrons in non-thermal discharges mostly provide excitation of lower vibrational levels, which determine vibrational temperature. The formation of the highly excited and chemically active molecules depends on a variety of different processes in plasma, but mostly on competition of VV-exchange processes and vibrational quanta losses in VT-relaxation. Similarly to the EEDFs define above, the physically clearest approach to the vibrational distribution functions is based on the Fokker-Planck equation. In the framework of this approach, the evolution of vibrational distribution functions is considered as diffusion and drift of molecules in the space of vibrational energies. The continuous distribution function $f(E)$ of molecules over vibrational energies E is considered in this case as the density in energy space. It is found from the continuity equation along the energy spectrum:

$$\frac{\partial f(E)}{\partial t} + \frac{\partial}{\partial E} J(E) = 0 \tag{2.163}$$

where $J(E)$ is the flux of molecules in the energy space, which includes all relaxation and reaction processes. Consider the contribution of VV- and VT relaxation processes as well as chemical reactions in one-component diatomic molecular gases. VT-relaxation of molecules with vibrational energy E and vibrational quantum $\hbar\omega$, considered as diffusion along the vibrational energy spectrum, is characterized by the flux:

$$j_{VT} = f(E)k_{VT}(E, E + \hbar\omega)n_0\hbar\omega - f(E + \hbar\omega)k_{VT}(E + \hbar\omega, E)n_0\hbar\omega \quad (2.164)$$

where n_0 is neutral gas density; $k_{VT}(E + \hbar\omega, E)n_0$ and $k_{VT}(E, E + \hbar\omega)n_0$ are frequencies of direct and reverse VT processes, with ratio $\exp(\hbar\omega/T_0)$. Expanding,

$$f(E + \hbar\omega) = f(E) + \hbar\omega\frac{\partial f(E)}{\partial E}$$

and defining $k_{VT}(E + \hbar\omega, E) \equiv k_{VT}(E)$, we can rewrite Equation (2.164) as:

$$j_{VT} = -D_{VT}(E)\left[\frac{\partial f(E)}{\partial E} + \tilde{\beta}_0 f(E)\right]. \quad (2.165)$$

We have introduced the diffusion coefficient D_{VT} of excited molecules in vibrational energy space, defined:

$$D_{VT}(E) = k_{VT}(E)n_0(\hbar\omega)^2 \quad (2.166)$$

which can be interpreted by taking into account: (1) the vibrational quantum $\hbar\omega$ in energy space corresponds to the mean free path in the conventional coordinate space; and (2) the quantum transfer frequency $k_{VT}(E)n_0$ in the energy space corresponds to frequency of collisions in coordinate space. The temperature parameter $\tilde{\beta}_0$ is:

$$\tilde{\beta}_0 = \frac{1 - \exp\left(-\frac{\hbar\omega}{T_0}\right)}{\hbar\omega} \quad (2.167)$$

which corresponds to an inverse temperature $\tilde{\beta}_0 = \beta_0 = 1/T_0$ if $T_0 > \hbar\omega$. The first term in Equation (2.165) can be interpreted as diffusion and the second term as drift in energy space. The ratio of diffusion coefficient and mobility at high temperature equals $\tilde{\beta}_0 = \beta_0 = 1/T_0$, which is in agreement with the Einstein relation. When VT-relaxation is the dominating process, kinetic Equation (2.163) can be written:

$$\frac{\partial f(E)}{\partial t} = \frac{\partial}{\partial E}\left\{D_{VT}(E)\left[\frac{\partial f(E)}{\partial E} + \frac{1}{T_0}f(E)\right]\right\}. \quad (2.168)$$

At steady-state conditions this Fokker-Planck kinetic equation yields, after integration, the quasi-equilibrium exponential Boltzmann distribution with temperature T_0: $f(E) \propto \exp(-E/T_0)$.

VV-exchange process involves two excited molecules (energies E, E'), which makes the VV-flux non-linear with respect to vibrational distribution function:

$$j_{VV}(E) = k_0 n_0 \hbar\omega \int_0^\infty \left[Q_{E+\hbar\omega,E}^{E',E'+\hbar\omega} f(E + \hbar\omega)f(E') - Q_{E,E+\hbar\omega}^{E'+\hbar\omega,E'} f(E)f(E' + \hbar\omega)\right] dE'. \quad (2.169)$$

Taking into account the defect of vibrational energy in the VV-exchange process $2x_e(E' - E)$, where x_e is the coefficient of anharmonicity and k_0 is the rate coefficient of gas-kinetic collisions, the VV-exchange probabilities Q in the integral (2.169) are related to each other by the detailed equilibrium equation:

$$Q_{E,E+\hbar\omega}^{E'+\hbar\omega,E'} = Q_{E+\hbar\omega,E}^{E',E'+\hbar\omega} \exp\left[-\frac{2x_e(E' - E)}{T_0}\right]. \tag{2.170}$$

The Treanor distribution function makes the VV-flux (Equation (2.169)) equal to zero. The Treanor distribution (see Section 2.4.5, also Figure 2.4) is therefore a steady-state solution of the Fokker-Planck kinetic equation. If VV-exchange is a dominating process and the vibrational temperature T_v exceeds the translational temperature T_0:

$$f(E) = B \exp\left(-\frac{E}{T_v} + \frac{x_e E^2}{T_0 \hbar\omega}\right), \quad \frac{1}{T_v} = \frac{\partial \ln f(E)}{\partial E}\Big|_{E\to 0}. \tag{2.171}$$

The exponentially parabolic Treanor distribution function, providing significant overpopulation of the highly vibrationally excited states, is illustrated in Figure 2.4. To analyze the complex VV-flux (Equation (2.169)), we divide it into linear $j^{(0)}$ and non-linear $j^{(1)}$ components:

$$j_{VV}(E) = j_{vv}^{(0)}(E) + j_{VV}^{(1)}(E). \tag{2.172}$$

The *linear VV flux-component* $j_{VV}^{(0)}(E)$ corresponds to non-resonant VV-exchange between a highly vibrationally excited molecule of energy E with the bulk of lower excited molecules ($0 < E' < T_v$):

$$j_{vv}^{(0)} = -D_{VV}(E)\left[\frac{\partial f(E)}{\partial E} + \left(\frac{1}{T_v} - \frac{2x_e E}{T_0 \hbar\omega}\right)(E)\right]. \tag{2.173}$$

The diffusion coefficient D_{VV} of excited molecules in the vibrational energy space, related to the non-resonant VV-exchange of a molecule of energy E with the bulk of low vibrational energy molecules, is given by:

$$D_{vv}(E) = k_{VV}(E) n_0 (\hbar\omega)^2. \tag{2.174}$$

Solution of the linear kinetic equation $j_{VV}^{(0)}(E) = 0$ with the flux (Equation (2.173)) gives the Treanor distribution function. The *non-linear flux-component* $j_{VV}^{(1)}(E)$ corresponds to the resonant VV-exchange between two highly vibrationally excited molecules ($|E - E'| \leq \delta_{VV}^{-1}$):

$$j_{VV}^{(1)} = -D_{VV}^{(1)} \frac{\partial}{\partial E}\left[f^2(E) E^2 \left(\frac{2x_e}{T_0} - \hbar\omega\frac{\partial^2 \ln f(E)}{\partial E^2}\right)\right]. \tag{2.175}$$

The diffusion coefficient $D_{VV}^{(1)}$ describing the resonance VV-exchange can be expressed:

$$D_{VV}^{(1)} = 3k_0\, n_0\, Q_{10}^{01}\, (\delta_{VV}\hbar\omega)^{-3}. \tag{2.176}$$

One solution of the non-linear kinetic equation $j_{VV}^{(1)}(E) = 0$ with flux (2.175) is again the Treanor distribution function. It is however not the only solution of the equation; another solution is a plateau-like vibrational distribution.

2.5.4   Vibrational distribution functions of excited molecules in plasma controlled by VV-exchange and VT-relaxation processes

Vibrational distributions in non-equilibrium plasma are mostly controlled by VV-exchange and VT-relaxation processes, while excitation by electron impact, chemical reactions, radiation and so on determine averaged energy balance and temperatures. The Fokker-Planck kinetic equation (2.163) gives at steady state $J(E) =$ const. At

$$E \to \infty : \frac{\partial f(E)}{\partial E} = 0, \quad f(E) = 0$$

and thereforeconst$(E) = 0$, which leads to:

$$j_{VV}^{(0)}(E) + j_{VV}^{(1)}(E) + j_{VT}(E) = 0. \tag{2.177}$$

With the fluxes described by Equations (2.173), (2.175) and (2.165), the Fokker-Planck equation for the vibrational distribution function controlled by VV- and VT-relaxation can be written:

$$D_{VV}(E) \left[\frac{\partial f(E)}{\partial E} + \frac{1}{T_v} f(E) - \frac{2x_e E}{T_0 \hbar \omega} f(E) \right]$$
$$+ D_{VV}^{(1)} \frac{\partial}{\partial E} \left[f(E)^2 E^2 \left(\frac{2x_e}{T_0} - \hbar \omega \frac{\partial^2 \ln f(E)}{\partial E^2} \right) \right] + D_{VT}(E) \left[\frac{\partial f(E)}{\partial E} + \tilde{\beta}_0 f(E) \right] = 0. \tag{2.178}$$

The first two terms are related to VV-relaxation and prevail at low vibrational energies; the third term is related to VT-relaxation and dominates at higher vibrational energies. Consider the case of the so-called weak excitation typical for most plasma-medical systems, when both the vibrational temperature ($< (5 - 10) \times 10^3$ K) and hence the Treanor factor are not very high:

$$\frac{x_e T_v^2}{T_0 \hbar \omega} < 1. \tag{2.179}$$

In this case, the non-linear term in the Fokker-Planck kinetic equation (2.178) can be neglected, and we then have:

$$\frac{\partial f(E)}{\partial E} (1 + \xi(E)) + f(E) \left(\frac{1}{T_v} - \frac{2x_e E}{T_0 \hbar \omega} + \xi(E) \tilde{\beta}_0 \right) = 0 \tag{2.180}$$

where $\xi(E)$ is an exponentially growing ratio of VT- to non-resonant VV-relaxation rates. Solution of the differential equation (2.180) gives the distribution:

$$f(E) = B \exp \left[-\frac{E}{T_v} + \frac{x_e E^2}{T_0 \hbar \omega} - \frac{\tilde{\beta}_0 - \frac{1}{T_v}}{2\delta_{VV}} \ln (1 + \xi(E)) \right]. \tag{2.181}$$

This distribution starts with a Treanor function with vibrational temperature at relatively low excitation energies, and finishes with a Boltzmann distribution with translational gas temperature at relatively high excitation energies.

2.5.5 Kinetics of population of electronically excited states in plasma

Transfer of electronic excitation energy in collisions of heavy particles is only effective for a limited number of specific electronically excited states. Even for high levels of electronic excitation, transitions between them are mostly due to collisions with plasma electrons at ionization degrees exceeding 10^{-6}. The population of highly electronically excited states in plasma $n(E)$, provided by energy exchange with electron gas, can also be described by the Fokker-Planck kinetic equation:

$$\frac{\partial n(E)}{\partial t} = \frac{\partial}{\partial E}\left[D(E)\left(\frac{\partial n(E)}{\partial E} - \frac{\partial \ln n^0}{\partial E}n(E)\right)\right] \tag{2.182}$$

where $n^0(E)$ is quasi-equilibrium population corresponding to electron temperature T_e, defined:

$$n^0(E) \propto E^{-\frac{5}{2}} \exp\left(-\frac{E_1 - E}{T_e}\right), \tag{2.183}$$

where E is the absolute value of electronic excitation energy (transition to continuum at $E = 0$) and E_1 is the ground state energy ($E_1 \geq E$).

The diffusion coefficient in energy space $D(E)$ can be defined:

$$D(E) = \frac{4\sqrt{2\pi}e^4 n_e E}{3\sqrt{mT_e}(4\pi\varepsilon_0)^2}\Lambda \tag{2.184}$$

where Λ is the Coulomb logarithm for the electronically excited state.

Taking boundary conditions as $y(E_1) = y_1$, $y(0) = y_e y_i$ (where $y(E) = n(E)/n^0(E)$ is a new variable and y_e and y_i are the electron and ion densities in plasma, divided by their equilibrium values), the steady-state solution of the kinetic equation can be written:

$$y(E) = \frac{y_1 \chi\left(\frac{E}{T_e}\right) + y_e y_i\left[\chi\left(\frac{E_1}{T_e}\right) - \chi\left(\frac{E}{T_e}\right)\right]}{\chi\left(\frac{E_1}{T_e}\right)} \tag{2.185}$$

where $\chi(x)$ is a special function, determined by the integral:

$$\chi(x) = \frac{4}{3\sqrt{\pi}}\int_0^x t^{\frac{3}{2}}\exp(-t)\,dt. \tag{2.186}$$

For electronically excited levels close to continuum ($E \ll T_e \ll E_1$), the relative population is:

$$y(E) \approx y_e y_i\left[1 - \frac{1}{2\sqrt{\pi}}\left(\frac{E}{T_e}\right)^{5/2}\right] + y_1\frac{1}{2\sqrt{\pi}}\left(\frac{E}{T_e}\right)^{5/2} \rightarrow y_e y_i. \tag{2.187}$$

This population of electronically excited states decreases exponentially with effective Boltzmann temperature T_e and has an absolute value corresponding to equilibrium with continuum $y(E) \rightarrow y_e y_i$. In the opposite case, the population of electronically excited states far from continuum ($E \gg T_e$) can be determined as:

$$y(E) = y_1 + y_e y_i \frac{4}{3\sqrt{\pi}} \exp\left(-\frac{E}{T_e}\right) \left(-\frac{E}{T_e}\right)^{3/2} \rightarrow y_1. \tag{2.188}$$

When electronic excitation energy is far from continuum, the population is also exponential with effective temperature T_e; in this case, the absolute value corresponds to equilibrium with ground state.

The Boltzmann distribution of electronically excited states with temperature T_e requires very high ionization degrees, that is, $n_e/n_0 \geq 10^{-3}$ (although domination of energy exchange with electron gas requires only $n_e/n_0 \geq 10^{-6}$). It is mostly due to the influence of some resonance transitions and the non-Maxwellian behavior of electron energy distribution function at the lower ionization degrees.

2.5.6   Macrokinetics of chemical reactions of vibrationally excited molecules

Macrokinetic rates of reactions of vibrationally excited molecules are self-consistent with influence of the reactions on vibrational distribution functions $f(E)$. These can be taken into account by introducing into the Fokker-Planck kinetic equation (2.177) an additional flux related to the reaction:

$$j_R(E) = -\int_E^\infty k_R(E') n_0 f(E') \, dE' = -J_0 + n_0 \int_0^E k_R(E') f(E') \, dE' \tag{2.189}$$

where $J_0 = -j_R(E = 0)$ is the total flux of molecules entering into the chemical reaction (total reaction rate $w_R = n_0 J_0$) and $k_R(E)$ is the microscopic reaction rate coefficient. For the weak excitation regime controlled by non-resonant VV and VT relaxation ($j_{VV}^{(0)} + j_{VT} + j_R = 0$), we have the Fokker-Planck kinetic equation:

$$\frac{\partial f(E)}{\partial E}(1 + \xi(E)) + f(E)\left(\frac{1}{T_v} - \frac{2x_e E}{T_0 \hbar \omega} + \tilde{\beta}_0 \xi(E)\right) = \frac{1}{D_{VV}(E)} j_R(E). \tag{2.190}$$

Solution $f(E)$ of the non-uniform linear equation (2.190) can be determined with respect to solution $f^{(0)}(E)$ of the corresponding uniform equation (Equation (2.181)):

$$f(E) = f^{(0)}(E)\left[1 - \int_0^E \frac{-j_R(E') \, dE'}{D_{VV}(E') f(E')(1 + \xi(E'))}\right]. \tag{2.191}$$

The function $-j_R(E)$ determines the flux of molecules in the energy spectrum which are going to react when they have enough energy ($E \geq E_a$). At relatively low energies ($E < E_a$) where chemical reactions can be neglected:

$$-j_R(E) = \int_{E_a}^\infty k_R(E') n_0 f(E') \, dE' = J_0 = \text{const.}$$

As a result, perturbation of the vibrational distribution function $f^{(0)}(E)$ at $E < E_a$ by chemical reaction can be written:

$$f(E) = f^{(0)}(E) \left[1 - J_0 \int_0^E \frac{dE'}{D_{VV}(E')f^{(0)}(E')(1 + \xi(E'))} \right]. \qquad (2.192)$$

At $E \geq E_a$, $j_R(E) \approx -k_R(E)n_0 f(E)\hbar\omega$ and Equation (2.192) can be solved:

$$f(E) \propto f^{(0)}(E) \exp \left[- \int_{E_a}^E \frac{k_R(E')\hbar\omega \, dE'}{D_{VV}(E')(1 + \xi(E'))} \right], \qquad (2.193)$$

which determines the fast exponential decrease of the vibrational distribution at $E \geq E_a$. Reaction rates of vibrationally excited molecules can be calculated in two specific limits of slow and fast reactions.

2.5.6.1   Fast reaction limit

The fast reaction limit implies that the chemical reaction is fast at $E \geq E_a$:

$$D_{VV}(E = E_a) \ll n_0 k_R(E + \hbar\omega)(\hbar\omega)^2, \qquad (2.194)$$

and the chemical process is limited by VV-diffusion to the threshold $E = E_a$. In this case, the distribution function $f(E)$ falls very fast at $E > E_a$, and we can assume $f(E = E_a) = 0$ in Equation (2.191), yielding

$$1 = \int_0^{E_a} \frac{-j_R(E') \, dE'}{D_{VV}(E')f^{(0)}(E')(1 + \xi(E'))}. \qquad (2.195)$$

This equation allows the total chemical process rate for the fast reaction limit to be found, taking into account that $-j_R(E) = J_0 = \text{const}$ at $E < E_a$:

$$w_R = n_0 J_0 = n_0 \left\{ \int_0^{E_a} \frac{dE'}{D_{VV}(E')f^{(0)}(E')[1 + \xi(E')]} \right\}^{-1}. \qquad (2.196)$$

As seen from Equation (2.196), the chemical reaction rate in this case is determined by frequency of VV-relaxation and by non-perturbed vibrational distribution $f^{(0)}(E)$. The rate described by Equation (2.196) is not sensitive to the characteristics of elementary chemical reaction.

2.5.6.2   Slow reaction limit

The slow reaction limit corresponds to inequality opposite to Equation (2.194). In this case, the population of highly reactive states $E > E_a$ by VV-exchange is faster than the elementary chemical reaction itself. The vibrational distribution function in this case is almost non-perturbed by the chemical reaction $f(E) \approx f^{(0)}(E)$, and the total macroscopic reaction rate coefficient can be found as:

$$k_R^{\text{macro}} = \int_0^\infty k_R(E') f(E') \, dE'. \qquad (2.197)$$

In the slow reaction limit, information about elementary chemical processes is explicitly presented in the formula for rate coefficient.

2.6 Plasma electrodynamics

2.6.1 Ideal and non-ideal plasmas

Plasma electrodynamics is a very important and widespread branch of plasma physics covering topics such as plasma sheaths, plasma oscillations and waves, propagation of electromagnetic waves in plasma, plasma instabilities, magneto-hydro-dynamics of plasma and collective and non-linear plasma phenomena. Only the most general aspects of plasma electrodynamics relevant to plasma-medical systems are discussed here.

  We consider first the general concepts of plasma ideality. The majority of plasmas are somewhat similar to gases from the point of view that electrons and ions move mostly in straight trajectories between collisions. This means that a potential energy $U \propto e^2/4\pi\varepsilon_0 R$, corresponding to the average distance between electrons and ions $R \approx n_e^{-1/3}$, is much lower than their kinetic energy (about T_e):

$$\frac{n_e e^6}{(4\pi\varepsilon_0)^3 T_e^3} \ll 1. \qquad (2.198)$$

Plasmas satisfying this condition are called ideal plasmas. Non-ideal plasma, corresponding to the inverse inequality (2.198), is not found in nature. Even the creation of such plasma in the laboratory is problematic. Interacting particles can be characterized by a non-ideality parameter, which is determined as the ratio of average potential energy of interaction between particles-neighbours and their average kinetic energy. The Coulomb coupling parameter for plasma electrons and ions is:

$$\Gamma_{e(i)} = \frac{e^2 n_{e(i)}^{1/3}}{4\pi\varepsilon_0 T_{e(i)}}. \qquad (2.199)$$

In most plasmas of interest, $\Gamma_{e(i)} \ll 1$ which means they are ideal. In a dusty plasma (or plasma with water droplets, which is of interest for plasma medicine) with particle or droplet density n_d, charge $Z_d e$, and temperature T_d, the non-ideality parameter (which is also called the Coulomb coupling parameter) is:

$$\Gamma_d = \frac{Z_d^2 e^2 n_d^{1/3}}{4\pi\varepsilon_0 T_d}. \qquad (2.200)$$

Because particles can be strongly charged ($Z_d \gg 1$), aerosol plasmas can be non-ideal.

2.6.2   Plasma polarization, Debye shielding of electric field in plasma

External electric fields induce plasma polarization, which prevents penetration of the field inside the plasma. Poisson's equation for space evolution of potential φ induced by the external field E is:

$$\operatorname{div}\vec{E} = -\Delta\varphi = \frac{e}{\varepsilon_0}(n_i - n_e). \tag{2.201}$$

Assuming a Boltzmann distribution for plasma electrons and neglecting ion motion, that is,

$$n_e = n_{e0}\exp\left(+\frac{e\varphi}{T_e}\right); \quad n_i = n_{e0}\exp\left(-\frac{e\varphi}{T_i}\right)$$

yields after expansion in a Taylor series ($e\varphi \ll T_e$):

$$\Delta\varphi = \frac{\varphi}{r_D^2}, \quad r_D = \sqrt{\frac{\varepsilon_0 T_e}{n_{e0}e^2}} \tag{2.202}$$

where r_D is the *Debye radius*, previously discussed regarding ambipolar diffusion (Section 2.4.7). In the one-dimensional (1D) case, Equation (2.202) becomes:

$$\frac{d^2\varphi}{dx^2} = \frac{\varphi}{r_D^2}$$

and describes Debye shielding of an external electric field from plasma boundary E_0 at $x = 0$ along the x axis ($x > 0$):

$$\vec{E} = -\nabla\varphi = \vec{E}_0\exp\left(-\frac{x}{r_D}\right). \tag{2.203}$$

The Debye radius gives the characteristic plasma size scale required for shielding of an external electric field. The same distance is necessary to compensate the electric field of a specified charged particle in plasma. In other words, the Debye radius indicates the scale of plasma quasi-neutrality. The Debye radius and plasma ideality are correlated; the non-ideality parameter Γ is related to the number of plasma particles in the Debye sphere N^D. For plasma consisting of electrons and positive ions:

$$N_{e(i)}^D = n_{e(i)}\frac{4}{3}\pi r_{De(i)}^3 \propto \frac{1}{\Gamma_{e(i)}^{3/2}}. \tag{2.204}$$

The number of electrons and ions in Debye sphere is usually large, that is, $N_{e(i)}^D \gg 1$, which confirms the conventional plasma ideality of $\Gamma \ll 1$.

2.6.3   Plasmas and sheaths

Plasma is assumed to be quasi-neutral ($n_e \approx n_i$) and to provide shielding of external electric fields and field around a specified charged particle, as discussed above. The characteristic plasma size should therefore exceed the Debye radius. Plasma is generally quasi-neutral ($n_e \approx n_i$). Near the walls, however, the quazi-neutrality

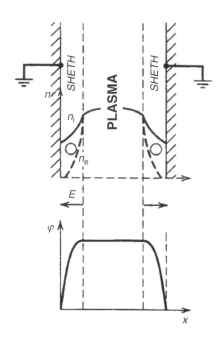

Figure 2.6   *Illustration of plasma and sheaths.*

of plasma is usually lost and the plasma creates positively charged thin layers called sheaths. An example of a sheath between plasma and zero-potential surfaces is illustrated in Figure 2.6.

Formation of the positively charged sheaths is due to the fact that electrons can move much faster than ions. For example, an electron thermal velocity (about $\sqrt{T_e/m}$) exceeds that of ions (about $\sqrt{T_i/M}$) by a factor of 1000. The fast electrons stick to the walls leaving the region near the walls (the sheath) for positively charged ions alone. The positively charged sheath results in the potential profile illustrated in Figure 2.6.

The bulk of plasma is quasi-neutral and hence iso-potential ($\varphi = \text{const}$) according to the Poisson equation (2.201). The positive potential falls sharply near discharge walls, providing a high electric field, acceleration of ions and deceleration of electrons. The sheaths play a significant role in various discharge systems engineered for surface treatment. Most of these systems operate at low gas temperatures ($T_e \gg T_0, T_i$) and low pressures, when the sheath can be considered as *collisionless*. In this case the basic 1D equation governing the DC sheath potential φ in the direction perpendicular to the wall can be obtained from the Poisson equation, energy conservation for the ions and Boltzmann distribution for electrons:

$$\frac{d^2\varphi}{dx^2} = \frac{en_s}{\varepsilon_0}\left[\exp\frac{\varphi}{T_e} - \left(1 - \frac{e\varphi}{E_{is}}\right)^{-1/2}\right] \tag{2.205}$$

where n_s is the plasma density at the sheath edge; $E_{is} = Mu_{is}^2/2$ is the initial energy of an ion entering the sheath (u_{is} is the corresponding velocity); and the potential is assumed to be zero ($\varphi = 0$) at the sheath edge ($x = 0$). Multiplying Equation (2.205) by $d\varphi/dx$ and then integrating the equation assuming boundary conditions $\varphi = 0$, $d\varphi/dx = 0$ at $x = 0$, allows us to find the electric field in the sheath, defined:

$$\left(\frac{d\varphi}{dx}\right)^2 = \frac{2en_s}{\varepsilon_0}\left[T_e\exp\left(\frac{e\varphi}{T_e}\right) - T_e + 2E_{is}\left(1 - \frac{e\varphi}{E_{is}}\right)^{1/2} - 2E_{is}\right]. \tag{2.206}$$

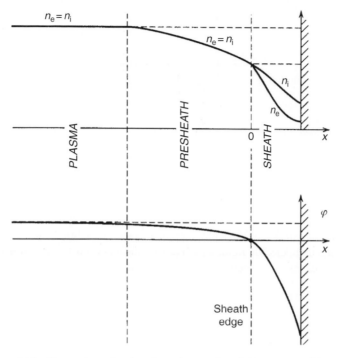

Figure 2.7 *Illustration of a sheath and a presheath in contact with a wall.*

A solution of Equation (2.206) can only exist if its right-hand side is positive. Expanding Equation (2.206) to the second order in a Taylor series shows that the sheath can exist only if the initial ion velocity exceeds the critical velocity, known as the Bohm velocity u_B:

$$u_{is} \geq u_B = \sqrt{T_e/M}. \tag{2.207}$$

The condition for the sheath existence (Equation (2.207)) is usually referred to as the *Bohm sheath criterion*. To provide ions with energy required to satisfy the Bohm criterion, there must be a quasi-neutral region (wider than the sheath, several r_D) with some electric field. This region, illustrated in Figure 2.7, is called the presheath. The minimum presheath potential (between bulk plasma and sheath) should be equal to:

$$\varphi_{\text{presheath}} \approx \frac{1}{2e} M u_B^2 = \frac{T_e}{2e}. \tag{2.208}$$

Balancing the ion and electron fluxes to the floating wall leads to the expression for the change of potential across the sheath in this case as:

$$\Delta\varphi = \frac{1}{e} T_e \ln \sqrt{\frac{M}{2\pi}m}. \tag{2.209}$$

The change of potential across the sheath is usually referred to as the floating potential. Because the ion-to-electron mass ratio M/m is large, the floating potential exceeds the potential across the presheath by a factor of 5–8. In the case of floating potential, numerical calculations give a typical sheath width of about a few Debye radii ($s \approx 3r_D$).

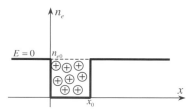

Figure 2.8   *Electron density distribution in plasma oscillations.*

2.6.4   Electrostatic plasma oscillations, Langmuir plasma frequency

The typical space-size characterizing plasma is the Debye radius, which is a linear size of electro-neutrality and shielding of external electric fields. A typical plasma timescale and typical time of plasma response to the external fields is determined by the plasma frequency, illustrated in Figure 2.8. We assume for 1D that all electrons at $x > 0$ are initially shifted to the right over the distance x_0, while heavy ions are not perturbed and remain at rest. This results in an electric field, which pushes the electrons back. If $E = 0$ at $x < 0$, this electric field acting to restore the plasma quasi-neutrality can be found at $x > x_0$ from the 1D Poisson equation as:

$$\frac{dE}{dx} = \frac{e}{\varepsilon_0}(n_i - n_e), \quad E = -\frac{e}{\varepsilon_0}n_{e0}x_0 (\text{at } x > x_0). \tag{2.210}$$

This electric field makes electrons, together with their boundary (at $x = x_0$), move back to the left (Figure 2.8) which results in electrostatic plasma oscillations:

$$\frac{d^2x_0}{dt^2} = -\omega_p^2 x_0, \quad \omega_p = \sqrt{\frac{e^2 n_e}{\varepsilon_0 m}} \tag{2.211}$$

where ω_p is the Langmuir frequency or plasma frequency, which determines the timescale of plasma response to external electric perturbations. This plasma frequency and the Debye radius are related by:

$$\omega_p \times r_D = \sqrt{2T_e/m}. \tag{2.212}$$

The time of plasma reaction to external perturbation ($1/\omega_p$) corresponds to a time required by a thermal electron (velocity $\sqrt{2T_e/m}$) to travel distance r_D required to shield the external perturbation. The plasma frequency ω_p (s^{-1}) depends only on plasma density, and numerically can be calculated as:

$$\omega_p = 5.65 \times 10^4 \sqrt{n_e} \tag{2.213}$$

where n_e is measured in units of cm^{-3}.

2.6.5   Penetration of slow-changing fields into plasma: Skin-effect in plasma

Consider the penetration of a low-frequency electromagnetic field ($\omega < \omega_p$) in plasma. In addition to Ohm's law ($\vec{j} = \sigma \vec{E}$), the Maxwell equation should be taken into account in this case:

$$\text{curl}\vec{H} = \vec{j} + \varepsilon_0 \frac{\partial \vec{E}}{\partial t}. \tag{2.214}$$

Assuming also that the frequency of the field is low with respect to plasma conductivity and that the displacement current (second current term in Equation (2.214)) can therefore be neglected, we can conclude that:

$$\operatorname{curl} \vec{H} = \sigma \vec{E}. \tag{2.215}$$

Taking the electric field from Equation (2.215) and substituting into another Maxwell equation, we have:

$$\operatorname{curl} \vec{E} = -\mu_0 \frac{\partial \vec{H}}{\partial t}, \tag{2.216}$$

which yields the differential relation for electromagnetic field decrease during penetration in plasma ($\omega < \omega_p$):

$$\frac{\partial \vec{H}}{\partial t} = -\frac{1}{\mu_0 \sigma} \operatorname{curl} \operatorname{curl} \vec{H} = -\frac{1}{\mu_0 \sigma} \nabla \left(\operatorname{div} \vec{H} \right) + \frac{1}{\mu_0 \sigma} \Delta \vec{H} = \frac{1}{\mu_0 \sigma} \Delta \vec{H}. \tag{2.217}$$

Equation (2.217) describes the decrease in plasma of amplitude of the low-frequency electric and magnetic fields with the characteristic space scale:

$$\delta = \sqrt{\frac{2}{\omega \mu_0 \sigma}}. \tag{2.218}$$

If this space scale δ is smaller than plasma sizes, then the external fields and currents are located only on the plasma surface layer with a penetration depth δ; this effect is known as the *skin effect*. The boundary layer, where the external fields penetrate and where plasma currents are located, is called the skin layer. The depth of the skin layer depends on the electromagnetic field frequency ($f = \omega/2\pi$) and plasma conductivity. For calculation of the skin-layer depth δ (cm), it is convenient to use:

$$\delta = \frac{5.03}{\sqrt{\sigma f}}. \tag{2.219}$$

2.6.6 Plasma magneto-hydrodynamics: Generalized Ohm's law, Alfven velocity and magnetic Reynolds' number

Plasma motion in a magnetic field induces the electric current $\vec{j} = en_e(\vec{v}_i - \vec{v}_e)$ (where n_e, $\vec{v}_e$ and $\vec{v}_i$ are quasi-neutral plasma density and electron and ion velocities, respectively), which together with the magnetic field influences the plasma motion. In two-fluid magneto-hydrodynamics, the *Navier-Stokes equation* includes the electron's mass m, pressure p_e, velocity and also takes into account friction between electrons and ions:

$$mn_e \frac{d\vec{v}_e}{dt} + \nabla p_e = -en_e \vec{E} - en_e \left(\vec{v}_e \vec{B} \right) - mn_e v_e (\vec{v}_e - \vec{v}_i) \tag{2.220}$$

where v_e is the frequency of electron collisions. A similar Navier-Stokes equation for ions includes the friction term with opposite sign. The first term in Equation (2.220) related to electron inertia can be neglected because of very low electron mass. Denoting the ion's velocity as $\vec{v}$ and plasma conductivity as $\sigma = n_e e^2 / m v_e$, Equation (2.220) can be written as the *generalized Ohm's law*:

$$\vec{j} = \sigma \left[\vec{E} + \left(\vec{v} \vec{B} \right) \right] + \frac{\sigma}{e n_e} \nabla p_e - \frac{\sigma}{e n_e} \left(\vec{j} \vec{B} \right). \tag{2.221}$$

This generalized Ohm's law differs from the conventional relation because it takes into account the electron pressure gradient and the $\vec{j} \vec{B}$ term related to the *Hall effect*. Solution of Equation (2.221) with respect to electric current is complicated because the current is present in two terms. The generalized Ohm's law can be simplified if plasma conductivity is high ($\sigma \to \infty$):

$$\vec{E} + \left(\vec{v} \vec{B} \right) + \frac{1}{e n_e} \nabla p_e = \frac{1}{e n_e} \left[\vec{j} \vec{B} \right]. \tag{2.222}$$

If electron temperature is uniform, the generalized Ohm's law can be rewritten:

$$\vec{E} = - \left(\vec{v}_e \vec{B} \right) - \frac{1}{e} \nabla \left(T_e \ln n_e \right). \tag{2.223}$$

Substituting Equation (2.223) into the Maxwell equation $\operatorname{curl} E = -\partial \vec{B} / \partial t$ gives ($\nabla \times \nabla(T_e \ln n_e) \equiv 0$):

$$\frac{\partial \vec{B}}{\partial t} = \operatorname{curl} \left(\vec{v}_e \vec{B} \right). \tag{2.224}$$

Thus two-fluid magneto-hydrodynamics shows that the magnetic field is frozen into the electron gas. The magneto-hydrodynamic effects are important when fluid dynamics are strongly coupled with the magnetic field. It demands that the 'diffusion' of the magnetic field is less than 'convection'. If the space-scale of plasma is L, it requires:

$$\frac{v B}{L} \gg \frac{1}{\sigma \mu_0} \frac{B}{L^2}, \quad \text{or } v \gg \frac{1}{\sigma \mu_0 L}. \tag{2.225}$$

These requirements of high conductivity and velocity are sufficient to make the magnetic field frozen in plasma, which is the most important feature of magneto-hydrodynamics. Using the concept of magnetic viscosity $D_m = 1/\sigma \mu_0$, the condition of magneto-hydrodynamic behavior described by Equation (2.225) becomes:

$$\operatorname{Re}_m = \frac{v L}{D_m} \gg 1 \tag{2.226}$$

where Re_m is the *magnetic Reynolds number*. Re_m is similar to the conventional Reynolds number, but kinematic viscosity has been replaced by magnetic viscosity. The plasma velocity v in magneto-hydrodynamic systems usually satisfies the approximate balance of dynamic and magnetic pressures:

$$\frac{\rho v^2}{2} \propto \frac{\mu_0 H^2}{2}, \quad \text{or } v \propto v_A = \frac{B}{\sqrt{\rho \mu_0}} \tag{2.227}$$

where $\rho = M n_e$ is plasma density. The characteristic plasma velocity v_A corresponding to the equality of the dynamic and magnetic pressures is called the *Alfven velocity*. The criterion of magneto-hydrodynamic behavior described by Equation (2.226) can be written in terms of the Alfven velocity:

$$\text{Re}_m = \frac{v_A L}{D_m} = B L \sigma \sqrt{\frac{\mu_0}{\rho}} \gg 1. \tag{2.228}$$

Magneto-hydrodynamic plasma behavior takes place not only at high conductivity and magnetic fields, but also at large sizes and low densities.

2.6.7 High-frequency conductivity and dielectric permittivity of plasma

High-frequency plasma conductivity and dielectric permittivity are important concepts to analyze the propagation of electromagnetic waves in plasma. One-dimensional electron motion in the electric field $E = E_0 \cos \omega t = \text{Re}(E_0 e^{i\omega t})$ can be described by the equation:

$$m \frac{du}{dt} = -eE - mu\nu_{en}, \tag{2.229}$$

where ν_{en} is the electron-neutral collision frequency and $u = \text{Re}(u_0 e^{i\omega t})$ is the electron velocity. The relation between the amplitudes of electron velocity and electric field is complex, and can be found from Equation (2.229) as:

$$u_0 = -\frac{e}{m} \frac{1}{\nu_{en} + i\omega} E_0. \tag{2.230}$$

The Maxwell equation $\text{curl}\, \vec{H} = \varepsilon_0 \partial \vec{E}/\partial t + \vec{j}$ allows the total current density to be written as the sum:

$$\vec{j}_t = \varepsilon_0 \frac{\partial \vec{E}}{\partial t} + \vec{j}. \tag{2.231}$$

The first current density component is related to displacement current, and its amplitude in complex form can be written $\varepsilon_0 i \omega E_0$. The second component corresponds to the conductivity current and has amplitude $-e n_e u_0$. The amplitude of the total current density is therefore:

$$j_{t0} = i \omega \varepsilon_0 E_0 - e n_e u_0. \tag{2.232}$$

Taking into account the complex electron mobility (Equation (2.230)), Equation (2.232) for the total current density and hence the Maxwell equation can be rewritten as:

$$j_{t0} = i \omega \varepsilon_0 \left[1 - \frac{\omega_p^2}{\omega(\omega - i\nu_{en})} \right] E_0, \quad \text{curl}\, \vec{H}_0 = i \omega \varepsilon_0 \left[1 - \frac{\omega_p^2}{\omega(\omega - i\nu_{en})} \right] \vec{E}_0 \tag{2.233}$$

where ω_p is the electron plasma frequency. Keeping Equation (2.233) in the form curl $\vec{H}_0 = i\omega\varepsilon_0\varepsilon E_0$, the complex dielectric permittivity of plasma can be introduced:

$$\varepsilon = 1 - \frac{\omega_p^2}{\omega(\omega - i\nu_{en})}. \tag{2.234}$$

The complex dielectric constant defined by Equation (2.234) can be rewritten in terms of its real and imaginary parts:

$$\varepsilon = \varepsilon_\omega - i\frac{\sigma_\omega}{\varepsilon_0\omega}. \tag{2.235}$$

The real component ε_ω is the high-frequency dielectric constant of the plasma:

$$\varepsilon_\omega = 1 - \frac{\omega_p^2}{\omega^2 + \nu_{en}^2} \tag{2.236}$$

and the imaginary component of Equation (2.234) corresponds to the high-frequency plasma conductivity:

$$\sigma_\omega = \frac{n_e e^2 \nu_{en}}{m(\omega^2 + \nu_{en}^2)}. \tag{2.237}$$

Expressions for the high-frequency dielectric permittivity and conductivity can be simplified into two cases: collisionless plasma and static limit. The collisionless plasma limit means $\omega \gg \nu_{en}$. For example, microwave plasma can be considered collisionless at low pressures (about 3 Torr and less). In this case:

$$\sigma_\omega = \frac{n_e e^2 \nu_{en}}{m\omega^2}, \quad \varepsilon_\omega = 1 - \frac{\omega_p^2}{\omega^2}. \tag{2.238}$$

Conductivity in collisionless plasma is proportional to the electron-neutral collision frequency, and the dielectric constant does not depend on the frequency ν_{en}. The ratio of conduction current to polarization current (which actually corresponds to displacement current) can be estimated as:

$$\frac{j_{conduction}}{j_{polarization}} = \frac{\sigma_\omega}{\varepsilon_0\omega|\varepsilon_\omega - 1|} = \frac{\nu_{en}}{\omega}. \tag{2.239}$$

The polarization current in collisionless plasma ($\nu_{en} \ll \omega$) greatly exceeds the conductivity current. In the opposite case for the static limit $\nu_{en} \gg \omega$, conductivity and dielectric permittivity are:

$$\sigma_\omega = \frac{n_e e^2}{m\nu_{en}}, \quad \varepsilon = 1 - \frac{\omega_p^2}{\nu_{en}^2}. \tag{2.240}$$

Conductivity in the static limit coincides with conventional DC conditions.

2.6.8 Propagation of electromagnetic waves in plasma

Electromagnetic wave propagation in plasma is described by the conventional wave equations:

$$\Delta \vec{E} - \frac{\varepsilon}{c^2} \frac{\partial^2 \vec{E}}{\partial t^2} = 0, \quad \Delta \vec{H} - \frac{\varepsilon}{c^2} \frac{\partial^2 \vec{H}}{\partial t^2} = 0. \tag{2.241}$$

Plasma peculiarities are related to the complex dielectric permittivity ε (Equations (2.234) and (2.235)). The dispersion equation for electromagnetic wave propagation in a dielectric medium:

$$\frac{kc}{\omega} = \sqrt{\varepsilon} \tag{2.242}$$

where c is the speed of light, is also valid in plasma with the complex dielectric permittivity ε. Assuming the electric and magnetic fields $\vec{E}, \vec{H}$ are proportional to $\exp(-i\omega t + i\vec{k}\vec{r})$ with real frequency ω, the wave number k should be complex:

$$k = \frac{\omega}{c}\sqrt{\varepsilon} = \frac{\omega}{c}(n + i\kappa) \tag{2.243}$$

where parameter n is the refractive index of electromagnetic wave; phase velocity is $v = \omega/k = c/n$; and the wavelength is $\lambda = \lambda_0/n$ (λ_0 corresponds to vacuum). The wave number κ characterizes the attenuation of electromagnetic waves in plasma: that is, the wave amplitude decreases e^κ times over the length $\lambda_0/2\pi$. Relations between refractive index and attenuation with high-frequency dielectric permittivity and conductivity are:

$$n^2 - \kappa^2 = \varepsilon_\omega, \quad 2n\kappa = \frac{\sigma_\omega}{\varepsilon_0 \omega}. \tag{2.244}$$

Solving this system of equations results in an explicit expression for the attenuation coefficient:

$$\kappa = \sqrt{\frac{1}{2}\left(-\varepsilon_\omega + \sqrt{\varepsilon_\omega^2 + \frac{\sigma_\omega^2}{\varepsilon_0^2 \omega^2}}\right)}. \tag{2.245}$$

Attenuation of the electromagnetic wave is determined by the plasma conductivity; if $\sigma_\omega \ll \varepsilon_\omega \varepsilon_0 \omega$, the electromagnetic field damping can be neglected. The explicit expression for the refractive index is:

$$n = \sqrt{\frac{1}{2}\left(\varepsilon_\omega + \sqrt{\varepsilon_\omega^2 + \frac{\sigma_\omega^2}{\varepsilon_0^2 \omega^2}}\right)}. \tag{2.246}$$

If conductivity is negligible, the refractive index $n \approx \sqrt{\varepsilon_\omega}$. Plasma polarization is negative for $\varepsilon_\omega < 1$, which means $n < 1$ at the low conductivity limit. Combining $n \approx \sqrt{\varepsilon_\omega}$ with Equation (2.238) leads, at low-conductivity conditions, to the dispersion equation for electromagnetic waves in a collisionless plasma:

$$\frac{k^2 c^2}{\omega^2} = 1 - \frac{\omega_p^2}{\omega^2}, \quad \text{or} \quad \omega^2 = \omega_p^2 + k^2 c^2. \tag{2.247}$$

Differentiation of Equation (2.247) relates phase and group velocities of electromagnetic waves in plasma:

$$\frac{\omega}{k} \times \frac{d\omega}{dk} = v_{\text{ph}} v_{\text{gr}} = c^2. \tag{2.248}$$

2.6.9   Plasma absorption and reflection of electromagnetic waves: Bouguer law and critical electron density

The energy flux of electromagnetic waves is determined by the Pointing-vector $\vec{S} = \varepsilon_0 c^2 [\vec{E} \times \vec{B}]$. Taking into account the relation between electric and magnetic fields $\varepsilon \varepsilon_0 E^2 = \mu_0 H^2$, damping of the electromagnetic oscillations in plasma can be written in the form of the *Bouguer law*:

$$\frac{dS}{dx} = -\mu_\omega S, \quad \mu_\omega = \frac{2\kappa\omega}{c} = \frac{\sigma_\omega}{\varepsilon_0 n c} \tag{2.249}$$

where μ_ω is an absorption coefficient (cm^{-1}). Energy flux S decreases by factor e over the length $1/\mu_\omega$. Product $\mu_\omega S$ is the electromagnetic energy dissipated per unit volume of plasma and corresponds to Joule heating:

$$\mu_\omega S = \varepsilon_0 c^2 \langle EB \rangle = \sigma \langle E^2 \rangle. \tag{2.250}$$

If the plasma ionization degree and absorption are relatively low, that is $n \approx \sqrt{\varepsilon} \approx 1$, the absorption coefficient can be simplified via Equations (2.249) and (2.237) to:

$$\mu_\omega = \frac{n_e e^2 n_e \nu_{\text{en}}}{\varepsilon_0 mc(\omega^2 + \nu_{\text{en}}^2)}. \tag{2.251}$$

This relation for electromagnetic wave absorption can be written numerically as:

$$\mu_\omega = 0.106 \times n_e \frac{\nu_{\text{en}}}{\omega^2 + \nu_{\text{en}}^2}. \tag{2.252}$$

At high frequencies ($\omega \gg \nu_{\text{en}}$), the absorption coefficient is proportional to the square of wavelength $\mu_\omega \propto \omega^{-2} \propto \lambda^2$; short electromagnetic waves can therefore propagate in plasma more readily.

If the plasma conductivity is not high, that is $\sigma_\omega \ll \omega \varepsilon_0 |\varepsilon|$, the electromagnetic wave propagates in plasma quite easily if the frequency is high enough. When the frequency decreases, however, the dielectric permittivity $\varepsilon_\omega = 1 - \omega_p^2/\omega^2$ becomes negative and the electromagnetic wave is unable to propagate. Negative values of dielectric permittivity make the refractive index equal to zero ($n = 0$) and attenuation coefficient $\kappa \approx \sqrt{|\varepsilon|}$. The penetration depth of electromagnetic wave in plasma for this case is:

$$l = \frac{\lambda_0}{2\pi \sqrt{|\varepsilon_\omega|}} = \frac{\lambda_0}{2\pi} \left| 1 - \frac{\omega_p^2}{\omega^2} \right|^{-1/2}, \tag{2.253}$$

which does not depend on conductivity and is not related to energy dissipation. Such non-dissipative stopping phenomenon is known as total electromagnetic wave reflection from plasma. The electromagnetic wave propagates from an area with low electron density to areas where plasma density is increasing. The electromagnetic wave frequency is fixed, but the plasma frequency increases together with electron density leading to a decrease of ε_ω. At the point when dielectric permittivity $\varepsilon_\omega = 1 - \omega_p^2/\omega^2$ becomes equal to zero, total reflection takes place. The total reflection of electromagnetic waves occurs when electron density reaches the *critical electron density*, which can be found from $\omega = \omega_p$ as:

$$n_e^{crit} = \frac{\varepsilon_0 m \omega^2}{e^2}; \quad n_e = 1.24 \times 10^4 f^2. \tag{2.254}$$

3

Selected Concepts in Biology and Medicine for Physical Scientists

As evident from the previous chapter, plasma is a complex medium having a variety of different properties that can potentially influence living systems. These influences can be purely physical, that is, directly related to temperature or electric fields for example, or biochemical. They can be subtle and selective affecting some biological molecules and cells more than others and, through them, initiating various cascades of biological consequences. This chapter will provide some background necessary to appreciate the structure of the biological world and living systems as the first step in learning possible consequences of applying plasma to such systems. Given the complexity of living systems and their molecular basis, the reader should be aware that the description below is a broad overview that relies on some oversimplification and may not contain the precision and detail expected of more specialized texts. This overview does not presume detailed prior knowledge of organic chemistry, biology or medicine and is intended as an introduction to the above topics. An important goal of such an introduction is not only to describe the key concepts, but to familiarize the reader with the relevant terminology.

3.1 Molecular basis of life: Organic molecules primer

3.1.1 Essential primer on bonds and organic molecules

Living systems, as we know them on our planet, consist primarily of organic molecules and water. Organic molecules, in turn, can be formed through activity of living systems (they may also be formed in other ways) and have a carbon backbone. Methane (CH_4) is an example of a simple organic molecule. Removing a hydrogen atom (H) from a methane molecule and linking several such units together can form a more complex organic molecule. When two methane molecules are so combined, the resultant molecule is ethane which has the chemical formula C_2H_6. Molecules made up of only H and C are known as hydrocarbons. The formulae and structural representations of several hydrocarbon molecules are provided in Figure 3.1.

In general, a number of smaller organic molecules acting as monomers can be joined together to form larger biopolymers (macromolecules). When two monomers join, a hydroxyl (OH) group is typically removed from one monomer and a hydrogen atom (H) is removed from the other, resulting in release of a water molecule. In this way, biopolymers are constructed by covalently bonding monomers in condensation reactions, where

Plasma Medicine, First Edition. Alexander Fridman and Gary Friedman.
© 2013 John Wiley & Sons, Ltd. Published 2013 by John Wiley & Sons, Ltd.

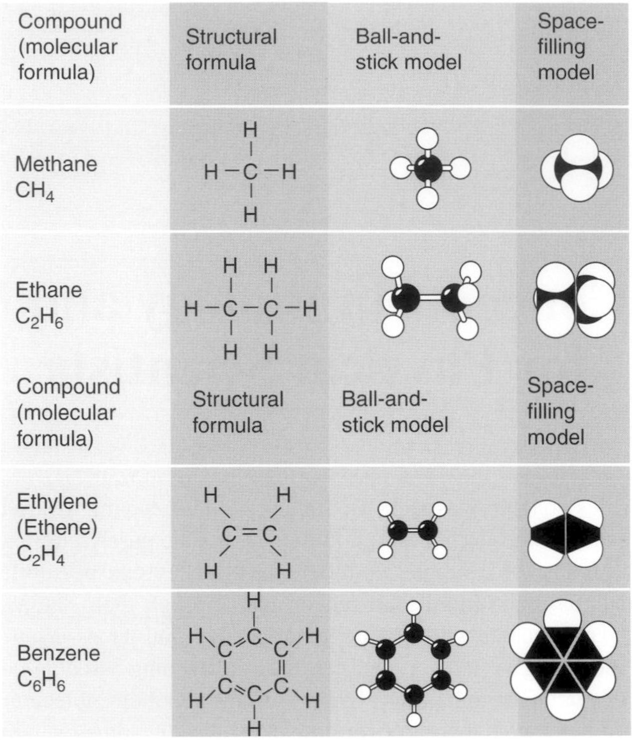

Compound (molecular formula)	Structural formula	Ball-and-stick model	Space-filling model
Methane CH₄	H–C–H with H above and below		
Ethane C₂H₆	H–C–C–H with H above and below each C		
Compound (molecular formula)	Structural formula	Ball-and-stick model	Space-filling model
Ethylene (Ethene) C₂H₄	C=C with H groups		
Benzene C₆H₆	ring structure		

Figure 3.1 *Formulae and structural representations of some hydrocarbon molecules.*

water is released from the monomers. In living systems, certain organic molecules called enzymes (these are mostly proteins) sometimes help carry out this condensation as well as the reverse reaction of biopolymer hydrolysis (that disassembles macromolecules into smaller organic subunits). This reverse reaction of hydrolysis (hydration) consumes water when it breaks down polymers, taking a hydroxyl (OH) group from water and attaching it to one of the organic subunits while attaching the remaining hydrogen (H) to another organic subunit.

Human societies as well as all living forms extract most of the energy required for all their activity from organic molecules primarily through oxidation processes. This energy can be viewed as being stored in various bonds within organic molecules. Table 3.1 below lists some of the bond energies. Energy of any given

Table 3.1 *Typical bond energies in organic molecules.*

Bond	Bond energy (kJ mole⁻¹)	Bond	Bond energy (kJ mole⁻¹)
H–H	432	C=O	799
O–H	460	C–C	347
C–H	410	C=C	611
C–O	360	N=O	623
O=O	494	O–O	142
S–S	250	S–H	360

bond can be affected by the presence of another bond to some extent, but energy that can be extracted from many organic molecules can be calculated with reasonable accuracy using energy per bond approximation.

For example, in the reaction of methane with molecular oxygen:

$$CH_4 + 2O_2 \rightarrow CO_2 + 2H_2O \tag{3.1}$$

based on the information given in Table 3.1, the initial reactants on the left-hand side have total bonding energy $U_i = -(4 \times 410) - (2 \times 494) = -2628$ kJ mole^{-1}, while the right-hand side has $U_f = -(2 \times 799) - (2 \times 460) = -2518$ kJ mole^{-1} of the total bonding energy. The right-hand side has lower 'bonding' energy; the reaction would therefore proceed under appropriate conditions (such as sufficiently high temperature needed to overcome an energy barrier) to lower this potential energy. About 800 kJ mole^{-1} of energy is produced in the process, usually in the form of heat.

While most of the bonds forming organic molecules are covalent (where, roughly speaking, atomic nuclei are attracted to the shared electron distribution that forms an electron pair), these bonds often have associated electric dipole moments. Different electronegativity of the bonding atoms is responsible for this dipole moment. Electronegativity, a concept originally introduced by Linus Pauling in the context of valence bond theory, describes the tendency of an atom to pull electrons in a covalent bond toward itself (although in reality is more complex because neighboring bonds can also affect each other and cannot always be considered independently). The most electronegative atom is fluorine, having an electronegativity value of 4 on the Pauling scale. The second-most electronegative atom is oxygen with a value of 3.44 on the Pauling scale, followed by chlorine and nitrogen with electronegativities of 3.16 and 3.04, respectively. Carbon has an electronegativity of 2.55 on the Pauling scale, which is slighter lower than sulfur (2.58) and somewhat larger than hydrogen (2.20) and phosphorus (2.19).

In defining his electronegativity scale, Linus Pauling proposed the following phenomenological relationship between electronegativity and bond energies (in kJ mole^{-1}):

$$U_{AB} \approx (U_{AA}U_{BB})^{0.5} + 180\,(\Delta\chi_{AB})^2 \tag{3.2}$$

where U_{AB} is the heteronuclear (having nuclei of different atoms) covalent bond energy of the atoms A and B, U_{AA} and U_{BB} are the corresponding homonuclear bond energies and $\Delta\chi_{AB}$ is the electronegativity difference. This relationship clearly reveals a correlation between energy release from bonds and change of bond polarities in a reaction. According to the above relationship, a bond becomes purely ionic when one of the atoms A or B does not form a stable covalent bond with itself, which is the case for metals such as sodium (Na), potassium (K) and others. Indeed, compounds such as NaCl and KCl are ionic and bond weakly compared to covalent bonds.

Although there are other methods of defining electronegativity in addition to Pauling's, all the different forms of relative electronegativity of two atoms reliably determine which of the atoms attracts electrons toward itself in a covalent bond. If two atoms in a bond have a relatively small electronegativity difference, the bond is non-polar or weakly polar (has no or weak dipole moment). On the other hand, a relatively large electronegativity difference results in a more polar bond. For example, bonds between carbon and hydrogen (C–H) are weakly polar (and bonding of other hydrogen atoms to the same carbon often further reduces the polarity of each such bond) in contrast to strongly polar bonds between hydrogen and oxygen.

As a note of caution, it should be mentioned that dipole moments of bonds may or may not combine to contribute to dipole moments of molecules. In a water molecule, due to the angle of c. 104° between the two O–H bonds, the dipole moments of the bonds contribute to a total dipole moment of the water molecule. This is the origin of a large dielectric constant of water. In an SF_6 molecule, despite strong polarity of the S–F bonds the molecule has no dipole moment because of the symmetry of the molecule.

Dipole moments of bonds in functional groups (clusters of atoms with a characteristic structure and function) in organic molecules play an important role in determining molecular properties. For example, since living cells are 70–90% water, the degree to which organic molecules interact with water strongly affects their function. Since electric dipoles can interact electrostatically, polar molecules with a sufficiently large dipole moment tend to surround themselves with polar water molecules in order to reduce their electrostatic energy. As a result, strongly polar functional groups, such as those containing C–O and O–H bonds, are typically hydrophilic and help organic molecules dissolve in water with relative ease. On the other hand, weakly polar (or non-polar) functional groups, such as in hydrocarbons containing only C–H bonds, are hydrophobic. Bond polarity in some molecules is so pronounced that this bond is virtually ionic (bond energy due to electronegativity contribution dominates the right-hand side of Equation (3.2)). As already noted, ionic bonds are weaker than covalent bonds and atoms can separate (bonds can ionize) in the presence of nearby water molecules, offering effective electric field screening. This is typically the case with relatively easily ionized carboxyl (COOH) functional group, the presence of which will typically make molecules hydrophilic. More selected functional groups and a classification of molecules are depicted in Figure 3.2.

The change in bond polarities during chemical reactions is the essence of what is commonly referred to as the oxidation reaction. In the language of electrochemistry, these are oxidation/reduction or redox processes. Oxidants in such redox processes are those atoms that are reduced by 'taking' electrons in a reaction that oxidizes atoms that donate the electrons. Redox reactions associated with formation of more polar bonds also lead, according to Pauling's relation (Equation (3.2)), to stronger and more stable bonds and therefore net energy release.

As an example, consider again the reaction of methane and oxygen described by Equation (3.1). As already discussed above, C–H bonds of methane are weakly polar. The bond in the oxygen molecule is completely non-polar. The products of the reaction are both highly polar molecules however because oxygen in carbon dioxide and in the water molecule is highly electronegative and 'takes' the electrons from carbon and hydrogen, respectively. The conversion of methane and oxygen molecules into carbon dioxide and water that produces excess energy therefore also produces net transfer of electrons from some atoms in the reactants to others. This electron transfer is often described as a change in the oxidation state of atoms (usually described by an integer indicating the number of electrons that shifted toward the more electronegative atom). Atoms that lose electrons (from which the electrons shift away as a result of the reaction) reach higher oxidation state or become more oxidized. In the above methane oxidation reaction for example, carbon 'had' all its electrons and an oxidation state of 0 in the methane molecule. It changed its oxidation state to $+4$ in the carbon dioxide molecule by losing all of its electrons to the more electronegative oxygen. At the same time, each hydrogen atom (which had all its electrons and an oxidation state of 0 in the methane molecule) lost an electron and changed its oxidation state to $+1$. Net increase in the oxidation state of all carbons and hydrogen atoms is balanced by the net reduction in the oxidation state of oxygen, which gains the electrons lost by carbon and hydrogen atoms.

In contrast to releasing energy in the form of heat, living systems often convert the energy of the redox reaction into other forms including electrostatic; this is the case in cellular respiration when electrons are transported across membranes of intracellular organelles (mitochondria). In fact, it appears that plasma treatment interacts with cells and tissues at least in part by directly or indirectly influencing redox reactions in cells and cellular respiration.

3.1.2   Main classes of organic molecules in living systems

In living systems there are four main classes of macromolecules that perform a variety of functions: (1) carbohydrates and sugars; (2) lipids (fats and oils); (3) polypeptides (proteins); and (4) nucleic acids.

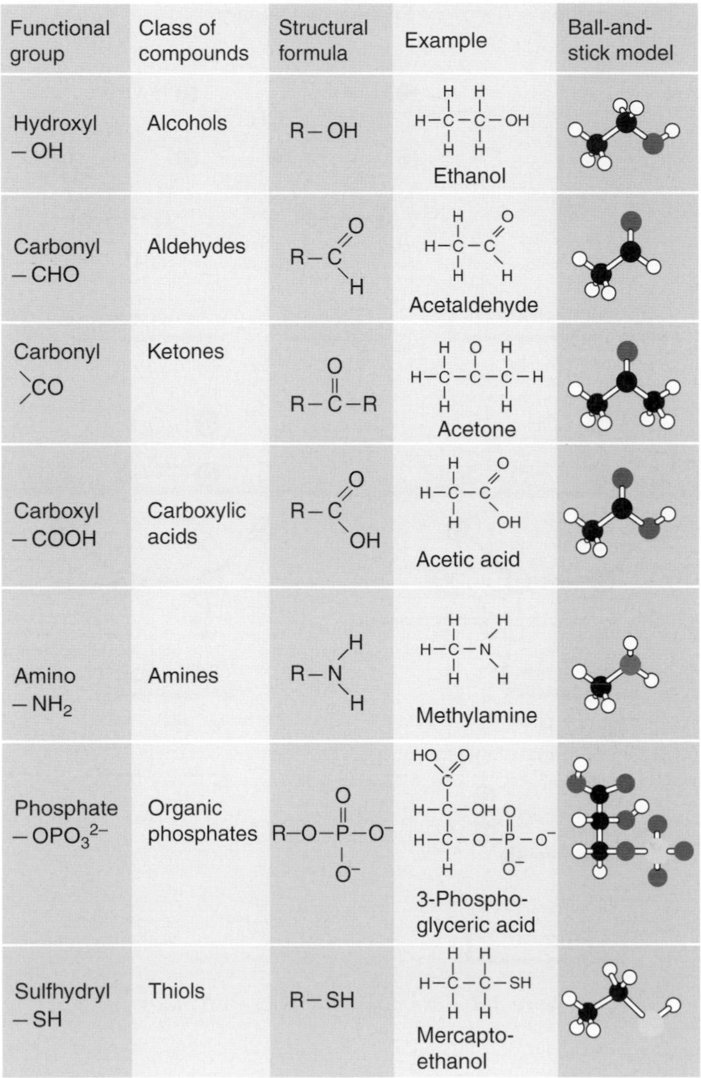

Functional group	Class of compounds	Structural formula	Example	Ball-and-stick model
Hydroxyl — OH	Alcohols	R — OH	Ethanol	
Carbonyl — CHO	Aldehydes	R — C (=O) H	Acetaldehyde	
Carbonyl >CO	Ketones	R — C (=O) — R	Acetone	
Carboxyl — COOH	Carboxylic acids	R — C (=O) OH	Acetic acid	
Amino — NH₂	Amines	R — N (H)(H)	Methylamine	
Phosphate — OPO₃²⁻	Organic phosphates	R — O — P(=O)(O⁻) — O⁻	3-Phospho-glyceric acid	
Sulfhydryl — SH	Thiols	R — SH	Mercapto-ethanol	

Figure 3.2 *Some classes of organic molecules and their relationship to specific functional groups.*

3.1.2.1 Carbohydrates

Carbohydrates (synonymous with saccharides in biochemistry) consist of carbon, oxygen and hydrogen where the number of hydrogen atoms is about twice as large as the number of oxygen atoms which, in turn, is about the same as the number of carbon atoms. Carbohydrates are used for a relatively short-term and intermediate-term energy storage (starch for plants and glycogen for animals). They are also employed as structural components in some cells (cellulose in the cell walls of plants and many protists and chitin in the exoskeleton of insects and other arthropods).

Sugars are structurally the simplest carbohydrates. They are the key building materials that make up other types of carbohydrates. Monosaccharides (see examples in Figure 3.3) are the simplest and smallest sugar

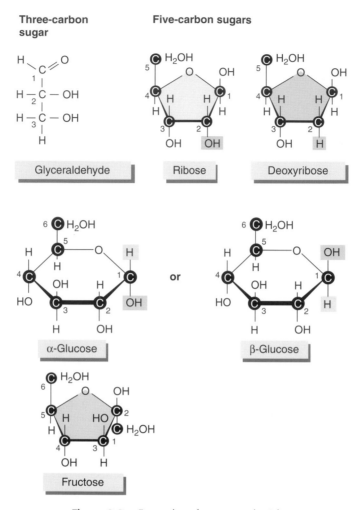

Figure 3.3   *Examples of monosaccharides.*

molecules with a formula $[CH_2O]_n$, where n is typically between 3 and 6. Some important monosaccharides include ribose ($C_5H_{10}O_5$), glucose ($C_6H_{12}O_6$), and fructose. We classify monosaccharides by the number of carbon atoms, the arrangement of atoms (molecules that have the same chemical formula but different atomic arrangements are called isomers) and the types of functional groups present in them. For example, glucose and fructose (illustrated in Figure 3.3) have the same chemical formula ($C_6H_{12}O_6$), but a different structure. Glucose has an aldehyde (internal hydroxyl shown as –OH) and fructose has a keto group (internal double-bond O, shown as =O). This functional group difference, as small as it seems, accounts for the greater sweetness of fructose as compared to glucose. In an aqueous solution, glucose tends to have two isomer structures, α- and β-, with an intermediate straight-chain form (shown in Figure 3.3).

Disaccharides are formed when two monosaccharides are bound together releasing water and using an oxygen atom to connect carbon on originally different subunits, a bond known as the ester bond. Sucrose, a common plant disaccharide, is composed of the monosaccharides glucose and fructose. Lactose, milk sugar, is a disaccharide composed of glucose and the monosaccharide galactose. The maltose that flavors

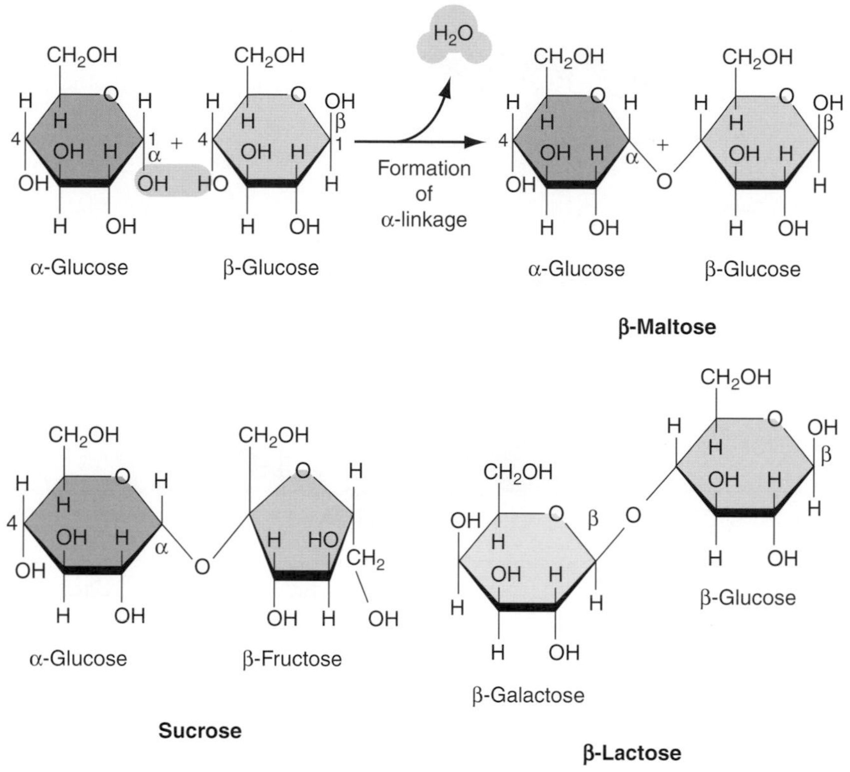

Figure 3.4 Examples of disaccharides.

a malted milkshake is also a disaccharide made of two glucose molecules, bound together as shown in Figure 3.4.

Polysaccharides are larger molecules composed of several individual monosaccharide units. Starch, a common plant polysaccharide, is made up of many glucose units forming one of two types of structures: amylose and amylopectin. Glycogen is another polysaccharide used as an animal long-term energy storage product that accumulates in the vertebrate liver. Cellulose, a biopolymer that forms the fibrous part of the cell wall, is also a polysaccharide. Cellulose is an important and easily obtained part of dietary fiber. As compared to starch and glycogen, which are each made up of mixtures of α and β glucose, cellulose (and the animal structural polysaccharide chitin) is made up of only β glucose. The three-dimensional structure of these polysaccharides is thus constrained into straight microfibrils by the uniform nature of its subunits, which resist the actions of enzymes (such as amylase) that breakdown energy-storing polysaccharides (such a starch).

3.1.2.2 Lipids

Lipids form the second important class of macromolecules. They are involved in long-term energy storage as well as structural (phospholipids in cell membranes) and signaling (hormones) functions. In contrast to carbohydrates, which contain a roughly equal number of carbon and oxygen atoms, lipids consist primarily of carbon and hydrogen atoms with a relatively few oxygen and other atoms. Remarkably, there is no universally accepted definition of the term 'lipid'. Some textbooks describe lipids as a group of naturally occurring

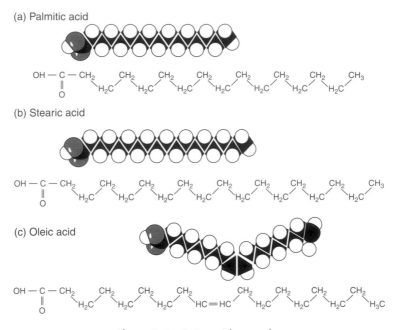

Figure 3.5   *Fatty acid examples.*

compounds which are not soluble in water (remember that C–H bonds and C–C bonds are weakly polar and non-polar), but are soluble in organic solvents such as hydrocarbons, chloroform, benzene, ethers and alcohols. Lipids include a diverse range of compounds such as fatty acids and their derivatives, carotenoids, terpenes, steroids and bile acids. Many of these compounds have little by way of structure or function that unites them. In fact, the above definition can be misleading since some of the substances that are now widely regarded as lipids may be almost as soluble in water as in organic solvents. One alternative definition is that lipids are fatty acids and their derivatives, and substances related biosynthetically or functionally to these compounds.

Fatty acids have 'tails' that are long hydrophobic hydrocarbon (consisting primarily of CH_2 units) chains and a 'head' that is a hydrophilic carboxyl (COOH) group. Some examples of fatty acids are shown in Figure 3.5. Fatty acids are the main component of soap for example, where their tails prefer to attach to oily dirt or bacterial particles and their heads are soluble in water to emulsify and wash away the oily dirt. When a fatty acid tail contains the maximum possible number of hydrogen molecules and no double bonds between carbon atoms, it is referred to as saturated (as in saturated by hydrogen). Palmitic and stearic acids in Figure 3.5 are saturated, while oleic acid is unsaturated.

Triglycerides, which are commonly known as fats and oils, are made from two kinds of molecules: glycerol (a type of alcohol with a hydroxyl group on each of its three carbons) and three fatty acids joined by dehydration synthesis as illustrated in Figure 3.6. Fats, having more of the saturated fatty acids, tend to remain solid around room temperature due to the fact that saturated fatty acids have relatively straight tails and can pack closer together in making a solid. Oils, on the other hand, are more liquid at around room temperature because they contain more unsaturated fatty acids, making close packing more difficult.

Fats and oils function as long-term energy storage materials. Animals convert excess sugars (beyond their glycogen storage capacities) into fats. Fats yield c. 9.3 kcal g^{-1}, while glycogens yield roughly 3.8 kcal g^{-1}. Most plants store excess sugars as starch, although some seeds and fruits have energy stored as oils (e.g. corn

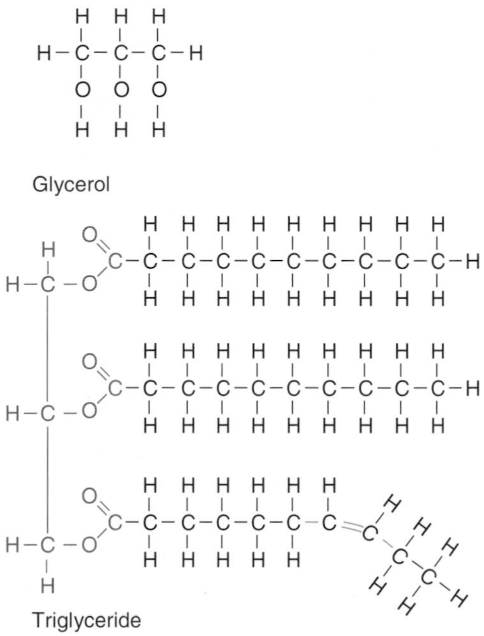

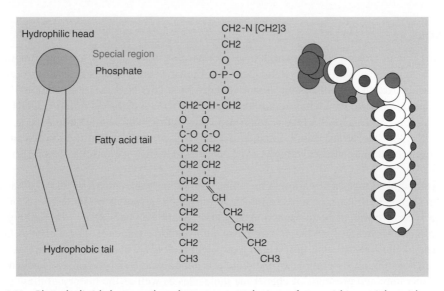

Figure 3.6   *Glycerol and formation of triglycerides from glycerol and fatty acids.*

oil, peanut oil, palm oil, canola oil and sunflower oil). Another use of fats is as insulators and cushions. The human body naturally accumulates some fats in the 'posterior' area. Sub-dermal ('under the skin') fat plays the role of insulation.

Phospholipids and glycolipids are also important examples of lipids. They are the key structural components of cell membranes. Phospholipids, shown in Figure 3.7, are made from glycerol, two fatty acids and, in place

Figure 3.7   *Phospholipids have a phosphate group replacing a fatty acid in a triglyceride molecule.*

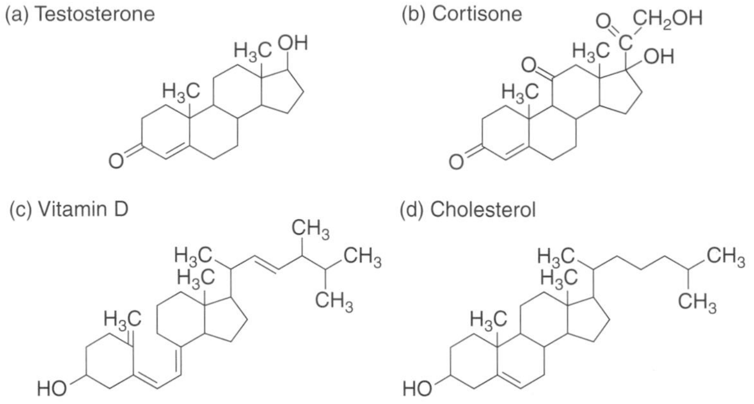

Figure 3.8   *Examples of (a) steroids; (b) vitamins; (c) hormones; and (d) cholesterol.*

of the third fatty acid, a phosphate group (PO_4^-) with some other molecule attached to its other end. The hydrocarbon tails of the fatty acids are still hydrophobic, but the phosphate group end of the molecule is hydrophilic because of the oxygen atoms with all of their pairs of unshared electrons. This means that phospholipids are soluble in both water and oil. Mammalian cell membranes are constructed as a double layer (bilayer) of phospholipids whose tails face each other in the interior of the membrane; phosphate group heads face water, which is on the outside as well as the inside of the cells.

Cholesterol, steroids and vitamins are another important type of lipids, which generally have a variety of biological roles ranging from structural to signaling. Cholesterol has the general structure consisting of three six-sided carbon rings side by side and a five-sided carbon ring, as illustrated in Figure 3.8. The central core of this molecule, consisting of four fused rings, is shared by all steroids including estrogen (estradiol), progesterone, corticosteroids such as cortisol (cortisone), aldosterone, testosterone and Vitamin D. In the various types of steroids, various other groups/molecules are attached around the edges of the carbon rings.

3.1.2.3   Proteins

Proteins play the primary role in control of all known biological systems. They also act as important structural elements in cell membranes and in extracellular tissues. Enzymes are proteins that act as organic catalysts (a catalyst is a chemical that promotes but is not changed by a chemical reaction). The building block of any protein is the amino acid, which has an amine group (NH_2) and a carboxyl group (COOH). The general structures of all amino acids, as well as the specific structures of the 20 biological amino acids, are illustrated in Figure 3.9. It is the side group (represented in the figure as *R*) that is unique to every amino acid. All known living things (including viruses, if they can be considered living) use various combinations of the same 20 amino acids.

Amino acids are linked together by joining the amino end of one molecule to the carboxyl end of another, as illustrated in Figure 3.10, forming a macromolecule referred to as polypeptide. Removal of water allows formation of a type of covalent bond known as a peptide bond. Polypeptide can be roughly viewed as a stretched-out protein. However, a protein is more than a polypeptide sequence. Its function is largely related to a three-dimensional (3D) structure into which a polypeptide folds. This folded structure of a polypeptide sequence is often not uniquely determined by the sequence alone and may depend on environmental conditions such as pH, temperature and presence of chaperone molecules as well as the history of these environmental conditions.

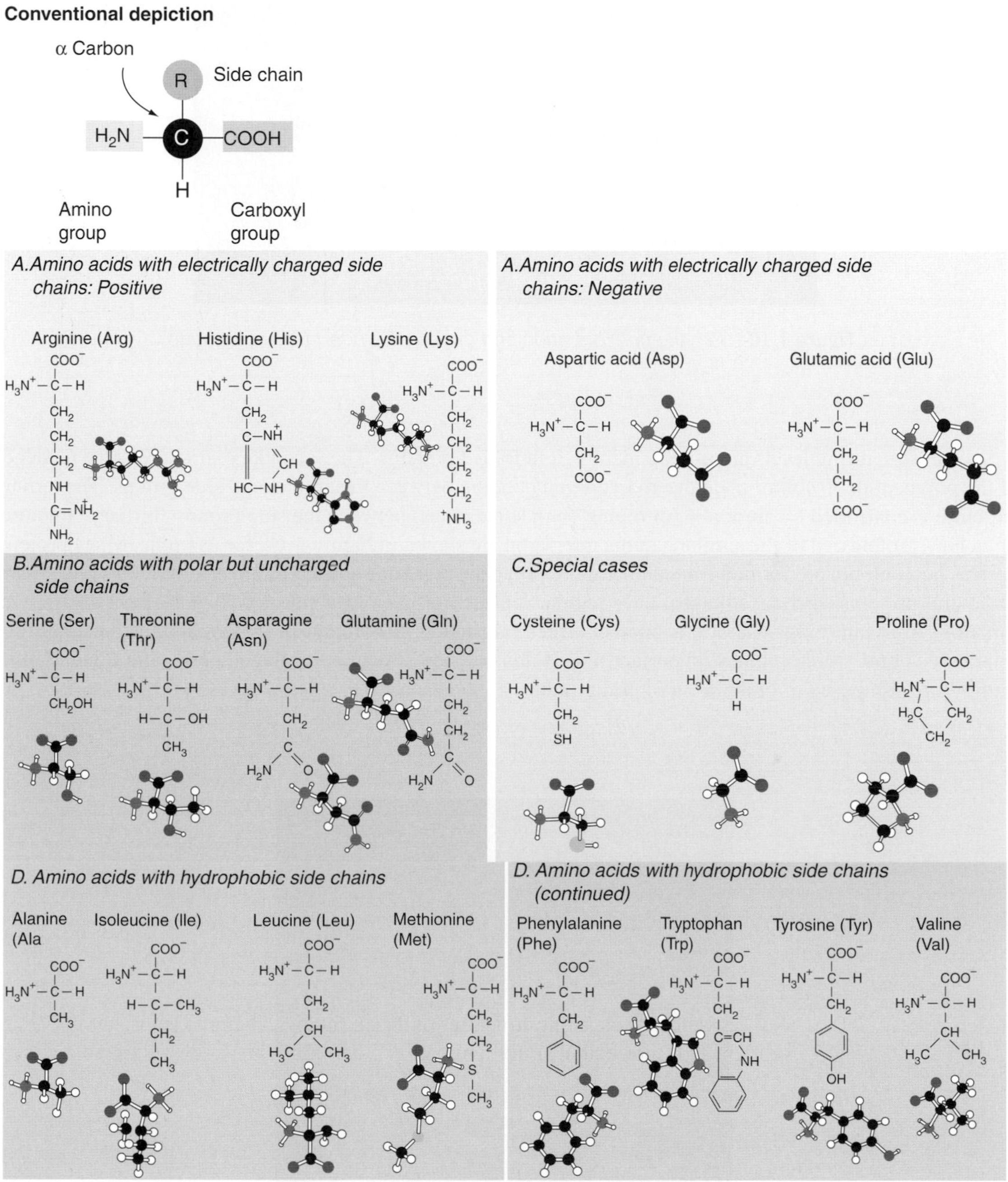

Conventional depiction

α Carbon

R   Side chain

H₂N — C — COOH

H

Amino group   Carboxyl group

A. Amino acids with electrically charged side chains: Positive

Arginine (Arg)   Histidine (His)   Lysine (Lys)

A. Amino acids with electrically charged side chains: Negative

Aspartic acid (Asp)   Glutamic acid (Glu)

B. Amino acids with polar but uncharged side chains

Serine (Ser)   Threonine (Thr)   Asparagine (Asn)   Glutamine (Gln)

C. Special cases

Cysteine (Cys)   Glycine (Gly)   Proline (Pro)

D. Amino acids with hydrophobic side chains

Alanine (Ala)   Isoleucine (Ile)   Leucine (Leu)   Methionine (Met)

D. Amino acids with hydrophobic side chains (continued)

Phenylalanine (Phe)   Tryptophan (Trp)   Tyrosine (Tyr)   Valine (Val)

Figure 3.9 *Structure of amino acids and 20 biological amino acids organized by properties of their side chains.*

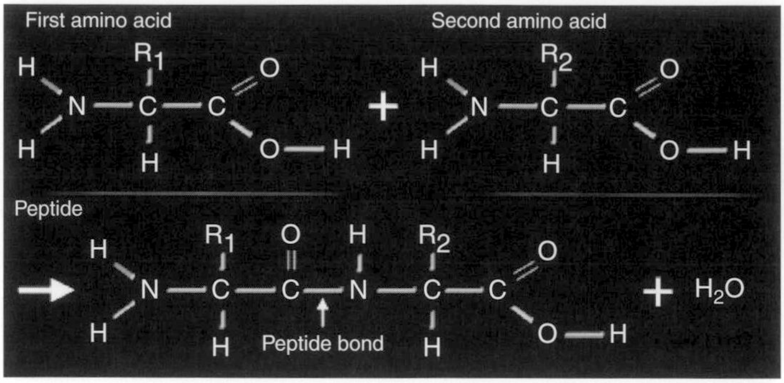

Figure 3.10   *Polypeptide formation and the peptide bond (see color plate).*

Several structuring levels are usually identified in folded proteins. The secondary structure is the tendency of the polypeptide to coil or pleat due to hydrogen bonding between the amino acid side groups. The tertiary structure is controlled by attraction (or repulsion in some cases) between the side groups. Tertiary structures of an HIV protein and of the similar gamma interferon are shown in Figure 3.11, for example. Many proteins such as hemoglobin are formed from one or more polypeptides. Such structures are referred to as quaternary. Structural proteins, such as collagen, have regular repeated primary structures. Collagens have a variety of functions in living things. Keratin is another structural protein. It is found in fingernails, feathers, hair and rhinoceros horns. Microtubules, important in cell division and structures of flagella and cilia (among other things), are composed of globular structural proteins.

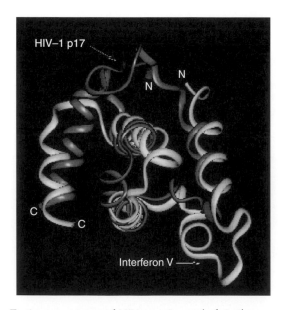

Figure 3.11   *Tertiary structures of HIV proteins and of similar gamma interferon.*

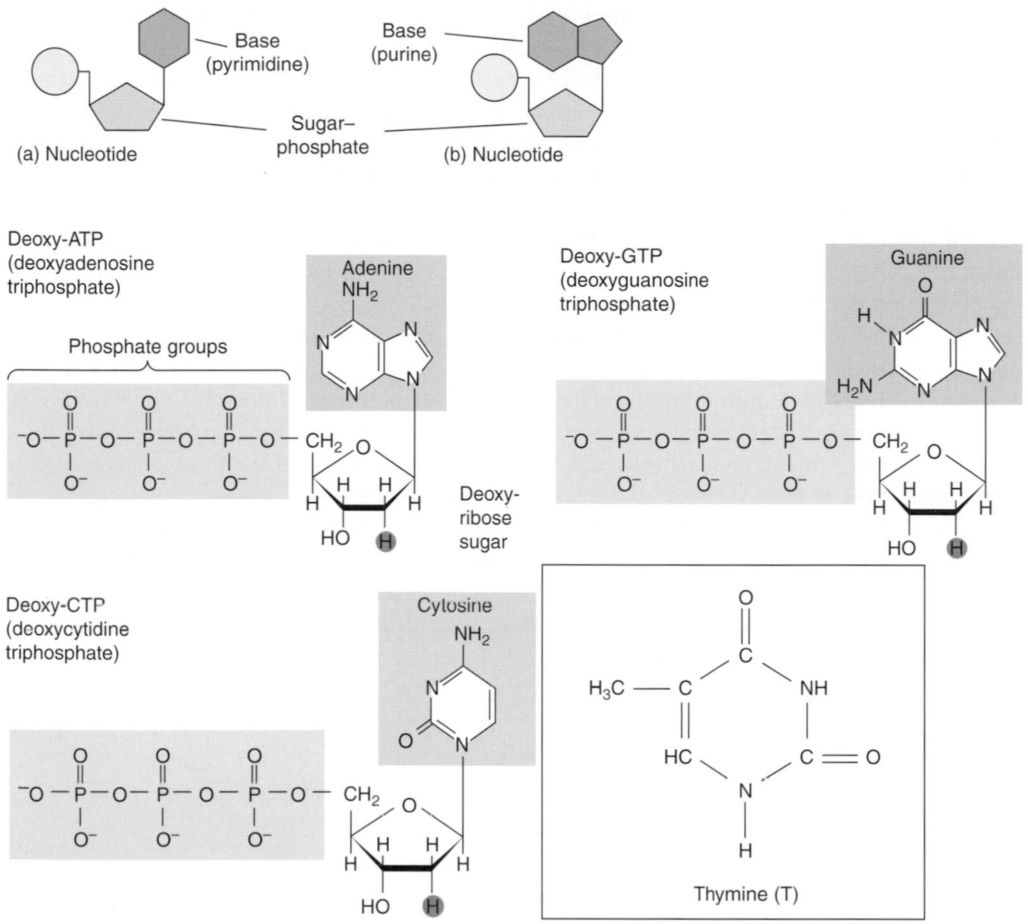

Figure 3.12   *Structure of nucleotides and the four nucleotides used as the basis for the DNA molecule structure: adenine (A), guanine (G), cytosine (C) and thymine (T). Only the nitrogenous base is shown for thymine.*

3.1.2.4   Nucleic acids

Nucleic acids are polymers composed of monomer units known as nucleotides. There are very few different types of nucleotides. The main functions of nucleotide-based molecules are information storage (deoxyribonucleic acid or DNA), protein synthesis (ribonucleic acid or RNA) and energy transfers (adenosine triphosphate or ATP and nicotinamide adenine dinucleotide or NAD^+). Nucleotides, as shown in Figure 3.12, consist of a sugar, a nitrogenous base and a phosphate. There are several common nitrogenous bases. Purines (adenine, guanine, xanthine and hypoxanthine) are double-ring structures, while pyrimidines (cytosine, thymine and uracil) are single-ringed.

DNA is probably the most well-known molecule based on nucleotides. It was first isolated by Friedrich Meischer in 1869 from fish sperm and the pus of open wounds. Since it came from nuclei, Meischer named this new chemical nuclein. Subsequently the name was changed to nucleic acid and lastly to deoxyribonucleic acid (DNA). In 1914, Robert Feulgen discovered that fuchsin dye-stained DNA, which was then found in the nucleus of all eukaryotic cells. During the 1920s, biochemist P.A. Levene analyzed the components of the DNA molecule. He found that, in addition to deoxyribose sugar and a phosphate group, it contained four out of five nitrogenous bases: cytosine, thymine, adenine and guanine (see Figure 3.12). Until about the 1940s, it was widely thought that protein (not DNA) was the carrier of hereditary information. DNA was not shown to be the carrier of genetic inheritance until the experiments of Alfred D. Hershey and Martha Chase in 1952. Using bacteriophage (virus which infects bacteria), a methodology pioneered by Max Delbruck and Salvador Luria in 1940s, Hershey and Chase showed that radioactively labeled DNA and not protein is passed to progeny during bacterial replication.

The actual structure of DNA polymer and the basic idea behind its replication (but not really a detailed mechanism of DNA replication) was specified by James Watson and Francis Crick in 1954 with the help of X-ray diffraction data from Rosalind Franklin and Maurice Wilkens. We now know that DNA consists of two strands, each containing sequences of complementary nucleotide bases (thymine/adenine or T/A and cytosine/guanine or C/G) bound to each other across the strands by hydrogen bonds (T/A requires two hydrogen bonds, while C/G requires three) and forming a double-helix coil structure as shown in Figure 3.13. Within a strand, different nucleotides are covalently bound through phosphate group (the 5′ end of the strand) oxygen on deoxyribose (the 3′ end) by releasing water molecule and two of the three phosphate groups on the original nucleotide.

RNA (Figure 3.14) was discovered after DNA. RNA, which is constructed using ribose sugar and can incorporate uracil as one of its nitrogenous bases, was found to be the molecule responsible for transcribing genetic information. It transports this information to the sites of protein construction and then actually helps to construct proteins according to this information. There are several different types of RNA including messenger

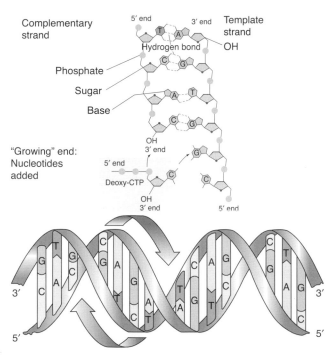

Figure 3.13   *Double helix structure of a DNA molecule. The so-called 5′ end is the phosphate group, while the 3′ end is the OH group on the deoxyribose. Double hydrogen bond holds the T and A nucleotides together, while a triple hydrogen bond holds the C and G nucleotides in adjacent strands of single-stranded DNA.*

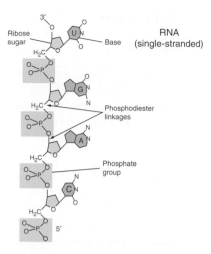

Figure 3.14 *Structure of single-stranded RNA. RNA molecules employ nucleotide Uracil, which is not employed by DNA.*

RNA (mRNA), which serves as the blueprint for construction of a protein copied from DNA, ribosomal RNA (rRNA) used for protein construction and transfer RNA (tRNA), which can be viewed as a vehicle for the proper amino acid delivery to the protein construction site at the right time.

Adenosine triphosphate, better known as ATP, is probably the most important energy-exchange vehicle in cells. It is known as a coenzyme because it is a non-protein molecule that often binds to proteins to create a functional enzyme. Structurally, ATP consists of the adenine nucleotide (ribose sugar, adenine base and phosphate group, PO_4^{2-}) plus two other phosphate groups. This molecule transfers energy in endergonic (energy-absorbing) reactions within the cell that create many other important molecules needed for living functions. The transferred energy is stored in the covalent bonds between phosphates, with the greatest amount of energy (approximately 30 kJ mole^{-1}) in the bond between the second and third phosphate groups. This covalent bond is known as a pyrophosphate bond. When ATP donates the energy of this bond, it is converted into adenosine diphosphate (ADP). The process of transfer of the phosphate group to a protein or another organic molecule is known as phosphorylation, while the removal of this group is called dephosphorylation. Enzymes that catalyze phosphorylation are often called kinases, while those that assist in dephosphorylation are known as phosphatases.

Nicotinamide adenine dinucleotide, abbreviated NAD$^+$, is a coenzyme found in all living cells that also plays an important role in cellular energy-exchange reactions. In cellular metabolism, NAD$^+$ is involved in redox (reduction and oxidation) reactions, carrying electrons from one reaction to another. The compound is a dinucleotide. It consists of two nucleotides joined through their phosphate groups; one nucleotide contains adenine base and the other nicotinamide. The coenzyme is therefore found in two forms in cells: NAD$^+$ is an oxidizing agent that accepts electrons from other molecules and becomes reduced. This reaction forms NADH, which can then be used as a reducing agent to donate electrons.

Nicotinamide adenine dinucleotide phosphate, abbreviated NADP$^+$ is another important coenzyme used in anabolic reactions such as lipid and nucleic acid synthesis, which employ NADPH (the reduced form of NADP$^+$) as a reducing agent. NADP$^+$ differs from NAD$^+$ in the presence of an additional phosphate group on the ribose ring that carries the adenine moiety. The NADPH system is also responsible for generating free radicals in immune cells. These radicals are used to destroy pathogens in a process called the respiratory burst.

3.2   Function and classification of living forms

3.2.1   What is life: Functionality of living forms

Defining a notion of life that is both useful and sufficiently general turns out to be notoriously difficult. In many ways, a lack of definition that clearly distinguishes the living system from its environment has been one of the key problems in the theory of evolution by natural selection as the unifying formalism and the organizing principle of all biology. Instead of a formal definition, the following functional characteristics of life are often mentioned.

- *Organization*: Living systems exhibit a hierarchy of organization levels being subdivided into communities of living organisms, with multicellular organisms being subdivided into cells, cells into organelles and organelles into molecules.
- *Homeostasis*: Living forms, from single cells to their communities to multicellular organisms, tend to maintain their internal environment including temperature, acidity levels, water concentrations and others within certain limits, despite time variations of various environmental parameters (such as temperature, acidity, etc.). Much metabolic energy goes toward this function.
- *Reproduction and heredity*: Living forms are produced by and from other existing living forms. The reproduction certainly involves the passing of some material, but may also involve some variation of this material. As a result, any given living form is rarely identical to its ancestor.
- *Growth and development*: All living forms, even single-celled organisms, grow and develop from the time of their first appearance. When first formed, cells are small and increase in size as they develop until maturity. Multicellular organisms pass through more complicated processes of differentiation and organogenesis.
- *Energy and mass flux*: Living forms exist in a state different from thermodynamic equilibrium usually maintained by a flux of energy and/or mass.

Organization and homeostatic properties of living systems are known among other types of self-organizing systems. The last three properties emphasize non-equilibrium nature of life. Specifically, it is on the basis of the last functional property that viruses are often excluded as living forms. Viruses, being relatively complex molecular assemblies that may include genetic and/or gene transcription molecules, are in or very near thermal equilibrium for most of their existence. However, the same argument can be made regarding some cells (particularly bacterial cells) that may remain dormant for a long time under some external conditions before coming alive in other conditions.

The difficulty of deciding if viruses are non-living or alive is only one example associated with the problem of defining the living system. However, the lack of a clear definition has not stopped substantial progress in life sciences whose practitioners often employ the principle "I know it when I see it", famously expressed by the US Supreme Court Justice Potter Stewart in connection with identifying obscenity.

3.2.2   Major classification of living forms

Organizing all living forms into categories is useful because it not only helps to recognize life in the absence of a uniformly accepted definition, but assists in systematic comparison of different manifestations of life. Using the system originally developed by Carl Linne (popularly known as Linneus), biologists often classify all living forms as one of five kingdoms: Monera, Protista, Fungi, Plantae and Animalia. Figure 3.15 illustrates what is now thought to be the relationship between the different kingdoms.

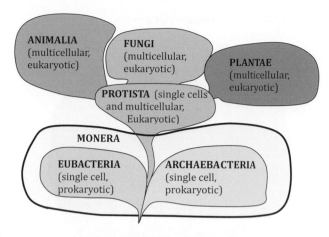

Figure 3.15   *Classification of living forms into five animal kingdoms.*

The subdivision of the kingdom of Monera into Eubacteria and Archaebacteria is somewhat dated. It is now clear that these two sub-kingdoms differ more than originally thought. To emphasize this, the names Bacteria and Archae are often employed today instead of Eubacteria and Archaebacteria. Both are single-cell microorganisms whose cells do not contain kernels (Greek 'karyose'). As a result, inhabitants of the Monera kingdom are referred to as prokaryotes. By 'kernel', biologists most often mean cell nucleus, which contains genetic material (molecules carrying information regarding the possible form a given living organism may assume). In addition, prokaryotes typically do not have other organelles such as mitochondria.

Protista were probably the first of the eukaryotic kingdoms to allow for compartmentalization and dedication of specific areas for specific functions. The chief importance of Protista is their role as a stem group for the remaining kingdoms: plants, animals, and fungi. Major groups within the Protista include the algae, euglenoids, ciliates, protozoa and flagellates.

3.3   Cells: Organization and functions

As mentioned above, life exhibits hierarchical organization. Atoms are organized into molecules, molecules into organelles, organelles into cells and so on. It seems, however, that all living things are composed of one or more cells as the most basic units of life, and the functions of a multicellular organism are a consequence of the types of cells it contains and how these cells are arranged and work together. Cells fall into two broad groups: prokaryotes and eukaryotes. Prokaryotic cells are smaller (as a general rule) and lack much of the internal compartmentalization and complexity of eukaryotic cells. No matter which type of cell we are considering, all known cells have certain features in common such as a cell membrane, DNA and RNA, cytoplasm and ribosomes.

The natural shapes of cells vary. For example, neurons can grow parts called axons that are often many centimeters long. Skeletal muscle cells can also be several centimeters long. Others such as parenchyma (a common type of plant cell) and erythrocytes (red blood cells) are much more equi-dimensional. Cells often change their shape when they attach to firm substrates or other cells. Some cells are encased in a rigid wall which constrains their shape, while others have a flexible cell membrane (and no rigid cell wall). The size of cells is also related to their functions. Eggs (or to use the Latin word, ova) are relatively large, often being the largest cells an organism produces. The large size of many eggs is related to the process of development

Anatomy of a cell

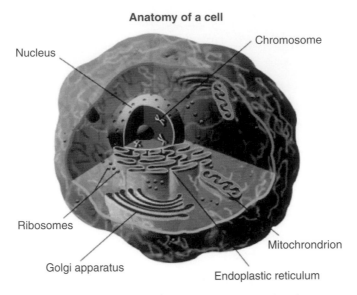

Figure 3.16   *Eukaryotic cell anatomy (see color plate).*

that occurs after the egg is fertilized, when the contents of the egg (now termed a zygote) are used in a rapid series of cellular divisions, each requiring tremendous amounts of energy. In general, cells range in size from small bacteria (about 1 micrometer) to unfertilized eggs produced by birds and fish.

3.3.1   Primary cell components

Just as larger living organisms differ from each other in structure and function, so do many cells. There are also many similarities between different types of cells. Figure 3.16 illustrates the typical structure of a eukaryotic cell, while Figure 3.17 shows the anatomy of a typical prokaryotic (bacterial in this case) cell.

3.3.1.1   The cell envelope: Membranes and walls

The cell membranc functions as a semi-permeable mechanically flexible barrier, allowing very few molecules across it while fencing the majority of organically produced chemicals inside the cell. Electron microscopy examinations of cell membranes have led to the development of the lipid bilayer model (also referred to as the fluid-mosaic model).

As illustrated in Figure 3.18, the most common molecule in the model is the phospholipid, which has a polar (hydrophilic) head and two non-polar (hydrophobic) tails. These phospholipids are aligned tail to tail so the non-polar areas form a hydrophobic region between the hydrophilic heads on the membrane surfaces, facing toward the inside and outside of the membrane. This membrane is termed a bilayer partly because freeze-fracture preparation for electron microscopy is able to split the membrane into two layers. The bilayer membrane is actually fluid-like in that various molecular structures in it can move as if in a fluid when these structures are not otherwise anchored in some way to other molecular structures within cells. In mammalian cells, these anchors and structural support are partly provided by the cytoskeleton filaments. Cholesterol is often an important component of mammalian cell membranes embedded in the hydrophobic areas of the inner

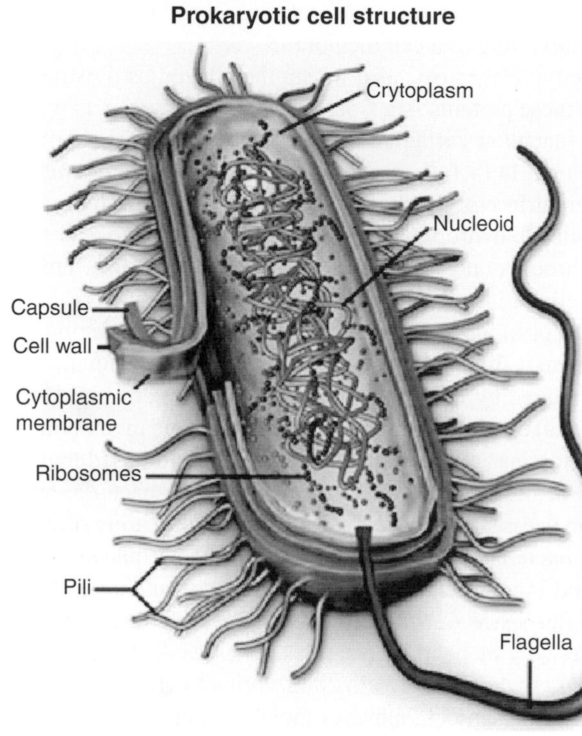

Prokaryotic cell structure

Crytoplasm

Nucleoid

Capsule

Cell wall

Cytoplasmic
membrane

Ribosomes

Pili

Flagella

Figure 3.17 *Anatomy of a bacterial cell (see color plate).*

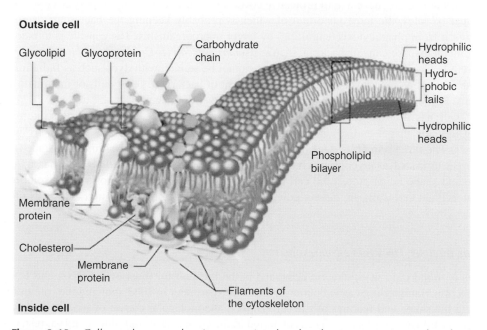

Figure 3.18 *Cell membrane and various associated molecular structures (see color plate).*

region (tail-tail region of the phospholipid bilayer). Most bacterial cell membranes do not contain cholesterol. Cholesterol partly aids in the flexibility of a cell membrane.

Cell membrane proteins are typically suspended within the bilayer, as illustrated in Figure 3.18, although the more hydrophilic areas of these proteins 'stick out' into the cell interior as well as outside the cell. These proteins function as gateways that allow certain molecules to cross into and out of the cell by moving through open areas of the protein channel. In fact, the proteins integrated into the membrane are sometimes known as gateway proteins. The outer surface of the membrane will tend to be rich in glycolipids, which have their hydrophobic tails embedded in the hydrophobic region of the membrane and their heads exposed outside the cell. These, along with carbohydrates attached to the integral proteins, are thought to function in the recognition of self, a sort of cellular identification system.

Most animal and animal-like protisan cells are enveloped only by cell membranes. Fluidity and flexibility differentiates these cell membranes from cell walls around many non-mammalian cell types. Many bacteria, fungi and plant cells are enveloped by cell walls that are complex semi-rigid structures. In gram-positive bacteria, for example, the cell wall envelopes the inner cytoplasmic lipid membrane. In gram-negative bacteria, the cell wall and the inner membrane are surrounded by an outer lipid-based membrane.

The chemical composition of the cell wall differs among various bacteria, but is substantially based on peptidoglycans. Peptidoglycan is essentially a polymer formed by repeating disaccharides interconnected by polypeptides. Gram-negative bacteria also have a relatively large periplasmic space or periplasm between the peptidoglycan cell wall and inner membrane, which may constitute up to 40% of the total cell volume. An equivalent but much smaller space outside the inner membrane also exists in Gram-positive species. Periplasm contains a loose network of peptidoglycan chains as well as a gel containing enzymes, nutrient binding and transport proteins, hydrolytic proteins and antibiotic resistance proteins. Some enzymes in the gel are involved in various biochemical pathways including peptidoglycan synthesis, electron transport (described in Section 3.3.4) and alteration of substances toxic to the cell. In some species the gel also contains beta-lactamase, an enzyme responsible for degrading penicillin.

Some species of bacteria have a third protective covering, a capsule made up of polysaccharides. Capsules play a number of roles, the most important of which is probably keeping the bacterium from drying out and protecting it from phagocytosis (engulfing) by larger microorganisms. The capsule is a major virulence factor in the major disease-causing bacteria, such as *Escherichia coli* (*E-Coli*) and *Streptococcus pneumoniae*. Capsules are relatively impermeable structures that cannot be stained with dyes such as India ink. Capsules can be in the form of a slime layer involved in attachment of bacteria to other cells or inanimate surfaces to form biofilms. Slime layers can also be used as a food reserve for the cell.

The cell wall has at least three major functions that include: (1) constraining the internal volume; (2) defining the cell shape; and (3) anchoring extracellular projections such as flagella, by which some bacteria move. The three primary shapes in bacteria are coccus (spherical), bacillus (rod-shaped) and spirillum (spiral). Bacteria such as mycoplasma have no cell wall and therefore have no definite shape. Constraining cell volume and shape helps protect the cell against rupturing when exposed to low salt solutions such as distilled water. Plant cell walls are mostly polysaccharides such as cellulose. Fungi cell walls often rely on chitin.

3.3.1.2   *The nucleus, DNA and chromosomes*

The nucleus is found only in eukaryotic cells, and is the location for most of the DNA and RNA. Danish biologist Joachim Hammerling carried out an important experiment in 1943 showing the role of the nucleus in controlling the shape and features of the cell. DNA is the molecular carrier of inheritance and, with the exception of plastid DNA (*cp*DNA and *m*DNA found in the chloroplasts and mitochondria, respectively), all DNA is restricted to the nucleus. RNA is formed in the nucleus using the DNA base sequence as a template, and moves out into the cytoplasm where it functions in the assembly of proteins. The nucleolus is an area of the nucleus (usually two nucleoli per nucleus) where ribosomes are constructed. The nuclear envelope is a double-membrane structure. Numerous pores occur in the nuclear envelope allowing RNA and other molecules to pass, but not DNA.

Chromatin is the combination of DNA and proteins that make up the contents of the nucleus of a cell. The primary functions of chromatin are: to package DNA into a smaller volume to fit in the cell's nucleus; to strengthen the DNA; to prevent its damage as much as possible; to control gene expression (a process that eventually leads to production of proteins the cell needs); and to control DNA replication and DNA repair, should damage occur. The primary protein components of chromatin are histones. Chromatin is found mostly in eukaryotic cells, while DNA-associated structure in prokaryotes is different and referred to as genophore. In eukaryotic cells chromatin is typically divided into separate chromosomes, which have linear strands of DNA containing anywhere from 10^5 to 10^9 nucleotides. In prokaryotic cells, the separate pieces of genetic material (which are also called chromosomes) are typically connected in a circle. Eukaryotic cells may also contain more than one type of chromosome. For example, mitochondria in most eukaryotes and chloroplasts in plants have their own small circular chromosomes.

The structure of chromatin depends on several factors, including the stage of the cell cycle. The majority of the time, the chromatin is structurally loose to allow DNA transcription and replication in preparation for cell division. The chromatin structure also exhibits local variations with more loosely packed DNA portions being actively transcribed. Epigenetic modifications of the structural proteins in chromatin are partly responsible for the local chromatin structure variations. Such chemical modifications include methylation and acetylation. As the cell prepares to divide, the chromatin packs more tightly in preparation for chromosome segregation. During this stage of the cell cycle, tight packing of chromatin makes the individual chromosomes more easily visible with an optical microscope.

There are typically three levels of chromatin organization. At the finest level, DNA wraps around histone proteins forming nucleosomes which might resemble a beads-on-a-string structure (euchromatin). Multiple histones wrap into their most compact form (heterochromatin), which is fiber of c. 30 nm diameter consisting of nucleosome arrays. These heterochromatin fibers pack into the chromosome during cell division. Unduplicated chromosomes are single linear strands, whereas duplicated chromosomes contain two identical copies (called chromatids) joined in a chromosomal region called the centromere.

There are also cells which do not follow this organization, however. For example, spermatozoa and avian red blood cells have more tightly packed chromatin than most eukaryotic cells. Some cells do not condense their chromatin into visible chromosomes for cell division.

Genetic information contained in chromosomes is often segregated between different chromosome sets, where some correspondence exists between certain chromosomes in different sets. This correspondence between chromosomes is often based on homology. Homologous chromosomes are those that carry alternate alleles. Allele is the term used to describe alternate forms of gene coding for alternate forms of functionally similar protein (for example genes that might code for some protein essential in the function of the heart or liver). Humans normally receive one set of homologous chromosomes from each parent.

Ploidy is the term referring to the number of chromosome sets. Diploid organisms have two (di) sets of chromosomes. Most human, animals and many plant cells are diploid. In humans, a single set consists of 23 chromosomes and most cells have a total of 46 chromosomes. Haploid organisms/cells have only one set of chromosomes. Organisms with more than two sets of chromosomes are termed polyploid.

3.3.1.3 *Cytoplasm and cytosol*

The cytoplasm is the material between the plasma membrane (cell membrane) and the nuclear envelope. The part of the cytoplasm that is outside all the organelles is called the cytosol. The cytosol is a gel-like complex mixture of cytoskeleton filaments (fibrous proteins), dissolved molecules and water that fills much of the volume of a cell. Cytoskeleton filaments maintain the shape of the cell, serve as organelle anchors and control movement of the cell and its internal reorganization. Three primary types of filaments are usually found in eukaryotic cells: microtubules, actin filaments (microfilaments) and intermediate filaments.

Microtubules are rope-like polymers of tubulin that can grow as long as 25 μm in length. The outer diameter of a microtubule is c. 25 nm. Microtubules are important for maintaining cell structure, providing platforms for intracellular transport, forming the spindle during mitosis (cell division process discussed in Section 3.3.3) as well as other cellular processes. There are many proteins that bind to the microtubules including motor proteins such as kinesin and dynein, severing proteins like katanin and other proteins important for regulating microtubule dynamics.

Actin filaments are the thinnest filaments of the cytoskeleton. Linear polymers of actin subunits are flexible and relatively strong, resisting buckling by pico-Newton range compressive forces and filament fracture by nano-Newton tensile forces. Microfilaments are highly versatile, playing the key role in cell movement and changes in cell shape. In inducing cell motility, one end of the actin filament elongates while the other end contracts, often as a result of interaction with myosin II molecular motor proteins. Additionally, they function as part of actomyosin-driven contractile molecular motors, wherein the *thin filaments* serve as tensile platforms for ATP-dependent pulling action of myosin in muscle contraction and uropod advancement.

Intermediate filaments (IFs) are a family of related proteins that share common structural and sequence features. They have an average diameter of c. 10 nm, which is between that of 7 nm actin filaments and 25 nm microtubules. Most types of intermediate filaments are cytoplasmic, but lamins are nuclear. IFs function as tension-bearing elements to help maintain cell shape and rigidity and to anchor in place several organelles, including the nucleus and desmosomes. In vertebrates, intermediate filament presence and composition is not only species-dependent, but it also varies with the tissue type. For example, most animal epithelial cells contain keratins, a diverse family of intermediate filaments consisting of more than 50 members, while mesenchymal and muscle cells are rich in the fibrous proteins, vimentin and desmin, respectively. Intermediate filaments found in neurons and glial cells include peripherin, neurofilaments and glial fibrillary acidic protein. A variety of associated proteins binds to intermediate filaments, either to improve stability (through cross-linking) or to provide attachment sites for other protein assemblies such as actin filaments and microtubules.

3.3.1.4   *Vacuoles and vesicles*

Vacuoles are single-membrane organelles that essentially form a cellular exterior within the boundaries of the cell membrane. Many cells will use vacuoles as storage compartments. Vesicles are much smaller than vacuoles and function as material transport vehicles both within and to the outside of the cell.

Many eukaryotic cell types also contain a form of vesicles known as *microbodies* or *peroxisomes*. They are involved in the breakdown of very long-chain and branched fatty acids for further catabolism (mechanism of generation of required materials from other nutrients). The name peroxisomes refers to generation of hydrogen peroxide within these vesicles during the oxidation of the fatty acids.

3.3.1.5   *Endoplasmic reticulum*

Endoplasmic reticulum is a mesh of interconnected membranes that performs functions involving protein synthesis and transport. Rough endoplasmic reticulum (rough ER) is so-named because of its rough appearance due to the numerous ribosomes (see the following section) that occur along the ER. Rough ER connects to the nuclear envelope through which the mRNA (the blueprint for proteins) travels to the ribosomes. Smooth ER lacks the ribosomes characteristic of rough ER and is thought to be involved in transport and a variety of other functions.

3.3.1.6   Ribosomes

Ribosomes are the sites of protein synthesis within the cytoplasm. They are not cell membrane bound and occur in both prokaryotes and eukaryotes. Eukaryotic ribosomes are slightly larger than prokaryotic ribosomes. Structurally, the ribosome consists of a small and larger subunit. Biochemically, the ribosome consists of rRNA and some 50 structural proteins. In eukaryotic cells ribosomes often cluster on the endoplasmic reticulum, in which case they resemble a series of factories adjoining a railroad line.

3.3.1.7   Golgi Apparatus

Golgi complexes are flattened stacks of membrane-bound sacs. Italian biologist Camillo Golgi discovered these structures in the late 1890s, although their precise role in the cell was not determined until the mid-1900s. Golgi function as a packaging plant, modifying vesicles produced by the rough ER. New membrane material is assembled in various cisternae (layers) of the Golgi.

3.3.1.8   Lysosomes

Lysosomes, discovered by the Belgian cytologist Christian de Duve in the 1960s, are relatively large vesicles formed by the Golgi. They are cellular organelles that contain acid hydrolase enzymes to breakdown waste materials and cellular debris and are found in animal cells; in yeast and plants the same roles are performed by lytic vacuoles. Lysosomes digest excess or worn-out organelles and food particles, and engulf viruses or bacteria that might end up within the cell. The membrane around a lysosome allows the digestive enzymes to work at the 4.5 pH they require. Lysosomes fuse with vacuoles and dispense their enzymes into the vacuoles, digesting their contents. They are created by the addition of hydrolytic enzymes to early endosomes from the Golgi apparatus. The name lysosome derives from the Greek words *lysis* (to separate) and *soma* (body). They are frequently nicknamed 'suicide-bags' or 'suicide-sacs' by cell biologists due to their autolysis.

The size of lysosomes varies over the range 0.1–1.2 μm. At pH 4.8, the interior of the lysosomes is acidic compared to the slightly alkaline cytosol (pH 7.2). The lysosome maintains this pH differential by pumping protons (H^+ ions) from the cytosol across the membrane via proton pumps and chloride ion channels. The lysosomal membrane protects the cytosol, and therefore the rest of the cell, from the enzymes within the lysosome. The cell is additionally protected from any lysosomal acid hydrolases that drain into the cytosol, as these enzymes are pH-sensitive and do not function well or at all in the alkaline environment of the cytosol. This ensures that cytosolic molecules and organelles are not lysed in case there is leakage of the hydrolytic enzymes from the lysosome.

3.3.1.9   Mitochondria

A mitochondrion (plural mitochondria) is a membrane-enclosed organelle found in most eukaryotic cells. These organelles range from 0.5 to 1.0 μm in size. Mitochondria are sometimes described as 'cellular power plants' because they generate most of the cell's supply of ATP used for chemical energy exchange (just like electricity generated by power plants is used in all kinds of devices requiring energy exchange). In addition to supplying cellular energy-exchange currency, mitochondria are involved in a range of other processes such as signaling, cellular differentiation, cell death as well as the control of the cell cycle and cell growth. Mitochondria have been implicated in several human diseases, including mitochondrial disorders and cardiac dysfunction, and may play a role in the aging process. The word mitochondrion comes from the Greek *mitos* (thread) and *chondrion* (granule).

The number of mitochondria in a cell varies widely by organism and tissue type. Many cells have only a single mitochondrion, whereas others can contain several thousand mitochondria. The organelle is composed of compartments that carry out specialized functions. These compartments include the outer membrane, the inter-membrane space, the inner membrane and the cristae and matrix. Mitochondrial proteins vary depending on the tissue and the species. In humans, 615 distinct types of proteins have been identified from cardiac mitochondria; 940 proteins encoded by distinct genes have been reported in rats. Although most of the DNA of a cell is contained in the cell nucleus, the mitochondrion has its own independent genome.

Interestingly, mitochondria are now believed to be originally derived from prokaryotes and incorporated into eukaryotic cells by the process of endosymbiosis. Endosymbiosis was originally proposed by the Russian botanist Konstantin Mereschkowski in 1905. It was largely ignored by a highly dogmatic approach to explanations of evolution of life by natural selection, which often claims gradual accumulation of genetic mutations as the sole innovative engine of evolution. It was probably the efforts of Lynn Margulis and other like-minded individuals in recent decades that helped revive the interest in symbiotic relationships, and endosymbiosis in particular, as one of the major sources of innovation in evolution of life. In particular, mitochondria provide substantial support for this hypothesis.

A mitochondrion contains DNA that is organized as several copies of a single, circular chromosome. The mitochondrial genome provides the codes for RNAs of ribosomes and the tRNAs necessary for the translation of messenger RNAs into protein. The circular structure is also found in prokaryotes, and the similarity is extended by the fact that mitochondrial DNA is organized with a variant genetic code similar to that of Proteobacteria. Mitochondria are also bounded by two membranes, similarly to some bacteria. The inner membrane folds into a series of cristae, which are the surfaces on which ATP is generated. The matrix is the area of the mitochondrion surrounded by the inner mitochondrial membrane. Ribosomes and mitochondrial DNA are found in the matrix.

One important mechanism of regulating mitochondrial activity in cells is calcium (Ca^{2+}) signaling (mitochondria is one of many possible targets of calcium signals in cells). In general, calcium performs a signaling role through binding to various organic molecules, which modulates their function. While a pathway for Ca^{2+} accumulation in mitochondria has long been established, its functional significance is only now becoming clear in relation to cell physiology. Accumulation of Ca^{2+} in mitochondria regulates mitochondrial metabolism and causes a transient depolarization of mitochondrial membrane potential. Mitochondria may act as a spatial Ca^{2+} buffer in many cells, regulating the local Ca^{2+} concentration in cellular microdomains. This regulates processes dependent on local cytoplasmic Ca^{2+} concentration, particularly the flux of Ca^{2+} through gated protein channels (discussed in Section 3.3.2) of the endoplasmic reticulum (ER) and the channels mediating Ca^{2+} influx through the plasma membrane.

Consequently, mitochondrial Ca^{2+} uptake plays a substantial role in shaping Ca^{2+} signals in many cell types. Under pathological conditions of cellular Ca^{2+} overload, particularly in association with oxidative stress, mitochondrial Ca^{2+} uptake may trigger pathological states that lead to cell death. Observations of intracellular calcium concentration variations as a result of non-thermal plasma treatment may therefore suggest one mechanism by which non-thermal plasma modulates mitochondrial activity, resulting in increased production of intracellular reactive oxygen species (ROS).

3.3.1.10 *Plastids*

Plastids are also membrane-bound organelles that only occur in plants and photosynthetic eukaryotes. Leucoplasts store starch, as well as some protein or oils. Chromoplasts store pigments associated with the bright colors of flowers and/or fruits. Like mitochondria, chloroplasts have their own DNA, called *cp*DNA. Chloroplasts of green algae (Protista) and plants (descendants of some of the green algae) are thought to have originated by endosymbiosis of a prokaryotic alga similar to living *Prochloron*.

3.3.2   Transport across cell membranes

Several researchers noted that transport of molecules across membranes changes following various types of plasma treatment. Knowledge of cell membrane transport mechanisms can therefore be important for developing an understanding of plasma-medical effects. It is known that cell membranes act as barriers to most, but not all, molecules. Development of a cell membrane that could allow some materials to pass while constraining the movement of other molecules was probably a major step in the evolution of the cell. Several transport mechanisms across cell membranes exist. Most of them can be divided into *passive* and *active* transport processes. Passive transport processes do not require cells to expend energy for the transport process, while active transport requires cellular energy expenditure. Although the concepts of passive and active transport are simple, it is not always easy to apply this classification because it is not always clear where the energy for transport comes from.

3.3.2.1   *Passive transport: diffusion*

Water (H_2O), carbon dioxide (CO_2), oxygen (O_2), hydrogen peroxide (H_2O_2) and nitric oxide (NO^*) (super-script * denotes a radical, discussed in Section 3.3.5) are among the few molecules that can cross the cell membrane by diffusion, a flow process driven by concentration gradients (occurring in the direction opposite to the concentration gradient, from higher to lower concentration). The ability of these molecules to diffuse is primarily due to their small size, lack of net charge in solution and absence of strong chemical reactivity with organic molecules. Animal cells typically produce carbon dioxide and consume oxygen through cellular metabolic processes. Carbon dioxide gradient (higher within cell, lower outside) and oxygen gradient (higher outside cell, lower inside) maintain the flow of these gases out and into the cell. By contrast, plant cells use CO_2 as a carbon source through photosynthesis while producing oxygen. Concentration gradients of these gases are consequently different across plant cell walls to across animal cell membranes.

Water is actually produced inside cells through aerobic metabolism. Flow of water out of the cell often occurs by osmosis, a phenomenon closely related to diffusion. In osmosis, a different concentration of some solute across the membrane drives the water preferentially in a direction that would tend to equalize solute concentration. For example, cells containing less salty water inside than outside will tend to lose water and shrink, while cells surrounded with less salty water will tend to uptake water and burst.

Larger molecules or even small ions are generally much less likely to diffuse across a lipid bilayer. However, limited diffusion even of these species can occur through 'flickering' pores in the lipid membrane, although this type of diffusion process is often ignored in most biology textbooks. Theoretical considerations suggest that pore flickering in a lipid bilayer occurs as a result of thermal fluctuation processes. Pores below certain sizes related to the Debye screening length in the cellular medium tend to close up quickly, while pores above this critical size could be more stable. The generation of such meta-stable pores can be viewed as a thermal activation process leading to pore flickering. The idea that pores exist (even if only transient) is supported by a number of experimental facts. For example, the penetration of larger molecules of various hydrodynamic sizes into various cells has been studied by a number of researchers showing that probability of penetration is consistent with the likelihood of existence of various pore dimensions. Phenomenon of electroporation, which can be viewed as increased probability of larger pores under external electric-field-induced cell membrane polarization, has been shown to reliably permit transfer of genetic material into cells that would otherwise not occur.

Facilitated diffusion is a passive form of the transport process that occurs in the direction opposite to the concentration gradient (from higher to lower concentration). In contrast to simple diffusion, facilitated diffusion requires the assistance of proteins integrated into various cell membranes (including intracellular organelle membranes). Facilitated diffusion is a process particularly important for transport ions, polar

(a) A channel protein

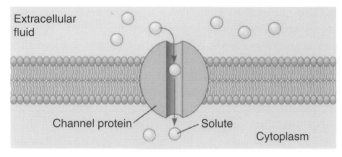

(b) A carrier protein

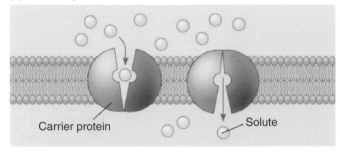

Figure 3.19   *Illustration of facilitated transport processes across lipid membranes (see color plate).*

molecules and larger non-polar molecules. Polar molecules and ions are largely repelled from polar surfaces of the phospholipids through usual colloidal-type (double-layer) interactions.

Two types of transmembrane proteins which facilitate diffusion can be distinguished: (1) those that restrict diffusion to particular species by size, shape, charge and other properties of the channel they form; and (2) those that change their conformation and work like a rotating doorway when they recognize the species they are meant to transport. These ideas are illustrated in Figure 3.19.

Channel-forming proteins are mostly used to facilitate ion transport (for this reason they are often called ion channels) and are often 'ligand-gated' (require binding of a small signaling molecule to open the protein channel). Voltage gating (electric field gating) and mechanical gating of channels is also common. Diffusion of larger polar and non-polar molecules typically occurs with the help of proteins that change their confirmation. These are called carrier, transporting or permeases proteins. Although carrier proteins are designed by nature to change their confirmation when binding to a target species, they can also restore their initial state (initial confirmation prior to binding) through thermal fluctuation.

3.3.2.2   *Active transport*

Active transport of ions and molecules often occurs in the direction of the gradient (that is, against diffusion direction) and requires the cell to spend chemical energy, usually through ATP molecules, to carry out the transport. Determining the use of energy by cells for transport experimentally is a non-trivial task. The problem is that cellular events such as production by the cell of appropriate molecules that are integrated into membranes to assist in the transport are also indirectly associated with any transport mechanism. Protein channels associated with facilitated diffusion may need to be renewed over time for example, and this

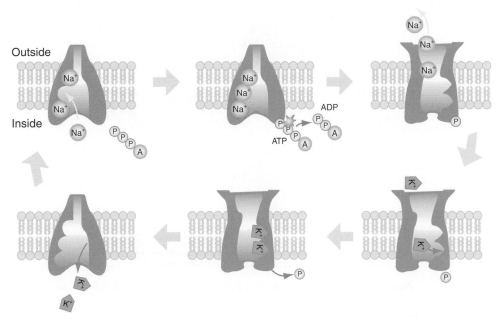

Figure 3.20 *Sodium-potassium pump transports three sodium ions to the exterior of the cell for every two potassium ions brought into the cell.*

requires cellular energy. One example where ATP molecules are directly involved in transport is protein pump, a transmembrane protein. In a protein pump, an ATP molecule donates energy to the transmembrane protein usually by letting it 'take' one phosphate group and becoming an ADP molecule in order to change the transmembrane protein confirmation after it binds some ions or molecules within it. This conformational change allows the bound species to be moved across a membrane from lower concentration to higher. Protein pumps are employed by bacteria to pump out anti-biotic molecules.

A good example of a protein pump in mammalian cells is the sodium-potassium pump illustrated in Figure 3.20. In the sodium-potassium pump, Na^+ is maintained at low concentrations inside the cell and K^+ is at higher concentrations. The reverse is the case on the outside of the cell. Nerve cells employ sodium-potassium pumping to establish conditions necessary for propagation of electrical impulses. In some cells, such as enterocytes found in human intestinal mucosa, sodium is then often brought back into cells together with glucose through sodium-dependent glucose co-transporter (SGLT). SGLT essentially uses concentration difference of sodium to pump glucose from lower concentration outside the cell to a higher one inside the cell. SGLT is sometimes referred to as a form of indirect active transport (because sodium gradients are established by use of ATP energy), although it can essentially be viewed as a facilitated diffusion mechanism.

Another interesting mechanism of what is often viewed as active transport is a phosphotransferase system that bacteria such as *Salmonella* and *E-Coli* use to bring glucose and other forms of sugars inside the cell. This system was probably first discovered by Saul Roseman in 1964. Beginning with ATP, it involves a sequence of phosphate group transfers that eventually transform glucose (or another sugar) into glucose-6-phosphate while the glucose is being transferred through a transmembrane protein (Enzyme II C, for example). Once glucose is converted into glucose-6-phosphate, it cannot go back out of the cell using the same route. The phosphotransferase system is an example of where molecules are modified in the transport process, circumventing the need to work against the concentration-gradient-driven flow. Although the energy

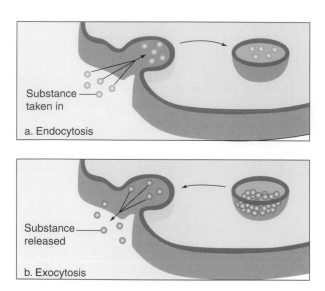

Figure 3.21   *Illustration of (a) endocytosis and (b) exocytosis.*

for this transport does come from ATP-donated phosphate groups, it can be argued that glucose is a source of cell-produced ATP and is therefore the energy source for its own transport. However, this energy is delivered through the cell rather than directly. For this reason, this mechanism is appropriately classified as active.

Another mechanism of transport often viewed as active is vesicle-mediated endocytosis/exocytosis, illustrated in Figure 3.21. Endocytosis is a process by which cells absorb molecules (such as proteins) by engulfing them. It is used primarily for large molecules. The process opposite to endocytosis is exocytosis (removal of large molecules from the cells). Endocytosis can be subdivided into several different types, depending on the molecular mechanism by which it is mediated. Clathrin-mediated endocytosis occurs through small (approximately 100 nm in diameter) vesicles that have a morphologically characteristic crystalline coat made up of a complex of proteins that are mainly associated with the cytosolic protein clathrin.

Clathrin-coated vesicles (CCVs) are found in virtually all cells and form domains of the plasma membrane referred to as clathrin-coated pits. Coated pits can concentrate large extracellular molecules that have different receptors responsible for the receptor-mediated endocytosis of ligands, for example, low-density lipoprotein, transferrin, growth factors, antibodies and many others. Caveolae are the most commonly reported non-clathrin-coated plasma membrane buds, which exist on the surface of many (but not all) cell types. They consist of the cholesterol-binding protein caveolin with the bilayer enriched in cholesterol and glycolipids. Caveolae are small (c. 50 nm in diameter) flask-shape pits in the membrane that resemble the shape of a cave (hence the name caveolae). They can constitute up to a third of the plasma membrane area of the cells of some tissues, being especially abundant in smooth muscle, type I pneumocytes, fibroblasts, adipocytes and endothelial cells. Uptake of extracellular molecules is also believed to be specifically mediated via receptors in caveolae.

Macropinocytosis, which usually occurs from highly ruffled regions of the plasma membrane, is the invagination of the cell membrane to form a pocket. This pocket then pinches off into the cell to form a vesicle (0.5–5 μm in diameter) filled with a large volume of extracellular fluid and molecules. The filling of the pocket occurs in a non-specific manner. The vesicle then travels into the cytosol and fuses with other vesicles such as endosomes and lysosomes.

A somewhat related mechanism is phagocytosis, the process by which cells bind and internalize particulate matter larger than around 0.75 μm in diameter such as small-sized dust particles, cell debris, microorganisms (bacteria) and even apoptotic cells. Like macropinocytosis, phagocytosis involves the uptake of larger membrane areas and typically forms vesicles inside the cell.

Cellular energy is apparently required for the process of endocytosis and exocytosis at many stages, although the exact mechanisms are not yet well understood. From a physical point of view, substantial deformation of the cell membrane clearly requires an input of energy due to a substantial increase of surface energy. Transport of vesicles that form during the endocytosis into the cell also requires energy. Both vesicle formation and transport energy comes from the cell. Some drug and gene delivery technologies employ lipid vesicles that are capable of fusing with the cell membrane. In this case, energy for the transport may actually come from the external source (surface energy of the drug- or gene-carrying vesicle).

3.3.3 Cell cycle and cell division

Plasma-medical research often reports on various effects of plasma treatment of cell cultures. However, cells in such cultures are rarely in the same state; as a result, they may show different susceptibility to plasma treatment. Some understanding of cell cycle and cell division mechanisms is therefore important for plasma medicine.

3.3.3.1 *The cell cycle*

Despite differences between prokaryotes and eukaryotes, there are several common features in their cell division processes. Replication of the DNA must occur. Segregation of the 'original' and its 'replica' must follow. Cytokinesis, the process that occurs after cell division where one cell splits off from its sister cell, ends the cell division process.

The cell cycle is the sequence of growth, DNA replication, growth and cell division that all cells go through. Beginning after cytokinesis, the daughter cells are quite small and low on ATP (energy-exchange molecule). They acquire ATP and increase in size during the G1 phase of Interphase. Most cells are observed in Interphase, the longest part of the cell cycle. After acquiring a sufficient size and amount of ATP, the cells then undergo DNA synthesis (replication of the original DNA molecules, making identical copies including one 'new molecule' eventually destined for each new cell), which occurs during the S phase. Since the formation of new DNA is an energy-draining process, the cell undergoes a second growth and energy (ATP) acquisition stage, the G2 phase. The energy acquired during G2 is used in cell division.

Regulation of the cell cycle is accomplished in several ways. Some cells divide rapidly (beans, for example, take c. 20 hours for the complete cycle, while bacteria take c. 20 minutes). Others such as nerve cells lose their capability to divide once they reach maturity. Some cells such as liver cells retain but do not normally utilize their capacity for division. Liver cells will divide if part of the liver is removed. The division continues until the liver reaches its former size. Cancer cells are those which undergo a series of rapid divisions such that the daughter cells divide before they have reached 'functional maturity'. Environmental factors such as changes in temperature and pH and declining nutrient levels lead to declining cell division rates. When cells stop dividing, they stop usually at a point late in the G1 phase known as the R point (for restriction).

3.3.3.2 *Prokaryotic cell division*

Prokaryotes are much simpler in their organization than eukaryotes. There are a great many more organelles in eukaryotes and more chromosomes. The usual method of prokaryote cell division is termed binary fission. The prokaryotic chromosome is a single DNA molecule and, after replication, attaches each copy to a different part of the cell membrane. When the cell begins to pull apart, the replicate and original chromosomes are separated. Following cell splitting (cytokinesis), there are then two cells of identical genetic composition (except for the rare chance of a spontaneous mutation). One consequence of this asexual method of reproduction is that all organisms in a colony are genetic equals. Moreover, all organisms in a colony are frequently of the same phenotype (not only the same genetically, but express the same proteins in very similar amounts).

3.3.3.3 *Eukaryotic cell division: Mitosis and meiosis*

Due to their increased numbers of chromosomes and organelles and diversity of cells, particularly within larger life forms, eukaryotic cell division is more complicated than prokaryotic although the same processes of replication, segregation and cytokinesis still occur. There are two primary types of eukaryotic cell division, mitosis and meiosis, depending on the two primary cell types of somatic (or vegetative, non-reproductive) and gametic (reproductive such as sperm and egg) cells. Mitosis is the process of forming genetically identical (except for occasional mutations) daughter cells by replicating and dividing the original chromosomes as well as separating the dividing cells. Most cells in the human body are produced by mitosis. Meiosis, on the other hand, is the process by which gametic cells are created. In meiosis, only some of the genetic material goes into created gametic cells. Specifically, only one complete set of chromosomes is transmitted to the created gametic cells. Meiosis reduces the number of sets of chromosomes by half, so that when gametic recombination (fertilization) occurs, the ploidy of the parents will be re-established.

Mitosis generally begins with tighter packing of chromatin, sometimes referred to as chromosome coiling, followed by disintegration of the nuclear membrane and formation of two spindles and associated spindle fibers in the cell. The spindle fibers from different spindles attach to the two chromatid stands of the duplicated chromosomes (kinetochore is the area on the centromere where the microtubules of the cell's spindle attach). The chromosomes then align and the chromatids separate, essentially doubling the chromosome number. Biogenesis of additional cellular organelles, such as mitochondria (which has its own set of genes requiring reproduction) and ribosomes, appears to occur during the reproduction of the nuclear genetic material. After that the cell also divides some organelles (such as mitochondria, ribosomes and some endoplasmic reticulum) between its progenies.

In meiosis, cells actually divide twice. In the first meiotic division the number of cells doubles, but the number of chromosomes remains the same. In this process, homologous chromosomes are first paired and a cross-over between them occurs. In the cross-over process, some of the alleles exchange place on the homologous chromosomes. After that, spindle fibers form, attach to the chromosomes and align them. The alignment pattern is random, sorting out which of the homologous chromosomes will go into different daughter cells. Separation of the homologous chromosomes follows. The second meiotic division is similar to mitotic division, including replication of the genetic material.

3.3.4   Cellular metabolism

Living cells constantly consume material and energy. They also dispose of various materials, perform mechanical work and generate heat. From a thermodynamic point of view, these cellular metabolic processes are what keep cells from thermal equilibrium and from the resulting death. From a biochemical point of view, cellular metabolism is the collection of chemical transformations that take place within a cell through which energy and molecular components are provided for all essential cellular processes, including the synthesis of new molecules and the breakdown and removal of others.

Different forms of life and different cells use different nutrients as sources of carbon, nitrogen, phosphorus, sulfur and various metals. Autotrophs such as plants and cyanobacteria, for example, obtain carbon from inorganic sources such as carbon dioxide (CO_2), while heterotrophs such as animals take carbon from organic molecules, often consuming other living forms for this purpose. Sources of energy for different life forms and cells can be light, as in the case of plants that carry out photosynthesis or chemical compounds (inorganic and organic). All sequences of metabolic chemical reactions (metabolic pathways) are redox processes that involve transfer of electrons between different atoms. Organisms can be classified according to primary electron donors in these pathways and electron acceptors. When the final electron acceptor is oxygen, the metabolic pathways is said to be aerobic; it is anaerobic otherwise.

Given the large variety of nutrients and sources of energy, it is natural that a large number of metabolic pathways exist. In fact, each cell type often has a large number of metabolic pathways available to it. The choice of a particular pathway depends on abundance of available nutrients in the environment and on other environmental factors. In some cases, once an organism chooses a particular pathway it may continue to employ it even when this pathway becomes sub-optimal (produces less energy and needed chemical compounds for a given amount of nutrients). This results in existence of multiple possible stable metabolic phenotypes for the same set of environmental conditions. The choice of a particular metabolic pathway can even be inherited (this is a form of so-called epigenetic inheritance or inheritance that is over and above genetic).

Metabolic pathways consist of catabolism and anabolism. The catabolic part of a pathway breaks down organic matter to harvest energy and basic construction materials. The anabolic part of a pathway uses net energy and materials provided through catabolism and the environment as input to construct components of cells such as lipids, proteins and nucleic acids.

3.3.4.1 *Important coenzymes in metabolic pathways*

Non-protein organic molecules (often based on nucleotides or vitamins) that bind to enzymes in order to help catalyze particular reactions are called coenzymes. Inorganic molecules or ions that perform similar functions are often called cofactors. The central coenzyme in metabolism is ATP or adenosine triphosphate. As described previously, it is an adenine nucleotide with an additional two phosphate groups. Removal of one phosphate group turns ATP into ADP.

Phosphorylation and dephosphorylation (transfers of one phosphate group) is the primary mechanism by which ATP regulates activity of many enzymes and proteins, transferring energy in the process. With the transfer of a phosphate group to a protein, effective transfer of an electron also occurs. In other words, phosphorylation is a redox process. ATP is oxidized during this process and the protein is reduced. In terms of energy ATP releases c. 30 kJ mole^{-1}, not all of which is gained by the phosphorylated protein. (The difference is dissipated in heat, which is always the commission price of energy transfers.) Enzymes use this energy to carry out various conformational changes (as in electrically driven mechanical machines) during transport and construction of various cellular materials. Phosphorylation and dephosphorylation processes are themselves assisted by enzymes called kinases and phosphatases that are specific to being used with particular enzymes/proteins. Within a protein, phosphorylation can occur on several amino acids. Phosphorylation on serine is the most common, followed by threonine. Tyrosine phosphorylation is relatively rare. Experimentally, a relatively reliable way of detecting protein conformational changes due to phosphorylation is through binding of fluorescent antibodies. Antibodies are immunoglobulin proteins produced by the immune system (typically in animals) to recognize specific target protein conformations.

In catabolic processes ATP is not usually produced from scratch, but rather created by phosphorylation of ADP. By way of an example, consider how many ATP molecules can be (and actually are) produced from ADP in the process of glucose catabolism:

$$C_6H_{12}O_6 + 6O_2 \rightarrow 6CO_2 + 6H_2O. \tag{3.3}$$

The above reaction releases c. 2870 kJ mole^{-1} of energy (estimate based on bond energies provided in Table 3.1; glucose bonds in Figure 3.3 actually yield 3005 kJ mole^{-1}). This much energy is sufficient to convert 95 ADP molecules into ATP. The standard aerobic process of glucose catabolism (consisting of glycolysis, cytochrome pathway and the Krebs cycle, see Section 3.3.4.4) converts 38 ADP molecules into ATP. The rest of the energy is irreversibly lost as heat. This energy conversion process is therefore c. 40% efficient. By comparison, typical natural-gas-to-electricity (the analog of ATP as the energy-exchange vehicle in human societies) conversion systems, engineered by humans, yield efficiencies of c. 25%.

Nicotinamide adenine dinucleotide (NAD) is another important coenzyme in metabolism. It is involved in redox reactions, carrying electrons from one reaction to another. The coenzyme is therefore found in two forms in cells: NAD$^+$ and NADH. NAD$^+$ is an oxidizing agent; it accepts electrons from other molecules and becomes reduced. This reaction forms a higher-energy NADH molecule, which can then be used as a reducing agent to donate electrons. The reaction can be written:

$$NAD^+ + H^+ + 2e^- \rightarrow NADH \tag{3.4}$$

There exist hundreds of separate types of dehydrogenases (enzymes) designed to remove electrons from their substrates (the molecules they act on) and reduce NAD$^+$ to NADH. This reduced form of the coenzyme is then a substrate for any of the reductases in the cell that need to reduce their substrates. Electron transfer reactions are the main function of NAD$^+$. Energy release in oxidation of NADH is more than five times larger (c. 158 kJ mole^{-1}) than the energy of ADP phosphorylation. Another coenzyme related to NADH is NADPH (having an additional phosphate group). While the NAD$^+$/NADH form is more important in catabolic reactions, NADP$^+$/NADPH is used in a similar fashion primarily in anabolic reactions. Another somewhat similar pair is FAD/FADH$_2$. FADH$_2$ can be oxidized to FAD to release energy. In contrast to NAD$^+$ which binds relatively loosely to enzymes, FAD often binds tightly. For this reason, it is often referred to as a prosthetic group rather than a coenzyme.

Redox processes involving NADH, NADPH and FADH$_2$ are all associated with significantly larger energy differences compared to ADP phosphorylation. The ability of ATP to transport energy in smaller quanta clearly offers greater flexibility in biochemical activation processes. At the same time, this energy quantum is not too small to be 'lost' in thermal fluctuation. Indeed, as already mentioned, phosphorylation of ADP involves c. 30 kJ mole^{-1}, which is c. 250 meV. At room temperature, the average energy transferred back and forth to a thermal reservoir in thermal fluctuations is c. 25 meV, which is c. 10 times lower than the energy of ADP phosphorylation. ADP phosphorylation and energy transferred by ATP to proteins during protein phosphorylation is therefore unlikely to be lost in thermal fluctuations. ATP probably strikes an ideal balance by transferring sufficiently small energy quanta, but not too small to be swamped by thermal fluctuations. This enables enzymes activated through ATP to truly function as mechanical machines at the molecular scale.

3.3.4.2   *Main stages of catabolism in eukaryotes*

As illustrated in Figure 3.22, most eukaryotic cells have three main stages of metabolism: digestion, glycolysis and oxidation.

Digestion involves the breakdown of more complex molecules including carbohydrates, fats and proteins into smaller molecular units that can be transported in some way into cells. Fats are broken down into fatty acids and glycerides. Proteins are separated into amino acids and carbohydrates are broken into simple sugars such as glucose or fructose. This typically involves a variety of digestion enzymes that are secreted into the extracellular environment and do not provide energy (in the form of ATP) directly to the cell.

Glycolysis is a set of processes that are employed by all known organisms, eukaryotic or prokaryotic, that convert simple sugars (such as glucose) to pyruvate, phosphorylating some ADP to ATP in the process and

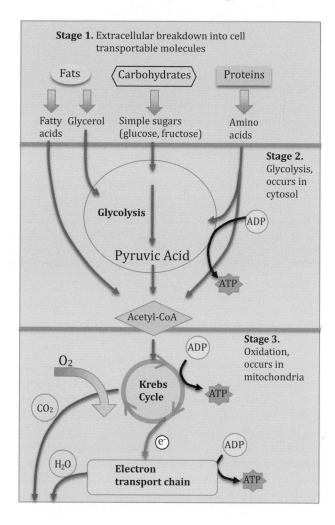

Figure 3.22   *Three main stages of metabolism.*

reducing NAD$^+$ to NADH. In eukaryotes, glycolysis occurs in the cytoplasm. The overall reaction can be summarized:

$$\text{Glucose} + 2\text{ADP} + 2\text{P}_i + 2\text{NAD}^+ \rightarrow 2\,(\text{Pyruvic acid}) + 2\text{NADH} + 2\text{ATP} \qquad (3.5)$$

where P_i is the phosphate group (inorganic). Under normal circumstances, most cells do not use amino acids to generate energy. In some situations however amino acids can be employed in the process called gluconeogenesis, which is a process of production of glucose from other metabolites. The primary carbon skeletons used for gluconeogenesis are derived from pyruvate, lactate and glycerol and the amino acids alanine and glutamine. Subsequently, glucose generated in this fashion (with the use of amino acids and glycerol) can be run through normal catabolic processes.

Two major pathways generally exist once pyruvic acid is generated: anaerobic (oxygen is not used as the electron acceptor) and aerobic (oxygen is used as the acceptor of electrons).

3.3.4.3   *Anaerobic pathways*

Under anaerobic conditions (in the absence of oxygen) pyruvic acid can be routed into one of two major pathways: alcohol fermentation and lactic acid (and other similar compounds) fermentation. In alcohol fermentation the pyruvic acid is first converted to acetaldehyde using a series of enzymes, releasing carbon dioxide. Acetaldehyde is then converted to ethanol using energy released by oxidation of NAHD, while pyruvic acid acts as the electron acceptor.

Most animals, including humans, lack genetic information and therefore all the appropriate enzymes to ferment alcohol in their own bodies. Yeast, on the other hand, can convert glucose to pyruvic acid via the glycolysis pathways, then go further and convert pyruvic acid into ethanol. Humans convert pyruvic acid following glycolysis into lactic acid in muscles when oxygen becomes depleted. This process also employs energy released from oxidation of NADH where pyruvic acid acts as an electron acceptor. This lactic acid causes the muscle stiffness. The stiffness goes away after a few days since the cessation of strenuous activity allows aerobic conditions to return to the muscle, and the lactic acid can be converted into ATP via the normal aerobic respiration pathways.

3.3.4.4   *Aerobic respiration*

When oxygen is present (aerobic conditions), most organisms will undergo two more steps – Krebs cycle and electron transport chain – to phosphorylate a large number of additional ADP into ATP. In eukaryotes these processes occur in the mitochondria, while in prokaryotes they occur in the cytoplasm.

Pyruvic acid is first altered in the transition reaction by removal of a carbon and two oxygens (which form carbon dioxide). When the carbon dioxide is removed, energy is given off and NAD^+ is converted into the higher-energy form NADH. Coenzyme A attaches to the remaining acetyl group forming acetyl-CoA complex (CoA is released later), as illustrated in Figure 3.22. Acetyl-CoA complex can also be generated in the process of oxidation of very long-chained fatty acids within peroxisomes. In this case a set of special enzymes is employed to oxidize the fatty acid, eventually transferring electrons to the oxygen molecule, generating hydrogen peroxide in the process and generating acetyl-CoA complex. Shorter fatty acids can be broken down in mitochondria, producing some ATP in the process and also producing acetyl-CoA.

Production of acetyl-CoA complex is a prelude to the Krebs cycle that is also known as the citric acid cycle. This cycle produces several higher-energy molecules from acetyl and releases carbon dioxide:

$$\text{Acetyl group} + 3NAD^+ + FAD + ADP + P_i \rightarrow 2CO_2 + 3NADH + FADH_2 + ATP. \qquad (3.6)$$

While the Krebs cycle occurs in the matrix of the mitochondrion, the electron transport system (ETS) chemicals are embedded in the membranes known as the cristae. The Krebs cycle completely oxidized the carbons in the pyruvic acids, producing a small amount of ATP and reducing NAD^+ and FAD into higher-energy forms. In the ETS those higher-energy forms are cashed in by phosphorylating a large number of ADP into ATP.

Cytochromes are molecules that pass the electrons along the ETS chain. In the process, H^+ ions are pumped out of the mitochondria creating an electric potential difference across the mitochondrial membrane and a gradient of proton ions across the membrane. This gradient and the flux of the proton ions back into the mitochondria through special channels is the process employed by these channels to phosphorylate ADP and produce additional ATP, and is known as the chemiosmotic coupling. The electrons that are passed along the mitochondrial ETS are captured in the end by oxygen:

$$2e^- + 2H^+ + O_2 \rightarrow 2H_2O. \tag{3.7}$$

While the majority of electrons captured at the end of this process results in the production of water molecules, some of the electrons are captured directly by the oxygen molecules producing a radical known as superoxide:

$$e^- + O_2 \rightarrow O_2^-. \tag{3.8}$$

The superoxide produced through cellular respiration in this way is the main source of highly reactive free oxygen radicals, which can play a variety of important roles in cells.

3.3.5 Reactive species in cells and living organisms

> They are in the sunlight. They float in the smog.
> They are present in food and increase when you jog.
> They form in the liver from all kinds of drugs,
> from solvents, from nasty things used to kill bugs.
> They will damage your heart. They'll make you grow old,
> and they will mutate your genome (we're told).
> Their causes, reactions, and guessed repercussions
> Are turning up often in journal discussions.
> The least stable molecules vanish so quick
> that publishing on them is no minor trick.
> But those with a half-life that lets them react
> with bio-components will also attract
> spectroscopists, spin trappers, and all of the rest
> who'll show you the things at which others only guess.

This small verse written by an unknown poet (probably a spectroscopist) expresses a common frustration about a number of highly reactive or unstable molecules and ions that play important roles in many cellular and extracellular processes. These are mostly oxygen, nitrogen, hydrogen and carbon (and sometimes chlorine) containing molecules or ions that will be jointly referred to as the reactive species. Some of these reactive species can be classified as radicals, while others fall apart to produce radicals or are in a state of electronic excitation (the electrons in molecules are in a higher energy state). Different reactive species have different lifetimes and different reaction rate constants with different organic molecules. They are frequently converted into each other and cannot be easily isolated. The exact role of each reactive molecule or ion in various cellular processes is often unclear as a result. At the same time, their collective importance is unquestionable. It has been shown in particular that plasma treatment of tissues and cells can influence intracellular concentrations of various reactive species. For this reason, in this section we will endeavor to provide a relatively brief and introductory overview of the reactive species in cells and their effects.

3.3.5.1 *Radicals*

Radicals (sometimes also called free radicals) are traditionally defined in chemistry as molecules or ions whose electrons have a non-zero total spin. Some radicals can form though homolytic breaking (scission) of

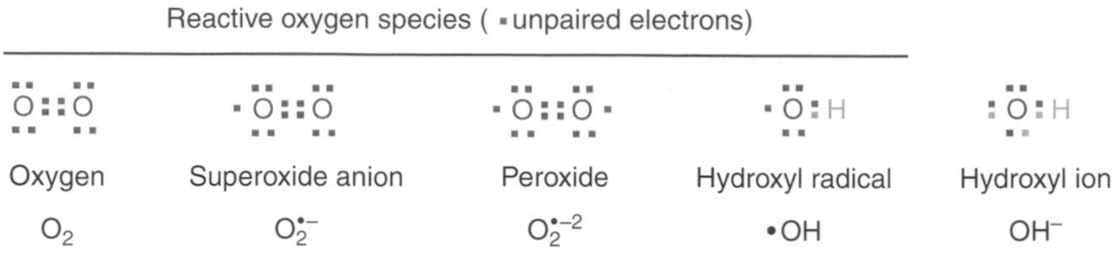

Reactive oxygen species (·unpaired electrons)

Oxygen	Superoxide anion	Peroxide	Hydroxyl radical	Hydroxyl ion
O_2	$O_2^{\bullet-}$	$O_2^{\bullet-2}$	$\bullet OH$	OH^-

Figure 3.23 *Radicals containing only oxygen and hydrogen and comparison with non-radicals.*

a covalent bond, when one of the electrons in the bond goes to each of the bonding species and each of the bonding species remains electrically neutral.

One example of an uncharged highly active radical is the hydroxyl radical OH*. Due to its high reactivity with organic molecules, it has a typical lifetime of the order nanoseconds in organic solutions (this depends on the concentration of the organic molecules). Some free radicals involving oxygen and hydrogen as well as some non-radical species are depicted in Figure 3.23. A much less reactive electrically neutral radical that can diffuse across cell membrane is nitric oxide (NO). As a result of its lower reactivity than hydroxyl, nitric oxide is often involved in transmitting signals across cellular membranes (e.g. acting as a neurotransmitter).

Organic molecules commonly have groups where a carbon atom is bound to hydrogen. Such molecules can be denoted by RH, where R is called the organic conjugate or residue. In some reactions, hydrogen (not hydrogen ion, but hydrogen atom) can be abstracted from this molecule forming an organic radical R*. Organic radicals often play a role in various chain reactions. Peroxidation reactions, for example, can be self-propagating reaction chains. Once they are initiated, as for example by:

$$RH + OH^* \rightarrow R^* + H_2O, \tag{3.9}$$

peroxidation will often propagate by the reactions:

$$R^* + O_2 \rightarrow ROO^* \tag{3.10}$$

and

$$ROO^* + RH \rightarrow R^* + ROOH. \tag{3.11}$$

Radicals such as ROO* are called peroxy radicals and molecules such as ROOH are called organic hydroperoxides (see the following section).

Ionic radicals can form by accepting or donating an electron. A common example of such a radical is superoxide O_2^-. This radical is much less reactive with organic molecules than hydroxyl, although it does react with organic molecules such as quinones (as an oxidizer) and metalloproteins. However, superoxide can directly or indirectly be responsible for production of other more reactive species.

3.3.5.2 *Reactive species*

Reactive species are not necessarily radicals. Moreover, some molecules traditionally defined as radicals are in fact very stable molecules and are not highly reactive. For example, although rarely viewed as a radical, diatomic oxygen molecule with its electrons in their ground state (triplet state) is a radical according to the traditional definition of radicals. An oxygen molecule is rather unusual in that the ground state of its electron system is one where the two electrons in the covalent bond have the same spin. As a result, this form of oxygen has non-zero total spin. In fact, the ground state oxygen molecule is sometimes called di-radical because it has two unpaired electron spins, as opposed to most radicals that have only one unpaired electron spin. Despite the fact that it is a radical, ground state oxygen does not readily react with most organic molecules because most covalent bonds in organic molecules involve a pair of electrons with a net zero spin (singlet state). Roughly speaking, conservation of spin quantum number (which can be viewed as conservation of angular momentum) would require a molecule in a singlet state to enter a triplet transition state in order for a reaction with the triplet oxygen to occur. The extra energy required for such a transition (energy barrier) is sufficient to prevent direct reaction at ambient temperatures. This is the reason oxygen does not kill us when we breath it. At higher temperatures, or in the presence of suitable catalysts, the reaction of organic molecules with the triplet oxygen proceeds much more readily. This is why fuels often need to burn to produce energy.

Singlet oxygen, often denoted as $^1O_2$, is a di-oxygen molecule where the bond-forming electrons have opposite spin (singlet state) and zero net spin. However, the energy of the singlet oxygen bonding is c. 94 kJ mole^{-1} (roughly 1 eV or one electron-volt per molecule) greater than the bond energy of the ground state (triplet) oxygen. Singlet oxygen is an example of a non-radical electronically excited highly reactive species. Reactivity of singlet oxygen is significantly higher than that of superoxide or triplet state oxygen, partly due to the fact that most bonds in organic molecules are in a singlet state and no longer require transition to the triplet state in order to react. Singlet oxygen will often react directly with lipids, for example, breaking the carbon-carbon double bonds (particularly when they are weakened by the adjacent alkyl groups). The reaction rates of singlet oxygen with amino acids mostly depend on the presence of double bonds or the presence of electron-rich sulfur atom in the amino acid. In practice, therefore, singlet oxygen reacts primarily with five amino acids: tryptophan, histidine, tyrosine, methionine and cysteine.

The category of reactive species also includes molecules that can split apart spontaneously or with the assistance of catalysts to produce radicals. Peroxides are examples of such molecules, and are molecules with the structure R-O-O-R$'$ where R and R$'$ are some atoms or organic molecular groups. When both R and R$'$ are hydrogen atoms, the peroxide is the hydrogen peroxide H-O-O-H or H_2O_2. Molecules of the type R-O-O-H are known as hydroperoxides. When R and R$'$ are both organic, the resulting molecule is called an organic peroxide.

The peroxide bond O-O is rather weak (see Table 3.1) and, when it breaks, the resulting two electrically neutral molecules are radicals. The presence of the peroxy radical O-O-H often makes organic molecules unstable. Break-up and radical generation from benzoyl peroxide, which is a compound frequently used to treat acne, is shown as an example in Equation (3.12) below, where a dot symbol denotes unbalanced spin in the radical.

$$(3.12)$$

Lipid peroxidation is a major mechanism of toxicity in living systems. Abstraction of hydrogen by radicals and attachment of peroxy group occurs most easily in polyunsaturated fatty acid chains at carbon locations that are adjacent to carbon double bonds. For example, if RH in Equation (3.11) is an oleic fatty acid molecule, the peroxy group is likely to form at the 8th carbon from the left in the formula:

$$OHCO - (CH_2)_6 - CH_2 - CH = CH - (CH_2)_7 - CH_3 + ROO^* \rightarrow$$
$$OHCO - (CH_2)_6 - CHOOH - CH = CH - (CH_2)_7 - CH_3 + R^*.$$

$$(3.13)$$

Breaking of the peroxide bond results in the alkoxy radical and a hydroxyl radical:

$$OHCO - (CH_2)_6 - CHOOH - CH = CH - (CH_2)_7 - CH_3 \rightarrow$$
$$OHCO - (CH_2)_6 - CHO^* - CH = CH - (CH_2)_7 - CH_3 + OH^* \tag{3.14}$$

followed by scission of the carbon bond on either side of the oxygen atom location in the alkoxy radical and formation of aldehydes (organic molecules with aldehyde groups HCO), for example:

$$OHCO - (CH_2)_6 - CHO^* - CH = CH - (CH_2)_7 - CH_3 \rightarrow$$
$$\left(OHCO - (CH_2)_6^*\right) + (HCO - CH = CH - (CH_2)_7 - CH_3). \tag{3.15}$$

The radical product of the above break-up can recombine with hydroxyl, also producing aldehyde. The lipid peroxidation process exemplified above essentially breaks down longer polyunsaturated fatty acids, producing aldehydes.

3.3.5.3　*Cellular sources of reactive species*

Among many reactive species that can exist within cells, reactive oxygen species (ROS) appear to have particular importance. These typically include O_2^-, OH^*, $^1O_2$ and H_2O_2. Most of the natural pathways for generation of intracellular ROS begin with the capture of an electron by molecular oxygen and formation of O_2^-. Electron transport chains such as those discussed above in mitochondria are probably responsible for the greatest amount of the superoxide produced under normal circumstances in cells converting of the order 1–2% of oxygen used by mitochondria into this radical. Another important mechanism that has been the subject of considerable investigation over the past decade was historically discovered in connection with cells such as neutrophils, which 'consume' bacteria through the process of phagocytosis and destroy it with the help of ROS. In such cells, superoxide is produced in the phagosomes (vesicles within the cells containing the offending foreign material) through NADPH oxidase, a phagosome membrane-bound enzyme complex that oxidizes NADPH:

$$NADPH + 2O_2 \leftrightarrow NADP^+ + 2O_2^- + H^+. \tag{3.16}$$

It turns out that the same process of generation of superoxide occurs in many other cell types, using a family of different plasma membrane-bound enzymes that have been called NOX and DUOX. It also appears to be possible that the same type of membrane-bound enzymes generate superoxide through oxidation of NADH, rather than only NADPH. In contrast to mitochondrial processes, where the primary effect is metabolic and the generation of superoxide appears to be an accompanying effect, superoxide production appears to be the primary function of the NOX and DUOX family of proteins.

　　Another known source of superoxide is xanthine oxidase (XO) enzyme that is used to help oxidize hypoxanthine into xanthine and xanthine into uric acid. This enzyme plays an important role in the catabolism of purines. The active site of XO is composed of a molybdopterin unit with the molybdenum atom also coordinated by terminal oxygen, sulfur atoms and a terminal hydroxide. In the reaction with xanthine to form uric acid, an oxygen atom is transferred from molybdenum to xanthine, whereby several intermediates are probably involved. The reformation of the active molybdenum center occurs by the addition of water.

　　Superoxide radicals created by one of the mechanisms described above can generate much more reactive hydroxyl radicals and singlet oxygen. This occurs during the spontaneous dismutation reaction (reaction where some of the oxygen is oxidized and some is reduced) of superoxide with production of hydrogen peroxide, and subsequent reaction of the superoxide with hydrogen peroxide:

$$2O_2^- + 2H^+ \rightarrow H_2O_2 + {}^1O_2 \tag{3.17}$$

and

$$O_2^- + H_2O_2 \rightarrow {}^1O_2 + OH^* + OH^-. \tag{3.18}$$

In addition, the presence of some multivalent metals in cells, such as Cu and Fe, catalyzes further production of hydroxyl radicals from hydrogen peroxide, according to the reaction proposed by Haber and Weiss in 1932:

$$Fe^{2+} + H_2O_2 \rightarrow Fe^{3+} + OH^* + OH^-. \tag{3.19}$$

This reaction is sometimes called the Fenton reaction or Fenton mechanism in honor of HJH Fenton. He first observed the oxidation of tartaric acid by hydrogen peroxide in the presence of ferrous iron ions (but did not actually propose a mechanism for the oxidation).

Given the high reactivity of hydroxyl radicals (its reactivity is limited only by the frequency of its encounters with organic molecules) and of the singlet oxygen, it is not surprising that several enzymes exist, allowing cells to bypass the generation of these ROS from superoxide (at least to some extent). One such enzyme is the superoxide dismutase (SOD), which catalyzes the reaction:

$$2O_2^- + 2H^+ \rightarrow H_2O_2 + O_2. \tag{3.20}$$

SODs are a family of metalloproteins where multivalent metals are co-factors that activate the enzyme. These metals are usually Cu and Zn (together), Mn, Fe or Ni. Copper- or zinc-based SOD occurs mostly in eukaryotes, while iron-, manganese- or nickel-based SOD occur most often in bacteria and mitochondria. Catalase enzyme helps dispose of the hydrogen peroxide, avoiding the production of the hydroxyl radicals by catalyzing the reaction:

$$2H_2O_2 \rightarrow 2H_2O + O_2. \tag{3.21}$$

In fact, catalase enzyme also typically involves iron as the center of the so-called heme group of the enzyme. The heme group is a non-protein molecular groups that consists of an iron (Fe) ion held in a heterocyclic ring, known as a porphyrin. It is the functional group that carries oxygen bound to iron in hemoglobin, a protein complex in blood (red blood cells) that carries oxygen to various tissues in the body.

Reactive nitrogen species (RNS) are those reactive molecules which originate from the nitric oxide radical NO*, just as ROS typically originates from superoxide. The most active among these species are nitrogen dioxide NO_2^* radical and peroxynitrite $ONOO^-$. The primary known mechanism of nitric oxide production in cells is oxidation of amino acid L-arginine catalyzed by an enzyme from a family of enzymes known as NOS or nitric oxide synthase. This catalytic reaction is:

$$2\,(\text{L-arginine}) + 3NADPH + 2H^+ + 4O_2 \rightarrow 2\,(\text{Citrulline}) \rightarrow 3NADP^+ + NO^*. \tag{3.22}$$

The above reaction is a two-stage process with each stage oxidizing a different amount of NADPH. Several different types of NOS are known, but all of their activities require five cofactors ensuring that electrons passing from NADPH to FAD to FMN (flavin mononucleotide) to heme group and, finally, to the oxygen molecule. The first to be identified was NOS1 or nNOS in neuron NOS cells. Endothelial NOS cells have a related enzyme, which is known as eNOS or NOS3 because it was the third to be identified. The activities of these two types of enzymes are dependent on the intracellular calcium Ca^{2+} concentration. Calcium concentration regulates the binding of calmodulin to specific nNOS and eNOS domains initiating electron transfer from the flavins to the heme moieties. In contrast, calmodulin remains tightly bound to the so-called inducible NOS (often referred to as iNOS or NOS2), making the activity of this enzyme relatively insensitive to intracellular calcium concentrations. Inducible NOS are known to become activated through inflammatory processes and may be responsible for several orders of magnitude of greater production of NO* than nNOS and eNOS. Activity of NOS generally appears to be affected by oxidative stress and concentration NO* can provide a negative feedback. Bacterial NOS (bNOS) has been shown to be coupled to transcription of SOD in bacteria.

Nitric oxide is not a highly reactive radical, particularly when compared to hydroxyl, but also with singlet oxygen and many organic peroxy radicals. Relatively weak reactivity permits diffusion of NO* through cytosol and often across cell membranes, which is critical to the signaling functionality of this molecule in biology. Given that nitrogen's electronegativity on the Pauling scale is 3.04 and oxygen's is 3.44, which is not a large difference, NO* can be considered hydrophobic. That makes its diffusion through cell membrane lipids often faster than diffusion of oxygen. While hydroxyl and peroxy radicals can often abstract hydrogen in carbon-hydrogen bonds, nitric oxide is usually incapable of doing so. Nitric oxide does bind strongly to the iron in heme groups, which is critical to the biological activity that involves activating guanylate cyclase and slowing mitochondrial respiration by binding to cytochrome-c oxidase (one of the key molecules in the mitochondrial electron transport chain). Similarly to molecular oxygen (triplet), nitric oxide also reacts with radicals quickly. However, in contrast to the di-oxygen molecule which has two unpaired spins and whose reaction with other radicals often results in radical chains or unstable molecules, NO* often terminates a chain reaction when it reacts with other radicals. For example, it can convert thiyl radicals into nitrosothiols by acting as a chain-terminating agent:

$$RS^* + NO^* \rightarrow RS - NO,$$

where R is some organic group and S is the sulfur atom.

Nitric oxide can act as both an oxidizing and a reducing agent, depending on its molecular environment. Recalling that nitrogen is more electronegative than hydrogen and less electronegative than oxygen, a typical sequence of the oxidation states of nitrogen species that may originate from nitric oxide is:

$$NO_3^- (+5) \leftrightarrow NO_2^* (+4) \leftrightarrow NO_2^- (+3) \leftrightarrow NO^* (+2) \leftrightarrow HNO (+1) \leftrightarrow$$
$$NH_2OH (-1) \leftrightarrow N_2H_2 (-2) \leftrightarrow NH_3 (-3) \tag{3.23}$$

where the number in parenthesis next to the compounds denotes the nitrogen oxidation state.

The two most active reactive nitrogen species produced from nitric oxide are probably peroxynitrite $ONOO^-$ and nitrogen dioxide NO_2^* radical, both of which are the major reactive products of nitric oxide with oxygen. Peroxynitrite, which is a structural isomer of the nitrate NO_3^- ion (in the nitrate all oxygen atoms are bound to nitrogen, while in peroxynitrite two oxygen atoms are nitrogen bound and one is oxygen bound), is produced primarily in the reaction with superoxide:

$$NO^* + O_2^- \rightarrow ONOO^-. \tag{3.24}$$

The rate of the above reaction is quite high, reported to be around 3×10^9 M^{-1} s^{-1}, and is comparable to the rate of superoxide dismutation by SOD, which is around 2×10^9 M^{-1} s^{-1}. This means that, at sufficiently high concentrations, nitric oxide can compete effectively for superoxide with SOD. Peroxynitrite can react directly with many important biomolecules, particularly those containing sulfhydryl (SH) groups and transition metal centers (iron, copper, etc.). This includes many important heme-group-containing enzymes that play an important role in electron transfer processes such as cytochrome c, hemoglobin, SOD and NOS as well as proteins containing cysteine and methionine amino acids. Peroxynitrite also causes direct DNA strand cleavage by oxidizing deoxyribose.

Indirect oxidative effects of peroxynitrite occur through several mechanisms. One is its decomposition resulting in the production of hydroxyl radicals, a moderately reactive nitrogen dioxide radical and nitric acid:

$$2ONOO^- + 2H^+ \leftrightarrow 2ONOOH \rightarrow NO_2^* + OH^* + HNO_3. \tag{3.25}$$

It has been demonstrated that hydroxyl radicals can in some cases be produced more effectively by this process than by the Fenton mechanism. Lipid peroxidation and DNA damage associated with peroxynitrite are likely to occur through this mechanism of hydroxyl radical generation. Nitrogen dioxide radical NO_2^* is capable of abstracting a hydrogen from some of the weaker C–H bonds (allylic hydrogen that is bound to a carbon which is adjacent to a carbon-carbon double bond). Protein tyrosine is a major target of NO_2^* in biological systems. Tyrosine nitration involves covalent protein modification resulting from the addition of a nitro ($-NO_2$) group adjacent to the hydroxyl group on the aromatic ring of tyrosine residues. Reactions with NO_2^* also result in nitrated lipids. It should be mentioned that decomposition of peroxynitrite is only one of several mechanisms that generate nitrogen dioxide radicals. Another mechanism is the direct reaction of nitric oxide with oxygen:

$$2NO^* + O_2 \rightarrow 2NO_2^*. \tag{3.26}$$

Peroxynitrite reacts fairly rapidly (reaction rate constant of around 4×10^4 M^{-1} s^{-1}) with carbon dioxide:

$$ONOO^- + CO_2 \rightarrow ONOOCO_2^-. \tag{3.27}$$

Nitrosoperoxycarbonate ($ONOOCO_2^-$) then homolyzes to produce two radicals:

$$ONOOCO_2^- \rightarrow CO_3^{*-} + NO_2^* \tag{3.28}$$

with probability of c. 0.3 or decomposes into:

$$ONOOCO_2^- \rightarrow CO_2 + NO_3^- \tag{3.29}$$

with probability of c. 0.6. Thus, under typical physiological conditions, when concentration of CO_2 is around 1 mM, carbon dioxide is the primary scavenger of peroxynitrite and Equation (3.28) is likely the primary producer of nitrogen dioxide in cells. Moreover, the carbonate radical CO_3^{*-} produced in this process can also be fairly reactive. It has been shown to abstract electrons from aromatic amino acids (tyrosine and tryptophan). It can cause oxidative damage to DNA bases, particularly to guanine as it is the easiest to oxidize. However, its reactions with sulfur-containing methionine and cysteine are less efficient and hydrogen atom abstraction by carbonate radicals is generally very slow. Carbonate does not react with lipid membranes very well either, due to its hydrophilic nature.

3.3.5.4   *Signaling functions of reactive species*

In 1956, soon after the discovery of oxygen free radicals in biological matter, Denham Harman hypothesized that they may be formed as by-products of enzymatic reactions. He described them as a Pandora's box of evils that may account for cellular damage, mutagenesis, cancer and the degenerative process of biological aging. The science behind biological effects of free radicals entered a new phase at the end of 1970s; publications of several papers established that ROS likely play important signaling functions in cells. In this section, after a very brief introduction to cellular signaling, we discuss some suppositions and facts explaining why and how reactive species are involved in signaling processes.

Cell signaling systems can be viewed as communication networks that control cellular functions and coordinate activities of different cells. Molecular carriers of signals between cells are often called 'first messengers', while intracellular signal carriers can be referred to as 'second messengers'. In contrast to telephony systems that establish communication channels between specific users or end points, cellular communications (of the biological kind) resemble a set of broadcasting systems where the broadcast messages can be targeted to a specific set of receivers, accepted by these receivers, interpreted and then re-broadcast, possibly using a different set of chemical carriers. An intracellular broadcast signal might be initiated, for example, at the cell membrane when a specific membrane receptor (usually a transmembrane protein) binds to a specific extracellular ligand (this could be a protein, hormone, a small molecule or an ion).

For example, a ligand called epidermal growth factor (EGF) binding to the EGF receptor (EGFR) on the extracellular side of the cell membrane usually leads to phosphorylation of the EGFR on the intracellular side, activating binding to the intracellular adaptor protein (GRB2 protein helps mediate protein–protein interactions). This couples to other proteins propagating the signal further inside the cell along several different pathways to different eventual targets. One of these pathways is called the mitogen-activated protein kinase (MAPK) pathway. The MAPK protein is a protein kinase (an enzyme) that phosphorylates various other proteins such as the transcription factor MYC, which leads to an altered gene transcription and, ultimately, cell cycle progression. Intracellular signals are usually passed by the mechanism of protein-protein interactions, often involving phosphorylation and other redox events and through diffusion of small molecules and ions.

It is natural for reactive species to participate in regulatory cell signaling. Generation of superoxide during cellular metabolism, for example, can be viewed as a signal of aerobic metabolic activity. Reduction of the amount of superoxide produced may signal, for example, reduction in the availability of oxygen or in the availability of glucose. Either way, cells will likely need to adjust their behavior, possibly switching to a different metabolic pathway or by enhancing glucose transport across cell membrane. Given that all cellular activities require energy and materials generated through metabolism, it might be expected that ROS, whose production is initiated through metabolic activity, might play an important signaling role.

There are various examples of growth factors, cytokines or other ligands that trigger ROS production in non-phagocytic cells through their corresponding membrane receptors. It has also been confirmed that intracellular ROS production can be triggered by extracellular messages using membrane-bound NADPH oxidase isoforms as the mechanism. ROS generation by the cardiovascular NADPH oxidase isoforms can be induced by hormones or hemodynamic forces, for example. Angiotensin II, a peptide hormone usually associated with blood vessel constriction, increases NADPH-driven superoxide production in cultured vascular smooth muscle cells and fibroblasts. Thrombin, platelet-derived growth factor (PDGF) and tumor necrosis factor-α (TNF-α) stimulate NADPH oxidase-dependent superoxide production in vascular smooth muscle cells. Interleukin-1, TNF-α and platelet-activating factor all increase NADPH-dependent O_2^- generation in fibroblasts. Mechanical forces have also been shown to stimulate NADPH oxidase activity in endothelial cells, and re-oxygenation stimulates NADPH oxidase activity in cardiac myocytes (heart muscle cells).

It would be difficult to detect small variations in the concentration of reactive species as a signal if their normal or 'background' concentrations were too high. Maintenance of reactive species homeostasis is therefore essential not only to minimize the damaging effects of the reactive species, but also to achieve a good signal-to-noise ratio. This homeostasis is usually maintained through scavenging of intracellular reactive species by various molecules called antioxidants.

Antioxidants come in several varieties. Some are water soluble, while others are lipophilic and tend to reside in lipid membranes. Some antioxidants are molecules that can become radicals relatively easily when exposed to reactive species, but do not easily convert other organic molecules into radicals. Radicals of such antioxidants readily combine with other organic radicals to terminate radical chain reactions. A well-known example of such antioxidant includes Vitamin E, which is in a class of lipophilic molecules. Although intracellular amino acids are not as easily oxidized by reactive species, their large concentration within cells may also buffer cells against high levels of reactive species. Consistent with this hypothesis, it has been noted that proteases (enzymes which degrade proteins into amino acids) appear to be over-expressed in cells during periods of increased oxidative stress.

An important class of antioxidants is based on oxidation of thiol groups (S–H) and generation of disulfide bond (S–S, sometimes called bridges). The bond of sulfur to hydrogen is weaker (see Table 3.1) than C–H or O–H bonds, and is relatively easily broken when sulfur donates electrons in redox reactions with reactive species. At the same time the disulfide bond is also relatively weak, making disulfide-containing compounds good candidates for reduction. This type of redox cycling is exactly what occurs in many cysteine residues containing proteins. A common and important redox cycling antioxidant mechanism in many eukaryotic cells is based on glutathione (most prokaryotes lack this particular antioxidant mechanism).

Glutathione (GSH) is a tripeptide that contains an unusual peptide linkage between the amine group of cysteine (which is attached by normal peptide linkage to a glycine) and the carboxyl group of the glutamate side-chain. One of the glutathione functions is to reduce disulfide bonds formed within cytoplasmic proteins to cysteines by serving as an electron donor. In donating an electron, glutathione itself becomes reactive, but readily reacts with another reactive glutathione to form glutathione disulfide (GSSG, also called L-glutathione). Such a reaction is possible due to the relatively high concentration of glutathione in cells (up to 5 mM in the liver, for example). In the process, GSH is converted to its oxidized form, GSSG. Once oxidized, glutathione can be reduced back by an enzyme glutathione reductase, using NADPH as an electron donor. Provision of cysteine is the rate-limiting factor in glutathione synthesis by the cells, since this amino acid is relatively rare in foodstuffs. Furthermore, if released as the free amino acid, cysteine appears to be toxic and spontaneously catabolized in the gastrointestinal tract and blood plasma. Raising GSH levels through direct supplementation of glutathione is difficult as it does not readily enter most cells from outside. Intracellular GSH concentrations can be raised by administration of certain supplements that serve as GSH precursors. N-acetylcysteine, commonly referred to as NAC, is the most bioavailable precursor of glutathione as it enter cells and provides the needed cysteine.

Many enzymes appear to be direct receivers of ROS signals. Some examples include guanylyl cyclase, phospholipase C, phospholipase A2 and phospholipase D. Ion channels also appear to be the ROS signaling targets, including calcium channels. Sulfurhydryl groups (R-S-H) and transition-metal-containing proteins appear to be the primary targets (receivers) of ROS signals. There are two sulfur-containing amino acids – cysteine and methionine – but only cysteine contains thiols. Through thiols, cysteine plays an important role in controlling protein confirmation by oxidation of thiols and formation of disulfate bridges (S–S bond) between different parts of a protein. Overall oxidative stress state (concentration of various ROS) can therefore modulate thiol/disulfate redox state of many proteins. This has been suggested as the mechanism by which levels of ROS affect human insulin receptor kinase activity and p38 MAPK signaling pathways described above, for example.

ROS and a pro-oxidative shift of the intracellular thiol/disulfide redox state appear to increase the overall signal strength in cellular broadcasting networks. There has been a growing body of evidence suggesting that this occurs by modulation of protein kinases and protein phosphatases (enzymes that phosphorylate and de-phosphorylate proteins) activities directly. Recall that protein phosphorylation/dephosphorylation processes play critical roles in regulating many cellular metabolic processes in eukaryotes. They also govern many signal transduction pathways. Among the extracellular signals, growth-factor-dependent protein tyrosine kinases and protein tyrosine phosphatases are of critical importance to mitogenesis, cell adhesion, cell differentiation, oncogenic transformation and apoptosis, to name a few. Target proteins are typically phosphorylated at specific transduction sites (usually at serine/threonine or tyrosine residues) by one or more protein kinases and the phosphates are removed by specific protein phosphatases. It is known that changing the activity of protein kinases, phosphatases or both can regulate the likelihood of phosphorylation at a particular site. The overall effect of reactive species on the signaling networks exerted by activation of protein kinases and phosphatases might be expected considering the fact that all protein tyrosine phosphatases, for example, have a conserved (as for many cells and species) cysteine residue in their catalytic domain; for complete activity, this catalytic domain needs to be in the fully reduced form. We can therefore make a reasonable argument that conditions of oxidative stress will lead to changes in the activity of several signaling enzymes.

In summary, although reactive species in general and ROS in particular have been initially viewed as a damaging by-products of metabolic activity, it is now clear that these species play a vital role in a large number of signaling mechanisms in cells. Several possible mechanisms by which the reactive species influence cell signaling have been suggested. This remains a highly active area of research in living systems.

3.4   Overview of anatomy and physiology

Anatomical language often uses terms related to position within the human body. Table 3.2 below summarizes some of the essential terms, while Figure 3.24 illustrates them.

3.4.1   Tissues

3.4.1.1   *Organization of the animal body*

Animals are multicellular heterotrophs whose cells lack cell walls. They developed external or internal skeletons to provide support, skin to prevent or lessen water loss, muscles that allowed them to move in

Table 3.2   *Anatomical position terms.*

Position	Description
Anterior	In front of another organ or structure, located toward the front from the coronal plane
Posterior	Behind another organ or structure, located toward the back from the coronal plane
Superior	Above another organ or structure, located above the mid-section of the body
Inferior	Below another organ or structure, located below the mid-section of the body
Deep	Located further away from the outer surface of the body
Superficial	Located closer to the outer surface of the body
Medial	Located closer to the median plane
Lateral	Located further away from the median plane
Proximal	Closer to the point of origin, such as the point of departure of a limb from the trunk
Distal	Further from the point of origin, such as the tip of a finger
Ipsilateral	The same side of the body (from the median plane)
Contralateral	Opposite side of the body (across the median plane)

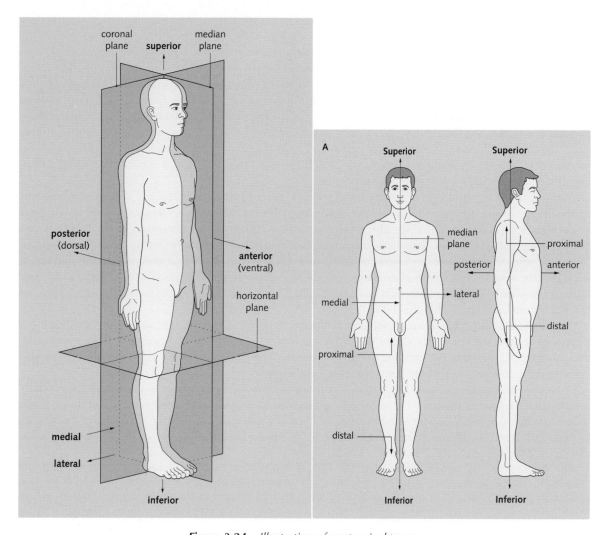

Figure 3.24   *Illustration of anatomical terms.*

search of food, brains and nervous systems for integration of stimuli and internal digestive systems for processing of food. Organs in animals are composed of a number of different tissue types, which are in turn composed of cells. The four basic types of animal tissue are: epithelial tissue, connective tissue, muscle tissue and nervous tissue.

3.4.1.2   *Epithelial tissue*

Epithelial tissue covers body surfaces, lines bodily cavities and forms most glands. Its name is derived from Greek with 'epi' meaning on or upon and 'theli' meaning tissue. Many epithelial tissues are avascular or poorly vascularize and receive nourishment primarily through diffusion. Exocrine and endocrine epithelial tissues, on the other hand, are highly vascular.

The structure of epithelial tissues often resembles a brick wall. Cells in the epithelial tissues are densely packed, connecting to each other through different types of junctions that play material and signal transmission roles. These junctions can be classified as tight junctions, adherens junctions, desmosomes, hemidesmosomes and gap junctions. Tight junctions involve a pair of transmembrane proteins fused on outer plasma membrane. Adherens junctions are protein layers on the inside plasma membrane which attaches both protein and microfilaments. Desmosomes attach to the microfilaments of cytoskeleton made up of keratin protein. Hemidesmosomes resemble desmosomes on a section, and are made up of the integrin (a transmembrane protein). Cells in epithelial tissues are typically supported by a basement membrane, which is a thin layer of protein fibers that acts as a scaffold for epithelium growth and regeneration after injuries. The basement membrane also acts as a selectively permeable membrane that determines which substances will be able to enter the epithelium more or less through diffusion.

Three different types of epithelium are usually identified: squamous, cuboidal and columnar. Squamous epithelial cells have the appearance of thin, flat plates. They fit closely together in tissues providing a relatively smooth surface. The shape of the nucleus in these cells usually corresponds to the cell form and helps to identify the type of epithelium. Squamous cells tend to have horizontally flattened oval-shaped nuclei. Typically, squamous epithelia are found lining surfaces that utilize diffusive transport of nutrients and gases, such as in the alveolar epithelium of the lungs. Specialized squamous epithelia also form the lining of cavities such as the blood vessels (endothelium) and pericardium (a double-walled sac that contains the heart) and the major cavities found within the body.

Cuboidal epithelium cells are roughly cuboidal in shape, appearing square in cross-section. Each cell has a spherical nucleus at the center. Cuboidal epithelium is commonly found in secretive or absorptive tissue, for example, the pancreas (an exocrine gland) and the (absorptive) lining of the kidney tubules. They also constitute the germinal epithelium that covers the female ovary.

Columnar epithelial cells are elongated and column-shaped. Their nuclei are elongated and are usually located near the base of the cells. Columnar epithelium forms the lining of the stomach and intestines. Some columnar cells are specialized for sensory reception such as in the nose, ears and the taste buds of the tongue. Goblet cells (unicellular glands) are found between the columnar epithelial cells of the duodenum (in the stomach). They secrete mucus, which acts as a lubricant.

Any epithelium can be simple or stratified. Simple epithelium has only a single cell layer. Stratified epithelium has more than one layer of cells. It is therefore found mostly where the body linings have to withstand mechanical or chemical insult such that layers can be abraded and lost without exposing sub-epithelial layers. In stratified epithelium, most cells are bound to each other by desmosomes. In keratinized epithelia, the most apical (exterior) layers of cells are dead and lose their nucleus and cytoplasm. Instead they contain a tough resistant protein called keratin. Mammalian skin contains the keratinized epithelium, making it waterproof. The lining of the esophagus is an example of a non-keratinized or 'moist' stratified epithelium. Transitional epithelia are found in tissues that stretch. It can appear to be stratified cuboidal when the tissue is not stretched or stratified squamous when the organ is distended and the tissue stretches. It is sometimes called the urothelium, since it is almost exclusively found in the bladder, ureters and urethra.

The primary functions of epithelial tissues are: (1) to protect the tissues that lie beneath it from radiation, desiccation, toxins and physical trauma; (2) the regulation and exchange of chemicals between the underlying tissues and a body cavity; and (3) the secretion of hormones, sweat, mucus, enzymes and other products. There are two major types of secretion organs or glands: endocrine glands and exocrine glands. Endocrine glands secrete their product into the extracellular space where it is rapidly taken up by the blood vascular system. The exocrine glands secrete their products into a duct that then delivers the product to the lumen of an organ or onto the free surface of the epithelium.

In addition to the above primary functions, epithelial tissues can be employed for sensing of the extracellular environment. Some epithelial cells, for example, have cilia (typically a cylindrical protuberance from the

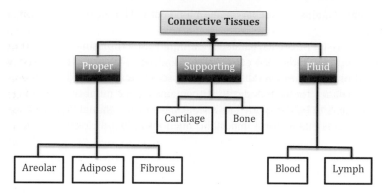

Figure 3.25   *Classification of different connective tissues.*

cellular body) that provide chemosensation, thermosensation and mechanosensation of the extracellular environment. These cells are often arranged in sheets forming a tube or tubule with cilia projecting into the lumen.

3.4.1.3   Connective tissue

Connective tissues serve many purposes in the body. This includes binding different organs in the body, providing rigidity and supporting the body, forming blood and lymphatic fluid that distributes essential materials throughout the body and storing nutrients such as fats. In contrast to epithelial tissue, a non-cellular matrix typically separates cells in connective tissues from one another. This matrix may be solid (as in a bone), soft (as in loose connective tissue) or liquid (as in blood). Cells, fibers and extracellular matrix of the connective tissue in one way or another are all embedded in a fluid.

Classification of different connective tissues is depicted in Figure 3.25. Areolar connective tissue is probably the most widespread throughout the body. Cells found in areolar tissue are: fibroblasts that create elastin and collagen fibers in the matrix; macrophage cells (considered to be part of the immune system) that help engulf foreign matter such as bacterial particles; and mast cells that also perform important immune system functions including production of histamine and heparin. Adipose tissue contains mostly specialized cells that store lipids (fat) and form protective padding on the body. Fibrous tissue consists primarily of fibroblasts and forms tendons and ligaments.

Supporting connective tissue, as the name implies, is tasked with construction of a skeleton. The matrix of cartilage is composed of chondrin. The cells lie in the matrix singly or in groups of two or four surrounded by fluid-filled spaces. The cartilage may be elastic, whose matrix has yellow fibers as in pinna of ear. Cartilage forms the embryonic skeleton of vertebrates and the adult skeleton of sharks and rays. It also occurs in the human body in the ears, tip of the nose and at joints such as the knee and between bones of the spinal column. The cartilage can be calcified where matrix is deposited with calcium salts as in the head of long bones.

Bones are of two types: spongy and compact. In spongy bone, osteocytes (bone cells) are irregularly arranged. Such bones are found at the ends of the other longer bones. In the compact bones, cells are arranged in circles or lamellae around a central canal known as the Haversian canal. Bone matrix is composed of ossein and contains salts of calcium, phosphorus and magnesium. The matrix in mammalian long bones (such as thigh bone) is arranged in concentric rings. The osteocytes typically lie on the lamellae (concentric rings in the matrix). Osteocytes give out branched processes that join with those of the adjoining cells. Some bones have a central cavity that contains a tissue producing blood cells. The substance contained in the bone cavity is called bone marrow.

Blood is a connective tissue of cells separated by a liquid (blood plasma) matrix. Two primary types of cells occur: red blood cells (erythrocytes) carry oxygen and white blood cells (leukocytes) function in the immune system. Blood plasma transports dissolved glucose, wastes, carbon dioxide and hormones, as well as regulating the water balance for the blood cells. It contains a large number of proteins such as fibrinogen, albumin, globulin and others to be transported to various parts of the animal body for various purposes.

Platelets are cell fragments that perform important functions in blood clotting.

3.4.1.4 *Muscle tissue*

Muscle tissue facilitates movements by contraction of individual muscle cells (sometimes referred to as muscle fibers). Three types of muscle fibers occur in animals (the only taxonomic kingdom to have muscle cells): skeletal (striated), smooth and cardiac.

Skeletal muscle cell can be anywhere from a few hundred micrometers to a few centimeters long. The longer cells are multinucleated, with the nuclei located just under the plasma membrane. Most of the cell is occupied by striated, thread-like myofibrils. Each myofibril is divided into sections along its length called sarcomere. A sarcomere is the functional unit of a skeletal muscle cell containing filaments of actin and myosin. Actin filaments serve as anchors, relative to which motion of myosin fibers occurs to cause contraction of the cell. Each sarcomere has thick and thin filaments. These skeletal muscle cells are typically anchored at the ends of the muscle to bones and have nerve endings that connect to the cells in order to signal movement. The skeletal muscles function in conjunction with the skeletal system for voluntary muscle movements.

Smooth muscle fibers lack the banding, although actin and myosin still occur. These cells usually actuate involuntary movements and/or autonomic responses (such as breathing, secretion, ejaculation, birth, peristaltic motion of the intestines and certain reflexes). Smooth muscle fibers consist of spindle-shaped cells. These fibers are components of structures in the digestive system, reproductive tract and blood vessels.

Cardiac muscle fibers are a type of striated muscle found only in the heart. The cell has a bifurcated (or forked) shape, usually with the nucleus near the center of the cell. The cardiac muscle cells are usually connected to each other by intercalated disks.

3.4.1.5 *Neural tissue*

Neural tissue (also called nervous tissue) is specialized for the conduction of electrical impulses that convey information or instructions from one region of the body to another. About 98% of neural tissue is concentrated in the brain and spinal cord, the control centers for the nervous system. Neural tissue integrates sensory input received by the body, controls movements and appears to be responsible for what we call thoughts and emotions. Nervous tissue is composed of two main cell types: neurons and glial (neuroglia) cells. Some parts of neurons are also enveloped by special cells referred to as Schwann cells.

Each neuron (Figure 3.26) has a cell body, so-called cell *processes* that include a single axon and many dendrites. Nerve cell processes are often less than a micron (1 μm) in diameter. However, the *length* of axons and dendrites is astonishingly great, far greater than ordinary cellular dimensions. Dendrites may extend several millimeters away from the cell body, into a volume the size of a pea. Axon length may exceed a meter (for many sensory and motor axons), and commonly extends for several centimeters. As a simple consequence of this cellular geometry, the cell body of a neuron may comprise less than one percent of the entire cell volume. From this, it may be deduced that the bulk of nervous tissue consists of nerve cell processes rather than nerve cell bodies.

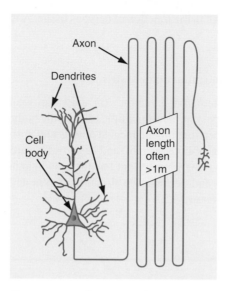

Figure 3.26   *Illustration of a neuron cell.*

The axon is a process that is specialized for conducting signals from one nerve cell to another. Each nerve cell has one and only one axon. Typical axons have relatively few branches, except near the terminal end. The diameter of an axon is relatively uniform along its entire length. The terminal branches of an axon make *synaptic contacts* onto other nerve cells or with peripheral effectors such as muscles and glands. The electrical signals of a nerve travel along axons away from the cell body and toward synaptic contacts (synapses) at the axonal terminal.

Axonal nerve signals, called *action potentials*, are electrical pulses initiated at the axon hillock, the site where an axon arises from the cell body. Action potentials are maintained by axons and do not decline in strength as they travel along the axon. Axons may be covered by layers of fatty myelin (myelinated) to increase the speed of signal conduction. Myelin is formed by the support cells (Schwann cells in the peripheral nerve system, oligodendroglia in the central nervous system) wrapping around the axons. An axon from one neuron communicates its signal at the synapse to another cell by neurotransmitters, usually small molecules secreted by one cell and binding to another. Neurotransmission can also be electrical, with ions passing directly from one cell into another. Each synapse has a presynaptic side at the axon terminal from which a neurotransmitter is released. The neurotransmitters are usually stored in *synaptic vesicles* within the presynaptic terminal. They are released in response to changes in membrane potential associated with the arrival of action potentials.

Dendrites are processes that specialize in receiving and integrating signals from other nerve cells. A nerve cell typically has several dendrites, each with numerous branches. The diameter of dendrites decreases away from the cell body. Dendrites typically receive synaptic contacts from axons of many other nerve cells. Synapses often occur on tiny *dendritic spines*. Nerve signals that travel along dendrites toward the cell body are called *synaptic potentials* and arise at synapses. Synaptic potentials usually fade as they propagate.

Glial cells are the most numerous cells within the central nervous system. The name 'glia' means 'glue', indicating that these cells fill the space between neurons. This name partly reflects enduring ignorance of glial cell functions. The small nuclei of glial cells may be readily observed in any section of central nervous tissue. The two most common types of glia, *oligodendroglia* and *astroglia*, both have extensive cytoplasmic processes and are intimately involved in the function of nervous tissue. A third glial type, *microglia*, functions similarly to macrophages.

3.4.2   The body covering: The integumentary system

The integumentary system is an alternative name for skin and various parts derived from skin such as hair, nails and glands. The integumentary system has multiple roles in the body homeostasis including protection, temperature regulation, sensory reception, biochemical synthesis and absorption. Skin is the largest organ in the human body by weight (12–15% of body weight) and has a surface area of 1–2 m. It is continuous organ, but structurally distinct from mucous membranes that line the mouth, anus, urethra and vagina. Two distinct layers occur in the skin: epidermis and dermis. The structure of skin and its various derivatives (processes originating from skin) is shown in Figure 3.27.

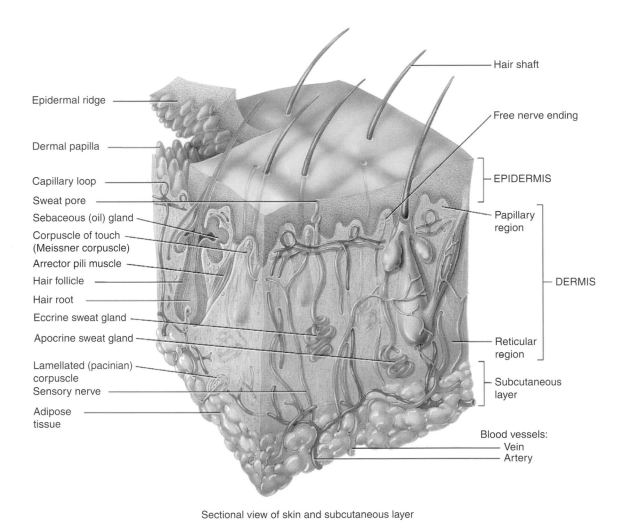

Sectional view of skin and subcutaneous layer

Figure 3.27   *Structure of skin and various processes originating from it (see color plate).*

3.4.2.1 Epidermis

The epidermis is the thinner avascular (no blood vessels) layer of the skin varying from 50 μm on the eyelids to 1.5 mm on the soles of feet. It is typically made up of 4 cell types:

1. *keratinocytes*: cells that produce keratin protein, a fibrous protein that helps protect the epidermis;
2. *melanocytes*: cell that produce the brown pigment melanin, a polyacetylene derivative that has excellent photo-protectant properties (transforms UVB into heat) and is the reason for long-lasting brown tan;
3. *Langerhan cells*: which participate in immune response; and
4. *Merkel cells*: responsible for the sense of touch.

Five distinct sub-layers are typically identified in the epidermis, as follows.

1. *Stratum corneum*: the outermost layer, made of 25–30 layers of dead flat keratinocytes. Lamellar granules (oblong 100–300 nm lipid-containing secretory organelles found in keratinocytes) provide water-repellent action and are continuously shed and replaced.
2. *Stratum lucidum*: only found in the fingertips, palms of hands and soles of feet. This layer is made up of 3–5 layers of flat dead keratinocytes.
3. *Stratum granulosum*: made up of 3–5 layers of keratinocytes, site of keratin formation and lamellar granule production.
4. *Stratum spinosum*: appears covered in thorn-like spikes that provide strength and flexibility to the skin.
5. *Stratum basale*: the deepest layer, made up of a single layer of cuboidal or columnar cells. Cells produced here constantly divide and move up to the apical (top) surface.

3.4.2.2 Dermis

The dermis is a connective tissue layer under the epidermis, and contains nerve endings, sensory receptors, blood capillaries, hair follicles, glands and elastic fibers. The dermis contains three main cell types:

1. *adipocytes*: lipid-storing cells;
2. *macrophages*: cells of the immune system; and
3. *fibroblasts*: typical cells of the connective tissues that produce fibers such as collagen and elastin.

There are two main divisions of the dermal layer as follows.

1. *Papillary region (Dermal papillae)*: the superficial layer of the dermis is made up of loose areolar connective tissue with elastic fibers. It contains finger-like structures that invade the epidermis and capillaries or Meissner corpuscles that respond to touch.
2. *Reticular region of the dermis*: the deeper region of the dermis is made up of dense irregular connective and adipose tissue containing sweat glands, sebaceous (oil) glands and blood vessels.

3.4.2.3 Follicles and glands

Hair follicles are lined with cells that synthesize the proteins that form hair. A sebaceous gland (that secretes the oily coating of the hair shaft), capillary bed, nerve ending and small muscles are typically associated with each hair follicle. If the sebaceous glands becomes plugged and infected, it becomes a skin blemish (or pimple). Sweat glands are normally open to the surface through skin pores. Eccrine glands are a type of sweat gland linked to the sympathetic nervous system, and occur all over the body. Apocrine glands are the other type of sweat gland, which are larger and occur in the armpits and groin areas. These produce a solution that bacteria act upon to produce 'body odor'.

3.4.2.4 Hair and nails

Hair, scales, feathers, claws, horns and nails are animal structures derived from skin. The hair shaft extends above the skin surface, while the hair root extends from the surface to the base or hair bulb located in the dermis layer. Genetics controls several features of hair: baldness, color and texture. Nails consist of highly keratinized, modified epidermal cells. The nail arises from the nail bed, which is thickened to form a lunula (or little moon). Cells forming the nail bed are linked together to form the nail.

3.4.2.5 Homeostasis

Skin homeostatic functions include maintenance of water balance, regulation of body temperature, sensory reception, synthesis of vitamins and hormones and absorption of nutrients. The skin's primary functions are to serve as a barrier to the entry of microbes and viruses and to prevent water and extracellular fluid loss. Acidic secretions from skin glands also retard the growth of fungi. Melanocytes form a second type of barrier: protection from the damaging effects of ultraviolet radiation. When a microbe penetrates the skin (or when the skin is breached by a cut), the inflammatory response occurs.

Heat and cold receptors are located in the skin. When the body temperature rises, the hypothalamus (brain region) sends a nerve signal to the sweat-producing skin glands, causing them to release c. 1–2 L of water per hour to cool the body. The hypothalamus also causes dilation of the blood vessels of the skin, allowing more blood to flow. When body temperature falls, the sweat glands constrict and sweat production decreases.

Water loss occurs in the skin by two routes: evaporation and sweating. In hot weather, up to 4 liters per hour can be lost by these mechanisms. Skin damaged by burns is less effective at preventing fluid loss, often resulting in a life-threatening problem if not treated.

Sensory receptors in the skin include those for pain, pressure (touch) and temperature. Deeper within the skin are Meissner's corpuscles, which are especially common in the tips of the fingers and lips and are very sensitive to touch. Pacinian corpuscles respond to pressure. Interestingly, there are more cold-temperature receptors than hot-temperature receptors.

Skin cells synthesize melanin and carotenes, which give the skin its color. The skin also assists in the synthesis of vitamin D. Children lacking sufficient vitamin D develop bone abnormalities known as rickets.

The skin is selectively soluble to fat-soluble substances such as vitamins A, D, E and K, as well as steroid hormones such as estrogen. These substances enter the bloodstream through the capillary networks in the skin. Patches have been used to deliver a number of therapeutic drugs in this manner including estrogen, scopolamine (motion sickness), nitroglycerin (heart problems) and nicotine (for those trying to quit smoking).

3.4.3 Circulatory system

Living things must be capable of transporting nutrients, wastes and gases to and from cells within their bodies. Sponges are the simplest animals, yet even they have a transport system; seawater is their medium of transport. It is propelled in and out of the sponge by ciliary action (collective pumping action of hair-like cilia). Simple animals, such as the hydra and planaria, lack specialized organs such as hearts and blood vessels. Instead, they use their skin and digestive tract as exchange points for materials and essential nutrients diffuse in and out. This, however, limits the size an animal can attain. To become larger, animals need specialized organs and organ systems. A circulatory system (often called cardiovascular system in humans) is a system animals developed to deliver oxygen and nutrients to organs and cells deep inside the body and remove carbon dioxide and metabolic wastes.

The main components of the circulatory system of humans and larger animals are:

1. *blood*: a connective tissue of liquid plasma and cells used as a gas and material exchange medium;
2. *blood vessels*: including arteries, capillaries and veins that serve as conduits for delivery of blood to all tissues and exchange of gases and some cells and materials between blood and tissue; and
3. *heart*: a muscular pump designed to move the blood.

Several types of circulatory systems can be found among animals. Open circulatory system is common to mollusks and arthropods, for example. In this system a fluid (called hemolymph or hemolymph) in a cavity called the hemocoel bathes the organs directly with oxygen and nutrients; there is no distinction between blood and interstitial fluid (such as lymph fluid in humans). Muscular movements by the animal during locomotion can facilitate hemolymph movement, but diverting flow from one area to another is limited. Some open circulatory systems, particularly in arthropods, do employ a heart for pumping the hemolymph. When the heart relaxes, hemolymph fluid is drawn back toward the heart through open-ended pores (ostia). Hemolymph is typically composed of water, inorganic salts (mostly Na^+, Cl^-, K^+, Mg^{2+} and Ca^{2+}) and organic compounds (mostly carbohydrates, proteins and lipids). The primary oxygen transporter molecule is hemocyanin. Hemocytes are free-floating cells located within the hemolymph, which play a role in the arthropod immune system.

Most vertebrates (animals with a vertebrae) and a few invertebrates have a closed circulatory system. Closed circulatory systems have the blood enclosed at all times within vessels of different size and wall thickness. In this type of system, blood is pumped by a heart through the vessels and does not normally fill larger body cavities. Iron containing hemoglobin protein, which binds and transports oxygen, causes the vertebrate blood to turn red in the presence of oxygen. Hemoglobin is estimated to transport c. 98% of oxygen used by various tissues in humans, while the remaining oxygen is dissolved in blood and various extracellular fluids. Vertebrates also have a secondary circulatory system – the lymphatic system – which collects fluid (lymph) and some cells and returns them to the cardiovascular system. The lymphatic system is considered by some to be part of the circulatory system and, by others, to be part of the immune system.

3.4.3.1 Blood

Blood normally makes up c. 7–8% of the human body weight and takes up c. 5 L by volume. It distributes oxygen, essential amino acids and glucose as well as lipids and many signaling molecules throughout the body, while getting rid of carbon dioxide, ammonia and other waste products. In addition, it plays a vital role in the function of the immune system and in maintaining a relatively constant body temperature. Blood is a highly specialized connective tissue having four primary components: red blood cells, platelets, white blood cells and blood plasma.

Red blood cells, or erythrocytes, can vary substantially in size and composition among vertebrates. In most vertebrates they are larger than 10 μm and have typical cellular organelles. In humans and many mammals (with several exceptions), mature red blood cells are much smaller, shaped as flexible biconcave disks and lack a cell nucleus and most other organelles. Human erythrocytes have diameters of 6–8 μm and thicknesses of around 2 μm. About 2.4 million new erythrocytes are produced per second in human bodies. The cells actually develop in the bone marrow, a flexible tissue found inside bones, and circulate for c. 100–120 days in the body before being destroyed by macrophages (white blood cells considered part of the immune system). Each circulation takes c. 20 seconds. Approximately a quarter of the cells in the human body are red blood cells, making up c. 40–50% of the total blood volume.

Erythropoiesis is the development process in which new erythrocytes are produced, developing from committed stem cells to mature erythrocytes in c. 7 days. These cells have nuclei and other organelles – such as mitochondria and endoplasmic reticulum – during the early phases of erythropoiesis, but extrude them during development as they mature providing more space for hemoglobin. As a result of not containing mitochondria, red blood cells use none of the oxygen they transport. Instead they produce the energy-exchange carrier ATP by the glycolysis of glucose and lactic acid fermentation of the resulting pyruvate (see Section 3.3.4.3 on anaerobic metabolism). Because they lack nuclei and other organelles, mature red blood cells do not contain DNA and cannot synthesize any RNA; consequently, they cannot divide and have limited repair capabilities. This also means that no virus can evolve to target mammalian red cells.

The primary function of red blood cells is oxygen transport, which is probably the main reason why erythrocytes consist mainly of hemoglobin. Recall that hemoglobin is a metalloprotein containing heme groups whose iron atoms temporarily bind to oxygen molecules (O_2) and release them throughout the body. Oxygen can easily diffuse through the cell membrane of the red blood cells. Hemoglobin in the erythrocytes also carries some of the waste product carbon dioxide back from the tissues, although most waste carbon dioxide is transported back to the pulmonary capillaries of the lungs as bicarbonate (HCO_3^-) dissolved in the blood plasma. The color of erythrocytes is due to the heme group of hemoglobin. The blood plasma alone is beige colored and relatively transparent, but the red blood cells change color depending on the state of the hemoglobin. When combined with oxygen, the resulting oxy-hemoglobin is scarlet; without oxygen, deoxy-hemoglobin is of a dark red/burgundy color, appearing bluish through the vessel wall and skin. Pulse oximetry takes advantage of this color change to directly measure the arterial blood oxygen saturation.

Erythrocytes appear to have other functions in addition to oxygen transport. It has been recently demonstrated that erythrocytes can also synthesize nitric oxide enzymatically in the same way as endothelial cells. Exposure of erythrocytes to physiological levels of shear stress activates nitric oxide synthase and export of nitric oxide, which may contribute to the regulation of vascular tonus. Erythrocytes can also produce hydrogen sulfide, a signaling gas that acts to relax vessel walls. It is believed that the cardioprotective effects of garlic are due to erythrocytes converting the sulfur compounds of garlic into hydrogen sulfide.

Red blood cells are very specific to every individual. Specifically, red blood cells have inherited antigenic substances on their surfaces. These antigens may be proteins, carbohydrates, glycoproteins or glycolipids, depending on the blood group system. Some of these antigens are also present on the surface of other types of cells of various tissues. Several of these red blood cell surface antigens can stem from one allele (or very closely linked genes) and collectively form a blood group system. Blood types are inherited and represent contributions from both parents. A total of 30 human blood group systems are now recognized by the International Society of Blood Transfusion (ISBT). Blood type is particularly important during blood transfusion. Patients should ideally receive their own blood or type-specific blood products to minimize the chance of a transfusion reaction, which may include various immunologic responses due to the red blood cells not being recognized by the immune system and by other red blood cells. The ABO system (blood types A, B, AB and O) is the most important blood-group system in human-blood transfusion. The Rh system is the second most significant blood-group system in human-blood transfusion.

Platelets or thrombocytes are small, irregularly shaped clear cell fragments containing no nucleus or DNA and reaching 2–3 μm in diameter. They are released, in the same way as vesicles, from relatively large precursor megakaryocytes cells located in the bone marrow. The average lifespan of a platelet is normally just 5–9 days.

The primary function of platelets is probably in cessation of bleeding and loss of blood through clot formation. The process of clot formation usually begins with platelet activation, often triggered by molecules released when blood vessels are injured such as collagen, tissue factor (TF) or von Willebrand factor (vWF). Platelet activation can also be triggered by thrombin molecules usually formed with the help of tissue factor or by molecular or cellular aggregates having effectively negatively charged surfaces. Activated platelets change in shape to become more spherical, while forming a kind of projection (pseudopod) on their surface. Activated platelets tend to adhere to each other, forming platelets aggregates. The aggregates are often held together using various molecules in blood such as fibrinogen and vWF. The most abundant platelet aggregation receptor is glycoprotein IIb/IIIa (gpIIb/IIIa), which is a calcium-dependent receptor for fibrinogen, fibronectin, vitronectin, thrombospondin and vWF. Recent evidence also suggests that ROS play a significant role in platelet activation and aggregation, most likely through transmembrane proteins responsible for cell adhesion referred to as integrins.

Platelets are also a natural source of growth factors. Growth factors released by platelets include platelet-derived growth factor (PDGF), a potent chemotactic agent, and TGF beta, which stimulates the deposition of extracellular matrix. Both of these growth factors have been shown to play a significant role in the repair and regeneration of connective tissues. Other healing-associated growth factors produced by platelets include basic fibroblast growth factor, insulin-like growth factor 1, platelet-derived epidermal growth factor and vascular endothelial growth factor. Local application of these factors in increased concentrations through platelet-rich plasma (PRP) has been used as an adjunct to wound healing for several decades.

White blood cells, also known as *leukocytes*, are somewhat larger than erythrocytes, have a nucleus and lack hemoglobin. These cells constitute less than 1% of the blood's volume. They are made from stem cells in bone marrow. There are five types of leukocytes, important components of the immune system, discussed in the following section.

Blood plasma is the liquid component of the blood. Mammalian blood consists of a liquid (plasma) and a number of cellular and cell fragment components. Plasma is c. 60% of a volume of blood; cells and fragments are 40%. Plasma has 90% water and 10% dissolved materials including proteins, glucose, ions, hormones and gases. It acts as a buffer, maintaining pH near 7.4. Plasma contains nutrients, wastes, salts and proteins.

3.4.3.2 *System of blood vessels*

Different designs of vascular systems exist among different animals. Fish, for example, pump blood from the heart to their gills where the gas exchange occurs, and then on to the rest of the body. On the other hand, mammals pump blood to the lungs for gas exchange, then back to the heart for pumping out to the systemic circulation. The human cardiovascular system contains two main circulation loops. The *pulmonary circulation* loops through the lungs where blood is oxygenated (hemoglobin takes up the oxygen) and the *systemic circulation* passes through the rest of the body to provide oxygenated blood. Both of these circulation systems pass through the heart.

Vessels through which blood exits the heart are called *arteries*, while vessels through which blood enters the heart are referred to as *veins*. Arteries and veins are blood conduits that have fairly large diameters (the largest over 10 mm). Combined, they do not have sufficiently large distribution and surface area to supply oxygen and nutrients and collect carbon dioxide and molecular waste throughout all the tissue in the body. As illustrated in Figure 3.28, a system of smaller blood vessels called *arterioles* and *venules* branch out from arteries and veins, respectively, into blood *capillaries* that can be as small as c. 5–10 μm in diameter. Blood capillaries form a distribution known as a capillary bed to exchange material with all the bodily tissues.

Being the vessels through which blood exits the heart, arteries constitute the high-pressure part of the cardiovascular system. Arterial pressure varies between peak pressure during heart contraction, called the *systolic pressure*, and the minimum or *diastolic pressure* between contractions when the heart expands and refills. Arterial walls are able to expand and contract in response to this pressure.

Three layers are usually identified in the microscopic structure of the arterial walls. The innermost layer in contact with the blood is called *intima* and consists primarily of endothelial cells. The intermediate layer, *media*, is made up of smooth muscle cells. The outermost layer, *tunica adventitia*, is the connective tissue layer. Larger arteries (those over 10 mm in diameter) are typically more elastic containing a greater proportion

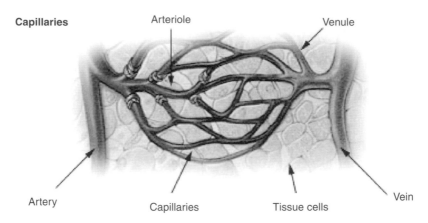

Capillaries   Arteriole   Venule

Artery   Capillaries   Tissue cells   Vein

Figure 3.28   *Arteries, arterioles, veins, venules and capillaries form the vascular structures that provide nutrition to tissues.*

of elastic fibers in their walls, while the smaller arteries (those between 0.1 and 10 mm diameter) have a greater proportion of muscle cells in their walls.

Arteries are more than just passive conduits for the blood flow. The muscular walls of the artery help the heart pump the blood. When the heart contracts, the artery expands as it fills with blood. When the heart relaxes the artery muscles contract, exerting a force strong enough to push the blood along. The primary artery in the systemic loop is called *aorta*. It carries oxygenated blood to all distant parts of the body. The pulmonary artery sends the blood from the heart toward the lungs, and is the only artery that carries oxygen-poor blood (this blood has already traveled through many parts of the body). Gas exchange occurs in the lungs allowing carbon dioxide to diffuse out and oxygen to diffuse in. Coronary arteries deliver oxygenated blood, food, etc. to the heart muscles.

Veins are similar to arteries but, because they transport blood at a lower pressure, they are not as strong. Like arteries, veins have three layers. The layers are thinner however, containing less tissue. Veins receive blood from the capillaries after the exchange of oxygen and carbon dioxide has taken place. The veins therefore transport waste-rich blood back to the lungs and heart. It is important that the waste-rich blood does not experience any backflow that could impede material exchange. In the veins, the backflow is prevented by valves that are located inside the veins. These valves are formed as folds in the intima layer and function like one-way gates. The valves in the veins also help deal with movement of blood along the veins against the force of gravity. Veins of the systemic loop return deoxygenated blood back to the heart through the primary vein called vena cavae. In contrast, the pulmonary vein coming back from the lungs to the heart carries oxygenated blood.

In contrast to arteries, capillaries are thin-walled blood vessels. In the capillary, the wall is only one endothelial cell layer thick. Capillaries in most of the body have small pores between the cells of the capillary wall, allowing materials to flow in and out of capillaries as well as permitting the passage of white blood cells. Blood vessels in the brain behave differently however, creating the blood–brain barrier (BBB). BBB is a mechanism of separation of circulating blood and the brain extracellular fluid. Capillary walls in the brain have tight junctions around the capillaries that do not exist in normal circulation. Endothelial cells restrict the diffusion of microscopic objects (e.g. bacteria) and large or hydrophilic molecules into the cerebrospinal fluid (CSF), while allowing the diffusion of small hydrophobic molecules (O_2, CO_2, hormones). Cells of the barrier actively transport metabolic products such as glucose across the barrier using specific proteins. The extensive network of capillaries in the human body is estimated at between 50 000 and 60 000 miles long.

It is important to note that although blood is supplied to all parts of the body, all capillary beds do not contain blood at all times. Blood is diverted to the parts of the body that need it most at a particular time. For instance, when you eat a meal blood is diverted from other parts of your body to the digestive tract to aid in digestion and nutrient absorption.

Blood flow in capillary beds is controlled is by the mechanism of microcirculation where the blood passes from arteries through arterioles into capillaries, then into venules and finally into the veins. Some organs such as liver, spleen and bone marrow contain highly porous (sometimes referred to as the open pore vessels) called sinusoids instead of typical capillaries that have less penetrable pores covered by a diaphragm-like tissue. Control of blood flow is achieved partly by pre-capillary sphincters typically positioned between the arterioles and capillaries. These sphincters contain muscle fibers that allow them to contract or relax, decreasing or increasing blood flow permitted through the capillaries. When the sphincters are closed, the blood can often bypass the capillaries moving directly from the arterioles to the venules through larger vessels called thoroughfare channels. It is believed that one of the mechanisms contributing to blood pressure dysfunction is related to misregulation of blood flow through capillaries by the nervous systems.

3.4.3.3   *The heart*

The heart is a muscular structure that contracts in a rhythmic pattern to pump blood. Hearts have a variety of forms: chambered hearts in mollusks and vertebrates; tubular hearts of arthropods; and aortic arches of annelids. Insects use accessory hearts to boost or supplement actions of the main heart. Fish, reptiles and amphibians have lymph hearts that help pump lymph back into the veins. The basic vertebrate heart, such as that in fish, has two chambers. An auricle is the chamber of the heart where blood is received from the body. A ventricle pumps the blood it received through a valve from the auricle out to the gills through an artery.

Amphibians have a three-chambered heart: two atria emptying into a single common ventricle. Some species have a partial separation of the ventricle to reduce the mixing of oxygenated (coming back from the lungs) and deoxygenated blood (coming in from the body). Two-sided or two-chambered hearts allow pumping at higher pressures, and the addition of the pulmonary loop allows blood to go to the lungs at lower pressure yet still go to the systemic loop at higher pressures. A four-chambered heart, along with the pulmonary and systemic circuits that completely separate oxygenated from deoxygenated blood, provides for the higher metabolic rates needed by warm-blooded birds and mammals.

The human heart is a two-sided, four-chambered structure with muscular walls as illustrated in Figure 3.29. It contains one atrium and one ventricle for each circulation loop. Since both loops have a systemic and a pulmonary circulation, there are four chambers in total: left atrium, left ventricle, right atrium and right ventricle. The right atrium is the upper chamber of the right side of the heart. The blood returned to the right atrium is deoxygenated (poor in oxygen) and passed into the right ventricle to be pumped through the pulmonary artery to the lungs for re-oxygenation and removal of carbon dioxide. The left atrium receives newly oxygenated blood from the lungs through the pulmonary vein, which is passed into the strong left ventricle to be pumped through the aorta to the different organs of the body. Atrioventricular (AV) valves called *tricuspid* and *mitral* separate each atrium from each ventricle. Semilunar (also known as arterial) valves, called pulmonary and aortic, separate each ventricle from its connecting artery. A thick wall of muscle called the septum separates the right the left sides of the heart. Physicians commonly refer to the right atrium and right ventricle together as the *right heart* and to the left atrium and ventricle as the *left heart*.

The human heart is enclosed in a double-walled protective sac called the pericardium. The double membrane of pericardium encloses pericardial fluid, which nourishes the heart and prevents mechanical shocks. The superficial part of this sac is called the fibrous pericardium. The outer wall of the human heart is composed of three layers. The outer layer is called the epicardium or visceral pericardium since it is also the inner wall of the (serous) pericardium. The middle layer is called the myocardium and is composed of muscle which

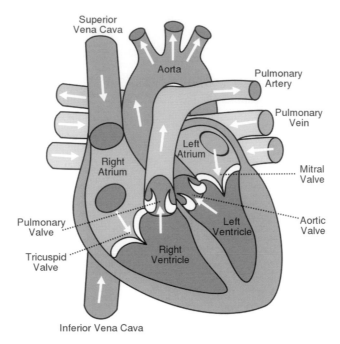

Figure 3.29   *Diagram of the human heart: the view from the front.*

contracts. The inner layer is called the endocardium and is in contact with the blood that the heart pumps. It also merges with the inner lining (endothelium) of blood vessels and covers heart valves.

The heart beats or contracts approximately 70–100 times per minute. It will undergo over 3 billion contraction cycles during an average lifetime. Each contraction cycle can be viewed as a propagating pulse. In fact, each such pulse is initiated as an electrical pulse at the sinoatrial (SA) node near the right atrium. Modified muscle cells contract, sending a signal to other muscle cells to contract. The signal spreads to the AV node, from where it is carried by specialized tissue called the Purkinje fibers. These then transmit the electric charge to the myocardium, causing the ventricles to contract simultaneously. The AV node is sometimes called the pacemaker since it keeps the heartbeat regular. Heartbeat is also controlled by nerve messages originating from the autonomic nervous system.

Cardiac muscle cells are serviced by a system of coronary arteries that are part of the systemic loop. During strenuous physical activity, the flow through these arteries is up to five times normal flow. Blocked flow in coronary arteries can result in death of heart muscle, leading to a heart attack. Blockage of coronary arteries is usually the result of gradual buildup of lipids and cholesterol in the inner wall of the coronary artery. Occasional chest pain, angina pectoralis, can result during periods of stress or physical exertion. Angina indicates oxygen demands are greater than the capacity to deliver it, and that a heart attack may occur in the future. Heart muscle cells that die are not replaced since heart muscle cells in adults do not divide. Heart disease and coronary artery disease are the leading causes of death in the United States.

3.4.3.4   *The lymphatic system*

Water and blood plasma are forced from the capillaries into intracellular spaces. This interstitial fluid transports materials between cells. Most of this fluid is collected in the capillaries of a secondary circulatory system known as the lymphatic system; the fluid in this system is therefore referred to as lymph. The lymphatic system is composed of lymph vessels, lymph nodes and organs. The functions of this system include the absorption of excess fluid and its return to the blood stream, absorption of fat (in the villi of the small intestine) and the immune system functions.

Lymph vessels are closely associated with the circulatory system vessels. Larger lymph vessels are similar to veins. Lymph capillaries are scatted throughout the body. Contraction of skeletal muscle causes movement of the lymph fluid through valves. Lymph organs include the bone marrow, lymph nodes, spleen and thymus. Bone marrow contains tissue that produces lymphocytes. B-lymphocytes (B-cells) mature in the bone marrow and T-lymphocytes (T-cells) mature in the thymus gland. Other white blood cells such as monocytes and leukocytes are produced in the bone marrow. Lymph nodes are areas of concentrated lymphocytes and macrophages along the lymphatic veins.

The spleen is similar to the lymph node, except that it is larger and filled with blood. The spleen serves as a reservoir for blood, and filters or purifies the blood and lymph fluid that flows through it. If the spleen is damaged or removed, the individual is more susceptible to infections. The thymus secretes a hormone, thymosin, which causes pre-T-cells to mature (in the thymus) into T-cells.

3.4.4 Immune system

An immune system can be viewed as a collection of molecules and cells whose primary function is to protect the living organism against infectious agents, parasites and materials foreign to the body. The job of an immune system consists of two major tasks. One is to recognize an invader, distinguishing it from 'self' cells and molecules native to the living organism. The second is to destroy the offending invader. The recognition that a particular living or material object is an invader occurs because this object carries or emits molecules referred to as *immunogens* (immune response generators) and/or *antigens* (antibody generators), which are discussed later in this section.

Even simple unicellular organisms such as bacteria possess elementary immune systems consisting of enzymes that protect the cells against bacteriophage (bacterial virus) infections. Basic immune mechanisms also existed in ancient eukaryotes and remain in their modern descendants, such as plants and insects. These mechanisms include phagocytosis (the process of engulfing and destroying foreign materials), antimicrobial peptides and the complement system (discussed below). Jawed vertebrates, including humans, have even more sophisticated defense mechanisms, including the ability to adapt over time to recognize specific pathogens and mount a faster and more specific defense.

In humans, the organs associated with the immune system are called the lymphoid organs. These organs are involved in producing cells and molecules that play various roles in the immune system. Lymphoid organs, as illustrated in Figure 3.30, include:

- adenoids (two glands located at the back of the nasal passage);
- appendix (a small tube that is connected to the large intestine);
- blood vessels (the arteries, veins and capillaries through which blood flows);
- bone marrow (the soft, fatty tissue found in bone cavities);
- lymph nodes (small organs shaped like beans, which are located throughout the body and connect via the lymphatic vessels);
- lymphatic vessels (a network of channels throughout the body that carries lymphocytes to the lymphoid organs and bloodstream);
- Peyer's patches (lymphoid tissue in the small intestine);
- spleen (a fist-sized organ located in the abdominal cavity);

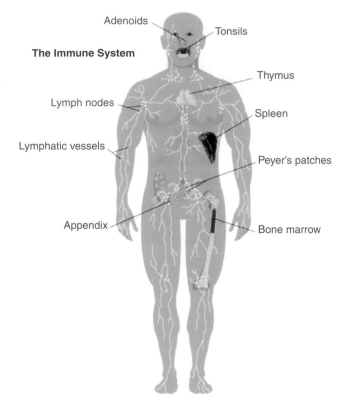

The Immune System

Figure 3.30   *Illustration of the main components of the lymphoid system.*

- thymus (two lobes that join in front of the trachea behind the breast bone); and
- tonsils (two oval masses in the back of the throat).

The immune system is often viewed as consisting of two main components: *innate* and *acquired* (*adaptive*). The responses of the innate immune system to invaders is genetically determined and does not change over time to adapt to specific invaders. Molecules and cells of this system have the task of recognizing a living or material object as an invader, attacking it and presenting information about the invader to the acquired immune system, typically using so-called *antigen-presenting* cells. Many cells in the body can present antigen to the cells of the adaptive immune system but certain cells of the innate system can be viewed as professional antigen presenters, capable of capturing the antigens and presenting them on a more massive scale. Presenting antigens to the adaptive immune system requires incorporation of these antigens into special molecular complexes known as major histocompatibility complex (MHC), which some of the more active immune cells are designed to recognize. Professional antigen-presenting cells are also equipped with special immune-stimulatory receptors that can enhance activation of the acquired immune system.

In contrast to the innate immune system, the *acquired immune system* changes over time by adapting its responses to specific invaders it has encountered in one way or another in the past. The ability to remember past invaders, mobilize an appropriately skilled army of cells and molecules quickly and mount an attack that minimizes collateral damage is essential, given the ability of many invasive bacteria and viruses to multiply quickly. Cells and molecules of the adaptive system develop a high degree of specialization and readiness accordingly.

Antibodies are the primary molecular mediators of the adaptive immune system. Antibodies are special molecules produced by cells of the adaptive immune system, which constitute so-called humoral immunity. These molecues, also called immunoglobulins or Igs, are soluble glycoproteins (molecules where sugar chains attached to amino acid residues), have molecular weights of 150–900 kDa (kilo-Daltons) and constitute the *gamma globulin* part of the blood proteins. Many of the antibodies are Y-shaped molecules. Each tip of the 'Y' of an antibody contains a paratope (a structure analogous to a lock) that is specific for one particular epitope (similarly, analogous to a key) on an antigen, allowing these two structures to bind together with precision. Using this binding mechanism, an antibody can *tag* a microbe or an infected cell for attack by other parts of the immune system, or can neutralize its target directly.

Antibodies can inactivate antigens directly by: (1) complement fixation where these proteins attach to an antigen surface and effectively destroy regions around the antigen leading to cell lysis; (2) neutralization where binding to specific antigen sites prevents attachment of cell or virus to other cells; (c) agglutination where antibody attachment causes clumping of cells or viruses, making them much less mobile; or (d) precipitation where attachment of antibodies force insolubility and settling out of solution.

It has been estimated that humans generate about 10 billion different antibodies, each capable of binding a distinct epitope of an antigen. Although the general structure of all antibodies is very similar, the paratope region is extremely variable. It allows millions of antibodies with slightly different tip structures, or antigen binding sites, to exist. This region is known as the *hypervariable region*. Each of these variants can bind to a different target, known as an antigen. This enormous diversity of antibodies allows the immune system to recognize an equally wide variety of antigens. The large and diverse population of antibodies is generated by random combinations of a set of gene segments that encode different antigen binding sites (*paratopes*). This is followed by random mutations in this area of the antibody gene, which create further diversity.

Antibodies are a very important tool in molecular biology and biochemistry today. They can be produced artificially in a relatively pure form from clonally identical immune cells, known as monoclonal antibodies. Such antibodies can be selected to attach very specifically to various proteins, protein fragments or cells that might be under investigation. Attaching such antibodies to fluorescent, magnetic or radioactive labels allows the movement and location of molecules or cells in tissues and in separation devices (such as those based on electrophoresis) to be investigated.

3.4.4.1 *Innate immune system*

Surface and mucosal barriers can be viewed as the first layer of protection offered by the innate immune system. The first and, arguably, most important barrier is the skin, which most bacteria and particulates cannot easily penetrate unless there is already a cut in the skin. Lungs can expel pathogens mechanically by ciliary action as the tiny hairs move in an upward motion. Coughing and sneezing abruptly ejects both living and non-living things from the respiratory system. The flushing action of tears, saliva and urine also force out pathogens, as does the sloughing off of skin. Sticky mucus in respiratory and gastrointestinal tracts traps many microorganisms. Acidic skin secretions can inhibit bacterial growth. Hair follicles secrete sebum that contains lactic and fatty acids, both of which inhibit the growth of some pathogenic bacteria and fungi (although acne does develop in the hair follicle region). Areas of the skin not covered by hair, such as the palms and soles of the feet, are most susceptible to fungal infections, partly due to the lack of secretions. Saliva, tears, nasal secretions and perspiration contain lysozyme, an enzyme that destroys Gram positive bacterial cell walls causing cell lysis. Vaginal secretions are also slightly acidic (after the onset of menses). Spermine and zinc in semen destroy some pathogens. Lactoperoxidase is a powerful enzyme found in mother's milk. The stomach is a formidable obstacle insofar as its mucosa secretes hydrochloric acid ($0.9 < $ pH $ < 3.0$, very acidic) and protein-digesting enzymes that kill many pathogens. The stomach can destroy drugs and other chemicals, making oral delivery of drugs a challenge.

Normal flora consisting of the microbes, mostly bacteria, that live in and on the body can be considered part of the immune barrier system. It is often argued that animals (and humans in particular) have co-evolved with their microbial flora. As a result, certain microorganisms are not immunogenic (do not initiate immune responses) when presented in certain regions of the body. The same microorganisms may be immunogenic in other areas of the body. It is estimated that humans have c. 10^{13} cells in their bodies and 10^{14} bacteria, most of which live in the large intestine. There are 10^3–10^4 microbes per square centimeter on the skin (*Staphylococcus aureus*, *Staph. epidermidis*, *diphtheroids*, *streptococci*, *Candida*, etc.). Various bacteria live in the nose and mouth. Lactobacilli live in the stomach and small intestine. The upper intestine has c. 10^4 bacteria per gram of tissue. The large bowel has 10^{11} per gram of tissue, of which 95–99% are anaerobes (do not use oxygen for metabolism). The urogenital tract is lightly colonized by various bacteria and diphtheroids. After puberty, the vagina is colonized by *Lactobacillus aerophilus* that ferment glycogen to maintain acidity.

Normal flora fills almost all of the available ecological niches in the body and produce bacteriocidins, defensins, cationic proteins and lactoferrin, all of which work to destroy other bacteria that compete for their niche in the body. For example, when antibiotics such as clindamycin kill some of the bacteria in the human intestinal tract, *Clostridium difficile* can overgrow resulting in pseudomembranous colitis, a rather painful condition where the inner lining of the intestine cracks and bleeds.

For a similar reason, the overuse of antimicrobial agents on skin can result in increased growth of less benign microorganisms. The resident bacteria can also become problematic when they invade spaces in which they were not meant to be. As an example, staphylococcus living on the skin can cause serious infection if it gains entry into the body through small cuts/nicks.

The *complement system* is a biochemical cascade that attacks surfaces of foreign cells and materials. It contains over 20 different proteins that complement the effects of antibodies (discussed below in this section). Proteins of the complement system typically bind to carbohydrates on the surfaces of microbes. This triggers a rapid progression of further binding events. The speed of the response is a result of autocatalytic signal amplification that occurs following sequential proteolytic activation of complement molecules. These proteases (enzymes that degrade proteins) also activate protease activity of other complement molecules. The cascade results in the production of peptides that attract immune cells, increase vascular permeability and opsonize (coat) the surface of a pathogen, marking it for destruction. Molecules of the complement system can also kill cells directly by disrupting their plasma membrane or by enhancing the ability of cells to kill the invader.

Cellular barriers are the key part of the innate immune system. Some of these cells are more commonly found in circulation, while others occur in different tissues. Many of these cells are phagocytes. Phagocytes (phagocyte cells) attract to foreign bodies and abnormal cells through chemotaxis. They then adhere to and engulf offending cells and foreign material, releasing a variety of oxidizing agents which are mostly various reactive species (ROS and RNS). In the process of such attack, the phagocytes themselves die. Their 'corpses', pockets of damaged tissue and fluid-form pus, are often observed in infected tissues or in the initial stages of wound healing.

In the above discussion of blood cells (Section 3.4.3.1) it was mentioned that white blood cells (also known as leukocytes) have an important role in the immune system. Leukocytes consist of granulocytes (three major cell types) and monocytes as follows.

1. Granulocytes, also known as myeloid cells. Three primary types of granulocytes exist.
    (a) Neutrophils are phagocytes that are the most abundant of the circulating leukocytes (estimates indicate 50–60% of all leukocytes). They are non-dividing, have an average size of 12–15 μm in blood and are relatively short-lived (1–4 day lifespan) cells. They have few mitochondria if any and

get most of their energy from stored glycogen. They are produced in bone marrow and can migrate from the blood vessels into interstitial tissue toward the infection site. Neutrophils provide the major defense against bacteria and are the first on the scene to fight infection. They are followed by the wandering macrophages (discussed below) c. 3–4 hours later.

(b) Eosinophils are acid-loving phagocyte cells (c. 15 μm in diameter) responsible for attacking multi-cellular parasites such as worms and are also involved in allergic reactions and asthma. They release major basic protein (MBP), cationic proteins, perforins and oxygen metabolites, all of which work together to burn holes in parasites. About 1–6% of the leukocytes are eosinophils. They are produced in bone marrow and are often found in the medulla and the junction between the cortex and medulla of the thymus. They also occur in the lower gastrointestinal tract, ovary, uterus, spleen and lymph nodes, but not in the lung, skin and esophagus (where their presence is associated with disease). Eosinophils are relatively long living. They persist in the circulation for c. 8–12 hours, but survive in tissues for an additional 8–12 days.

(c) Basophils occur more readily in non-acidic environments in the body. They have a relatively low abundance in blood (around 0.3% of leukocytes). They are usually implicated in responses to ectoparasites (often larger parasites such as ticks that invade tissue surfaces) and allergic responses due to the release of histamine.

2. Monocytes are phagocyte cells that circulate in the blood until they receive the signal to extravasate (exit circulation) into the peripheral tissue. Once in the tissue, they can mature into several different cells including macrophages and dendritic cells (discussed below). Monocytes constitute c. 3–8% of leukocytes, are produced in bone marrow and are stored primarily in the spleen.

Cellular barriers of the innate immune system also include the following.

• Mast cells, closely related to basophil granulocytes. While basophils leave the bone marrow already mature, the mast cells circulate in an immature form, maturing once in a tissue site. Two types of mast cells are usually recognized: those from connective tissue and a distinct set of mucosal mast cells. Mast cells are present in most tissues characteristically surrounding blood vessels and nerves, and are especially prominent near the boundaries such as the skin, mucosa of the lungs, digestive tract and in the mouth, conjunctiva and nose. Mast cells play a key role in the inflammatory process. When activated by the presence of antibodies (it has a particularly high affinity for a relatively rare immunoglobuline E) a mast cell rapidly releases its characteristic *granules* and various hormonal mediators into the tissue. These granules contain a variety of molecules such as: proteases that cleave peptide bonds in proteins; neurotransmitters (such as serotonin); molecules that increase permeability of blood capillary walls (histamine); anticoagulants (such as heparin) which presumably permits better blood flow into affected tissues and the delivery of other immune cells and molecules; and various small signaling molecules (such as cytokines).

• The main function of dendritic cells (DCs) is to process antigen material and present it on their surface to other cells of the immune system. That is, dendritic cells function as antigen-presenting cells. They act as messengers between the innate and adaptive immunity. Dendritic cells are present in tissues in contact with the external environment, such as the skin (where there is a specialized dendritic cell type called Langerhans cells) and the inner lining of the nose, lungs, stomach and intestines. They can also be found in an immature state in the blood. Once activated to present antigens, they migrate to the lymph nodes where they interact with the T-cells and B-cells to initiate and shape the adaptive immune response. They grow branched projections at certain development stages, the *dendrites* that give the cell its name ('dendron' being Greek for tree). While similar in appearance, these are distinct structures from the dendrites of neurons.

- Macrophages, which are phagocyte cells derived from monocytes, are concentrated in the lungs, liver (Kupffer cells), lining of the lymph nodes and spleen, brain microglia, kidney mesoangial cells, synovial A cells and osteoclasts. Macrophages are relatively long-living cells, depend on mitochondria for energy and are best at attacking dead cells and pathogens capable of living within cells. Once a macrophage phagocytizes a cell, it places some of its proteins (called epitopes) on its surface. These surface markers serve as an alarm to other immune cells that then infer the form of the invader. An important function of macrophages is therefore also antigen presentation for communication with the adaptive immune system.

3.4.4.2   *Adaptive immune system*

The cells of the adaptive immune system are a type of leukocyte (white blood cell or WBC) called a *lymphocyte*. B-cells and T-cells are the major types of lymphocytes. The human body has c. 2 trillion lymphocytes, constituting 20–40% of WBCs. Their total mass has been estimated to be about the same as the mass of the brain or liver. The peripheral blood contains 20–50% of circulating lymphocytes, while the rest move within the lymphatic system.

B-cells and T-cells are derived from the same multipotent (having multiple potential phenotypes) hematopoietic stem cells, and are morphologically indistinguishable from one another until after they are activated. B-cells play a large role in the humoral immune response, producing antibodies; some of these remain on the B-cell membrane while others are released into body. T-cells are intimately involved in cell-mediated immune responses, which means that they primarily attack the invaders and pass information regarding the invader and the type of antigens it contains. In nearly all vertebrates, B-cells and T-cells are produced by stem cells in the bone marrow. T-cells travel to and develop in the thymus, from which they derive their name. About 10% of plasma cells will survive to become long-lived antigen-specific memory B-cells. Already primed to produce specific antibodies, these cells can be called upon to respond quickly if the same pathogen re-infects the host (while the host experiences few, if any, symptoms). This memory of infection is the reason behind vaccination.

3.4.5   Digestive system

Single-celled organisms can directly take in nutrients from their outside environment. Multicellular animals, with most of their cells removed from direct contact with the outside environment, have developed specialized structures for obtaining and breaking down their food. Absorptive feeders such as tapeworms absorb nutrients directly through their body wall. Filter feeders, such as oysters and mussels, collect small organisms and particles from the surrounding water. Fluid feeders, such as aphids, pierce the body of a plant or animal and withdraw fluids. Animals, for the most part, ingest their food as large, complex molecules that must be broken down into smaller molecules (monomers) that can then be distributed throughout the body of every cell. This vital function is accomplished by a series of specialized organs that comprise the digestive system.

The digestive system uses mechanical and chemical methods to break food down into nutrient molecules that can be absorbed into the blood. Once in the blood, the food molecules are routed to every cell in the animal's body. There are two types of animal body plans as well as two locations for digestion to occur. Sac-like plans are found in many invertebrates that have a single opening for food intake and the discharge of wastes. Vertebrates, the animal group humans belong to, use the tube-within-a-tube plan with food entering through one opening (the mouth) and wastes leaving through another (the anus).

Where the digestion of the food takes place is variable. Some animals use intracellular digestion, where food is taken into cells by phagocytosis with digestive enzymes being secreted into the phagocytic vesicles. This type of digestion occurs in sponges, coelenterates (corals, hydras and their relatives) and most protozoans.

Extracellular digestion occurs in the lumen (or opening) of a digestive system, with the nutrient molecules being transferred to the blood or some other body fluid.

3.4.5.1 Stages of the digestive process

For the most part, food consists of various organic macromolecules such as starch, proteins and fats. These molecules are polymers made of individual monomer units. Breaking these large molecules into smaller components involves: (1) movement that propels food through the digestive system; (2) secretion that releases digestive juices in response to a specific stimulus; (3) digestion that breaks down food into molecular components small enough to cross the plasma membrane; (4) absorption that passes the molecules into the body's interior and their passage throughout the body; and (5) elimination that removes undigested food and wastes.

3.4.5.2 Components of the digestive system

The human digestive tract is a coiled, muscular tube (6–9 m long when fully extended) stretching from the mouth to the anus. Several specialized compartments occur along the length of the digestive tract: mouth, pharynx, esophagus, stomach, small intestine, large intestine and anus. As well as the digestive tract, the digestive system includes accessory digestive organs that are connected to the main tract by a series of ducts: salivary glands, parts of the pancreas, the liver and gall bladder (biliary system). Figure 3.31 illustrates the primary components of the human digestive system.

3.4.5.3 The mouth and pharynx

Mechanical breakdown begins in the mouth by chewing (teeth) and actions of the tongue. In the mouth, teeth, jaws and the tongue begin the mechanical breakdown of food into smaller particles. Most vertebrates (except birds who usually have a hardened bill instead of teeth) have teeth for tearing, grinding and chewing food. The tongue manipulates food during chewing and swallowing; mammals have taste buds clustered on their tongues.

The chemical breakdown of starch occurs by production of salivary amylase from the salivary glands. Salivary glands secrete salivary amylase, an enzyme that begins the breakdown of starch into glucose. This mixture of food and saliva is then pushed into the pharynx by the tongue where an involuntary swallowing reflex prevents food from entering the lungs and directs food and saliva into the esophagus. The esophagus is a muscular tube whose wave-like muscular contractions (peristalsis) propel food past the gastroesophageal sphincter into the stomach. Mucus moistens food and lubricates the esophagus. Heartburn results from irritation of the esophagus by gastric juices that leak through this sphincter. This gastroesophageal reflux can lead to a pre-cancerous condition associated with the change in the lining of the esophagus known as the Barrett's esophagus, which is often treated today using hot plasma ablation.

3.4.5.4 The stomach

During a meal, the stomach gradually fills to a capacity of c. 1 L in an average-sized adult, from an empty capacity of c. 50–100 mL. At the price of discomfort, the stomach can distend to hold c. 2 L or more.

Epithelial cells line the inner surface of the stomach and secrete c. 2 L of gastric juices per day. Gastric juice contains hydrochloric acid, pepsinogen and mucus, all ingredients important in digestion. Secretions are controlled by the nervous (smells, thoughts and caffeine) and endocrine signals. Hydrochloric acid does not directly function in digestion. Its primary function is to help kill microorganisms and to lower the pH

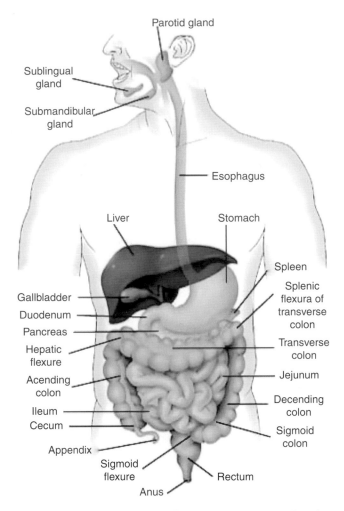

Figure 3.31   *Illustration of the digestive system (see color plate).*

to c. 1.5–2.5, which helps activate pepsinogen. Pepsinogen is an enzyme that starts protein digestion; it is activated by cleaving off a portion of the molecule, producing the enzyme pepsin that splits off fragments of peptides from a protein molecule during digestion in the stomach.

There are several mechanisms that protect the stomach lining from food-degrading processes within the stomach. Bicarbonate ions reduce acidity near the cells lining the stomach. Tight junctions link the epithelial stomach-lining cells together, further reducing or preventing stomach acids from passing. Pepsin is inactivated when it comes into contact with the mucus.

The stomach also mechanically churns the food. Chyme, the mix of acid and food in the stomach, leaves the stomach and enters the small intestine. Smaller molecules, such as alcohol and aspirin, are absorbed through the stomach lining into the blood. Carbohydrate digestion, begun by salivary amylase in the mouth, continues in the food bolus as it passes to the stomach. The bolus is broken down into acid chyme in the lower third of the stomach, allowing the stomach's acidity to inhibit further carbohydrate breakdown. Food is mixed in the lower part of the stomach by peristaltic waves that also propel the acid-chyme mixture against the pyloric sphincter. Increased contractions of the stomach push the food through the pyloric sphincter and into the small intestine as the stomach empties over a 1–2 hour period.

3.4.5.5   Ulcers

Peptic ulcers result when various lining-protecting mechanisms fail. Bleeding ulcers result when tissue damage is so severe that bleeding occurs into the stomach. Perforated ulcers, where a hole has formed in the stomach wall, are life-threatening situations. At least 90% of all peptic ulcers are caused by *Helicobacter pylori*. Other factors, including stress and aspirin, can also produce ulcers.

3.4.5.6   The small intestine

The small intestine is the major site for digestion and absorption of nutrients. It is a tube that is up to 6 m long and is 2–3 cm wide. The upper part, the duodenum, is the most active in digestion. Secretions from the liver and pancreas are used for digestion in the duodenum. Epithelial cells of the duodenum secrete watery mucus. The pancreas secretes digestive enzymes and stomach-acid-neutralizing bicarbonate. The liver produces bile, which is stored in the gall bladder before entering the bile duct into the duodenum.

Digestion of carbohydrates, proteins and fats continues in the small intestine. Starch and glycogen are broken down into maltose by small intestine enzymes. Proteases are enzymes secreted by the pancreas that continue the breakdown of protein into small peptide fragments and amino acids. Bile emulsifies fats, facilitating their breakdown into progressively smaller fat globules until they can be acted upon by lipases. Bile contains cholesterol, phospholipids, bilirubin and a mix of salts. Fats are completely digested in the small intestine, unlike carbohydrates and proteins.

Most absorption of nutrients occurs in the duodenum and jejeunum (middle third of the small intestine). The inner surface of the intestine has circular folds that more than triple the surface area for absorption. Folding in the small intestine plus villi dramatically increase the surface area of this tube, aiding absorption of nutrients. Villi have cells producing intestinal enzymes that complete the digestion of peptides and sugars. Sugars and amino acids go into the bloodstream via capillaries in each villus. Glycerol and fatty acids go into the lymphatic system.

Maltose, sucrose and lactose are the main carbohydrates present in the small intestine and are absorbed by the microvilli. Starch is broken down into two-glucose units (maltose) elsewhere. Enzymes in the cells convert these disaccharides into monosaccharides that then leave the cell and enter the capillary. Lactose intolerance results from a genetic lack of the enzyme lactase produced by the intestinal cells.

Digested fats are not very soluble. Bile salts surround fats to form micelles that can pass into the epithelial cells. The bile salts return to the lumen to repeat the process. Fat digestion is usually completed by the time the food reaches the ileum (lower third) of the small intestine. Bile salts are in turn absorbed in the ileum and are recycled by the liver and gall bladder. Fats pass from the epithelial cells to the small lymph vessel that also runs through the villus.

3.4.5.7   The liver and gall bladder

The liver produces and sends bile to the small intestine via the hepatic duct. Bile contains bile salts which emulsify fats, making them susceptible to enzymatic breakdown. In addition to digestive functions, the liver plays several other roles including: (1) detoxification of blood; (2) synthesis of blood proteins; (3) destruction of old erythrocytes and conversion of hemoglobin into a component of bile; (4) storage of glucose as glycogen and its release when blood sugar levels drop; and (5) production of urea from amino groups and ammonia.

The gall bladder stores excess bile for release at a later time. We can live without our gall bladders. The drawback, however, is a need to be aware of the amount of fats in the food we eat when the stored bile of the gall bladder is no longer available.

Glycogen is a polysaccharide made of chains of glucose molecules. Starch is the storage form of glucose in plants, while animals use glycogen for the same purpose. Low glucose levels in the blood cause the release of hormones, such as glucagon, that travel to the liver and stimulate the breakdown of glycogen into glucose which is then released into the blood (raising blood glucose levels). When no glucose or glycogen are available, amino acids are converted into glucose in the liver. The process of deamination removes the amino groups from amino acids. Urea is formed and passed through the blood to the kidney for export from the body. In this sense, liver can be viewed as a component of the excretory system. The hormone insulin promotes the take-up of glucose into liver cells and its formation into glycogen.

3.4.5.8   *Liver diseases*

Jaundice occurs when the characteristic yellow tint to the skin is caused by excess hemoglobin breakdown products in the blood, a sign that the liver is not properly functioning. Jaundice may occur when liver function has been impaired by obstruction of the bile duct and by damage caused by hepatitis.

Hepatitis A, B and C are all viral diseases that can cause liver damage. As for any viral disease, the major treatment efforts focus on the symptoms and not the removal of the viral cause. Hepatitis A is usually a mild malady indicated by a sudden fever, malaise, nausea, anorexia and abdominal discomfort. Jaundice follows for several days. The virus causing hepatitis A is primarily transmitted by fecal contamination, although contaminated food and water can also promote transmission. A rare disease in the US, hepatitis B is endemic in parts of Asia where hundreds of millions of individuals are possibly infected.

Hepatitis B virus (HBV) may be transmitted by blood and blood products as well as sexual contact. The blood supply in developed countries has been screened for the virus that causes this disease for many years, and transmission by blood transfusion is rare. The risk of HBV infection is high among sexually active homosexual men, but is also transmitted heterosexually. Effective vaccines are available for the prevention of hepatitis B infection. Individuals with chronic hepatitis B have an increased risk of developing primary liver cancer and may develop cirrhosis of the liver. Although this type of cancer is relatively rare in the US, it is the leading cause of cancer death in the world due to the virus causing it being endemic in eastern Asia.

Hepatitis C affects approximately 170 million people worldwide and 4 million in the US. The virus is transmitted primarily by blood and blood products. Most infected individuals have either received blood transfusions prior to 1990 (when screening of the blood supply for the hepatitis C virus began) or have used intravenous drugs. Sexual transmission can occur between monogamous couples (rare) but infection is far more common in those with more sexual partners. In rare cases, hepatitis C causes acute disease and even liver failure. About 20% of individuals with hepatitis C who develop cirrhosis of the liver will also develop severe liver disease. Cirrhosis caused by hepatitis C is presently the leading cause of the need for liver transplants in the US. Individuals with cirrhosis from hepatitis C also bear increased chances of developing primary liver cancer. All current treatments for hepatitis C employ various preparations of the potent antiviral interferon alpha. Not all patients who have the disease are suitable for treatment, so infected individuals are urged to consult their physician.

Cirrhosis of the liver commonly occurs in alcoholics, who place the liver in a stress situation due to the amount of alcohol to be broken down. Cirrhosis can cause the liver to become unable to perform its biochemical functions. Chemicals responsible for blood clotting are synthesized in the liver as is albumin, the major protein in blood. The liver also makes or modifies bile components. Blood from the circulatory system passes through the liver, so many of the body's metabolic functions occur primarily there including the metabolism of cholesterol and the conversion of proteins and fats into glucose. Cirrhosis is a disease resulting from damage to liver cells due to toxins, inflammation and other causes. Liver cells regenerate in an abnormal pattern primarily forming nodules that are surrounded by fibrous tissue. Changes in the structure of the liver can decrease blood flow, leading to secondary complications. Cirrhosis has many causes including alcoholic liver disease, severe forms of some viral hepatitis, congestive heart failure, parasitic infections (for example schistosomiasis) and long-term exposure to toxins or drugs.

3.4.5.9   The pancreas

The pancreas sends pancreatic juice, which neutralizes the chyme, to the small intestine through the pancreatic duct. In addition to this digestive function, the pancreas is the site of production of several hormones, such as glucagon and insulin. The pancreas contains exocrine cells that secrete digestive enzymes into the small intestine and clusters of endocrine cells (the pancreatic islets). The islets secrete the hormones insulin and glucagon, which regulate blood glucose levels.

Blood glucose levels rise after a meal, prompting the release of insulin which causes cells to take up glucose and liver and skeletal muscle cells to form the carbohydrate glycogen. As glucose levels in the blood fall, further insulin production is inhibited. Glucagon causes the breakdown of glycogen into glucose, which in turn is released into the blood to maintain glucose levels within a homeostatic range. Glucagon production is stimulated when blood glucose levels fall, and inhibited when they rise.

Diabetes results from inadequate levels of insulin. Type I diabetes is characterized by inadequate levels of insulin secretion, often due to a genetic cause. Type II usually develops in adults from both genetic and environmental causes. Loss of response of targets to insulin rather than lack of insulin causes this type of diabetes. Diabetes may cause impairment in the functioning of the eyes, circulatory system, nervous system and failure of the kidneys. It is the second leading cause of blindness in the US, for example. Treatments might involve daily injections of insulin, oral medications such as metformin, monitoring of blood glucose levels and a controlled diet.

The fifth leading cause of cancer death in the US is from pancreatic cancer, which is nearly always fatal. Scientists estimate that 25 000 people may die from this disease each year. Standard treatments today are ineffective.

3.4.5.10   The large intestine

The large intestine is made up of the colon, cecum, appendix and rectum. Material in the large intestine is mostly indigestible residue and liquid. Movements are due to involuntary contractions that shuffle contents back and forth and propulsive contractions that move material through the large intestine. The large intestine performs three basic functions in vertebrates: (1) recovery of water and electrolytes from digested food; (2) formation and storage of feces; and (3) microbial fermentation. The large intestine supports an amazing flora of microbes; these produce enzymes that can digest many of the molecules that are indigestible by vertebrates.

Secretions in the large intestine are in the form of an alkaline mucus that protects epithelial tissues and neutralizes acids produced by bacterial metabolism. Water, salts and vitamins are absorbed and the remaining contents in the lumen form feces (mostly cellulose, bacteria and bilirubin). Bacteria such as *E. coli* in the large intestine produce vitamins (including vitamin K) that are absorbed.

3.4.6   The nervous system

Multicellular animals must monitor and maintain a constant internal environment as well as monitor and respond to an external environment. In many animals, these two functions are coordinated by two integrated organ systems: the nervous system and the endocrine system (Section 3.4.7).

The three basic functions performed by nervous system are:

1. receiving sensory input from internal and external environments;
2. integrating the input; and
3. responding to stimuli.

3.4.6.1   *Sensory input and output*

Receptors are parts of the nervous system that sense changes in the internal or external environments. Sensory input can be in many forms, including pressure, taste, sound, light, blood pH or hormone levels, that are converted to a signal and sent to the brain or spinal cord. The barrage of sensory input is integrated and a response is generated in the sensory centers of the brain or in the spinal cord. The response is typically a motor output, which is a signal transmitted to organs that can convert the signal into some form of action such as movement, changes in heart rate, release of hormones and so on.

3.4.6.2   *Divisions of the nervous system*

The nervous system monitors and controls almost every organ system through a series of positive and negative feedback loops. The central nervous system (CNS) includes the brain and spinal cord. The peripheral nervous system (PNS) connects the CNS to other parts of the body, and is composed of nerves (bundles of neurons).

Not all animals have highly specialized nervous systems. Those with simple systems tend to be either small and very mobile or large and immobile. Large, mobile animals typically have highly developed nervous systems. It appears that the evolution of nervous systems was important in the evolution of body size and mobility.

Bilaterally symmetrical animals have a body plan that includes a defined head and a tail region. Development of bilateral symmetry is associated with cephalization, the development of a head with the accumulation of sensory organs at the front end of the organism. Flatworms have neurons associated into clusters known as ganglia, which in turn form a small brain. Vertebrates have a spinal cord in addition to a more developed brain. The vertebrate nervous system is divided into a number of parts. The CNS includes the brain and spinal cord while the PNS consists of all the body nerves. Motor neuron pathways are of two types: somatic (skeletal) and autonomic (smooth muscle, cardiac muscle and glands). The autonomic system is subdivided into the sympathetic and parasympathetic systems.

3.4.6.3   *Peripheral nervous system*

The PNS contains only nerves and connects the brain and spinal cord (CNS) to the rest of the body. The axons and dendrites are surrounded by a white myelin sheath in the PNS. Cell bodies are in the CNS or ganglia, which are collections of nerve cell bodies. Cranial nerves in the PNS take impulses to and from the brain (CNS), while spinal nerves take impulses to and from the spinal cord. There are two major subdivisions of the PNS motor pathways: somatic and autonomic.

Two main components of the PNS are sensory (afferent) pathways that provide input from the body to the CNS and motor (efferent) pathways that carry signals to muscles and glands (effectors). Most sensory input carried in the PNS remains below the level of conscious awareness. Input that does reach the conscious level contributes to perception of our external environment.

3.4.6.4   *Somatic nervous system*

The somatic nervous system (SNS) includes all nerves controlling the muscular system and external sensory receptors. External sense organs (including skin) are receptors. Muscle fibers and gland cells are effectors. The reflex arc is an automatic, involuntary reaction to a stimulus. When the doctor taps your knee with a rubber hammer, she/he is testing your reflex (or knee-jerk). The reaction to the stimulus is involuntary, with the CNS being informed but not consciously controlling the response. Examples of reflex arcs include balance, the blinking reflex and the stretch reflex. Sensory input from the PNS is processed in the CNS and responses from the CNS are sent by the PNS to the organs of the body. Motor neurons of the somatic system are distinct from those of the autonomic system. Inhibitory signals cannot be sent through the motor neurons of the somatic system.

3.4.6.5   *Autonomic nervous system*

The autonomic nervous system (ANS) controls muscles in the heart and the smooth muscle in internal organs such as the intestine, bladder and uterus. It consists of motor neurons that control internal organs, and has two subsystems: (1) the sympathetic nervous system, which is involved in the fight or flight response; and (2) the parasympathetic nervous system, which is involved in relaxation. Each of these subsystems operates in the reverse of the other (antagonism). Both systems innervate the same organs and act in opposition to maintain homeostasis. For example, when you are scared the sympathetic system causes your heart to beat faster. The parasympathetic system reverses this effect. Motor neurons in this system do not reach their targets directly (as do those in the somatic system), but rather connect to a secondary motor neuron (which in turn innervates the target organ).

3.4.6.6   *Central nervous system*

The CNS is composed of the brain and spinal cord, and is surrounded by bone-skull and vertebrae. Fluid and tissue also insulate the brain and spinal cord. The brain is composed of three parts: the cerebrum (seat of consciousness), the cerebellum and the medulla oblongata (these latter two are part of the unconscious brain).

The medulla oblongata is closest to the spinal cord, and is involved with the regulation of heartbeat, breathing, vasoconstriction (blood pressure) and reflex centers for vomiting, coughing, sneezing, swallowing and hiccupping. The hypothalamus regulates homeostasis. It has regulatory areas for thirst, hunger, body temperature, water balance and blood pressure, and links the nervous system to the endocrine system. The midbrain and pons are also part of the unconscious brain. The thalamus serves as a central relay point for incoming nervous messages.

The cerebellum is the second largest part of the brain after the cerebrum. Its functions include muscle coordination, maintenance of normal muscle tone and posture and coordination of balance.

The conscious brain includes the cerebral hemispheres, which are separated by the *corpus callosum*. In reptiles, birds and mammals, the cerebrum coordinates sensory data and motor functions. The cerebrum governs intelligence and reasoning, learning and memory. While the cause of memory is not yet definitely known, studies on slugs indicate learning is accompanied by a synapse decrease. Within the cell, learning involves change in gene regulation and increased ability to secrete transmitters.

3.4.6.7   *The brain*

The structure of the human brain is illustrated in Figure 3.32. During embryonic development the brain first forms as a tube. The anterior end enlarges into three hollow swellings that form the brain, and the posterior

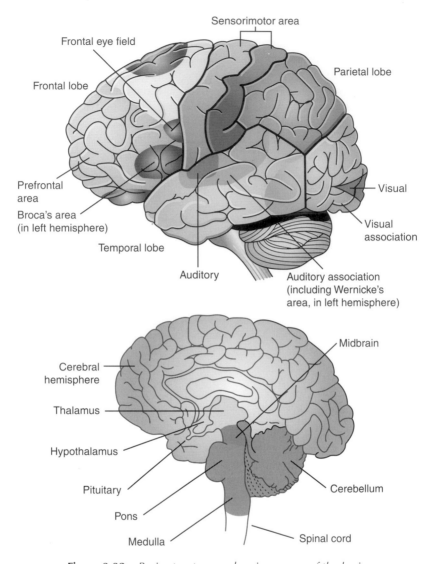

Figure 3.32   *Brain structure and various areas of the brain.*

end develops into the spinal cord. Some parts of the brain have changed little during vertebrate evolutionary history. Vertebrate evolutionary trends include: increase in brain size relative to body size; subdivision and increasing specialization of the forebrain, midbrain and hindbrain; and growth in relative size of the forebrain, especially the cerebrum which is associated with increasingly complex behavior in mammals.

3.4.6.8   *The brain stem and midbrain*

The brain stem is the smallest and, from an evolutionary viewpoint, the oldest and most primitive part of the brain. The brain stem is continuous with the spinal cord and is composed of the hindbrain and midbrain. The medulla oblongata and pons control heart rate, constriction of blood vessels, digestion and respiration. The midbrain consists of connections between the hindbrain and forebrain. Mammals use this part of the brain mostly for eye reflexes.

3.4.6.9   The cerebellum

The cerebellum is the third part of the hindbrain, but it is not considered part of the brain stem. Functions of the cerebellum include fine motor coordination and body movement, posture and balance. This region of the brain is enlarged in birds and controls muscle action needed for flight.

3.4.6.10   The forebrain

The forebrain consists of the diencephalon (comprising the thalamus and hypothalamus) and cerebrum. The thalamus acts as a switching center for nerve messages and the hypothalamus is a major homeostatic center, having both nervous and endocrine functions.

The cerebrum, the largest part of the human brain, is divided into left and right hemispheres connected to each other by the corpus callosum. The hemispheres are covered by a thin layer of gray matter known as the cerebral cortex, the most recently evolved region of the vertebrate brain. Fish have no cerebral cortex, and amphibians and reptiles have only rudiments of this area.

The cortex in each hemisphere of the cerebrum is between 1 and 4 mm thick. Folds divide the cortex into four lobes: occipital, temporal, parietal and frontal. No region of the brain functions alone, although major functions of various parts of the lobes have been determined.

The occipital lobe (back of the head) receives and processes visual information. The temporal lobe receives auditory signals, processing language and the meaning of words. The parietal lobe is associated with the sensory cortex and processes information about touch, taste, pressure, pain and heat and cold. The frontal lobe has three functions: motor activity and integration of muscle activity, speech and thought processes.

Most people who have been studied have their language and speech areas on the left hemisphere of their brain. Language comprehension is found in Wernicke's area. Speaking ability is in Broca's area; damage to Broca's area causes speech impairment but not impairment of language comprehension. Lesions in Wernicke's area impair the ability to comprehend written and spoken words but not speech. The remaining parts of the cortex are associated with higher thought processes, planning, memory, personality and other human activities.

3.4.6.11   The spinal cord

The spinal cord runs along the dorsal side of the body and links the brain to the rest of the body. Vertebrates have their spinal cords encased in a series of (usually) bony vertebrae that comprise the vertebral column. The gray matter of the spinal cord consists mostly of cell bodies and dendrites. The surrounding white matter is made up of bundles of inter-neuronal axons (tracts). Some tracts are ascending (carrying messages to the brain), while others are descending (carrying messages from the brain). The spinal cord is also involved in reflexes that do not immediately involve the brain.

3.4.6.12   Senses

Input to the nervous system is in the form of our five senses: pain, vision, taste, smell and hearing. Vision, taste, smell and hearing input are the special senses. Pain, temperature and pressure are known as somatic senses. Sensory input begins with sensors that react to stimuli in the form of energy that is transmitted into an action potential and sent to the CNS.

3.4.6.13   *Sensory receptors*

Sensory receptors are classified according to the type of energy they can detect and respond to. Mechanore-ceptors are sensitive to mechanical excitations such as those needed for hearing and sensed during stretching. Photoreceptors are sensitive to light. Chemoreceptors are responsible mainly for smell as well as internal sensors in the digestive and circulatory systems. Thermoreceptors sense changes in temperature. Electroreceptors detect electrical currents from the surrounding environment.

   Mechanoreceptors vary greatly in the specific type of stimulus and duration of stimulus/action potentials. The most adaptable vertebrate mechanoreceptor is the hair cell, present in the lateral line of fish. In humans and mammals hair cells are involved with detection of sound, gravity and providing balance.

3.4.6.14   *Hearing*

Hearing involves the actions of the external ear, eardrum, ossicles and cochlea. In hearing, sound waves in air are converted into vibrations of a liquid then into movement of hair cells in the cochlea. Finally they are converted into action potentials in a sensory dendrite connected to the auditory nerve. Very loud sounds can cause violent vibrations in the membrane under hair cells, causing a shearing or permanent distortion to the cells and resulting in permanent hearing loss.

3.4.6.15   *Orientation and gravity*

Orientation and gravity are detected at the semicircular canals. Hair cells along three planes respond to shifts of liquid within the cochlea, providing a three-dimensional sense of equilibrium. Calcium carbonate crystals can shift in response to gravity, providing sensory information about gravity and acceleration.

3.4.6.16   *Photoreceptors detect vision and light sensitivity*

The human eye can detect light in the 400–700 nm range of the electromagnetic spectrum, that is, the visible light spectrum. Light with wavelengths within the range 100–400 nm is termed ultraviolet (UV) light. Light with wavelengths between 700 nm and 1 mm is termed infrared (IR) light.

3.4.6.17   *Eye*

In the eye, two types of photoreceptor cells (rods and cones) are clustered on the retina or back portion of the eye. These receptors apparently evolved from hair cells. Rods detect differences in light intensity, while cones detect color. Rods are more common in a circular zone near the edge of the eye. Cones occur in the center (or fovea centralis) of the retina. Light reaching a photoreceptor causes the breakdown of the chemical rhodopsin, which in turn causes a membrane potential that is transmitted to an action potential. The action potential transfers to synapsed neurons that connect to the optic nerve. The optic nerve connects to the occipital lobe of the brain.

   Humans have three types of cones, each sensitive to a different color of light: red, blue and green. Opsins are chemicals that bind to cone cells and make those cells sensitive to light of a particular wavelength (or color). Humans have three different form of opsins coded by three genes on the X chromosome. Defects in one or more of these opsin genes can cause color blindness, which is more common in males.

3.4.7   The endocrine system

The nervous system coordinates rapid and precise responses to stimuli using electrical signals (action potentials). The endocrine system maintains homeostasis and long-term control using chemical signals. The endocrine system works in parallel with the nervous system to control growth and maturation along with homeostasis.

3.4.7.1   *Hormones*

The endocrine system is a collection of glands that secrete chemical messages referred to as hormones. These signals are passed through the blood to arrive at a target organ, which has cells possessing the appropriate receptor. Exocrine glands (not part of the endocrine system) secrete products that are passed outside the body. Sweat glands, salivary glands and digestive glands are examples of exocrine glands. Hormones are grouped into three classes based on their structure: steroids, peptides and amines.

3.4.7.2   *Steroids*

Steroids are lipids derived from cholesterol, and testosterone is the male sex hormone. Estradiol, similar in structure to testosterone, is responsible for many female sex characteristics. Steroid hormones are secreted by the gonads (sexual organs), adrenal cortex and placenta.

Most hormones are peptides (short chains of amino acids). They are secreted by the pituitary, parathyroid, heart, stomach, liver and kidneys. Amines are derived from the amino acid tyrosine and are secreted from the thyroid and the adrenal medulla. Solubility of the various hormone classes varies. Steroid hormones are derived from cholesterol by a biochemical reaction series. Defects along this series often lead to hormonal imbalances with serious consequences. Once synthesized, steroid hormones pass into the bloodstream; they are not stored by cells, and the rate of synthesis controls them.

Peptide hormones are synthesized as precursor molecules and processed by the endoplasmic reticulum and Golgi, where they are stored in secretory granules. When needed, the granules are dumped into the bloodstream. Different hormones can often be made from the same precursor molecule by cleaving it with a different enzyme. Amine hormones (notably epinephrine) are stored as granules in the cytoplasm until needed.

The endocrine system uses negative feedback to regulate physiological functions. Negative feedback regulates the secretion of almost every hormone. Cycles of secretion maintain physiological and homeostatic control, and can range from hours to months in duration.

3.4.7.3   *Mechanisms of hormone action*

The endocrine system acts by releasing hormones that in turn trigger actions in specific target cells. Receptors on target cell membranes bind only to one type of hormone. More than 50 human hormones have been identified, all acting by binding to receptor molecules. The binding of hormone changes the shape of the receptor, causing the response to the hormone. There are two mechanisms of hormone action on all target cells.

Non-steroid hormones (water soluble) do not enter the cell but bind to plasma membrane receptors, generating a chemical signal (second messenger) inside the target cell. Five different second messenger chemicals, including cyclic AMP, have been identified. Second messengers activate other intracellular chemicals to produce the target cell response.

The second mechanism involves steroid hormones, which pass through the plasma membrane and act in a two-step process. Once inside the cell, steroid hormones bind to the nuclear membrane receptors producing an activated hormone-receptor complex. The activated hormone-receptor complex binds to DNA and activates specific genes, increasing production of proteins.

3.4.7.4   *Endocrine-related problems*

Primary problems include overproduction of a hormone, underproduction of a hormone and non-functional receptors that cause target cells to become insensitive to hormones. The pituitary gland (often called the master gland) is located in a small bony cavity at the base of the brain. A stalk links the pituitary to the hypothalamus, which controls release of pituitary hormones. The pituitary gland has two lobes: the anterior and posterior lobes. The anterior pituitary is glandular. The hypothalamus contains neurons that control releases from the anterior pituitary. Seven hypothalamic hormones are released into a portal system connecting the hypothalamus and pituitary, and cause targets in the pituitary to release eight hormones.

Growth hormone (GH) is a peptide anterior pituitary hormone essential for growth. GH-releasing hormone stimulates release of GH while the GH-inhibiting hormone suppresses the release of GH. The hypothalamus maintains homeostatic levels of GH. Cells under the action of GH increase in size (hypertrophy) and number (hyperplasia). GH also causes increase in bone length and thickness by deposition of cartilage at the ends of bones. During adolescence, sex hormones cause replacement of cartilage by bone, halting further bone growth even though GH is still present. Too little or two much GH can cause dwarfism or gigantism, respectively.

Hypothalamus receptors monitor blood levels of thyroid hormones. Low blood levels of thyroid-stimulating hormone (TSH) cause the release of TSH-releasing hormone from the hypothalamus, which in turn causes the release of TSH from the anterior pituitary. TSH travels to the thyroid where it promotes production of thyroid hormones, which in turn regulate metabolic rates and body temperatures.

Gonadotropins and prolactin are also secreted by the anterior pituitary. Gonadotropins (which include the follicle-stimulating hormone or FSH and luteinizing hormone or LH) affect the gonads by stimulating gamete formation and production of sex hormones. Prolactin is secreted near the end of pregnancy and prepares the breasts for milk production.

The posterior pituitary stores and releases hormones into the blood. Antidiuretic hormone (ADH) and oxytocin are produced in the hypothalamus and transported by axons to the posterior pituitary where they are dumped into the blood. ADH controls water balance in the body and blood pressure. Oxytocin is a small peptide hormone that stimulates uterine contractions during childbirth.

3.4.7.5   *The adrenal glands*

Each kidney has an adrenal gland located above it, which is divided into an inner medulla and an outer cortex. The medulla synthesizes amine hormones, while the cortex secretes steroid hormones. The adrenal medulla consists of modified neurons that secrete two hormones: epinephrine and norepinephrine. Stimulation of the cortex by the sympathetic nervous system causes release of hormones into the blood to initiate the fight-or-flight response. The adrenal cortex produces several steroid hormones in three classes: mineralocorticoids, glucocorticoids and sex hormones. Mineralocorticoids maintain electrolyte balance. Glucocorticoids produce a long-term slow response to stress by raising blood glucose levels through the breakdown of fats and proteins; they also suppress the immune response and inhibit the inflammatory response.

3.4.7.6   The thyroid gland

The thyroid gland is located in the neck. Follicles in the thyroid secrete thyroglobulin, a storage form of thyroid hormone. The thyroid-stimulating hormone (TSH) from the anterior pituitary causes conversion of thyroglobulin into thyroid hormones T4 and T3. Almost all body cells are targets of thyroid hormones.

The thyroid hormone increases the overall metabolic rate, regulates growth and development as well as the onset of sexual maturity. Calcitonin is also secreted by large cells in the thyroid, which plays a role in the regulation of calcium.

3.4.7.7   The pancreas

The pancreas contains exocrine cells that secrete digestive enzymes into the small intestine and clusters of endocrine cells (the pancreatic islets). The islets secrete the hormones insulin and glucagon, which regulate blood glucose levels. Blood glucose levels rise after a meal, prompting the release of insulin. This causes cells to take up glucose and liver and skeletal muscle cells to form the carbohydrate glycogen. As glucose levels in the blood fall, further insulin production is inhibited. Glucagon causes the breakdown of glycogen into glucose, which in turn is released into the blood to maintain glucose levels within a homeostatic range. Glucagon production is stimulated when blood glucose levels fall, and inhibited when they rise.

Diabetes results from inadequate levels of insulin or its reception in tissues (type I and type II).

3.4.7.8   Other chemical messengers

Interferons are proteins released when a cell has been attacked by a virus. They cause neighboring cells to produce antiviral proteins. Once activated, these proteins destroy the virus. Prostaglandins are fatty acids that behave in many ways like hormones. They are produced by most cells in the body and act on neighboring cells. Pheromones are chemical signals that travel between organisms rather than between cells within an organism. Pheromones are used to mark territory, signal prospective mates and communicate. The presence of a human sex attractant/pheromone has not been established conclusively.

3.4.7.9   Biological cycles

Biological cycles ranging from minutes to years occur throughout the animal kingdom. Cycles involve hibernation, mating behavior, body temperature and many other physiological processes. Rhythms or cycles that show cyclic changes on a daily (or even a few hours) basis are known as circadian rhythms. Many hormones such as ACTH-cortisol, TSH and GH show circadian rhythms. The menstrual cycle is controlled by a number of hormones secreted in a cyclical fashion. Thyroid secretion is usually higher in winter than in summer. Childbirth is hormonally controlled, and is highest between the hours of 0200 and 0700. Internal cycles of hormone production are controlled by the hypothalamus, specifically the suprachiasmic nucleus (SCN). According to one model, the SCN is signaled by messages from the light-detecting retina of the eyes. The SCN signals the pineal gland in the brain to signal the hypothalamus, and so on.

3.4.8   The muscular and skeletal systems

The single-celled protozoan ancestors of animals had their weight supported by water and were able to move by cilia or other simple organelles. The evolution of large and more complex organisms (animals) necessitated the development of support and locomotion systems. Animals use their muscular and skeletal systems for support, locomotion and maintaining their shape.

3.4.8.1   *Skeletal system*

The skeleton helps transmit that movement. Skeletons can be a fluid-filled body cavity, an exoskeleton or an internal skeleton.

Hydrostatic skeletons consist of fluid-filled closed chambers. Internal pressures generated by muscle contractions cause movement as well as maintain the shape of the animals, such as the sea anemone and worms. The sea anemone has one set of longitudinal muscles in the outer layer of the body, and a layer of circular muscles in the inner layer of the body. The anemone can elongate or contract its body by contracting one or the other set of muscles.

Exoskeletons are characteristic of the phylum arthropoda. Exoskeletons are hard segments that cover the muscles and visceral organs. Muscles for movement attach to the inner surface of the exoskeleton. Exoskeletons restrict the growth of the animal, thus it must shed its exoskeleton (or molt) to form a new one that has room for growth. The bulk and weight of the exoskeleton and associated mechanical problems limits the size animals can attain. Spiders use a combination of an exoskeleton for protection and fluid pressure for movement.

Vertebrates have developed an internal mineralized (in most cases) endoskeleton composed of bone and/or cartilage. Muscles are on the outside of the endoskeleton. Cartilage and bone are types of connective tissue. Sharks and rays have skeletons composed entirely of cartilage; other vertebrates have an embryonic cartilage skeleton progressively replaced by bone as they mature and develop. Some areas of the human body retain cartilage in the adult, however (e.g. in joints and flexible structures such as the ribs, trachea, nose and ears). The human skeletal system is depicted in Figure 3.33.

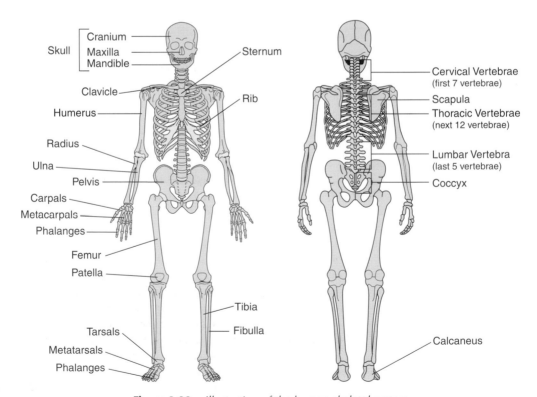

Figure 3.33   *Illustration of the human skeletal system.*

The axial skeleton consists of the skull, vertebral column and rib cage. The appendicular skeleton contains the bones of the appendages (limbs, wings or flippers/fins) and the pectoral and pelvic girdles.

The human skull or cranium has a number of individual bones tightly fitted together at immovable joints. At birth, many of these joints are not completely sutured together as bone. This leads to a number of 'soft spots' or fontanels, which do not join completely until the age of 14–18 months.

The vertebral column has 33 individual vertebrae separated from each other by a cartilage disk. These disks allow a certain flexibility to the spinal column, although the disks deteriorate with age and produce back pain. The sternum is connected to all the ribs except the lower pair. Cartilage allows for the flexibility of the rib cage during breathing.

The arms and legs are part of the appendicular skeleton. The upper bones of the limbs are single: humerus (arm) and femur (leg). Below a joint (elbow or knee), both limbs have a pair of bones (radius and ulna in the arms; tibia and fibula in legs) that connect to another joint (wrist or ankle). The carpals make up the wrist joint; the tarsals are in the ankle joint. Each hand or foot ends in five digits (fingers or toes) composed of metacarpals (hands) or metatarsals (feet).

Limbs are connected to the rest of the skeleton by collections of bones known as girdles. The pectoral girdle consists of the clavicle (collar bone) and scapula (shoulder blade). The humerus is joined to the pectoral girdle at a joint and is held in place by muscles and ligaments. A dislocated shoulder occurs when the end of the humerus slips out of the socket of the scapula, stretching ligaments and muscles. The pelvic girdle consists of two hipbones that form a hollow cavity: the pelvis. The vertebral column attaches to the top of the pelvis and the femur of each leg attaches to the bottom. The pelvic girdle in land animals transfers the weight of the body to the legs and feet. Pelvic girdles in fish, which have their weight supported by water, are primitive; land animals have more developed pelvic girdles. Pelvic girdles in bipeds are recognizably different from those of quadrupeds.

Although bones vary greatly in size and shape, they have certain structural similarities. Bones have cells embedded in a mineralized (calcium) matrix and collagen fibers. Compact bone forms the shafts of long bones and can be found on the outer side of the bone. Spongy bone forms the inner layer.

Compact bone has a series of Haversian canals around which concentric layers of bone cells (osteocytes) and minerals occur. New bone is formed by the osteocytes. The Haversian canals form a network of blood vessels and nerves that nourish and monitor the osteocytes.

Spongy bone occurs at the ends of long bones and is less dense than compact bone. The spongy bone of the femur, humerus and sternum contains red marrow, in which stem cells reproduce and form the cellular components of the blood and immune system. Yellow marrow at the center of these bones is used to store fats. The outer layer of the bones is known as the periosteum. The inner layer of the periosteum forms new bone or modifies existing bone to meet new conditions. It is rich in nerve endings and blood and lymphatic vessels. When fractures occur, the pain is carried to the brain by nerves running through the periosteum.

Endochondral ossification is the process of converting the cartilage in embryonic skeletons into bone. Cartilage is deposited early in development into shapes resembling the bones-to-be. Cells inside this cartilage grow and begin depositing minerals.

The spongy bone forms first, and osteoblasts attach and lay down the mineral portions of spongy bone. Osteoclasts remove material from the center of the bone, forming the central cavity of the long bones. The perichondrium, a connective tissue, forms around the cartilage and begins forming compact bone while the above changes are occurring. Blood vessels form and grow into the perichondrium, transporting stem cells into the interior. Two bands of cartilage remain as the bone develops, one at each end of the bone. During childhood, this cartilage allows for growth and changes in the shape of bones. Eventually the elongation of the bones stops and the cartilage is all converted to bone.

There are three types of joints: immovable, partly movable and synovial. Immovable joints such as those connecting the cranial bone, have edges that tightly interlock. Partly movable joints allow some degree of flexibility and usually have cartilage between the bones (vertebrae is an example). Synovial joints permit the greatest degree of flexibility and have the ends of bones covered with a connective tissue filled with synovial fluid (hip is an example).

The outer surface of the synovial joints contains ligaments that strengthen joints and hold bones in position. The inner surface (the synovial membrane) has cells producing synovial fluid that lubricates the joint and prevents the two cartilage caps on the bones from rubbing together. Some joints also have tendons (connective tissue linking muscles to bones). Bursae are small sacs filled with synovial fluid that reduce friction in the joint; the knee joint contains 13 bursae.

3.4.8.2   Skeletal muscle system

Vertebrates move by the actions of muscles on bones. Tendons attach many skeletal muscles across joints, allowing muscle contraction to move the bones across the joint. Muscles generally work in pairs to produce movement; when one muscle flexes (or contracts) the other relaxes, a process known as antagonism.

Muscles have both electrical and chemical activity. There is an electrical gradient across the muscle cell membrane and the outside is more positive than the inside. Stimulus causes an instantaneous reversal of this polarity, causing the muscle to contract which produces a twitch or movement. Muscles contract by shortening each sarcomere. The sliding filament model of muscle contraction has thin filaments on each side of the sarcomere sliding past each other until they meet in the middle. Myosin filaments have club-shaped heads that project toward the actin filaments. Myosin heads attach to binding sites on the actin filaments; they swivel toward the center of the sarcomere, detach and then reattach to the nearest active site of the actin filament. Each cycle of attachment, swiveling and detachment shortens the sarcomere by 1%. Hundreds of such cycles occur each second during muscle contraction.

Calcium ions are required for each cycle of myosin-actin interaction. Calcium is released into the sarcomere when a muscle is stimulated to contract, and uncovers the actin binding sites. When the muscle no longer needs to contract, the calcium ions are pumped from the sarcomere and back into storage.

Neuromuscular junctions are the point where a motor neuron attaches to a muscle. Acetylcholine is released from the axon end of the nerve cell when a nerve impulse reaches the junction. A wave of electrical changes is produced in the muscle cell when the acetylcholine binds to receptors on its surface. Calcium is released from its storage area in the cell's endoplasmic reticulum. An impulse from a nerve cell causes calcium release and brings about a single, short muscle contraction called a twitch. Skeletal muscles are organized into hundreds of motor units, each of which is a motor neuron and a group of muscle fibers. A graded response to a circumstance will involve controlling the number of active motor units. While individual muscle units contract as a unit, the entire muscle can contract on a graded basis due to their organization into motor units.

3.4.9   The respiratory system

A sufficient supply of oxygen is required for the aerobic respiratory machinery of Krebs cycle and the electron transport system to efficiently convert stored organic energy into energy trapped in ATP. Carbon dioxide is also generated by cellular metabolism and must be removed from the cell. There must be an exchange of gases: carbon dioxide leaving the cell and oxygen entering. Single-celled organisms exchange gases directly across their cell membrane. However, the slow diffusion rate of oxygen limits the size of the organism. Simple animals that lack specialized exchange surfaces have flattened, tubular or thin-shaped body plans, which appear to be the most efficient for gas exchange.

Large animals cannot maintain gas exchange by diffusion across their outer surface. They have developed a variety of respiratory surfaces that all increase the surface area for exchange, thus allowing for larger bodies. A respiratory surface is covered with thin, moist epithelial cells that allow oxygen and carbon dioxide to exchange. Those gases can only cross cell membranes when they are dissolved in water or an aqueous solution; respiratory surfaces therefore must be moist.

Functions of the respiratory system include:

- movement of an oxygen-containing medium so it contacts a moist membrane overlying blood vessels;
- diffusion of oxygen from the medium into the blood;
- transport of oxygen to the tissues and cells of the body;
- diffusion of oxygen from the blood into cells; and
- carbon dioxide following a reverse path.

The respiratory system in humans includes the lungs, pathways connecting them to the outside environment and structures in the chest involved with moving air in and out of the lungs. Lungs are ingrowths of the body wall and connect to the outside by a series of tubes and small openings. Lung breathing probably evolved about 400 million years ago, and is not the sole property of vertebrates.

The human respiratory system is illustrated in Figure 3.34. Air enters the body typically through the nose, is warmed, filtered and passed through the nasal cavity. Air passes the pharynx, which has the epiglottis to prevent food from entering the trachea. The upper part of the trachea contains the larynx. The vocal cords are two bands of tissue that extend across the opening of the larynx. After passing the larynx, the air moves into the bronchi that carry air in and out of the lungs.

In humans, the trachea divides into the two main bronchi that enter the roots of the lungs. The lungs are large, lobed, paired organs in the chest, also known as the thoracic cavity that also accommodates the heart. The diaphragm forms the bottom of the thoracic cavity. Having entered each lung, the bronchi continue to divide and, after multiple divisions, give rise to bronchioles. Bronchi are reinforced to prevent their collapse and are lined with ciliated epithelium and mucus-producing cells. The bronchial tree continues branching until it reaches the level of terminal bronchioles, which lead to alveolar sacs. Alveolar sacs comprise clusters of alveoli, like individual grapes within a bunch. The individual alveoli are tightly wrapped in blood vessels and it is here that gas exchange actually occurs. Only c. 0.2 μm separate the alveoli from the capillaries due to the extremely thin walls of both structures.

Although similar in appearance, the two lungs are not identical. Both are separated into lobes by fissures, with three lobes on the right and two on the left. The lobes are further divided into segments and then into lobules, hexagonal divisions of the lungs that are the smallest subdivision visible to the naked eye. The connective tissue that divides lobules is often blackened in smokers. The medial border of the right lung is nearly vertical, while the left lung contains a cardiac notch. The cardiac notch is a concave impression molded to accommodate the shape of the heart. Each lobe is surrounded by a pleural cavity, which consists of two pleurae. The parietal pleura lies against the rib cage, and the visceral pleura lies on the surface of the lungs. In between the pleura is pleural fluid. The pleural cavity helps to lubricate the lungs, as well as providing surface tension to keep the lung surface in contact with the rib cage.

Ventilation is the mechanics of breathing in and out. When you inhale, muscles in the chest wall contract, lifting the ribs and pulling them outward. The diaphragm at this time moves downward, enlarging the chest cavity. Reduced air pressure in the lungs causes air to enter the lungs. Exhaling reverses these steps.

Carbon dioxide concentration in metabolically active cells is much greater than in capillaries, so carbon dioxide diffuses from the cells into the capillaries. Water in the blood combines with carbon dioxide to form bicarbonate. This removes the carbon dioxide from the blood, so diffusion of even more carbon dioxide from the cells into the capillaries continues yet still manages to 'package' the carbon dioxide for eventual passage out of the body. In the alveoli capillaries, bicarbonate combines with a hydrogen ion (proton) to form carbonic acid, which breaks down into carbon dioxide and water. The carbon dioxide then diffuses into the alveoli and out of the body with the next exhalation.

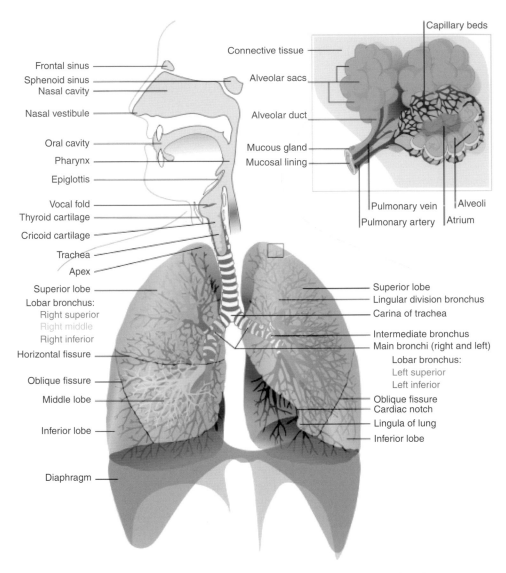

Figure 3.34   *Illustration of the human respiratory system (see color plate).*

Muscular contraction and relaxation controls the rate of expansion and constriction of the lungs. These muscles are stimulated by nerves that carry messages from the part of the brain that controls breathing, the medulla. Two systems control breathing: an automatic response and a voluntary response. Both are involved in holding your breath.

Although the automatic breathing regulation system allows you to breathe while you sleep, it sometimes malfunctions. Apnea involves stoppage of breathing for as long as 10 seconds, which for some individuals can occur as often as 300 times per night. This failure to respond to elevated blood levels of carbon dioxide may result from viral infections of the brain, tumors or it may develop spontaneously. A malfunction of the breathing centers in newborns may result in sudden infant death syndrome (SIDS).

As altitude increases, atmospheric pressure decreases. Above 10 000 feet decreased oxygen pressures cause loading of oxygen into hemoglobin to drop off, leading to lowered oxygen levels in the blood. The result can be mountain sickness (nausea and loss of appetite) which does not result from oxygen starvation but rather from the loss of carbon dioxide due to increased breathing in order to obtain more oxygen.

3.4.10 The excretory system

Cells produce water and carbon dioxide as by-products of the metabolic breakdown of sugars, fats and proteins. Inorganic chemical groups such as nitrogen, sulfur and phosphorous must be stripped from the large molecules to which they were formerly attached as part of preparing them for energy conversion. The continuous production of metabolic wastes establishes a steep concentration gradient across the plasma membrane, causing wastes to diffuse out of cells and into the extracellular fluid.

Single-celled organisms have most of their wastes diffuse out into the outside environment. Multicellular organisms, and animals in particular, must have a specialized organ system to concentrate and remove wastes from the interstitial fluid into the blood capillaries and eventually deposit that material at a collection point for removal entirely from the body.

Excretory systems regulate the chemical composition of body fluids by removing metabolic wastes and retaining the proper amounts of water, salts and nutrients. Although components of this system in vertebrates are often cited to include the kidneys, liver, lungs and skin, skin actually plays a relatively minor role acting primarily to secrete aqueous solutions during temperature regulation rather than to excrete waste. Being the major part of the respiratory system, lungs excrete primarily gaseous waste (carbon dioxide). Not all animals use the same routes or excrete their wastes in the same way as humans. Excretion applies to metabolic waste products that cross a plasma membrane. Elimination is the removal of feces.

3.4.10.1 Key functions of excretory system

One of the key functions of the excretory system is removal of nitrogen wastes, a by-product of protein metabolism. Amino groups are removed from amino acids prior to energy conversion. The NH_2 (amino group) combines with a hydrogen ion (proton) to form ammonia (NH_3). Ammonia is usually very toxic and excreted directly by marine animals. Terrestrial animals usually need to conserve water. Ammonia is converted to urea, a compound the body can tolerate at higher concentrations than ammonia. Birds and insects secrete uric acid that they make through large energy expenditure but little water loss. Amphibians and mammals secrete urea that they form in their liver. Amino groups are turned into ammonia which in turn is converted to urea, dumped into the blood and concentrated by the kidneys.

The excretory system is also responsible for regulating water balance in various body fluids. Osmoregulation refers to the state aquatic animals are in: they are surrounded by freshwater and must constantly deal with the influx of water. Animals such as crabs have an internal salt concentration very similar to that of the surrounding ocean. Such animals are known as osmoconformers, as there is little water transport between the inside of the animal and the isotonic outside environment. Marine vertebrates, however, have internal concentrations of salt that are about one-third of the surrounding seawater. They are said to be osmoregulators.

Osmoregulators face two problems: prevention of water loss from the body and prevention of salts diffusing into the body. Fish deal with this by passing water out of their tissues through their gills by osmosis and salt through their gills by active transport. Cartilaginous fish have a greater salt concentration than seawater, causing water to move into the shark by osmosis which is used for excretion. Freshwater fish must prevent water gain and salt loss. They do not drink water, and have their skin covered by thin mucus. Water enters and leaves through the gills and the fish excretory system produces large amounts of dilute urine. Terrestrial animals use a variety of methods to reduce water loss including living in moist environments, developing impermeable body coverings and the production of more concentrated urine. Water loss can be considerable: a person in an environment of $100°F$ loses 1 liter of water per hour.

3.4.10.2   *The liver*

The breakdown of amino acids into sugars, glycogen or fats releases nitrogen. This nitrogen is changed by the liver into urea, which is carried to the kidneys by the blood stream. The urea is eliminated from the body from the kidneys. The liver also changes hemoglobin from dead red blood cells into bile, which is used in the small intestine to breakdown fats.

3.4.10.3   *The urinary system*

The urinary system is made up of the kidneys, ureters, bladder and urethra. Waste is filtered from the blood and collected as urine in each kidney. Urine leaves the kidneys by ureters, and collects in the bladder. The bladder can distend to store urine that eventually leaves through the urethra. The nephron is the kidney's basic functional unit.

The nephron contains a special network of capillaries and other structures, which help filter the blood under the action of arterial pressure. Fluids and solutes are returned to the capillaries that surround the nephron tubule. The nephron has three functions:

- filtration of water and solutes from the blood;
- tubular reabsorption of water and conserved molecules back into the blood; and
- tubular secretion of ions and other waste products from surrounding capillaries into the distal tubule and, ultimately, into urinary tract.

Nephrons filter c. 125 mL of bodily fluids per minute (roughly equivalent to the entire fluid component of the body 16 times each day). Nephrons produce c. 180 L of filtrate in a 24 hour period; c. 179 L are reabsorbed and the remaining 1 L forms urine. In some cases, excess wastes crystallize as kidney stones. They grow and can become a painful irritant that may require surgery or ultrasound treatments. Some stones are small enough to be forced into the urethra; others are too large to be passed this way.

Water reabsorption is controlled by the antidiuretic hormone (ADH) in negative feedback. ADH is released from the pituitary gland in the brain. Dropping levels of fluid in the blood signal the hypothalamus to cause the pituitary to release ADH into the blood. This acts to increase water absorption in the kidneys, which puts more water back in the blood. When too much fluid is present in the blood, sensors in the heart signal the hypothalamus to cause a reduction of the amounts of ADH in the blood. This increases the amount of water absorbed by the kidneys, producing larger quantities of more dilute urine.

Aldosterone, a hormone secreted by the kidneys, regulates the transfer of sodium from the nephron to the blood. When sodium levels in the blood fall, aldosterone is released into the blood causing more sodium to pass from the nephron to the blood. This causes water to flow into the blood by osmosis. Renin is released into the blood to control aldosterone.

4

Major Plasma Disharges and their Applicability for Plasma Medicine

Specific plasma-medical technologies require application of specific plasma sources, that is, specific electric discharges. Various electric discharges generate plasmas with very different parameters including: temperatures and densities of electron, ions and neutrals; different densities of excited and active atoms and molecules; and different intensities of UV and other types of radiation. In classifying electric discharges applied in plasma chemistry, it is convenient to distinguish between thermal (e.g. arc) and non-thermal (e.g. glow) discharges; high-pressure (e.g. arc, corona, dielectric barrier discharges or DBDs) and low-pressure (e.g. glow) discharges; electrode discharges (e.g. arc, glow) and electrodeless discharges (e.g. inductively coupled plasma or ICP radio frequency or RF and microwave); direct current (DC) and non-DC discharges; and self-sustained and non-self-sustained discharges. All these groups of discharges have their own niche of biological and medical applications. For example, thermal plasma discharges are widely applied for blood cauterization and for surgical cutting and ablation of tissue; atmospheric-pressure cold discharges are widely applied for sterilization, healing of wounds and diseases; and low-pressure discharges (especially low-pressure RF) are widely used for tissue engineering. Detailed consideration of all major types of electric discharges can be found in Fridman and Kennedy (2004, 2011). In this chapter, the major plasma discharges will be discussed with a focus on specifics critical for their plasma-medical applications.

4.1 Electric breakdown and steady-state regimes of non-equilibrium plasma discharges

4.1.1 Townsend mechanism of electric breakdown, Paschen curves

Consider breakdown in a plane gap d by DC voltage V corresponding to electric field $E = V/d$. Occasional primary electrons near the cathode provide low initial current i_0. The primary electrons drift to the anode, ionizing the gas and generating avalanches. The ionization in avalanches is usually described by the Townsend ionization coefficient α, indicating production of electrons per unit length along the electric field:

$$\frac{dn_e}{dx} = \alpha n_e,$$

Plasma Medicine, First Edition. Alexander Fridman and Gary Friedman.
© 2013 John Wiley & Sons, Ltd. Published 2013 by John Wiley & Sons, Ltd.

where the number density of electrons n_e in the x direction is defined:

$$n_e(x) = n_{e0} \exp(\alpha x).$$

The Townsend ionization coefficient is related to the ionization rate coefficient $k_i(E/n_0)$ and electron drift velocity v_d as:

$$\alpha = \frac{v_i}{v_d} = \frac{1}{v_d} k_i \left(\frac{E}{n_0} \right) n_0 = \frac{1}{\mu_e} \frac{k_i(E/n_0)}{E/n_0}, \quad (4.1)$$

where v_i is ionization frequency and μ_e is electron mobility, inversely proportional to pressure. The Townsend coefficient α is usually presented as similarity parameter α/p, depending on reduced electric field E/p. Each primary electron generated near a cathode produces $\exp(\alpha d) - 1$ positive ions moving back to the cathode. These ions lead to the extraction of $\gamma[\exp(\alpha d) - 1]$ electrons from the cathode due to secondary electron emission characterized by the Townsend coefficient γ (Section 2.1.6, Equation (2.47)). Typical γ values are 0.01–0.1. Taking into account current of primary electrons i_0 and electron current due to the secondary electron emission from the cathode, the total electronic part of the cathode current i_{cath} is:

$$i_{cath} = i_0 + \gamma i_{cath}[\exp(\alpha d) - 1]. \quad (4.2)$$

Total current in the external circuit is equal to electronic current at the anode, where ion current is absent. The total current can be found as $i = i_{cath} \exp(\alpha d)$, which leads to the *Townsend formula*:

$$i = \frac{i_0 \exp(\alpha d)}{1 - \gamma[\exp(\alpha d) - 1]}. \quad (4.3)$$

The current in the gap is non-self-sustained as long as the denominator of Equation (4.3) is positive. When the electric field and Townsend coefficient α become high enough, the denominator of Equation (4.3) tends to zero and the transition to self-sustained current takes place. This is called the *Townsend breakdown mechanism*:

$$\gamma[\exp(\alpha d)1] = 1, \quad \text{where} \quad \alpha d = \ln \left(\frac{1}{\gamma} + 1 \right). \quad (4.4)$$

The similarity parameters α/p and E/p are related semi-empirically according to Equation (4.1) as:

$$\frac{\alpha}{p} = A \exp \left(-\frac{B}{E/p} \right), \quad (4.5)$$

where parameters A and B for different gases at $E/p = 30 - 500$ V (cm Torr)$^{-1}$ are given in Table 4.1.

The combination of Equations (4.4) and (4.5) gives the formula for calculation of breakdown voltage and breakdown-reduced electric field as functions of an important similarity parameter pd:

$$V = \frac{B(pd)}{C + \ln(pd)}, \quad \frac{E}{p} = \frac{B}{C + \ln(pd)} \quad (4.6)$$

Table 4.1 *Parameters A and B for calculation of the Townsend coefficient α.*

Gas	A	B	Gas	A	B
Air	15	365	N_2	10	310
CO_2	20	466	H_2O	13	290
H_2	5	130	He	3	34
Ne	4	100	Ar	12	180
Kr	17	240	Xe	26	350

where

$$C = \ln A - \ln \ln \left(\frac{1}{\gamma} + 1 \right)$$

is almost constant. The breakdown voltage dependence on the similarity parameter pd is usually referred to as the *Paschen curve* (see Figure 4.1). These curves have a minimum corresponding to the easiest breakdown conditions, which can be found from Equation (4.6) as:

$$V_{min} = \frac{2.72 \times B}{A} \ln \left(1 + \frac{1}{\gamma} \right), \quad \left(\frac{E}{p} \right)_{min} = B, \quad (pd)_{min} = \frac{2.72}{A} \ln \left(1 + \frac{1}{\gamma} \right). \tag{4.7}$$

Reduced electric field E/p required for breakdown (Equation (4.6)) decreases only logarithmically with pd. Breakdown of larger gaps is less sensitive to the secondary electron emission and cathode material, which explains the E/p reduction with pd. This reduction of breakdown electric field in electronegative gases is limited by electron attachment processes, characterized by the Townsend coefficient β which is defined:

$$\beta = \frac{v_a}{v_d} = \frac{1}{v_d} k_a \left(\frac{E}{n_0} \right) n_0 = \frac{1}{\mu_e} \frac{k_a(E/n_0)}{E/n_0}. \tag{4.8}$$

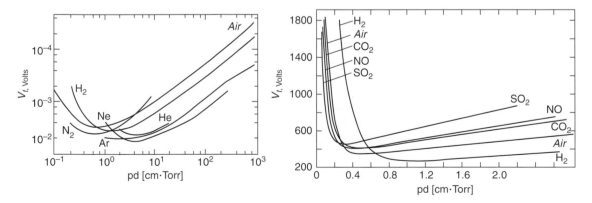

Figure 4.1 *Paschen curves for different gases.*

Table 4.2   *Townsend breakdown electric fields for centimeter-size gaps at atmospheric pressure.*

Gas	E	Gas	E	Gas	E
Air	32	O_2	30	N_2	35
H_2	20	Cl_2	76	CCl_2F_2	76
CSF_8	150	CCl_4	180	SF_6	89
He	10	Ne	1.4	Ar	2.7

In this relation, $k_a(E/n_0)$ and ν_a are the attachment rate coefficient and frequency with respect to an electron. The Townsend coefficient β characterizes electron losses due to attachment per unit length:

$$\frac{dn_e}{dx} = (\alpha - \beta)n_e, \qquad n_e(x) = n_{e0}\exp[(\alpha - \beta)x]. \tag{4.9}$$

Similarly to α, the Townsend coefficient β is an exponential function of the reduced electric field (although not as strong). Ionization rate therefore exceeds attachment at high electric fields, and the β coefficient can be neglected with respect to α in this case (short gaps, Equation (4.6)). When the gaps are relatively large (≥ 1 cm at 1 atm), the Townsend breakdown electric fields in electronegative gases become almost constant and limited by attachment processes. The breakdown electric fields at high pressures and long gaps for electronegative and non-electronegative gases are presented in Table 4.2.

4.1.2   Streamer or spark breakdown mechanism

The Townsend quasi-homogeneous breakdown mechanism can be applied only for relatively low pressures and short gaps ($pd < 4000$ Torr cm at atmospheric pressure and $d < 5$ cm). Another breakdown mechanism, called spark or streamer, takes place in larger gaps at high pressures. The sparks provide breakdown in a local narrow channel without direct relation to electrode phenomena. Sparks are also primarily related to avalanches, but in large gaps the avalanches cannot be considered as independent. The spark breakdown at high pd and considerable overvoltage develops much faster than the time necessary for ions to cross the gap and provide the secondary emission.

The mechanism of spark (or streamer) breakdown is based on the concept of *streamer*, which is a thin ionized channel that rapidly propagates between electrodes along the positively charged trail left by an intensive primary avalanche. This avalanche also generated photons, which in turn initiate numerous secondary avalanches in the vicinity of the primary avalanche. Electrons of the secondary avalanches are pulled by the strong electric field into the positively charged trail of the primary avalanche, creating the rapidly propagating streamer between electrodes.

The qualitative change occurs when the charge amplification in the avalanche $\exp(\alpha x)$ becomes large and the created space charge leads to considerable electric field $\vec{E}_a$, which should be added to the external field $\vec{E}_0$. The electrons are in the head of the avalanche while the positive ions remains behind, creating a dipole with the characteristic length $1/\alpha$ and charge $N_e \approx \exp(\alpha x)$. For the breakdown field of about 30 kV cm^{-1} in atmospheric-pressure air, the α-coefficient is about 10 cm^{-1} and characteristic ionization length can be estimated as $1/\alpha \sim 0.1$ cm. Transverse avalanche size can also be estimated as $1/\alpha \sim 0.1$ cm; the maximum electron density in an avalanche is therefore 10^{12}–10^{13} cm^{-3}.

The external electric field distortion due to the space charge of the dipole is shown in Figure 4.2. In front of the avalanche combine to make the total field stronger, which accelerates ionization. On the contrary, between the separated charges or 'inside' the avalanche, the total electric field is lower than the external field,

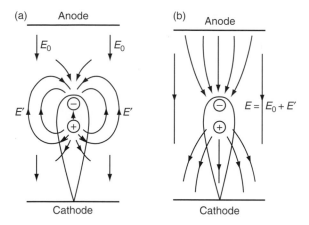

Figure 4.2   *Electric field distribution in an avalanche: external and space charge fields are shown (a) separately, and (b) combined.*

which slows down the ionization. The electric field of the charge $N_e \approx \exp(\alpha x)$ on the distance about the avalanche radius reaches the value of the external field $\vec{E}_0$ at some critical value of αx. Numerically, during the 1-cm-gap breakdown in air, the avalanche radius is about $r_A = 0.02$ cm and the critical value of αx when the avalanche electric field becomes comparable with E_0 is $\alpha x = 18$. As soon as the avalanche head reaches the anode, electrons flow into the electrode and it is mostly the ionic trail that remains in the gap.

Electric field distortion due to the space charge is illustrated in Figure 4.3. Total electric field is due to the external field, the ionic trail and also the ionic charge 'image' in the anode. A strong primary avalanche is able to amplify the external electric field and form a streamer. When the streamer channel connects the electrodes, the current may be significantly increased to form the spark. The avalanche-to-streamer transformation takes place when the internal field of an avalanche becomes comparable with the external field. If gap is short, the transformation occurs only when the avalanche reaches the anode.

Such a streamer grows from anode to cathode and is called the *cathode-directed* or *positive streamer*. If the gap and overvoltage are large, the avalanche-to-streamer transformation can take place far from anode

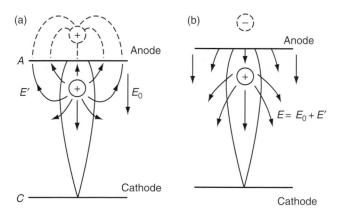

Figure 4.3   *Electric field distribution when the avalanche reaches the anode: external and space charge fields are shown (a) separately, and (b) combined.*

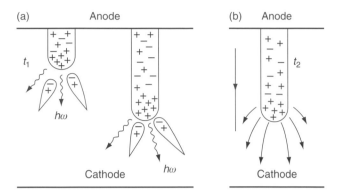

Figure 4.4   *Illustration of a cathode-directed streamer: (a) propagation, and (b) electric field near the streamer head.*

and the *anode-directed* or *negative streamer* grows toward both electrodes. The mechanism of formation of a cathode-directed streamer is illustrated in Figure 4.4. High-energy photons emitted from the primary avalanche provide photo-ionization in the vicinity, which initiates the secondary avalanches. Electrons of the secondary avalanches are pulled into the ionic trail of the primary avalanche and create a quasi-neutral plasma channel. The cathode-directed streamer starts near the anode, where the positive charge and electric field of the primary avalanche is the highest.

The streamer looks like a thin conductive needle growing from the anode. The electric field at the tip of the 'anode needle' is very high, and provides high electron drift and streamer growth velocities of about 10^8 cm s^{-1}. The diameter of the streamer channel is 0.01–0.1 cm and corresponds to the maximum size of a primary avalanche head $1/\alpha$. Plasma density in the streamer also corresponds to the maximum electron density in the head of the primary avalanche $10^{12}/10^{13}$ cm^{-3}. Specific energy input in a streamer channel is small during the short period (~30 ns) of the streamer growth between electrodes. In molecular gases it is about 10^{-3} eV mole^{-1}, which corresponds to heating to about 10 K. The anode-directed streamer occurs if the primary avalanche becomes strong before reaching the anode. Such a streamer growing in two directions is illustrated in Figure 4.5.

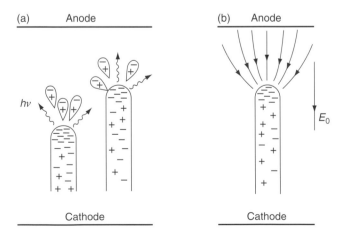

Figure 4.5   *Illustration of an anode-directed streamer: (a) propagation and (b) electric field near the streamer head.*

The mechanism of the anode-directed streamer is similar to that of the cathode-directed streamers, but the secondary avalanches can also be initiated by some electrons moving in front of the primary avalanche. The anode-directed streamers can be effectively generated when voltage rise-time is very short (not usually more than several nanoseconds, which is shorter than characteristic breakdown time) and therefore significant overvoltage with respect to the Paschen curve can be achieved. At the significant overvoltage level, surface density of the anode-directed streamers is high which leads to their overlapping and, as a result, to uniformity of such nanosecond pulsed discharges. The uniform nanosecond pulsed discharges, and especially nanosecond pulsed DBD discharge, play an important role in applied plasma-medical systems, discussed in Chapters 6, 7, 9 and 10.

4.1.3   Meek criterion of streamer formation and streamer propagation models

A streamer occurs when the electric field of space charge in an avalanche E_a equals the external field E_0:

$$E_a = \frac{e}{4\pi\varepsilon_0 r_A^2} \exp\left[\alpha\left(\frac{E_0}{p}\right)d\right] \approx E_0. \tag{4.10}$$

Assuming avalanche head radius as $r_A \approx 1/\alpha$, formation of a streamer in the gap (d) can be presented as a requirement for the avalanche amplification parameter αd to exceed the critical value:

$$\alpha\left(\frac{E_0}{p}\right)d = \ln\left(\frac{4\pi\varepsilon_0 E_0}{e\alpha^2}\right) \approx 20, \quad N_e = \exp(\alpha d) \approx 3 \times 10^8. \tag{4.11}$$

This criterion of streamer formation is known as the *Meek breakdown condition* ($\alpha d \geq 20$). Electron attachment mitigates their multiplication in avalanches and increases the electric field required for streamer formation. The ionization coefficient α in the Meek breakdown condition should be replaced in electronegative gases by $\alpha - \beta$. However, when discharge gaps are short (in air, $d \leq 15$ cm), electric fields required by the Meek criterion are relatively high; therefore $\alpha \gg \beta$ and the attachment can be neglected.

Increasing distance d between electrodes in electronegative gases does not lead to a gradual decrease of the required electric field (Equation (4.11)). The minimal field required for streamer formation is due to ionization-attachment balance $\alpha(E_0/p) = \beta(E_0/p)$, which gives 26 kV cm^{-1} in air and 117.5 kV cm^{-1} in SF$_6$. Electric field non-uniformity has a strong influence on breakdown conditions and the transformation of an avalanche into a streamer. Voltage applied non-uniformly should provide intensive electron multiplication only near the electrode to initiate a streamer. Once the plasma channel is initiated, it grows mostly due to the high electric field of its own streamer tip. In the case of very long (about a meter and longer) non-uniform systems, average breakdown electric field can be as low as 2–5 kV cm^{-1}.

The breakdown threshold for non-uniform electric fields also depends on polarity. The threshold voltage in a long gap between a negatively charged rod and a plane is about twice as higher than for the case of positively charged rod. In the case of a rod anode, the avalanches approach the anode where the electric field becomes stronger, which facilitates the avalanche-to-streamer transition. The avalanche electrons also easily sink into the anode in this case, leaving the ionic trail near the electrode enhancing the electric field.

There are two streamer propagating models depending on the assumption regarding conductivity of the streamer channels. The model of *quasi-self-sustained streamers* assumes low conductivity of the streamer channel, which makes it self-propagating and independent of the anode. Photons initiate an avalanche, which then develops in the self-induced electric field of the positive space charge. To provide continuous and steady propagation of the self-sustained streamer, its positive space charge should be compensated by the negative charge of the avalanche head at the meeting point of the avalanche and streamer. A qualitatively different model assumes a streamer channel as an ideal conductor connected to an anode.

Table 4.3   *Typical parameters of a streamer DBD microdischarge.*

Lifetime	10–40 ns	Filament radius	50–100 μm
Electron avalanche duration	10 ns	Electron avalanche transported charge	0.01 nC
Cathode-directed streamer duration	1 ns	Cathode-directed streamer charge transfer	0.1 nC
Plasma channel duration	30 ns	Plasma channel charge transfer	1 nC
Microdischarge remnant duration	1 ms	Microdischarge remnant charge	>1 nC
Peak current	0.1 A	Current density	0.1–1 kA cm^{-2}
Electron density	$10^{14} - 10^{15}$ cm^{-3}	Electron energy	1–10 eV
Total transported charge	0.1–1 nC	Reduced electric field	$E/n = (1\text{-}2)(E/n)_{\text{Paschen}}$
Total dissipated energy	5 μJ	Gas temperature	Close to average, about 300 K
Overheating	5 K		

The *ideally conducting streamer channel* is considered as an anode elongation in the direction of external electric field E_0 with the shape of an ellipsoid. The propagation velocity is determined by electron drift in the appropriate electric field E_m on the tip of the streamer with length l and radius r:

$$\frac{E_m}{E_0} = 3 + \left(\frac{l}{r}\right)^{0.92}, \quad \text{where} \quad 10 < \frac{l}{r} < 2000. \tag{4.12}$$

4.1.4   Streamers and microdischarges

Although streamers are elements of the spark breakdown, their visual observation is often related to dielectric barrier (DBD) and some corona discharges. The DBD gap (from 0.1 mm to 3 cm) usually includes one or more dielectric layers located in the current path between metal electrodes. Typical frequency is 0.05–100 kHz and voltage is about 10 kV at atmospheric pressure. In most cases, DBDs are not uniform and consist of numerous microdischarges built from streamers and distributed in the discharge gap. Electrons in the conducting plasma channel established by the streamers dissipate from the gap in about 40 ns, while slowly drifting ions remain there for several microseconds (Table 4.3). Deposition of electrons from the conducting channel onto the anode dielectric barrier results in charge accumulation and prevents new avalanches and streamers nearby until the cathode and anode are reversed. Usual DBD operation frequency is around 20 kHz; the voltage polarity reversal therefore occurs within 25 μs. After the voltage polarity reverses, the deposited negative charge facilitates formation of new avalanches and streamers in the same spot. As a result, a multi-generation family of streamers is formed that is macroscopically observed as a bright spatially localized *filament*.

It is important to distinguish the terms streamer and microdischarge. An initial electron starting from some point in the discharge gap (or from a cathode or a dielectric that covers the cathode) produces secondary electrons by direct ionization and develops an electron avalanche. If the avalanche is big enough (Meek condition, see above), the cathode-directed streamer is initiated. The streamer bridges the gap in a few nanoseconds and forms a conducting channel of weakly ionized plasma. Intensive electron current flows through this plasma channel until the local electric field collapses. Collapse of the local electric field is caused by the charges accumulated on the dielectric surface and ionic space charge (ions are too slow to leave the gap for the duration of this current peak). The group of local processes in the discharge gap initiated by avalanche and developed until electron current termination is usually called *microdischarge*.

After electron current termination there is no longer an electron-ion plasma in the main part of the microdischarge channel, but high levels of vibrational and electronic excitation in channel volume along with charges deposited on the surface and ionic charges in the volume allow us to separate this region from the rest of the volume, referring to it as *microdischarge remnant*. Positive ions (or positive and negative ions in the case of electronegative gas) of the remnant slowly move to electrodes resulting in low and very long ($\sim$10 μs for 1 mm gap) falling ion current.

The microdischarge remnant facilitates the formation of a new microdischarge in the same spot as the polarity of the applied voltage changes; this is why it is possible to see single filaments in a DBD. If microdischarges were to form at a new spot each time the polarity changed, the discharge would appear uniform. A filament in a DBD is a group of microdischarges that form on the same spot each time polarity is changed. The fact that the microdischarge remnant is not fully dissipated before formation of the next microdischarge is called the *memory effect*. Typical characteristics of DBD microdischarges in a 1 mm gap in atmospheric air are summarized in Table 4.3.

Charge accumulation on the surface of the dielectric barrier reduces the electric field at the location of a microdischarge. It results in current termination within just several nanoseconds after breakdown. The short duration of microdischarges leads to very low overheating of the streamer channel. Principal microdischarge properties for most frequencies do not depend on characteristics of external circuit, but only on gas composition, pressure and the electrode configuration. An increase of power leads to the generation of a larger number of microdischarges per unit time, which simplifies DBD scaling.

4.1.5   Interaction of streamers and microdischarges

The mutual influence of microdischarges in DBD are related to their electric interaction with residual charges left on the dielectric barrier, as well as the influence of excited species generated in one microdischarge on the formation of another microdischarge. The interaction of streamers and microdischarges is responsible for the DBD microdischarge patterns, which may have significant influence on the performance of plasma-medical treatment of living tissues.

Consider the propagation of streamers from anode to cathode. The resulting plasma channels have a net positive charge because electrons leave the gap much faster than ions. The residual positive charge (together with the deposited negative charge in the case of dielectric surface) influences the formation of nearby families of avalanches and streamers and, therefore, the formation of neighboring microdischarges. The positive charge (or dipole field in the case of deposited negative charge) intensifies electric field in the cathode area of the neighboring microdischarge and decreases electric field in the anode area. Since the avalanche-to-streamer transition depends mostly on near-anode electric field (from which new streamers originate), the formation of neighboring microdischarges is actually prevented and microdischarges effectively repel each other.

The *quasi-repulsion between microdischarges* leads to formation of short-range order that is related to a characteristic repulsion distance between microdischarges. Observation of this cooperative phenomenon depends on several factors, including the number of the microdischarges and the operating frequency. For example, when the number of microdischarges is not large enough (when average distance between microdischarges is larger than characteristic interaction radius), no significant microdischarge interaction is observed. Short voltage-rise time and significant overvoltage of the nanosecond pulsed discharges can lead to overlapping of the streamers and to plasma quasi-uniformity, which is especially important in plasma-medical systems focused on the safe treatment of living tissue.

4.1.6   Steady-state regimes of non-equilibrium electric discharges

Steady-state regimes of non-equilibrium discharges are provided by the balance of generation and losses of charged particles. The generation of electrons and positive ions is mostly due to volume ionization processes. To sustain the steady-state plasma, the ionization should be quite intensive; this usually requires electron temperature to be at least one-tenth that of ionization potential ($\sim$1 eV). Losses of charged particles can also be related to volume processes of recombination or attachment, also provided by diffusion of charged particles to the walls with further surface recombination. These two mechanisms of charge losses separate two different regimes of sustaining the steady-state discharge: the first controlled by volume processes and second by diffusion to the walls. If the ionization degree in plasma is relatively high and diffusion can be considered as ambipolar, the frequency of charge losses due to diffusion to the walls can be described as:

$$v_\mathrm{D} = \frac{D_\mathrm{a}}{\Lambda_\mathrm{D}^2}, \qquad (4.13)$$

where D_a is coefficient of ambipolar diffusion and Λ_D is characteristic diffusion length. Volume-related charge losses dominate, and hence non-equilibrium discharges are controlled by volume processes, when:

$$k_\mathrm{i}(T_\mathrm{e})n_0 \gg \frac{D_\mathrm{a}}{\Lambda_\mathrm{D}^2}. \qquad (4.14)$$

In this relation $k_\mathrm{i}(T_\mathrm{e})$ is the ionization rate coefficient and n_0 is neutral gas density. Criterion (4.14) actually restricts pressure, because $D_\mathrm{a} \propto 1/p$ and $n_0 \propto p$. When pressure is low, non-equilibrium discharges are controlled by diffusion to the walls and surface recombination. When pressure exceeds 10–30 Torr (the so-called range of moderate and high pressures), diffusion is relatively slow and the balance of charge particles is due to volume processes:

$$\frac{\mathrm{d}n_\mathrm{e}}{\mathrm{d}t} = k_\mathrm{i}n_\mathrm{e}n_0 - k_\mathrm{a}n_\mathrm{e}n_0 + k_\mathrm{d}n_0n_- - k_\mathrm{r}^\mathrm{ei}n_\mathrm{e}n_+, \qquad (4.15)$$

$$\frac{\mathrm{d}n_+}{\mathrm{d}t} = k_\mathrm{i}n_\mathrm{e}n_0 - k_\mathrm{r}^\mathrm{ei}n_\mathrm{e}n_+ - k_\mathrm{r}^\mathrm{ii}n_+n_-, \qquad (4.16)$$

$$\frac{\mathrm{d}n_-}{\mathrm{d}t} = k_\mathrm{a}n_\mathrm{e}n_0 - k_\mathrm{d}n_0n_- - k_\mathrm{r}^\mathrm{ii}n_+n_- \qquad (4.17)$$

where n_+, n_- are concentrations of positive and negative ions and n_e, and n_0 are concentrations of electrons and neutral species. The rate coefficients k_i, k_a, k_d, k_r^ei, k_r^ii are related to the processes of ionization by electron impact, dissociative or other electron attachment, electron detachment from negative ions, electron-ion and ion-ion recombination, respectively. The rate coefficients of processes involving neutral particles (k_i, k_a, k_d) are expressed in Equations (4.15)–(4.17) with respect to the total gas density. If the moderate or high-pressure gas is not electronegative, the volume balance of electrons and positive ions can be reduced to the simple ionization-recombination balance. In electronegative gases, however, two qualitatively different self-sustained regimes can be achieved (at different effectiveness of electron detachment): one controlled by recombination and the other by electron attachment.

4.1.7   Discharge regime controlled by electron-ion recombination

In some plasma-medical systems, the destruction of negative ions by, for example, associative electron detachment is faster than ion-ion recombination:

$$k_d n_0 \gg k_r^{ii} n_+. \tag{4.18}$$

In this case, actual losses of charged particles are also due to the electron-ion recombination, in the same way as for non-electronegative gases. Such a situation can occur, in particular, in plasma processes of CO_2 and H_2O dissociation and NO-synthesis in air, when associative detachment processes:

$$O^- + CO \rightarrow CO_2 + e, \quad O^- + NO \rightarrow NO_2 + e, \quad O^- + H_2 \rightarrow H_2O + e \tag{4.19}$$

are very fast (about 0.1 μs at concentrations of CO, NO and H_2 molecules of about 10^{17} cm^{-3}). Electron attachment and detachment (Equation (4.17)) are in dynamic quasi-equilibrium in the recombination regime during time intervals sufficient for electron detachment ($t \gg 1/k_d n_0$). The concentration of negative ions can then be considered in dynamic quasi-equilibrium with electron concentration:

$$n_- = \frac{k_a}{k_d} n_e = n_e \varsigma. \tag{4.20}$$

Using the parameter $\varsigma = k_a / k_d$ and presenting quasi-neutrality as $n_+ = n_e + n_- = n_e(1 + \varsigma)$, Equations (4.15)–(4.17) can be simplified to a single kinetic equation for electron density:

$$\frac{dn_e}{dt} = \frac{k_i}{1 + \varsigma} n_e n_0 - (k_r^{ei} + \varsigma k_r^{ii}) n_e^2. \tag{4.21}$$

The parameter $\varsigma = k_a / k_d$ shows the detachment ability to compensate for electron losses due to attachment. If $\varsigma \ll 1$, the attachment influence is negligible and the kinetic equation (4.21) becomes equivalent to that for non-electronegative gases. The kinetic equation includes the effective rate coefficients of ionization $k_i^{eff} = k_i / (1 + \varsigma)$ and recombination $k_r^{eff} = k_r^{ei} + \varsigma k_r^{ii}$. Equation (4.21) describes electron density evolution to the steady-state magnitude of the recombination-controlled regime:

$$\frac{n_e}{n_0} = \frac{k_i^{eff}(T_e)}{k_r^{eff}} = \frac{k_i}{(k_r^{ei} + \varsigma k_r^{ii})(1 + \varsigma)}. \tag{4.22}$$

4.1.8   Discharge regime controlled by electron attachment

This regime takes place if the balance of charged particles is due to the volume processes (Equation (4.14)) and the discharge parameters correspond to the inequality (opposite to Equation (4.18)). The discharge regime controlled by electron attachment is typical for atmospheric-pressure cold discharges in air, especially important for plasma-medical applications. Negative ions produced by electron attachment go almost instantaneously into ion-ion recombination, and electron losses are mostly due to the attachment process. The steady-state solution (Equation (4.15)) for the attachment-controlled regime is:

$$k_i(T_e) = k_a(T_e) + k_r^{ei} \frac{n_+}{n_0}. \tag{4.23}$$

In the attachment-controlled regime, the electron attachment is usually faster than recombination and Equation (4.23) actually requires $k_i(T_e) \approx k_a(T_e)$. Typical functions $k_i(T_e)$ and $k_a(T_e)$ are shown in Figure 4.6 and have a

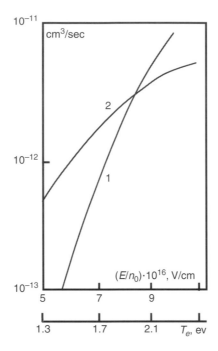

Figure 4.6 *Rate coefficients of ionization (1) and dissociative attachment (2) for CO_2.*

single crossing point T_{st}, which determines the steady-state electron temperature for a non-thermal discharge, self-sustained in the attachment-controlled regime.

4.1.9 Non-thermal discharge regime controlled by diffusion of charged particles to the walls

When pressure is relatively low and the opposite inequality to Equation (4.14) is valid, the balance of charged particles is provided by competition of ionization in volume and diffusion of charged particles to the walls, where they recombine on the surface. The balance of direct ionization by electron impact and ambipolar diffusion to the walls of long discharge chamber of radius R gives the following relation between electron temperature and pressure:

$$\left(\frac{T_e}{I}\right)^{1/2} \exp\left(\frac{I}{T_e}\right) = \frac{\sigma_0}{\mu_i p}\left(\frac{8I}{\pi m}\right)^{1/2}\left(\frac{n_0}{p}\right)(2.4)^2(pR)^2. \tag{4.24}$$

This is the *Engel-Steenbeck relation* for diffusion-controlled regime of non-equilibrium discharges. Here I is ionization potential, μ_i is ion mobility, σ_0 is electron-neutral gas-kinetic cross-section and m is the electron mass. If gas temperature is fixed (for example, at room temperature), the parameters $\mu_i p$ and n_0/p are constant and the Engel-Steenbeck relation can be written:

$$\sqrt{\frac{T_e}{I}} \exp\left(\frac{I}{T_e}\right) = C(pR)^2. \tag{4.25}$$

The constants in the Engel-Steenbeck relation depend on the type of gas, and are listed in Table 4.4.

Table 4.4 *Parameters of the Engel-Steenbeck relation.*

Gas	C	c	Gas	C	c
N_2	2×10^4	4×10^{-2}	Ar	2×10^4	4×10^{-2}
He	2×10^2	4×10^{-3}	Ne	4.5×10^2	6×10^{-3}
H_2	1.25×10^3	10^{-2}			

The universal relation between T_e/I and the similarity parameter cpR for the diffusion-controlled regime is presented in Figure 4.7. According to the Engel-Steenbeck curve, the electron temperature in the diffusion-controlled regimes decreases with growth of pressure and the radius of the discharge tube.

4.2 Glow discharge and its application to biology and medicine

4.2.1 Glow discharge structure

Glow discharge is the best-known and best-investigated type of non-thermal discharge. Most general features of cold plasma discharges can be understood by analyzing the glow discharge. The term 'glow' highlights that plasma of the discharge is luminous in contrast to the relatively low-power dark discharge. Glow discharge can be defined as the self-sustained continuous DC discharge with cold cathode, which emits electrons as a result of a secondary emission mostly induced by positive ions, depicted in Figure 4.8.

A distinctive feature of a glow discharge is the *cathode layer* with positive space charge, strong electric field and potential drop of about 100–500 V. The thickness of the cathode layer is inversely proportional to gas density. If the distance between the electrodes is large enough, a quasi-neutral plasma with low electric field, the so-called *positive column*, is formed between the cathode layer and the anode. The positive column of glow discharge is the most typical example of weakly ionized non-equilibrium low-pressure plasma, and is separated from the anode by an anode layer. The *anode layer* is characterized by a negative space charge, slightly elevated electric field and some potential drop. Typical glow discharge parameters are summarized in Table 4.5.

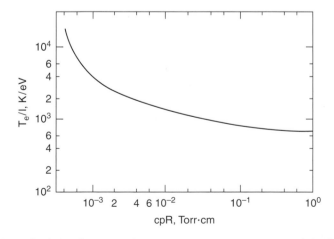

Figure 4.7 *Universal relation between electron temperature, pressure and discharge tube radius.*

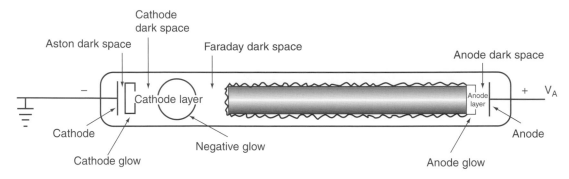

Figure 4.8   *General structure of a glow discharge in a long tube.*

Consider a pattern of light emission in a classical low-pressure discharge in a tube. A sequence of dark and bright luminous layers can be seen along such a discharge tube (Figure 4.9a). The typical size of the structure is proportional to electron mean free path $\lambda \propto 1/p$, and hence inversely proportional to pressure. It is therefore easier to observe the glow pattern at low pressures (of the scale centimeters at pressures of about 0.1 Torr).

Each layer of the glow pattern illustrated in Figure 4.9a has a particular name. Immediately adjacent to the cathode is a dark layer known as the *Aston dark space*. The relatively thin layer of the *cathode glow* follows, which is followed by the *cathode dark space*. The next zone is *negative glow*, which is sharply separated from the dark cathode space. The negative glow is gradually less and less bright toward the anode, becoming the *Faraday dark space*. The *positive column* then begins which is bright (but not as bright as the negative glow), uniform and long. Near the anode, the positive column is first transferred into *anode dark space* and finally into the narrow zone of the *anode glow*.

The glow pattern can be interpreted based on distribution of the discharge parameters shown in Figure 4.9b–g. Electrons are ejected from the cathode with insufficient energy (about 1 eV) for the excitation of atoms, which explains the Aston dark space. Electrons then obtain enough energy for electronic excitation from the electric field, which provides the cathode glow. Further acceleration of electrons in the cathode dark space leads mostly to ionization and not to electronic excitation; this explains the low level of radiation and the increase of electron density in the cathode dark space. Slowly moving ions have a higher concentration

Table 4.5   *Parameters of a conventional low-pressure glow discharge.*

Parameter of a glow discharge	Typical values
Discharge tube radius	0.3 – 3 cm
Discharge tube length	10–100 cm
Plasma volume	c. 100 cm^3
Gas pressure	0.03–30 Torr
Voltage between electrodes	100–1000 V
Electrode current	10^{-4}–0.5 A
Power level	around 100 W
Electron temperature in positive column	1–3 eV
Electron density in positive column	10^9–10^{11} cm^{-3}

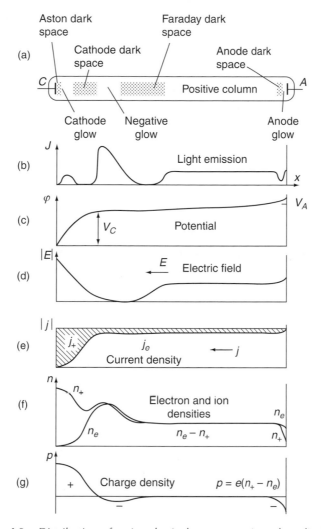

Figure 4.9 Distribution of major physical parameters in a glow discharge.

in the cathode layer and provide most of the current. High electron density at the end of the cathode dark space results in a decrease of the electric field, electron energy and ionization rate, but leads to intensification of radiation. This explains the transition to the brightest layer, the negative glow. Electron energy decreases further from the cathode, resulting in a transition from the negative glow to the Faraday dark space. There plasma density decreases and an electric field develops, establishing the positive column. Average electron energy in the positive column is about 1–2 eV, which provides light emission. The cathode layer structure remains the same if electrodes are moved closer, while the positive column shrinks. The positive column can be extended to a long length, connecting the electrodes. The anode repels ions and pulls out electrons from the positive column, which creates the negative space charge and leads to an increase of the electric field in the anode layer. A reduction of the electron density explains the anode dark space, while the electric field increase explains the anode glow.

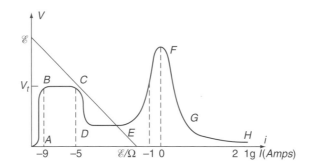

Figure 4.10   *Current voltage characteristic of DC discharges.*

4.2.2   Current-voltage characteristics of DC discharges

If the voltage between the electrodes exceeds the critical threshold value V_t necessary for breakdown, a self-sustained discharge can be ignited. The general current-voltage characteristic of such a discharge is illustrated in Figure 4.10 for a wide range of currents I. The electric circuit also includes an external ohmic resistance R, which results in Ohm's law usually being referred to as the *load line*:

$$\text{EMF} = V + RI \tag{4.26}$$

where EMF is electromotive force and V is voltage on the discharge gap.

The intersection of the current-voltage characteristic and the load line gives current and voltage in a discharge. If the external ohmic resistance is high and the current in the circuit is low (about $10^{-10} - 10^{-5}$ A), electron and ion densities are negligible and perturbation of the external electric field in plasma can be neglected. Such discharge is known as the *dark Townsend discharge*. The voltage necessary to sustain this discharge does not depend on current and coincides with the breakdown voltage. The dark Townsend discharge corresponds to the plateau BC in Figure 4.10. An increase of the EMF or a decrease of the external resistance R leads to a growth of the current and plasma density, which results in significant restructuring of the electric field. It leads to a reduction of voltage with current (interval CD in Figure 4.10) and to the transition from dark to glow discharge. This very low-current glow discharge is called the sub-glow discharge. Further EMF increase or R reduction leads to the lower-voltage plateau DE on the current-voltage characteristic, corresponding to the normal glow discharge existing over a large range of currents $10^{-4}/0.1$ A.

The current density on the cathode is fixed for normal glow discharges. An increase of the total discharge current is provided by growth of the cathode spot through which the current flows. When the current is so high that no more free surface is left on the cathode, further current growth requires a voltage increase to provide higher values of current density. Such a regime is called the abnormal glow discharge, which corresponds to the growing interval EF on the current-voltage characteristics (Figure 4.10). Further increase of current and voltage in the abnormal glow regime leads to higher power and to transition-to-arc discharge. The glow-to-arc transition usually takes place at currents of about 1 A.

4.2.3   Townsend dark discharge

A distinctive feature of the dark discharge is the smallness of its current and plasma density, which keeps the external electric field unperturbed and determined by Townsend breakdown condition (4.4). The relations

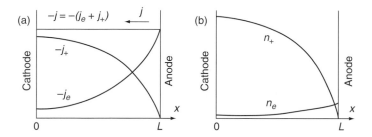

Figure 4.11 *(a) Current density and (b) electron/ion density distributions in a dark discharge.*

between electron (j_e), ion (j_+) and total (j) current densities in a dark discharge (distance between electrodes d_0 from cathode x and Townsend coefficient α) are:

$$\frac{j_e}{j} = \exp\left[-\alpha\left(d_0 - x\right)\right],$$

$$\frac{j_+}{j} = 1 - \exp\left[-\alpha\left(d_0 - x\right)\right], \tag{4.27}$$

$$\frac{j_+}{j_e} = \exp\left[-\alpha\left(d_0 - x\right)\right] - 1.$$

According to Equation (4.4), $\alpha d_0 = \ln[(\gamma + 1)/\gamma] \gg 1$ and therefore the ion current exceeds the electron current over the major part of the discharge gap (Figure 4.11a). The electron and ion currents become equal only near the anode ($j_e = j_+$ at $x = 0.85 d_0$). The difference in concentrations of electrons and ions is even stronger because of the additional big difference in electron and ion mobilities (μ_e, μ_+). Electron and ion concentrations become equal at a point very close to the anode (Figure 4.11b), where:

$$1 = \frac{n_+}{n_e} = \frac{\mu_e}{\mu_+}\frac{j_+}{j_e} = \frac{\mu_e}{\mu_+}\left[\exp \alpha\left(d_0 - x\right) - 1\right]. \tag{4.28}$$

Assuming $\mu_e/\mu_+ \approx 100$, electron and ion concentrations become equal at $x = 0.998$; almost the complete gap is charged positively. Dark-to-glow discharge transition at higher currents is due to growth of the positive space charge and distortion of the external electric field, which results in formation of the cathode layer. To describe the transition, the Maxwell equation can be used:

$$\frac{dE}{dx} = \frac{1}{\varepsilon_0}e\left(n_+ - n_e\right) \tag{4.29}$$

where $n_+ \approx j/e\mu_+ E \gg n_e$. Equation (4.29) therefore gives the following distribution of electric field:

$$E = E_c\sqrt{1 - \frac{x}{d}}, \quad d = \frac{\varepsilon_0 \mu_+ E_c^2}{2j}, \tag{4.30}$$

where E_c is the electric field at the cathode. The electric field decreases near the anode with respect to the external field and grows in the vicinity of the cathode, as is illustrated in Figure 4.12. Higher current densities lead to more distortion of the external electric field. The parameter d corresponds to a virtual point, where

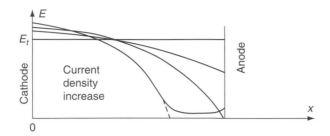

Figure 4.12   *Electric field evolution in a dark discharge.*

the electric field equals zero. This point is located far beyond the discharge gap ($d \gg d_0$) at the low currents typical for dark discharges. At higher current densities, the point of zero electric field reaches the anode ($d = d_0$). This critical current density is the maximum for the dark discharge, and corresponds to formation of the cathode layer and to the transition from dark to glow discharge:

$$j_{\max} = \frac{\varepsilon_0 \mu_+ E_c^2}{2d_0}. \tag{4.31}$$

In N_2 at pressure 10 Torr, inter-electrode distance 10 cm, electrode area 100 cm^2 and secondary emission coefficient $\gamma = 10^{-2}$, the maximum dark discharge current is $j_{\max} \approx 3 \times 10^{-5}$ A. The dependence of the Townsend coefficient α on the electric field is exponential and very strong. It leads to an interesting consequence: the typical voltage of a glow discharge is lower than that of a dark discharge (Figure 4.10). Growth of the positive space charge during the dark-to-glow transition results in redistribution of the initially uniform electric field: it becomes stronger near the cathode and lower near the anode. However, the increase of the exponential function $\alpha[E(x)]$ in the cathode side is more significant than its decrease in the anode side. The electric field non-uniformity therefore facilitates the breakdown conditions, which explains why the typical voltage of a glow discharge is usually lower than in a dark discharge.

4.2.4   Current-voltage characteristics of the cathode layer

The cathode layer is the most distinctive zone of a glow discharge. It provides its self-sustaining behavior and generates enough electrons to balance plasma current in the positive column. When voltage is applied to a discharge gap, the uniform distribution of electric field is not optimal; it is easier to sustain the discharge when a sufficiently high potential drop occurs near the cathode. The required electric field non-uniformity is provided by positive space charge formed near the cathode due to low ion mobility. A theory of the cathode layer has been developed by Engel and Steenbeck. Assuming zero electric field at the end of a cathode layer, neglecting ion current into a cathode layer from a positive column and considering the cathode layer as an independent system of length d, the Engel-Steenbeck theory gives the relations between the electric field E_c in the cathode layer, the cathode potential drop V_c and the length of cathode layer pd, which are similar to those describing breakdown of a gap, as:

$$V_c = \frac{B(pd)}{C + \ln(pd)}, \quad \frac{E_c}{p} = \frac{B}{C + \ln(pd)}, \tag{4.32}$$

where

$$C = \ln A - \ln \left[\ln \left(\frac{1}{\gamma} + 1 \right) \right]$$

and A and B are Townsend parameters (see Equation (4.5) and Table 4.1). V_c, E_c and similarity parameter pd depend on the discharge current density j, which is close to the ion current density because $j_+ \gg j_e$ near the cathode. To find out this dependence, the positive ion density $n_+ \gg n_e$ is first determined from the Maxwell equation:

$$n_+ \approx \frac{\varepsilon_0}{e} \left| \frac{dE(x)}{dx} \right| \approx \frac{\varepsilon_0 E_c}{ed}. \tag{4.33}$$

The total current density in the cathode vicinity is close to the current density of positive ions, which results in the Engel-Steenbeck current-voltage characteristics of the cathode layer:

$$j = en_+\mu_+ E \approx \frac{\varepsilon_0\mu_+ E_c^2}{d} \approx \frac{\varepsilon_0\mu_+ V_c^2}{d^3}. \tag{4.34}$$

The cathode potential drop V_c as a function of the similarity parameter pd corresponds to the Paschen curve for breakdown, thus the function $V_c(pd)$ has a minimum V_n. Taking into account Equation (4.34), the potential drop V_c as a function of current density j has also the same minimum point V_n. Relations between V_c, E_c, pd and j can be expressed using the dimensionless parameters:

$$\tilde{V} = \frac{V_c}{V_n}, \quad \tilde{E} = \frac{E_c/p}{E_n/p}, \quad \tilde{d} = \frac{pd}{(pd)_n}, \quad \tilde{j} = \frac{j}{j_n} \tag{4.35}$$

where electric field E_n/p and cathode layer length $(pd)_n$ correspond to the minimum point of the cathode voltage drop V_n. The subscript n represents the 'normal' regime of a glow discharge. All three 'normal' parameters E_n/p, $(pd)_n$ as well as V_n can be found from Equations (4.7) originally derived for electric breakdown as parameters of the Paschen curve. The corresponding value of the normal current density (in units of A cm^{-2} $Torr^{-2}$) can be derived from Equation (4.34) using the numeric formula constructed with similarity parameters:

$$\frac{j_n}{p^2} = \frac{1}{9 \times 10^{11}} \frac{(\mu_+ p) \times (V_n)^2}{4\pi \left[(pd)_n\right]^3}. \tag{4.36}$$

Relations between V_c, E_c, j and the cathode layer length pd can be expressed:

$$\tilde{V} = \frac{\tilde{d}}{1 + \ln \tilde{d}}, \quad \tilde{E} = \frac{1}{1 + \ln \tilde{d}}, \quad \tilde{j} = \frac{1}{\tilde{d} \left(1 + \ln \tilde{d}\right)^2}. \tag{4.37}$$

Voltage $\tilde{V}$, electric field $\tilde{E}$ and cathode layer length $\tilde{d}$ are presented in Figure 4.13 as functions of current density, which is referred to as the dimensionless current-voltage characteristic of a cathode layer. According to Equation (4.37), any current densities are possible in a glow discharge. In reality, a cathode layer 'prefers' to operate at the only value of current density (i.e. j_n) which corresponds to the minimum of the cathode

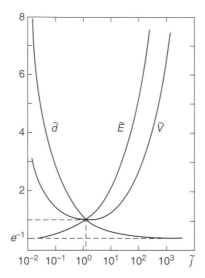

Figure 4.13   *Dimensionless parameters of a cathode layer.*

potential drop. It can be interpreted in terms of the *Steenbeck minimum power principle*. The current-conducting channel occupies a *cathode spot* with area $A = I/j_n$, which provides the normal current density. Other current densities are unstable, provided the cathode surface is big enough. A glow discharge with the normal cathode current density is referred to as the *normal glow discharge*. Normal glow discharges have fixed current density j_n (and corresponding fixed cathode layer thickness $(pd)_n$) and voltage V_n which, at room temperature, depend only on gas composition and cathode material. Normal glow discharge parameters are listed in Table 4.6.

A typical normal current density is 100 μA cm^{-2} at a pressure of about 1 Torr. The thickness of the normal cathode layer at this pressure is about 0.5 cm. The normal cathode potential drop is about 200 V and does not depend on pressure and temperature.

4.2.5   Abnormal, subnormal and obstructed regimes of glow discharges

An increase of the current in a normal glow discharge is provided by growth of the cathode spot at $j = j_n = \mathrm{const}$. As soon as the entire cathode is covered, further current growth results in an increase of current density over the normal value. This discharge is called an *abnormal glow discharge*. The abnormal glow discharge corresponds to the right-hand branches ($j > j_n$) in Figure 4.13. The current-voltage characteristic of the abnormal discharge $\tilde{V}(\tilde{j})$ is growing. It corresponds to the interval EF on the general current-voltage characteristic, Figure 4.10. According to Equation (4.35), when the current density is growing ($j \rightarrow \infty$) the cathode layer thickness decreases asymptotically to a finite value $\tilde{d} = 1/e \approx 0.37$ while the cathode potential drop and electric field grow as:

$$\tilde{V} = \frac{1}{e^{3/2}}\sqrt{\tilde{j}}, \quad \tilde{E} \approx \frac{1}{e^{1/2}}\sqrt{\tilde{j}}. \tag{4.38}$$

Actual growth of the current and cathode voltage are limited by the cathode overheating. Significant cathode heating at voltages of about 10 kV and current densities 10–100 A cm^{-2} results in transition of the abnormal glow discharge into an arc discharge. The normal glow discharge transition to a dark discharge takes place at

Table 4.6   *Normal current density, normal thickness of cathode layer and normal cathode potential drop for different gases and cathode materials at room temperature.*

Gas	Cathode material	Normal current density, j_n/p^2	Normal thickness of cathode layer, $(pd)_n$	Normal cathode potential drop, V_n
Air	Al	330	0.25	229
Air	Cu	240	0.23	370
Air	Fe	—	0.52	269
Air	Au	570	—	285
Ar	Fe	160	0.33	165
Ar	Mg	20	—	119
Ar	Pt	150	—	131
Ar	Al	—	0.29	100
He	Fe	2.2	1.30	150
He	Mg	3	1.45	125
He	Pt	5	—	165
He	Al	—	1.32	140
Ne	Fe	6	0.72	150
Ne	Mg	5	—	94
Ne	Pt	18	—	152
Ne	Al	—	0.64	120
H_2	Al	90	0.72	170
H_2	Cu	64	0.80	214
H_2	Fe	72	0.90	250
H_2	Pt	90	1.00	276
H_2	C	—	0.90	240
H_2	Ni	—	0.90	211
H_2	Pb	—	0.84	223
H_2	Zn	—	0.80	184
N_2	Pt	380	—	216
N_2	Fe	400	0.42	215
N_2	Mg	—	0.35	188
N_2	Al	—	0.31	180
O_2	Pt	550	—	364
O_2	Al	—	0.24	311
O_2	Fe	—	0.31	290
O_2	Mg	—	0.25	310

low currents (about 10^{-5} A) and starts with the *subnormal discharge*, which corresponds to the interval CD on the current-voltage characteristic (Figure 4.10). The size of the cathode spot at the low currents becomes comparable with the cathode layer thickness, which results in significant electron losses with respect to the normal glow; higher voltages are therefore required to sustain the discharge.

Another regime called an *obstructed glow* occurs at low pressures and narrow gaps when pd_0 is less than normal $(pd)_n$. The obstructed discharge corresponds to the left-hand branch of the Paschen curve, where voltage exceeds the minimum value V_n. The short inter-electrode distance in the obstructed discharge is not sufficient for effective multiplication of electrons, so the inter-electrode voltage should be greater than the normal voltage in order to sustain the mode.

4.2.6   Positive column of glow discharge

The positive column can be long, homogeneous and releasing most of the discharge power. Its physical function is simple: closing the electric circuit between cathode layer and anode. Non-equilibrium behavior of the discharge $(T_e \gg T)$ is controlled by heat balance, which can be illustrated as:

$$w = jE = n_0 c_p (T - T_0) \nu_T \tag{4.39}$$

where w is the power per unit volume; c_p is the specific heat; T and T_0 are the positive column and room temperatures; and ν_T is the cooling frequency for cylindrical discharge tube of radius R and length d_0, defined:

$$\nu_T = \frac{8}{R^2} \frac{\lambda}{n_0 c_p} + \frac{2u}{d_0}, \tag{4.40}$$

where λ is the coefficient of thermal conductivity and u is the gas velocity. The first term in Equation (4.40) is related to thermal conductivity and the second term describes convective heat removal. If heat removal is controlled by thermal conduction, the discharge power causing a doubling of temperature $(T - T_0 = T_0)$ is:

$$w = jE = \frac{8\lambda T_0}{R^2}. \tag{4.41}$$

The thermal conductivity coefficient λ does not depend on pressure and can be estimated as $\lambda \sim 3 \times 10^{-4}$ W (cm K)$^{-1}$. Specific discharge power as defined by Equation (4.41) therefore does not depend on pressure, and can be estimated as 0.7 W cm^{-3} for $R = 1$ cm. Higher powers result in higher gas temperatures and glow discharge contraction. Typical current density in the positive column with the heat removal controlled by thermal conduction is inversely proportional to pressure:

$$j = \frac{8\lambda T_0}{R^2} \frac{1}{(E/p)} \frac{1}{p}. \tag{4.42}$$

If $E/p = 3/10$ V (cm Torr)$^{-1}$, then $j \sim 100/p$ in units of mA cm^{-2}. The corresponding electron density in the positive column can be calculated from Ohm's law $j = \sigma E$:

$$n_e = \frac{w}{E^2} \frac{m \nu_{en}}{e^2} = \frac{w}{(E/p)^2} \frac{m \, k_{en}}{e^2 T_0} \frac{1}{p} \tag{4.43}$$

where ν_{en}, k_{en} are the frequency and rate coefficient of electron-neutral collisions. Numerically, $n_e \sim 3 \times 10^{11}/p$ (in units of cm^{-3}) in a positive column with conductive heat removal. The reduction of the plasma ionization degree with pressure is significant: $n_e/n_0 \propto 1/p^2$. Low pressures are therefore generally more favorable for sustaining the steady-state homogeneous non-thermal plasma.

4.2.7   Atmospheric pressure glow discharges and applications in plasma medicine

Glow discharges usually operate at low gas pressures. If the steady-state discharge cooling is controlled by conduction (the diffusive regime), the pressure increase is limited by overheating. The maximum current density and electron concentration of glow discharges proportionally decrease with pressure, and the ionization degree decreases as square of pressure, that is, $n_e/n_0 \propto 1/p^2$. Higher pressures can be reached by operating the glow discharges in fast flows. However, these pressures are also limited by plasma instabilities because their very short induction times should be comparable or longer than the gas residence times in the discharge.

Operating a stable, continuous, non-thermal glow discharge at atmospheric pressure is a challenging task. Organization of the glow discharges at atmospheric pressure is especially interesting for biomedical applications, however, keeping in mind the uniformity of the glow discharges and relatively low voltage required to sustain them. Atmospheric pressure glow (APG) discharges have been accomplished in several special discharge systems such as glow discharges in transonic and supersonic flows. Application of special aerodynamic techniques allows the uniform steady-state glow discharges to be sustained at atmospheric pressure and specific energy input up to 500 J g^{-1}. The glow discharge de-contraction at atmospheric pressure becomes possible due to suppression of the transverse diffusion influence on the temperature and current density distribution (see details in Fridman and Kennedy, 2011).

Another approach to organization of the glow discharge at atmospheric pressure is related to the use of special gas mixtures as working fluid, additionally elaborating some special types of electrodes. Such gas mixtures are assumed to be able to provide the necessary level of the ionization rate at relatively low values of reduced electric field E/p and are suitable to sustain glow discharge operation at atmospheric pressure.

Gases for such discharges usually include different inert gas mixtures. The reduced electric field necessary to sustain a glow discharge in inert gases can be 30 or more than times less than that for molecular gases; this is due to the absence of electron energy losses to vibrational excitation. Mixtures of helium or neon with argon or mercury are very effective for ionization when the Penning effect takes place. In this case, hundreds of volts are sufficient to operate a glow discharge at atmospheric pressure. Use of helium is also helpful to increase heat exchange and cooling of the systems.

The normal cathode current density is proportional to the square of pressure and becomes large at elevated pressures. For this reason, special types of electrodes should be applied, for example, fine wires or barrier discharge type electrodes. These are necessary to avoid overheating and correlate the current density on the cathode and in the discharge volume. While atmospheric pressure glow discharges are physically similar to traditional non-thermal atmospheric pressure discharges such as corona or DBD, their voltage can be much less. Generally, the glow mode of DBD can be operated at lower voltages (down to hundreds volts); streamers are avoided as the electric fields are below the Meek criterion and discharge operates in the Townsend ionization regime. Secondary electron emission from dielectric surfaces, which sustains the Townsend ionization regime, relies upon adsorbed electrons (with binding energy only about 1 eV) that were deposited during previous DBD excitation (high voltage) cycle.

If enough electrons 'survive' voltage switching time without recombining, they can trigger transition to the homogeneous Townsend mode of DBD. Survival of electrons and crucial active species between cycles or the DBD memory effect is critical for organization of APG, and depends on the properties of the dielectric surface as well as operating gas. In electronegative gasses, the memory effect is weaker because of attachment losses of electrons. If the memory effect is strong, the transition to Townsend mode can be accomplished and uniform discharge can be generated without streamers.

A streamer DBD is easy to produce, while organization of APG at the same conditions is not always possible. This can be explained by considering that the streamer discharge is not sensitive to the secondary electron emission from dielectric surface, while it is critical for operation of APG. Glow discharges usually undergo contraction with increase of pressure due to the thermal instability. The thermal instability in the DBD-APG is somewhat suppressed by using alternating voltage. Discharge therefore only operates when voltage is high enough to satisfy the Townsend criteria; the rest of time the discharge is idle, which allows the dissipation of heat and active species.

If time between excitation cycles is not enough for the dissipation then an instability will develop and discharge will undergo a transition to filamentary mode. The avalanche-to-streamer transition in the APG-DBD depends on the level of pre-ionization. Meek criterion is related to an isolated avalanche while, in the case of intensive pre-ionization, avalanches are produced close to each other and interact. If two avalanches occur close enough, their transition to streamers can be electrostatically prevented and discharge remains uniform. A modified Meek criterion of the avalanche-to-streamer transition can be obtained considering two simultaneously starting avalanches with maximum radius R, separated by distance L:

$$\alpha d - (R/L)^2 \approx \text{const}, \tag{4.44}$$

$$\alpha d \approx \text{const} + n_e^p R^2 d \tag{4.45}$$

where α is the Townsend coefficient and d is distance between electrodes.

In Equation (4.45), distance between avalanches is approximated using the pre-ionization density n_e^p. The constant in Equations (4.45) and (4.46) depends on gas: in air at 1 atm this constant equals 20. According to the modified Meek criterion, the avalanche-to-streamer transition can be avoided by increasing avalanche radius and by providing sufficient pre-ionization. The type of gas is important in the transition to the APG. Helium is relevant for the purpose: it has high-energy electronic excitation levels and no electron energy losses on vibrational excitation, resulting in higher electron temperatures at lower electric fields. Also, fast heat and mass transfer processes prevent contraction and other instabilities at high pressures.

The space-uniform APG can be effectively applied for the treatment of living tissue, especially when a small amount of oxygen in added to helium.

4.2.8 Resistive barrier discharge (RBD) as modification of APG discharges

The resistive barrier discharge (RBD) as a modification of the APG discharges can be operated with DC or AC (60 Hz) power supplies and is based on DBD configuration, where the dielectric barrier is replaced by highly resistive sheet (few $M\Omega$ cm^{-1}) covering one or both electrodes. The system can consist of a top wetted high-resistance ceramic electrode and a bottom electrode. The highly resistive sheet plays the role of distributed resistive ballast, which prevents high currents and arcing. If He is used and the gap distance is not too large (5 cm and below), a spatially diffuse discharge can be maintained in the system for several tens of minutes. If 1% of air is added to helium, the discharge forms filaments. Even when driven by a DC voltage, the current signal of RBD is pulsed with the pulse duration of few microseconds at a repetition rate of a few tens of kiloHertz. When the discharge current reaches a certain value, the voltage drop across the resistive layer becomes large to the point where the voltage across the gas became insufficient to sustain the discharge. The discharge extinguishes and current drops rapidly; voltage across the gas then increases to a value sufficient to reinitiate the RBD. The RBD discharge can be used to effectively sterilize water and other liquids, as well as for plasma-medical treatment of different liquids.

4.2.9 Atmospheric pressure micro glow-discharges

Scaling down with a constant similarity parameter (pd) should not change the properties of discharges significantly. Conventional non-equilibrium discharges are operated in the pd range of around 10 cm Torr, therefore organization of strongly non-equilibrium discharges at atmospheric pressure should be effective in the sub-millimeter sizes. Some specific new properties can be achieved by scaling down the plasma size to sub-millimeters:

1. Size reduction of non-equilibrium plasmas permits an increase of their power density to a level typical for the thermal discharges (because of intensive heat losses of the tiny systems)

2. At high pressures, volumetric recombination and especially three-body processes can go faster than diffusion losses, which results in a significant change of plasma composition. For example, the high-pressure microdischarges can contain significant amount of molecular ions in noble gases.
3. Sheathes (about 10–30 μm at atmospheric pressure) occupy a significant portion of the plasma volume.
4. Plasma parameters move to the 'left' side of the Paschen curve. The Paschen minimum is about $pd = 3$ cm Torr for some gases, therefore a 30 μm gap at 1 atm corresponds to the left side of the curve.

The specific properties can lead to the positive differential resistance of a microdischarge, which allows many discharges to be supported in parallel from a single power supply without using multiple ballast resistors.

Another important consequence is relatively high electron energy in the microplasmas, which are mostly strongly non-equilibrium. As an example, atmospheric DC micro glow-discharge has been generated between a thin cylindrical anode and a flat cathode. The discharge has been studied using an inter-electrode gap spacing in the range of 20 μm to 1.5 cm so that the influence of the discharge scale on plasma properties could be observed. Current-voltage characteristics, visualization of the discharge and estimations of the current density indicate that the discharge operates in the normal glow regime. Emission spectroscopy and gas temperature measurements using the second positive band of N_2 indicate that the discharge generates non-equilibrium plasma. For 0.4 mA and 10 mA discharges, rotational temperatures are 700 and 1550 K while vibrational temperatures are 5000 and 4500 K, respectively.

It is possible to distinguish a negative glow, Faraday dark space and positive column regions of the discharge (see Figure 4.14; significant investigations of the micro-glow discharge has been accomplished by a Drexel group led by Professors Farouk and Fridman, and especially by D. Staack). The radius of the column is about 50 μm and remains relatively constant with changes in the electrode spacing and discharge current. Such radius allows the heat generation and conductive cooling to be balanced to help prevent thermal instability and the transition to an arc. Generally, there is no significant change in the current-voltage characteristics of the discharge for different electrode materials or polarity. There are several notable exceptions to this for certain configurations as follows.

1. For a thin upper electrode wire (<100 μm) and high discharge currents, the upper electrode melts. This occurs when the wire is the cathode, indicating that the heating is due to energetic ions from the cathode sheath and not resistive heating.

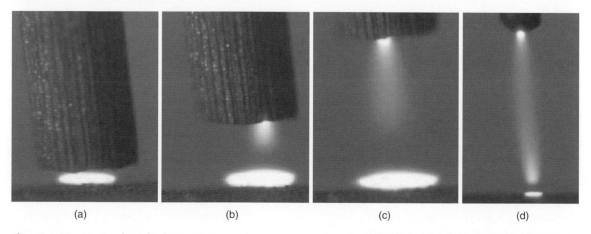

(a)   (b)   (c)   (d)

Figure 4.14 *Micro glow-discharge in atmospheric-pressure air with electrode spacing: (a) 0.1 mm, (b) 0.5 mm, (c) 1 mm and (d) 3 mm.*

2. For a medium-sized wire (~200 μm) as a cathode, the width of the negative glow increases as the current increases until it covers the entire lower surface of the wire. If the current is further increased, the negative glow 'spills over' the edge of the wire and begins to cover the side of the wire. This effect is similar to the transition from a normal glow to an abnormal glow in low-pressure glow discharges. However, there is no increase in the current density since the cathode area is not limited. For sufficiently large electrode wires, this effect does not occur.

3. In air discharges with oxidizable cathode materials, the negative glow moves around the cathode electrode leaving a trail of oxide coating behind until there is no clean surface within the reach of the discharge and the discharge extinguishes. For a small spacing of electrodes, the current-voltage characteristics are relatively flat, which is consistent with the normal discharge mode. For a normal glow discharge in air, the potential drop at the normal cathode sheath is around 270 V which occurs mostly in the positive column.

For larger electrode spacing, the current-voltage characteristics have a negative differential resistance dV/dI. This is due to the discharge temperature increase with gap length resulting in growth of conductivity. A short discharge loses heat through the thermal conductivity of electrodes. A long discharge cooling is not efficient because the thermal conductivity of the gas is much lower than that of metal electrodes; the temperature of the long discharge is therefore higher. Such behavior demonstrates a new property related to the size reduction to microscale. Diffusive heat losses can balance the increased power density only at elevated temperatures of a microdischarge, and the traditionally cold glow discharge becomes 'warm'. Table 4.7 summarizes the micro glow-discharge parameters corresponding to currents 0.4 and 10 mA.

The atmospheric-pressure DC microdischarge is a normal glow discharge thermally stabilized by its size and maintaining a high degree of vibrational-translational non-equilibrium.

The micron-sized precise micro glow-discharges or their arrays can be effectively used for direct microscale treatment of living tissues.

Table 4.7   *Micro glow-discharge parameters in air.*

Microdischarge and microplasma parameters	Microdischarge current (mA)	
	0.4	10
Electrode spacing (mm)	0.05	0.5
Microdischarge voltage (V)	340	380
Microdischarge power (W)	0.136	3.8
Diameter of negative glow (μm)	39	470
Positive column diameter (μm)	—	110
Electric field in the positive column (kV cm^{-1})	5.0	1.4
Translational gas temperature (K)	700	1550
Vibrational gas temperature (K)	5000	4500
Negative glow current density (A cm^{-2})	33.48	5.8
Positive column current density (A cm^{-2})	—	105
Reduced electric field E/n_0 (V cm^2)	4.8×10^{-16}	3×10^{-16}
Electron temperature T_e (eV)	1.4	1.2
Electron density n_e in negative glow (cm^{-3})	3×10^{13}	7.2×10^{12}
Electron density n_e in positive column (cm^{-3})	—	1.3×10^{14}
Ionization degree in negative glow	3×10^{-6}	15×10^{-7}
Ionization degree in positive column	—	3×10^{-5}

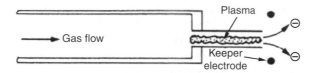

Figure 4.15 *Schematic of the Lidsky capillary hollow cathode.*

4.2.10   Hollow-cathode glow discharge and hollow-cathode APG microplasma

The *hollow-cathode discharge* (HCD) is an electron source based on intensive 'non-local' ionization in the negative glow zone of the glow discharges, where the electric field is not very high but quite a few electrons are very energetic. These energetic electrons were formed in the cathode vicinity and crossed the cathode layer with only a few inelastic collisions. They provide the non-local ionization and lead to electron densities in negative glow exceeding that in the positive column.

Imagine a glow discharge with a cathode arranged as two parallel plates and the anode to the side. If we gradually decrease the distance between the cathodes, the current at some point grows 100–1000 times without any substantial change of voltage. The effect occurs when the two negative glow regions overlap, accumulating energetic electrons from both cathodes.

Strong photoemission from the cathodes also contributes to the intensification of ionization. The hollow cathode can be arranged as a cylinder with the anode located further along the axis. The magnitude of the pressure should be selected such that the cathode layer thickness is comparable with the internal diameter of the hollow cylinder. The most traditional configuration of the system is the *Lidsky hollow cathode* (Figure 4.15), which is a narrow capillary-like nozzle operating with axially flowing gas and anode located about 1 cm downstream. The Lidsky hollow cathode is hard to initiate, but it provides electron current densities exceeding the limits of the Child law.

The HCD can be effectively generated in the microscale at atmospheric pressure. Similarly to the conventional hollow-cathode discharges, micro-HCDs are interesting for applications because of their ability to generate high-density plasma. While conventional HCDs are generated at low pressures and macroscale, the micro-HCDs can operate at atmospheric pressure in agreement with the (pD)-similarity. The micro-HCD can be effectively arranged in the form of special arrays. If (pD) is in the range 0.1–10 Torr cm, the discharge develops in stages.

At low currents, a 'pre-discharge' – which is a glow discharge with the cathode falling outside the hollow-cathode structure – is observed. As the current increases and the glow discharge starts its transformation into the abnormal glow with a positive differential resistance, a positive space-charge region moves closer to the hollow-cathode structure and can enter the cavity. After that, the positive space charge in the cavity acts as a virtual anode, resulting in the redistribution of the electric field inside the cavity. A potential well for electrons appears at the center of the cavity, forming a cathode sheath along the cavity walls. At this transition from the axial pre-discharge to a radial discharge, the sustaining voltage drops. Sometimes this transition is not very sharp; in this case a negative slope in the current-voltage characteristic (i.e. a negative differential resistance) appears, which is traditionally referred as the 'hollow-cathode mode'.

The micron-sized precise micro hollow-cathode glow-discharges or their arrays can be effectively used for direct microscale treatment of living tissues at atmospheric pressure.

4.3   Arc discharge and its medical applications

4.3.1   Major types of arc discharges

Arc discharges are the most conventional sources of thermal plasma, characterized by very high gas temperatures often exceeding 10 000 K. Plasma-medical applications of arc discharges are therefore mostly

Table 4.8   *Plasma parameters typical for thermal and not-completely-thermal arcs.*

Discharge plasma parameter	Thermal arc discharge	Non-completely-thermal arc
Gas pressure	0.1–100 atm	10^{-3}–10^2 Torr
Arc current (A)	30–30 000	1–30
Cathode current density (A cm^{-2})	10^4–10^7	10^2–10^4
Voltage (V)	10–100	10–100
Power per unit length (kW cm^{-1})	>1	<1
Electron density (cm^{-3})	10^{15}–10^{19}	10^{14}–10^{15}
Gas temperature	1–10 eV	300–6000 K
Electron temperature (eV)	1–10	0.2–2

focused on coagulation of blood (surgical cauterization), surgical ablation, removal and cutting of tissue. Arc discharges in air are able to generate significant amounts of nitrogen oxides, which also results in application of these thermal plasma sources in NO therapy.

As a rule, arcs are self-sustaining DC discharges with relatively low cathode fall voltage of about 10 V. Arc cathodes emit electrons by intensive *thermionic* and *field emission*. Arc cathodes receive a large amount of Joule heating from the discharge current and are therefore able to reach very high temperatures, which leads to evaporation and erosion of electrodes.

The classical type of arc discharge is the *voltaic arc*, which is the carbon electrode arc in atmospheric air. Arc discharges were first discovered in this form. Cathode and anode layer voltages in the voltaic arc are both about 10 V; the rest of the voltage corresponds to the positive column. Arcs can be sustained in thermal and non-completely-thermal regimes; cathode emission is thermionic in the non-completely-thermal regimes and field emission in thermal arcs. The reduced electric field E/p is low in thermal arcs and higher in non-completely-thermal arcs. The total arc voltage is low, sometimes only tens of volts. The ranges of plasma parameters are listed in Table 4.8.

The thermal arcs operating at high pressures are much more energy intensive, having higher currents and current densities and higher power per unit length. A variety of DC-discharges with low cathode-fall voltage are usually considered as arc discharges and classified by principal cathode and positive column mechanisms as follows.

− *Hot thermionic cathode arcs*: The entire cathode in such arcs has a temperature of 3000 K or above, which provides high current due to thermionic emission. The arc is stationary connected to the same fixed and quite large cathode spot. Current is distributed over a relatively large cathode area and its density is not so high (about 10^2–10^4 A cm^{-2}). Only special refractory materials such as carbon, tungsten, molybdenum, zirconium, tantalum and so on can withstand such high temperatures. The hot thermionic cathode can be heated to the high temperature not only by the arc current, but also by an external source of heating. Such cathodes are applied in low-pressure arcs and in thermionic converters in particular. The cathodes in such arc discharges are usually activated to decrease the temperature of thermionic emission.

− *Arcs with hot cathode spots*: If the cathode is made from lower-melting-point metals such as copper, iron, silver or mercury, the high temperature required for emission cannot be sustained permanently. Electric current flows in this case through hot spots, which appear to move fast and disappear on the cathode surface. Current density in the spots is extremely high (10^4–10^7 A cm^{-2}) which leads to intensive but local and short heating and evaporation of the cathode material, while the rest of the cathode actually stays cold. Mechanism of electron emission from the spots is thermionic field emission. Cathode spots appear not only on the low-melting-point cathodes, but also on refractory metals at low currents and low pressures.

- *Vacuum arcs*: This type of low-pressure arc, operating with cathode spots, is special because the gas-phase working fluid is provided by erosion and evaporation of the electrode material. This type of arc is of importance in high-current electrical equipment, high-current vacuum circuit breakers and switches.
- *High-pressure arc discharges*: The positive column of an arc plasma is in quasi-equilibrium at pressures exceeding 0.1–0.5 atm. Thermal arcs also operate at very high pressures exceeding 10 atm. The plasma in this case is so dense that most of the discharge power (80–90%) is converted into radiation.
- *Low-pressure arc discharges*: The positive column of arc discharges at low pressures $10^{-3}/1$ Torr consists of non-equilibrium plasma. Ionization degree in the non-thermal arcs is higher than in glow discharges because arc currents are much larger (see Table 4.8).

The current-voltage characteristic of continuous DC discharges in a wide range of currents is presented in Figure 4.10; The transition from glow to arc corresponds to the interval FG. Current density in abnormal glow increases, resulting in cathode heating and growth of thermionic emission which determines the glow-to-arc transition. The glow-to-arc transition is continuous in the case of the thermionic cathodes made from refractory metals, and takes place at currents of about 10 A. Cathodes made from low-melting-point metals provide the transition at lower currents (0.1–1 A). This transition is sharp, unstable and accompanied by the formation of hot cathode spots. When the current grows, the electric field and voltage decreases until a critical point of sharp voltage reduction, followed by an almost horizontal current-voltage characteristic. The transition is accompanied by a specific hissing noise, which is the formation of hot anode spots with intensive evaporation.

4.3.2   Cathode and anode layers of arc discharges and spots

The function of the cathode layer is to provide the high current necessary for arc operation. Electron emission from cathode in arcs is due to thermionic and field emission, which are much more intensive than ion-induced secondary electron emission which dominates in the cathode layers of glow. In the case of thermionic emission, ion bombardment provides cathode heating which then leads to escape of electrons from the surface. Secondary electron emission gives about $\gamma \approx 0.01$ electrons per ion, while thermionic emission can generate much more $\gamma_{\text{eff}} = 2 - 9$. The fraction of electron current S near the cathode in a glow discharge is very small $(\gamma/(\gamma + 1) \approx 0.01)$, but in the cathode layer of an arc we have:

$$S = \frac{\gamma_{\text{eff}}}{\gamma_{\text{eff}} + 1} \approx 0.7 \text{ to } 0.9. \tag{4.46}$$

Thermionic emission from the cathode provides most of electric current in the arc. Current in the positive column of both arc and glow discharges is almost completely provided by high-mobility electrons. In contrast to a glow, the direct electron-impact ionization in the cathode layer of an arc discharge should provide only a minor fraction of the total discharge current $1 - S \sim 10$–30%. The cathode voltage in arcs is therefore relatively low, similar to or even less than the ionization potential. Sufficient ion density is generated in the cathode layer to provide necessary cathode heating in the case of thermionic emission. Gas temperature near the cathode is the same as cathode surface temperature and about half that of the temperature in the positive column (Figure 4.16). Thermal ionization is therefore unable to provide the necessary ionization degree and additional non-thermal ionization is required.

This leads to an elevated electric field near the cathode, which stimulates electron emission by decrease of work function (*Schottky effect*) and contribution of field emission. Intensive ionization in the cathode vicinity leads to a high concentration of ions in the layer and to the formation of a positive space charge, which actually provides the elevated electric field. The distribution of arc parameters (temperature, voltage and electric field) along the discharge from cathode to anode is illustrated in Figure 4.16.

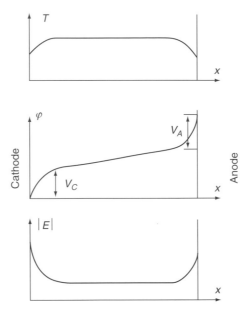

Figure 4.16   *Distributions of temperature, potential and electric field from cathode to anode.*

The structure of the cathode layer is illustrated in Figure 4.17. A large positive space-charge, with high electric field and most of the cathode voltage, is located in the narrow layer by the cathode. This layer is shorter than the mean free path and referred to as the *collisionless zone of the cathode layer*.

Between the narrow collisionless layer and positive column, the longer *quasi-neutral zone of cathode layer* is located. Electric field there is not so high, but ionization is intensive because electrons keep the high energy they received in the collisionless layer. Most of the ions carrying current and energy to the cathode are generated here. Electron and ion currents are constant in the collisionless layer, where there are no sources of charge particles. The electron fraction of current grows from $S \approx 0.7 - 0.9$ in the cathode layer to about 1 in the positive column. Plasma density in the cathode layer grows in the direction of the positive column. Electron j_e and ion j_+ components of the total current density j in the collisionless zone are:

$$j_e = Sj = n_e e v_e, \quad j_+ = (1 - S)j = n_+ e v_+. \tag{4.47}$$

Electron and ion velocities v_e, v_+ can be presented as a function of voltage V, assuming $V = 0$ at the cathode and $V = V_C$ at the end of the collisionless layer:

$$v_e = \sqrt{2eV/m}, \quad v_+ = \sqrt{2e(V_C - V)/M}. \tag{4.48}$$

where m and M are masses of electrons and ions. Based on Equations (4.47) and (4.48), Poisson's equation for the voltage V in the collisionless layer is:

$$-\frac{d^2 V}{dx^2} = \frac{e}{\varepsilon_0}(n_+ - n_e) = \frac{j}{\varepsilon_0 \sqrt{2e}} \left[\frac{(1-S)\sqrt{M}}{\sqrt{V_C - V}} - \frac{S\sqrt{m}}{\sqrt{V}} \right]. \tag{4.49}$$

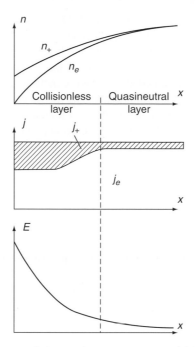

Figure 4.17   *Distributions of charge density, current and field in the cathode layer.*

Taking into account that

$$\frac{d^2V}{dx^2} = \frac{1}{2}\frac{dE^2}{dV},$$

Poisson's equation (4.49) can be integrated assuming $E \approx 0$ at $V = V_C$ as a boundary condition. This results in the relation between the electric field near cathode, current density and the cathode voltage drop V_C:

$$E_C^2 = \frac{4j}{\varepsilon_0\sqrt{2e}}\left[(1-S)\sqrt{M} - S\sqrt{m}\right]\sqrt{V_C}. \tag{4.50}$$

This relation can be rewritten by taking into account $S = 0.7$ to 0.9 in the numerical form:

$$E_C = 5 \times 10^3 A^{1/4}(1-S)^{1/2}(V_C)^{1/4}(j)^{1/2}, \tag{4.51}$$

where A is atomic mass of ions in a.m.u and E_C is measured in units of V cm^{-1}. For example, arc discharge in nitrogen ($A = 28$) at typical values of current density for hot cathodes, that is, $j = 3 \times 10^3$ A cm^{-2}, cathode voltage drop $V_C = 10$ V and $S = 0.8$ gives, according to Equation (4.51), the electric field near cathode $E_C = 5.7 \times 10^5$ V cm^{-1}. The length of the collisionless zone of the cathode layer can be expressed:

$$\Delta l = \frac{4V_C}{3E_C}. \tag{4.52}$$

Numerically, the length of the collisionless layer is about $\Delta l \approx 2 \times 10^{-5}$ cm. Arcs can be connected to the anode either by diffuse connection or by anode spots. The diffuse connection occurs on the large area of the anodes at current densities of about 10 A cm^{-2} The anode spots appear on small and non-homogeneous anodes and current densities within the spots are 10^4–10^5 A cm^{-2}. The number of spots grows with current and pressure. The anode spots can be arranged and moved in patterns.

The anode voltage drop consists of two components. The first is related to negative space charge near the anode, which repels ions. This small voltage drop (of about ionization potential) stimulates some additional electron generation to compensate for the absence of ion current in the region. The second voltage component is related to the arc discharge geometry. If the anode surface is smaller than the positive column, the current near the electrode should be provided only by electrons. It requires higher electric fields near the electrode and an additional anode voltage drop. Each electron brings to the anode an energy of about 10 eV, so the energy flux to the anode spot at 10^4–10^5 A cm^{-2} is about 10^5–10^6 W cm^{-2}. The temperature at the anode spots of vacuum metal arcs is about 3000 K and in the carbon arc about 4000 K.

The cathodes spots are the localized current centers which appear on the cathode when significant current should be provided, but the cathode cannot be heated enough as a whole. The most typical cause of the cathode spots is application of metals with relatively low melting point. The cathode spots can also be caused by low arc currents, which are only able to provide the necessary electron emission when concentrated within a small area. The cathode spots also appear at low gas pressures (<1 Torr) when metal vapor from the cathode provides atoms to generate positive ions, bringing their energy to the cathode to sustain the electron emission. To provide requited evaporation, current is concentrated in spots at pressures <1 Torr and currents 1–10 A; such spots even occur on refractory metals.

4.3.3   Positive column of high-pressure arcs

The Joule-heating power per unit length of positive column in high-pressure arcs is quite significant 0.2–0.5 kW cm^{-1}. Such plasma in molecular gases ($p \geq 1$ atm) is usually in quasi-equilibrium at any currents. In inert gases, the electron-neutral energy exchange is less effective and requires high currents and electron densities to reach quasi-equilibrium (Figure 4.18). Temperatures of electrons and neutrals can differ at low pressures ($p \leq 0.1$ Torr) and currents ($I \approx 1$ A). Electric field E is constant along the positive column and defines the voltage. The current-voltage characteristics are hyperbolic, which indicates that Joule heat per unit length $w = EI$ does not change significantly with current I. $w = EI$ grows with pressure due to intensification of heat transfer mostly related to radiation. The contribution of radiation increases proportionally to the square of plasma density, and hence grows with pressure. The arc radiation losses in atmospheric air are about 1%, but they become significant at pressures above 10 atm and for higher arc power. The highest level of radiation can be reached in Hg, Xe and Kr, which is applied in mercury and xenon lamps. Formulae for the calculation of plasma radiation are listed in Table 4.9.

The fraction of power conversion into radiation is high in mercury and xenon, even at low values of the Joule heat per unit length. Temperature distribution in a long cylindrical steady-state thermal plasma column stabilized by walls in a tube of radius R is described by the *Elenbaas-Heller equation*, assuming heat transfer across the positive column is provided by heat conduction with the coefficient $\lambda(T)$. According to the Maxwell equation curl $E = 0$ and the electric field in a long homogeneous arc column is constant across its cross-section. Radial distributions of electric conductivity $\sigma(T)$, current density $j = \sigma(T)E$ and Joule heating density $w = jE = \sigma(T)E^2$ are determined only by radial temperature distribution $T(r)$, therefore:

$$\frac{1}{r}\frac{d}{dr}\left[r\lambda(T)\frac{dT}{dr}\right] + \sigma(T)E^2 = 0. \tag{4.53}$$

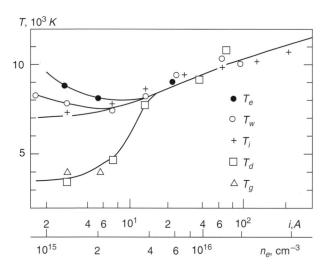

Figure 4.18   *Temperature separation in the positive arc column in Ar with an admixture of H_2 at p = 1 atm as a function of the current density or electron density. T_e is the electron temperature, T_w corresponds to the population of the upper levels, the ion temperature T_i is related to n_e by the Saha formula, T_g is the gas temperature and T_d corresponds to the population of the lower levels.*

Boundary conditions for Equation (4.53) are: $dT/dr = 0$ at $r = 0$, and $T = T_w$ at $r = R$. The experimentally controlled parameter is current I, defined:

$$I = E \int_0^R \sigma T(r) \, 2\pi r \, dr. \tag{4.54}$$

The Elenbaas-Heller equation (4.53) together with (4.54) permits calculations of the function $E(I)$, the current-voltage characteristic of the plasma column. This is determined by two material functions: electric conductivity $\sigma(T)$ and thermal conductivity $\lambda(T)$. To reduce the number of the material functions, it is convenient to introduce the *heat flux potential* $\Theta(T)$, where:

$$\Theta = \int_0^T \lambda(T) \, dT, \quad \lambda(T)\frac{dT}{dr} = \frac{d}{dr}\Theta. \tag{4.55}$$

Table 4.9   *Radiation power per unit length of positive column of arc discharges at different pressures and values of joule heat per unit length $w = EI$.*

Gas	Pressure (atm)	Radiation power per unit length (W cm^{-1})	w
Hg	≥ 1	$0.72(w-10)$	—
Xe	12	$0.88(w-24)$	>35
Kr	12	$0.72(w-42)$	>70
Ar	1	$0.52(w-95)$	>150

Using the heat flux potential, the Elenbaas-Heller equation can be simplified as follows:

$$\frac{1}{r}\frac{d}{dr}\left(r\frac{d\Theta}{dr}\right) + \sigma(\Theta)E^2 = 0, \tag{4.56}$$

but not analytically solved because of the non-linearity of the material function $\sigma(\Theta)$.

4.3.4   Steenbeck – Raizer 'channel' model of positive column of arc discharges

The Steenbeck-Raizer channel model is based on the strong dependence of plasma electric conductivity on temperature. At temperatures below 3000 K, plasma conductivity is low; it grows significantly only when the temperature exceeds 4000 K. The temperature decrease $T(r)$ from the axis to walls is gradual, while the conductivity change with radius $\sigma[T(r)]$ is sharp. According to the model, arc current is located mostly in a channel of radius r_0. Temperature and electric conductivity are considered as constant inside the arc channel, and equal to their maximum value on the discharge axis: T_m and $\sigma(T_m)$. The total arc current can then be expressed as:

$$I = E\sigma(T_m)\pi r_0^2. \tag{4.57}$$

Outside of the channel, that is, $r > r_0$, electric conductivity, current and Joule heating can be neglected. Based on the Elenbaas-Heller equation, the arc power per unit length can be expressed as a function of maximum temperature T_m on the axis of the arc channel, relevant heat conductivity and ionization potential I_i:

$$w = 8\pi\lambda_m(T_m)\frac{T_m^2}{I_i}. \tag{4.58}$$

The arc temperature as a function of specific discharge power w per unit length (with ionization potential I_i and thermal conductivity coefficient λ_m as parameters) is:

$$T_m = \sqrt{w\frac{I_i}{8\pi\lambda_m}}. \tag{4.59}$$

The temperature does not depend directly on tube radius, but only on specific power w. The principal parameter of an arc discharge is the current. Assuming that $\lambda = \text{const}$, $\Theta = \lambda T$, the conductivity in the arc channel is proportional to current:

$$\sigma_m = I\sqrt{\frac{I_i C}{8\pi^2 R^2 \lambda_m T_m^2}}. \tag{4.60}$$

Plasma temperature in the arc channel grows with current I, but only logarithmically:

$$T_m = \frac{I_i}{\ln\left(8\pi^2\lambda_m C T_m^2/I_i\right) - 2\ln(I/R)}. \tag{4.61}$$

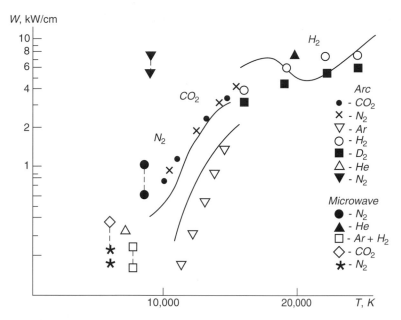

Figure 4.19   *Dissipated power per unit length versus maximum temperature for different types of thermal discharges (solid lines represent numerical calculations).*

The weak logarithmic growth of temperature in the channel with electric current leads to the similar weak logarithmic dependence on the electric current of the arc discharge power w per unit length:

$$w \approx \frac{\text{const}}{(\text{const} - \ln I)^2}.$$
(4.62)

Experimental data demonstrating the dependence are presented in Figure 4.19. Taking into account that $w = EI$, the decrease of the electric field in the positive column with current I is close to hyperbolic:

$$E = \frac{8\pi \lambda_m T_m^2}{I_i} \frac{1}{I} \approx \frac{\text{const}}{I(\text{const} - \ln I)^2}.$$
(4.63)

This explains the hyperbolic decrease of current-voltage characteristics typical for thermal arc discharges.

4.3.5   Configurations of arc discharges and their applicability to plasma medicine

Because of the very high temperatures of the arc discharges, their plasma-medical applications are mostly focused on coagulation of blood (surgical cauterization), surgical ablation, removal and cutting of tissue. The organization of arc discharges in air results in significant generation of nitrogen oxides, which is applied for different aspects of NO-therapy and plasma-NO-therapy. We now consider the major configurations of these thermal plasma sources.

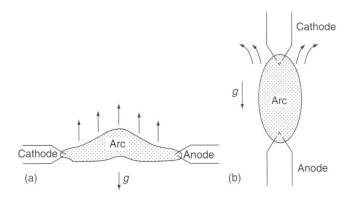

Figure 4.20   *Illustration of (a) horizontal and (b) vertical free-burning arc.*

4.3.5.1   *Free-burning linear arcs*

Free-burning linear arcs are the simplest axisymmetric electric arcs burning between two electrodes. They can be arranged in horizontal and vertical configurations (Figure 4.20). Sir Humphrey Davy first observed such a horizontal arc at the beginning of the 19th century. Buoyancy of hot gases in the horizontal arc leads to bowing up or 'arcing' of the plasma channel, which explains the term 'arc'. If the free-burning arc is vertical (Figure 4.20b), the cathode is usually placed at the top; buoyancy then provides more intensive cathode heating. If the arc length is shorter than the diameter (Figure 4.21), the discharge is referred to as the *obstructed arc*. The distance between electrodes in such discharges is typically about 1 mm; nevertheless, voltage exceeds the anode and cathode drops. The obstructed arcs are electrode-stabilized.

   The important advantage of linear free-burning arcs is their simplicity. They can be effectively applied as generators of nitrogen oxides for the purposes of NO-therapy and plasma-NO-therapy.

4.3.5.2   *Wall-stabilized linear arcs*

Wall-stabilized linear arcs are widely used for gas heating. A simple schematic of the wall-stabilized arc with a unitary anode is shown in Figure 4.22. The cathode is axial in this configuration, and the unitary anode is hollow and coaxial. The arc is axisymmetric and stable with respect to perturbations. If the arc channel is asymmetrically perturbed and approaches a coaxial anode, it leads to intensification of the discharge cooling and to a temperature increase on the axis (according to the Elenbaas-Heller equation). The increase of temperature results in displacement of the arc channel back on the axis of the discharge tube. The arc can attach the anode at any point along the axis. A better-defined discharge arrangement can be achieved in the so-called segmented wall-stabilized arc configuration; see Figure 4.23. The anode walls in this system are water-cooled and electrically segmented and isolated. Such a configuration provides a linear decrease of axial

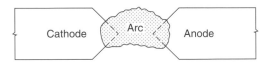

Figure 4.21   *Illustration of the obstructed electrode-stabilized arc.*

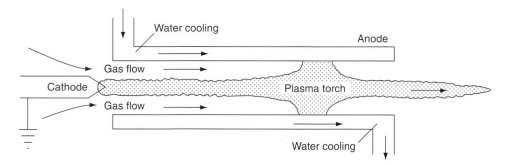

Figure 4.22 *Illustration of the wall-stabilized arc.*

voltage, and forces the arc attachment to the wider anode segment (the furthest from the cathode). The length of other segments is set small enough to avoid breakdown between them.

The above-described non-transferred arc jets are the engineering base for the thermal plasma-based cauterization devices, as well as plasma devices for tissue cutting and ablation.

4.3.5.3 *Transferred arcs*

Transferred arcs with water-cooled non-consumable cathodes are illustrated in Figure 4.24. The generation of electrons on the inner walls of the hollow cathodes is provided by field emission, which permits operation of the transferred arcs at the multi-megawatt power for thousands of hours. The electric circuit is completed by transferring the arc to an external anode (which is a conducting material) where the arc is to be applied. The arc root can move over the cathode surface, which further increases its lifetime.

4.3.5.4 *Flow-stabilized linear arcs*

For flow-stabilized linear arcs, the arc channel can be stabilized on the axis of the discharge chamber by radial injection of cooling water or gas. Such a configuration, illustrated in Figure 4.25, is usually referred to as the *transpiration-stabilized arc*. This discharge is similar to the segmented wall-stabilized arc, but the transpiration of cooling fluid through annular slots between segments increases the lifetime of the interior segments.

Another linear arc configuration providing high power of the discharge is the so-called *coaxial flow-stabilized arc*, illustrated in Figure 4.26. In this case the anode is located far from the main part of the plasma

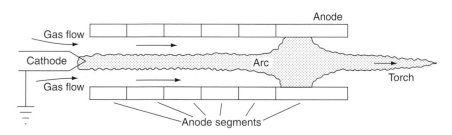

Figure 4.23 *Wall-stabilized arc with segmented anode.*

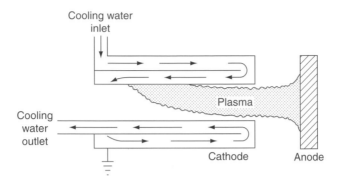

Figure 4.24   *Illustration of the transferred arc configuration.*

channel and cannot provide wall stabilization. Instead of a wall, the arc channel is stabilized by a coaxial gas flow moving along the outer surface of the arc. Such stabilization is effective with essentially no heating of the discharge chamber walls, if the coaxial flow is laminar. Similar arc stabilization can be achieved using a flow rotating around the arc column.

Different configurations of *vortex-stabilized arcs* are shown in Figure 4.27. The arc channel is stabilized by a vortex gas flow, which is introduced from a special tangential injector. The vortex gas flow cools the edges of the arc, and keeps the arc column confined to the axis of the discharge chamber. Applicability of the flow-stabilized arcs to plasma medicine is similar to that of the wall-stabilized arcs.

4.3.5.5   *Non-transferred arcs*

A non-linear wall-stabilized non-transferred arc is depicted in Figure 4.28. It consists of a cylindrical hollow cathode and coaxial hollow anode located in a water-cooled chamber and separated by an insulator. Gas flow blows the arc column out of the anode opening to heat a downstream material, which is assumed to be treated. In contrast to transferred arcs, the treated material is not assumed to operate as anode. Magnetic $\vec{I} \times \vec{B}$ forces cause the arc roots to rotate around electrodes (Figure 4.28), which provides longer-lifetime electrodes. The generation of electrons on the cathode is provided in this case by field emission.

An axisymmetric version of the non-transferred arc, usually referred to as the *plasma torch* or the *arc jet*, is illustrated in Figure 4.29. The arc is generated in a conical gap in the anode, and pushed out of this opening by gas flow. The heated gas flow forms a very-high-temperature arc jet which is sometimes at supersonic velocities.

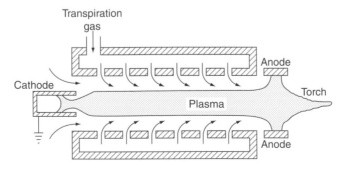

Figure 4.25   *Illustration of the transpiration-stabilized arc.*

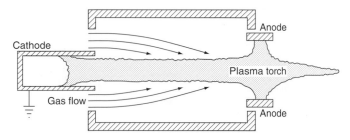

Figure 4.26   *Illustration of the coaxial flow-stabilized arc.*

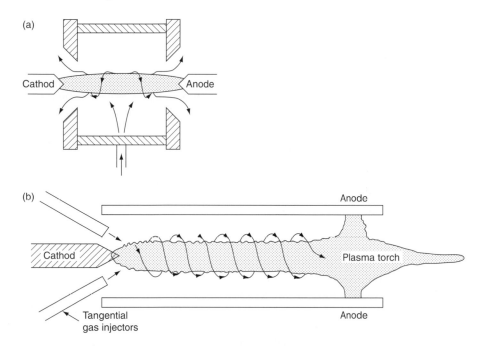

Figure 4.27   *Illustration of vortex-stabilized arcs: (a) side gas injection and (b) tangential gas injection.*

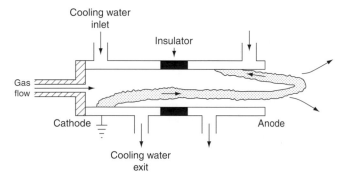

Figure 4.28   *Illustration of the non-transferred arc.*

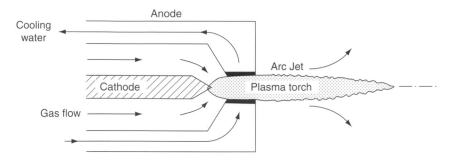

Figure 4.29   *Illustration of the plasma torch.*

4.3.5.6   *Magnetically stabilized rotating arcs*

Magnetically stabilized rotating arcs are illustrated in Figure 4.30. An external axial magnetic field provides $\vec{I} \times \vec{B}$ forces, which cause arc rotation and protect the anode from local overheating. Figure 4.30(a) shows magnetic stabilization of the wall-stabilized arc; Figure 4.30(b) shows a magnetically stabilized plasma torch. Magnetic stabilization is an essential supplement to the wall and aerodynamic stabilization. Protecting the electrodes from overheating (by all different means described above) also results in a decrease of the metal erosion and the following yield into the thermal plasma jet, a very important issue for all plasma-medical applications of arc discharges.

4.3.6   Gliding arc discharge as a powerful source of non-equilibrium plasma

Gliding arc is an auto-oscillating periodic discharge between at least two diverging electrodes submerged in gas flow (Figure 4.31). Self-initiated in the upstream narrowest gap, the discharge forms the plasma column connecting the electrodes. This column is dragged by the gas flow towards the diverging downstream section. The arc grows with the increase of inter-electrode distance until it extinguishes, but reignites itself at the minimum distance between the electrodes to start a new cycle. The gliding arc plasma can be either thermal or non-thermal, depending on the power and flow rate. Along with completely thermal and non-thermal modes, it is possible to operate the arc in the transitional regime when the discharge starts as thermal but, during the

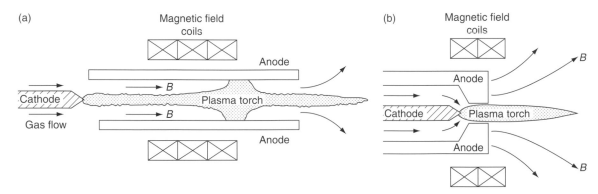

Figure 4.30   *Illustration of magnetically stabilized arcs: (a) open anode configuration and (b) restricted-flow anode configuration.*

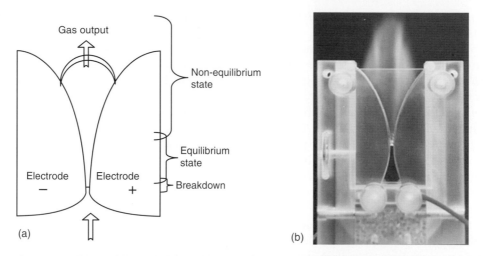

Figure 4.31   *(a) Illustration and (b) picture of gliding arc discharge.*

space and time evolution, becomes a non-thermal discharge. The powerful and energy-efficient transitional discharge combines the benefits of both equilibrium and non-equilibrium discharges.

Thermal plasma discharges have been designed for many diverse applications covering a wide range of operating power levels from less than 1 kW to over 50 MW. However, despite providing sufficient power levels, these do not appear to be well adapted to the purposes of plasma chemistry, biology and medicine, where selective treatment of reactants and high energy efficiency are required.

An alternative approach is the non-thermal discharge, which offers good selectivity and energy efficiency of chemical processes but is usually limited to low pressures and powers. In the two general types of discharge it is impossible to simultaneously keep a high level of non-equilibrium high electron temperature and high electron density, whereas a significant number of important prospective plasma-chemical and plasma-medical applications require high power for high reactor productivity and a high degree of non-equilibrium to support selective chemical processes at the same time. This can be achieved in the transitional regime of the gliding arc, which can be operated at atmospheric pressure or higher, with power at non-equilibrium conditions reaching up to 40 kW per electrode pair.

A simple electrical scheme of the DC gliding arc is shown in Figure 4.32. The high-voltage generator (up to 5 kV) is used in this scheme to ignite the discharge. The second generator is the power generator with

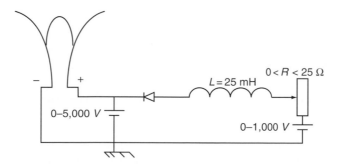

Figure 4.32   *Typical gliding arc discharge electrical scheme.*

voltages up to 1 kV and a total current I up to 60 A. A variable resistor $R = 0$–25 Ω is in series with a self-inductance $L = 25$ mH.

Initial breakdown at the shortest distance (1–2 mm) between the two electrodes begins the cycle of the gliding arc evolution. For atmospheric-pressure air and 1 mm distance between the electrodes, the breakdown voltage V_b is about 3 kV. After about 1 μs, low-resistance plasma is formed and the voltage between the electrodes falls.

The *equilibrium stage* occurs after formation of the plasma channel. The gas flow pushes the plasma column with velocity of about 10 m s^{-1}. The length l of the arc increases together with voltage; the power increases up to the maximum value P_{max} provided by the power supply. Electric current increases up to its maximum value of $I_m = V_0/R \approx 40$. During the quasi-equilibrium stage, gas temperature T_0 does not change significantly (remaining at around 3000 K).

The *non-equilibrium stage* begins when the length of the gliding arc exceeds its critical value l_{crit}. Heat losses from the plasma column begin to exceed the energy supplied by the source, and it is not possible to sustain the plasma in quasi-equilibrium. The plasma then rapidly cools down to $T_0 = 1000$–2000 K, while conductivity is sustained by high electron temperature $T_e = 1$ eV. After decay of the non-equilibrium discharge, a new breakdown takes place at the shortest distance between the electrodes and the cycle repeats. Assuming specific power w remains constant allows a description of the evolution of the current, voltage and power during the gliding arc quasi-equilibrium phase. Neglecting the self-inductance L, Ohm's law can be written:

$$V_0 = RI + wl/I \tag{4.64}$$

where V_0, R and I are open-circuit voltage of the power supply, external resistance and current. The arc current can be determined from Ohm's law as a function of the growing arc length l:

$$I = \frac{V_0 \pm \sqrt{V_0^2 - 4wlR}}{2R}. \tag{4.65}$$

The solution with + describes the steady state of the gliding arc column; the solution with $I < V_0/2R$ is unstable and corresponds to negative differential resistance ρ ($=dV/dJ$) of the circuit. The current described by Equation (4.65) decreases slightly during the quasi-equilibrium period while the arc voltage grow as Wl/I and the total arc power $P = wl$ increases almost linearly with length l. The quasi-equilibrium evolution of gliding arc is terminated when the arc length approaches the critical value:

$$l_{crit} = \frac{V_0^2}{4wR}, \tag{4.66}$$

and the square root in Equation (4.65) becomes equal to zero. The current decreases at this point to its minimal stable value of $I_{crit} = V_0/2R$, which is one-half of the initial current. Plasma voltage, electric field E and total power at the same critical point approach their maximum values, defined:

$$V_{crit} = V_0/2, \quad E_{crit} = w \overleftrightarrow{} /I_{crit}, \quad W_{crit} = V_0^2/4R. \tag{4.67}$$

Plasma resistance at the critical point becomes equal to the external resistance, and the maximum value of the discharge power equals one-half that of the maximum generator power. When the arc length exceeds the critical value $l > l_{crit}$, the heat losses wl continue to grow. Power from the power supply can no longer be increased, and the gas temperature rapidly decreases. Plasma conductivity can still be maintained by electron

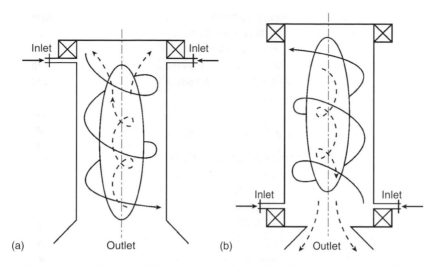

Figure 4.33   *Plasma stabilization by (a) forward and (b) reverse vortex flows.*

temperatures (≈ 1 eV) and stepwise ionization. The fast equilibrium-to-non-equilibrium transition is due to the increase of electric field $E = w/I$ and electron temperature T_e:

$$T_e = T_0 \left(1 + \frac{E^2}{E_i^2} \right), \tag{4.68}$$

where E_i corresponds to the transition from thermal to direct electron impact ionization. After the fast transition, the gliding arc continues evolution under the non-equilibrium conditions $T_e \gg T_0$. Up to 70–80% of the total power can be dissipated in the non-equilibrium plasma phase with $T_e \approx 1$ eV and $T_0 < 1000 - 1500$ K. Such gliding arc evolution is possible only if the electric field during the transition is high enough and the current is not too high.

The gliding arc discharges can be generated in various ways. An important option for plasma-medical applications is gliding arc stabilization in the reverse vortex (tornado) flow. This approach is opposite to the conventional *forward-vortex stabilization* (Figure 4.33a), where the swirl generator is placed upstream with respect to discharge and the rotating gas provides the walls with protection from the heat flux. Some reverse axial pressure gradient and central reverse flow appears due to fast flow rotation and strong centrifugal effect near the gas inlet, which becomes a bit slower and weaker downstream. The hot reverse flow mixes with incoming cold gas and increases heat losses to the walls, which makes the insulation of the discharge walls less effective.

More effective wall insulation is achieved by the *reverse-vortex stabilization* illustrated in Figure 4.33b. In this case, the outlet of the plasma jet is directed along the axis to the swirl generator side. Cold incoming gas moves first by the walls providing their cooling and insulation; after that it goes to the central plasma zone and becomes hot. In the case of reverse-vortex stabilization, the incoming gas entering the discharge zone from all directions except the outlet side, which makes it effective for gas heating/conversion efficiency and for protection of discharge walls. A picture of the gliding arc discharge trapped in the reverse vortex (tornado) flow is shown in Figure 4.34.

The non-equilibrium gliding arc tornado discharge is effectively applied for sterilization, and especially for sterilization of water. It is also applied for plasma activation of water and different solutions for their further plasma-medical usage (so-called plasma pharmacology).

Figure 4.34   *Non-equilibrium gliding arc discharge moving along a spiral electrode and stabilized in the reverse vortex 'tornado' flow.*

4.4   Radio-frequency and microwave discharges in plasma medicine

4.4.1   Generation of thermal plasma in radio-frequency discharges

Although arcs are the most conventional generators of thermal plasma, more expensive radio frequency (RF) discharges are also effective for application, especially when it is important to avoid impurities related to direct contact between plasma and electrodes. Thermal plasma generation in inductively-coupled plasma (ICP) discharges is illustrated in Figure 4.35. High-frequency electric current passes through a solenoid coil providing axial magnetic field, which induces the vortex electric field sustaining the RF-ICP discharge. The magnetic field in the ICP discharge is determined by current in the solenoid, while electric field according to the Maxwell equations is also proportional to the frequency. To achieve sufficient electric fields, the RF

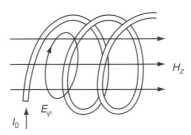

Figure 4.35   *Generation of the inductively coupled plasma.*

frequency 0.1–100 MHz is usually required. A dielectric tube is usually inserted inside the solenoid coil to sustain the thermal plasma in the gas of interest. To avoid interference with radio-communication systems, specific frequencies f were assigned for operation of the industrial RF-discharges; 13.6 MHz (wavelength 22 m) is used mostly.

The RF-ICP discharges are effective in sustaining thermal plasma at atmospheric pressure. Low electric fields are however not sufficient here for ignition of the discharges, which requires for example insertion of an additional rod-electrode heated and partially evaporated by the Foucault currents. RF-ICP discharge plasma should be coupled as a load with RF-generator. Electrical parameters of the plasma load, such as resistance and inductance, influence operation of the electric circuit as a whole and determine effectiveness of the coupling.

Relevant analysis of the ICP parameters can be accomplished in frameworks of the *metallic cylinder model*, where plasma is considered as a cylinder with the fixed conductivity σ $(\Omega\ \text{cm})^{-1}$ corresponding to the maximum temperature T_m on the axis. Thermal plasma conductivity is high and the skin effect prevents deep penetration of the electromagnetic field. Heat release related to the inductive currents is therefore localized in the relatively thin skin layer of the column. Thermal conductivity λ_m inside the cylinder provides the temperature plateau in the central part of the discharge where the inductive heating is negligible. Electromagnetic flux S_0 (W cm^{-2}) and power per unit surface of the long cylindrical ICP column can be calculated in frameworks of the metallic cylinder model as a function of current I_0 in the solenoid and number n of turns per unit length of the coil:

$$S_0 = 9.94 \times 10^{-2}(I_0 n)^2 \sqrt{\frac{f}{\sigma}}. \tag{4.69}$$

Plasma temperature in the metallic cylinder T_m can be expressed in terms of ionization potential I_i:

$$2\sqrt{2}\lambda_m \frac{T_m^2}{I_i}\sigma_m = I_0^2 n^2. \tag{4.70}$$

Conductivity σ_m strongly depends on temperature, while T_m and λ_m change only slightly; therefore $\sigma_m \propto (I_0 n)^2$. The Saha equation results in a weak logarithmic dependence of $T_m(I_0, n)$:

$$T_m = \frac{\text{const}}{\text{const} - \ln(I_0 n)}. \tag{4.71}$$

The dependence of ICP power per unit length on solenoid current and frequency can be shown as:

$$w \propto I_0 n \sqrt{\omega}. \tag{4.72}$$

Typical atmospheric-pressure thermal RF-ICP discharge ($f = 13.6$ MHz, tube radius 3 cm) has a temperature of about 10 000 K, thermal conductivity $\lambda_m = 1.4 \times 10^{-2}$ W $(\text{cm K})^{-1}$, electric conductivity $\sigma_m \approx 25$ $(\Omega\ \text{cm})^{-1}$ and skin layer $\delta \approx 0.27$ cm. The electromagnetic flux to sustain such plasma is $S_0 \approx 250$ W cm^{-2}, with corresponding solenoid current and number of turns $I_0 n \approx 60$ and magnetic field $H_0 \approx 6$ kA m^{-1}. Electric field on the external plasma boundary is about 12 V cm^{-1}, the density of circular current is 300 A cm^{-2}, total current per unit length of the column is about 100 A cm^{-1}, thermal flux potential is about 0.15 kW cm^{-1}, the distance between the effective plasma surface and discharge tube is $\Delta r \approx 0.5$ cm and finally, ICP-discharge power per unit length is about 4 kW cm^{-1}. The ICP discharge, where magnetic field is 'primary', has lower electric fields than the *capacitively-coupled plasma* (CCP) discharge, where the electric field is primary. For this reason, the ICP-discharges at moderate–high pressures usually generate thermal plasma, while the RF-CCP discharges in the same conditions can generate non-equilibrium plasma.

4.4.2   Atmospheric-pressure microwave discharges and their biomedical applications

In contrast to RF discharges, microwave plasma is sustained by the centimeter-range electromagnetic waves interacting with plasma in a quasi-optical way. The optical discharges are sustained by laser radiation with even shorter wavelengths. Thermal plasma generation in microwave and optical discharges is usually related to high-pressure systems. Microwave generators (in particular magnetrons) operating with power exceeding 1 kW in the giga-Hertz frequency range, are able to maintain the steady-state thermal microwave discharges at atmospheric pressure. Electromagnetic energy in the microwave discharges can be coupled with plasma in different ways. The most typical coupling is provided in waveguides, where the dielectric tube transparent to the electromagnetic waves (usually quartz) crosses the rectangular waveguide. Plasma is ignited and maintained inside the tube. Different modes of electromagnetic waves, formed in the rectangular waveguide, can be used to operate the microwave discharges. The most typical mode is H_{01}, where the electric field is parallel to the narrow walls of the waveguide and its maximum E_{max} (kV cm^{-1}) is related to power P_{MW} (kW) as:

$$E_{max}^2 = \frac{1.51 P_{MW}}{a_w b_w} \left[1 - \left(\frac{\lambda}{\lambda_{crit}} \right)^2 \right]^{-1/2} \tag{4.73}$$

where λ is wavelength; λ_{crit} is its maximum when propagation is still possible ($\lambda_{crit} = 2a_w$); and a_w and b_w are lengths of the waveguide walls in cm.

The H_{01} mode is convenient for microwave plasma generation because electric field in this case has its maximum in the center of the discharge tube. To provide plasma stabilization, gas flow is usually supplied tangentially. The waveguide dimensions are related to the microwave frequency: if $f = 2.5$ GHz ($\lambda = 12$ cm), the wide waveguide wall should be longer than 6 cm, usually 7.2 cm. The narrow waveguide wall is typically 3.4 cm long, the dielectric tube diameter 2 cm and the diameter of the plasma about 1 cm. When microwave power at atmospheric pressure is 1–2 kW, the temperature in molecular gases is c. 4000–5000 K, which is lower with respect to RF-ICP discharges.

The incident electromagnetic wave interacts with plasma, which results in partial dissipation and reflection of the electromagnetic wave. Typically about half of the incident microwave power is dissipated in plasma, about quarter is transmitted and another quarter is reflected. To increase the coupling effectiveness to 90–95%, the transmitted wave can be reflected back. Parameters of the thermal microwave discharge at frequency $f = 10$ GHz ($\lambda = 3$ cm) in atmospheric-pressure air are listed in Table 4.10.

The depth of microwave absorption in plasma l_ω grows significantly with temperature reduction; at $T = 3500$ K it reaches 2 cm. At lower temperatures, microwave plasma becomes transparent for the electromagnetic wave. High microwave power should be provided to sustain such plasma. The minimal temperature of the microwave discharges can be estimated from $l_\omega(T_{min}) = R_f$, where R_f is plasma radius. It is about 4200 K for the filament radius of 3 mm; the corresponding absorbed flux is $S_1 = 0.2$ kW cm^{-2} and minimal microwave energy flux to sustain the thermal plasma is about $S_0 = 0.25$ kW cm^{-2}. The maximum temperature is limited by reflection at high plasma conductivity. This effect is seen in Table 4.10: the reflection coefficient becomes very high $\rho = 81\%$ at $T_m = 6000$ K. For this reason, the quasi-equilibrium temperature of the microwave plasma in atmospheric air does not usually exceed 5000–6000 K.

Waveguide mode-converters can be applied for better coupling of microwave radiation with plasma; such a plasma torch is depicted in Figure 4.36. Microwave power is first delivered here by a rectangular waveguide,

Table 4.10 *Characteristics of atmospheric-pressure microwave discharge plasma in air at frequency* $f = 10$ *GHz ($\lambda = 3$ cm) as a function of plasma temperature.*

Plasma temperature, T_m	4500 K	5000 K	5500 K	6000 K
Electron density, n_e (10^{13} cm^{-3})	1.6	4.8	9.3	21
Plasma conductivity, σ_m (10^{11} s^{-1})	0.33	0.99	1.9	4.1
Thermal conductivity, λ (10^{-2} W cm^{-1} K^{-1})	0.95	1.1	1.3	1.55
Refractive index n of plasma surface	1.3	2.1	2.8	4.3
Plasma attenuation coefficient, κ	2.6	4.7	7.3	11
Depth of the microwave absorption layer, l_ω ($1/\mu_\omega$, 10^{-2} cm)	9.1	5.0	3.2	2.2
Energy flux absorbed, S_1 (kW cm^{-2})	0.23	0.35	0.56	1.06
Microwave reflection coefficient, ρ	0.4	0.65	0.76	0.81
Microwave energy flux to sustain plasma, S_0 (kW cm^{-2})	0.38	1.0	2.3	5.6

and then goes through a quartz window into an impedance-matching mode converter which couples the microwave power into a coaxial waveguide. The coaxial waveguide then operates as an arc jet: the center conductor of the waveguide forms one electrode and the other electrode comprises an annual flange on the outer coaxial electrode.

Another configuration of microwave discharge based on conversion of the rectangular waveguide mode into a circular mode is illustrated in Figure 4.37. A microwave plasma torch, presented in Figure 4.38, has been used by G. Morfill's group at the Max Planck Institute for sterilization of different surfaces, treatment of

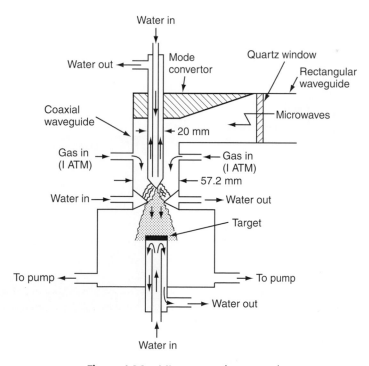

Figure 4.36 *Microwave plasma torch.*

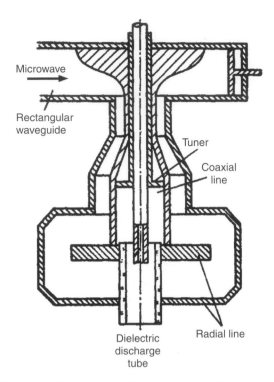

Figure 4.37   *Radial microwave plasmatron with rectangular-to-circular microwave mode converter.*

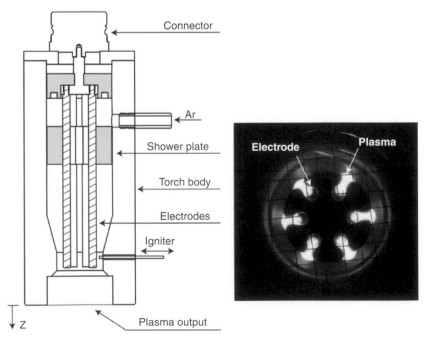

Figure 4.38   *Microwave plasma torch (left) and plasma output photo (right). Reproduced from New Journal of Physics: Plasma medicine: an introductory review – Vol 11, 115012, 2009, with permission from IOP Publishing Ltd.*

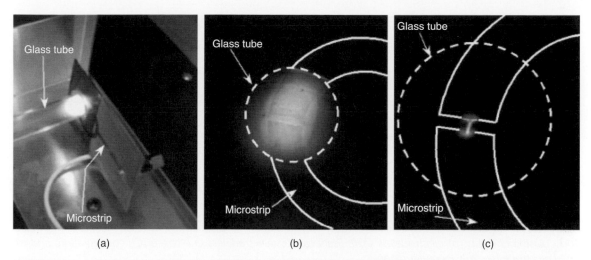

(a)   (b)   (c)

Figure 4.39 *Microwave discharge of Ar-plasma, power 1 W: (a) device operating at 9 Torr; (b) diffuse plasma at 20 Torr; (c) confined plasma at 760 torr. Copyright 2003 IEEE. Reprinted, with permission from Transactions on Plasma Science, Low-power microwave plasma source based on a microstrip split-ring resonator, vol 31, no 4, Iza and Hopwood.*

wounds and other medical applications. The relatively weak microwave discharge (2.45 GHz, about 100 W) has been generated in atmospheric-pressure argon flow (about 2 slm or standard litter per minute). The plasma-medical effect in this system is mostly due to energy transfer from electronically excited argon to air during its contact with the jet. It should be mentioned that while plasma temperature in this microwave discharge is rather high, the temperature at the tip of the jet is relatively low, permitting effective treatment of living tissues.

Microwave plasma microdischarges, developed by Hopwood and his group, also have important biomedical applications. This low-power microwave microplasma source is based on a microstrip split-ring 900 MHz resonator (Figure 4.39) and operates at pressures up to 1 atm; see Figure 4.39. Argon and air discharges can be self-started in the system with power less than 3 W. Ion density of 1.3×10^{11} cm^{-3} in argon at 400 mTorr can be produced with only 0.5 W power. Atmospheric discharges can also be sustained in argon with 0.5 W. The low power allows portable air-cooled operation of the system. This type of microplasma microwave source can be integrated into portable devices for applications such as bio-MEMS sterilization, small-scale biomaterial processing and microchemical and biochemical analysis.

4.4.3   Non-thermal RF discharges: CCP and ICP coupling

At low pressures, RF plasma is strongly non-equilibrium and cold. Electron-neutral collisions are less frequent while gas cooling by the walls is intensive, therefore $T_e \gg T_0$ similarly to glow discharges. The cold electrodeless RF discharges are widely applied in technologies of high-precision surface treatment, in particular in electronics, tissue engineering and treatment of polymers and biopolymers. The upper frequency limit of the discharges is due to wavelengths close to the system sizes (shorter wavelengths are referred to as microwaves). The lower RF limit is due to frequencies of ionization and ion transfer. Ion density in RF discharge plasmas and sheaths can usually be considered as constant during a period of electromagnetic field oscillation. Radio frequencies therefore usually exceed 1 MHz (sometimes smaller); the most industrially used frequency is 13.6 MHz.

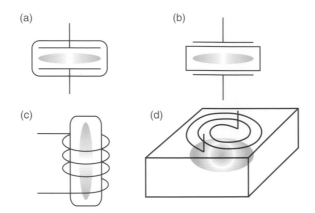

Figure 4.40   *Capacitively (CCP) and inductively (ICP) coupled RF discharges: (a) CCP discharge with electrode/plasma contact; (b) electrodeless CCP discharge; (c) electrodeless coil-based ICP configuration; and (d) spiral electrode-based ICP configuration.*

Non-thermal RF discharges can be subdivided into those of moderate and those of low pressure. In the moderate-pressure discharges (1–100 Torr), the electron energy relaxation length is smaller than characteristic system sizes and the electron energy distribution function (EEDF) is determined by the local electric field. In the low-pressure discharges ($pL < 1$ in inert gases), electron energy relaxation length is comparable to the discharge sizes and EEDF is determined by electric field distribution in the entire discharge.

Non-thermal RF discharges can be either capacitively (CCP) or inductively (ICP) coupled (Figure 4.40). The CCP discharges provide electromagnetic field by electrodes located either inside or outside of the chamber (Figure 4.40a and b). They stimulate an electric field, facilitating ignition. Electromagnetic field in the ICP discharges is induced by the inductive coil; discharge can be located either inside the coil (Figure 4.40c) or adjacent to the plane coil (Figure 4.40d). The ICP discharges primarily stimulate magnetic field, while the corresponding non-conservative electric field is relatively low. Non-thermal ICP is therefore usually generated at lower pressures when E/p is sufficient for ionization.

Coupling between the inductive coil and plasma can be interpreted as a voltage-decreasing transformer: the coil represents the primary multi-turn windings and the plasma represents the secondary single-turn winding. Effective coupling with RF power supply requires low plasma resistance. As a result, the ICP discharges are useful for reaching high currents, electric conductivities and electron densities. In contrast to that, CCP discharges are more convenient for provide higher electric fields. RF power supplies for plasma generation typically require an active load of 50–75 Ω.

To provide effective correlation between resistance of the leading line from the RF generator and the CCP or ICP discharge impedance, the special coupling circuit should be applied (Figure 4.41). The RF generator

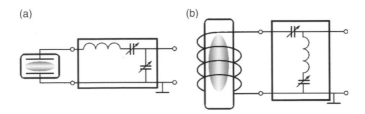

Figure 4.41   *Coupling circuits for (a) CCP and (b) ICP discharges.*

should not only be correlated with the RF discharge during its continuous operation, but also provide initially sufficient voltage for breakdown. To provide the ignition, the coupling electric circuit should form the AC resonance being in series with the generator and the discharge system during its idle operation. For this reason, the CCP coupling circuit includes inductance in series with the generator and discharge system (Figure 4.41a). Capacitance is variable as the design of variable inductance is more complicated. In the case of ICP discharge, a variable capacitance in the electric coupling circuit in series with generator and idle discharge system provides the necessary effect of initial breakdown.

4.4.4 Non-thermal RF-CCP discharges at moderate pressure regime

An external circuit in CCP discharges provides either a fixed voltage or fixed current, which is a typical choice for practical application. The fixed current regime can be represented simply by current density $j = -j_0 \sin(\omega t)$ between two parallel electrodes. It can be assumed that RF plasma density $n_e = n_i$ is high and the electron conductivity current exceeds the displacement current:

$$\omega \ll \frac{1}{\varepsilon_0} |\sigma_e| \equiv \frac{1}{\tau_e} \tag{4.74}$$

where τ_e is Maxwell time for electrons, that is, the time required for the charged particles to shield the electric field; $\omega \ll \nu_e$, where ν_e is electron collisional frequency; and σ_e is complex electron conductivity, defined:

$$\sigma_e = \frac{n_e e^2}{m(\nu_e + i\omega)}. \tag{4.75}$$

The inequality opposite to (4.74) is usually valid for ions because of their much higher mass $M \gg m$:

$$\omega \gg \frac{1}{\varepsilon_0} |\sigma_i| \equiv \frac{1}{\tau_i}, \quad \sigma_i = \frac{n_i e^2}{M(\nu_i + i\omega)}. \tag{4.76}$$

Ion conductivity current can therefore be neglected with respect to displacement current, and the ion drift motion during an oscillation period of electric field can be neglected. Ions are at rest and form the 'skeleton' of plasma, while electrons oscillate between the electrodes (Figure 4.42).

Electrons are present in the sheath of width L near an electrode for only part of the oscillation period called the plasma phase. Another part of the oscillation period (no electrons in the sheath) is referred to as the space-charge phase. The oscillating space charge creates electric field, which forms the displacement current and closes the circuit. The quasi-neutral plasma zone is called the positive column. The electric field of the space charge has oscillating as well as constant components, which is directed from plasma to electrodes. The constant component of the space-charge field provides faster ion drift to the electrodes than for the case of ambipolar diffusion. As a result, ion density in the space charge layers near electrodes is lower than in plasma.

Moderate-pressure CCP discharges assume energy relaxation lengths smaller than sizes of plasma zone and sheaths $\lambda_\varepsilon < L, L_p$, which corresponds to the pressure interval $1 - 100$ Torr. In this pressure range, $\omega < \delta \nu_e$, where δ is fraction of electron energy lost per electron collision and ν_e is the frequency of the electron collisions. In this regime, electrons lose and gain their energy during the time interval shorter than the period of electromagnetic RF oscillations. EEDFs and hence ionization and excitation rates are determined in the moderate-pressure discharges by local and instantaneous values of electric field. In particular, it results in

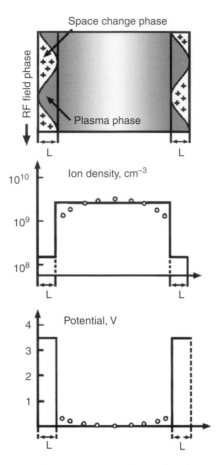

Figure 4.42   *Ion density and potential distribution and boundary layer oscillation in RF CCP-discharge.*

essential contribution of ionization in sheaths, where electric fields have maximum values and are able to provide maximum electron energies.

Moderate-pressure RF-CCP discharges can be sustained in two forms, referred to as α-discharge and γ-discharge (Levitsky, 1957). The main differences between the α- and γ-regimes are related to current density and luminosity distribution in the discharge gap. The α-discharge is characterized by low luminosity in the plasma volume. Brighter layers are located closer to electrodes, but layers immediately adjacent to the electrodes are dark. The γ-discharge occurs at much higher current density. In this regime, the discharge layers immediately adjacent to the electrodes are very bright, but relatively thin. The plasma zone is also luminous and separated from the bright electrode layers by dark layers similar to the Faraday dark space in glow discharges. In the γ-regime, ionization processes are due to electrons formed on the electrode by the secondary electron emission and accelerated in a sheath, which is similar to glow discharges. The α- and γ-discharges operate at normal current density similarly to glow discharges. The increase in current is provided by growth of the electrode area occupied by the discharge, while the current density remains constant. Normal current density in both regimes is proportional to frequency of electromagnetic oscillations. In the case of γ- discharges, normal current density exceeds current density of α-ischarges by a factor of >10.

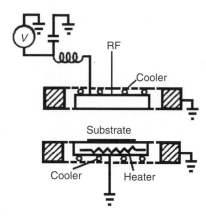

Figure 4.43 *Low-pressure CCP discharge.*

4.4.5 Low-pressure CCP RF discharges

Low-pressure RF discharges (≤ 0.1 Torr), widely used in electronics, the treatment of polymers and tissue engineering, are illustrated in Figure 4.43. The luminosity pattern is different from that of moderate pressure. The plasma zone is bright and separated from electrodes by dark pre-electrode sheaths. The discharge is usually asymmetric; a sheath located by the electrode where RF voltage is applied is about 1 cm thick while another is only about 0.3 cm. In contrast to the moderate-pressure discharges, plasma fills out the entire gap and the normal current density does not occur. Electron energy relaxation length λ_ε in the low-pressure RF-CCP discharges exceeds the typical sizes of plasma zone and sheaths ($\lambda_\varepsilon > L_p, L$).

The EEDF is not local, and is determined by the electric fields in the zone of about λ_ε. For He discharges, it requires $p(L_p, L) \leq 1$ Torr cm. Since $\omega \gg \delta v_e$ (where δ is average fraction of electron energy lost per collision and v_e is frequency of the electron collisions), the time of EEDF formation is longer than the period of RF oscillations which allows EEDF to be considered as stationary.

The electric field can be divided into constant and oscillating components. The constant component provides a balance of electron and ion fluxes and quasi-neutrality; it provides current and heating of the plasma electrons. Distribution of plasma density and electric potential φ (corresponding to the constant component) is depicted in Figure 4.42. Electric field in the space-charge sheath exceeds that in plasma; a sharp change of potential therefore occurs on the boundary of plasma and the sheath. Plasma electrons move between the sharp potential barriers. Electron heating in RF-CCP discharges of not-very-low pressures is due to electron-neutral collisions v_e.

Energy of the systematic oscillations, received by an electron from the electromagnetic field after a previous collision, can be transferred to chaotic electron motion during the next collision. If $\omega^2 \gg v_e^2$, electric conductivity and Joule heating are proportional to v_e and the discharge power should be low at low pressures. Experimentally, however, the discharge power in such conditions can be significantly higher because of the contribution of the *stochastic heating effect*, which provides heating even in a collisionless regime. For example, when the mean free path of electrons exceeds the sheath size, an electron arriving at the sheath is reflected by the space-charge potential of the sheath boundary in the same way as an elastic ball is bounced off a massive wall. If the sheath boundary moves from the electrode then the reflected electron receives energy; if the sheath boundary moves to the electrode and a fast electron 'catches the sheath up', the reflected electron loses its kinetic energy. Electron flux to the boundary moving from the electrode exceeds the other and so energy is mostly transferred to the fast electrons, which is the stochastic heating effect.

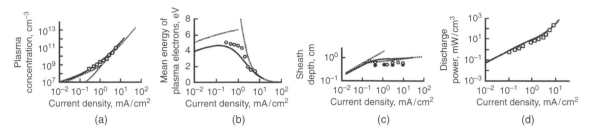

Figure 4.44   *(a) Plasma concentration, (b) electron energy, (c) sheath depth and (d) discharge power as a function of current density. Solid curve: exact calculation; dotted line: asymptotic formulae; points: observations (f = 13.56 MHz; L_0 = 6.7 cm; p = 0.03 Torr).*

Parameters of the low-pressure RF-CCP discharge in argon are provided in Figure 4.44 as functions of current. The RF-CCP discharges are usually generated in grounded metal chambers. One electrode is connected to the chamber wall and grounded and the other is powered, which makes the discharge asymmetric. The current between an electrode and the grounded metallic wall is capacitive in moderate pressure discharges and does not play any important role. In the low-pressure discharges, plasma occupies a bigger volume because of diffusion and some fraction of the discharge current goes from the loaded electrode to the grounded walls. As a result, current density in the sheath located near the powered electrode exceeds current density in the sheath related to the grounded electrode. The surface area of the powered electrode is also smaller, therefore current densities there are higher. Lower current density in the sheath corresponds to lower voltage.

The constant component of the voltage, which is the plasma potential with respect to electrode, is also lower at lower current density. A constant potential difference (called *auto-displacement voltage*) therefore occurs between the electrodes if they are covered by a dielectric layer or if capacitance is installed in the electric circuit (Figure 4.45) to avoid direct current between electrodes.

Low-frequency (< 100 kHz) RF-CCP discharges are also applied in plasma chemistry. In particular, nitrogen plasma (generated in such discharges) is effective in: polymer surface treatment which aims to promote adhesion of silver to polyethylene terephthalate; the treatment of polyester web to promote adhesion of gelatin-containing layers related to production of photographic film; and sputter deposition of metals.

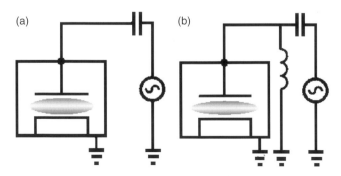

Figure 4.45   *RF discharge with (a) disconnected and (b) DC-connected electrodes.*

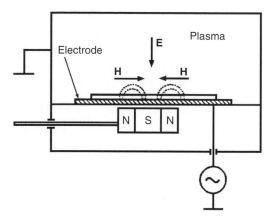

Figure 4.46 *RF-magnetron discharge.*

4.4.6 Low-pressure RF magnetron discharges for surface treatment

Sheath voltages in the RF-CCP discharges are relatively high, leading to low ion densities and high ion-bombarding energies which can be disadvantageous. The ion-bombarding energies cannot be varied independently of the ion flux in these discharges. RF magnetron discharges (referred to as magnetically enhanced reactive ion etchers or MERIEs) were developed to make the relevant improvements. In this case, relatively weak DC magnetic field (50–200 G) is imposed on the low-pressure RF-CCP discharge parallel to the powered electrode and perpendicular to the RF electric field and current. The RF magnetron allows the ionization degree to be increased at lower RF voltages and the energy of ions bombarding the powered electrode to be decreased. This discharge also allows the ion flux-intensifying etching to be increased. The RF magnetron discharges can be effectively sustained at much lower pressures (down to 10^{-4} Torr); a well-directed ion beam is therefore able to penetrate the sheath without any collisions. The ratio of the ion flux to the flux of active neutral species grows, which also leads to higher-quality etching.

The RF magnetron used for ion etching is depicted in Figure 4.46. The RF voltage is applied to the smaller lower electrode, where a sample is located. A rectangular samarium-cobalt magnet is placed under the powered electrode. The part of the electrode where the magnetic field is horizontal does not cover the entire electrode. A scanning device is used to move the magnet and to provide a horizontal magnetic field for the entire electrode.

The RF magnetron can be also generated in cylindrical systems with coaxial electrodes; the magnetic field in these systems is directed along the cylinder axis. Electrons oscillating with the sheath boundary additionally rotate in the RF magnetrons around the horizontal magnetic field lines with the cyclotron frequency.

When magnetic field and the cyclotron frequency are high enough, the amplitude of the electron oscillations along the RF electric field decreases significantly; the magnetic field 'traps' electrons. The cyclotron frequency plays the same role as the frequency of electron-neutral collisions: electrons become unable to reach the amplitude of their free oscillations in the RF electric field. The amplitude of electron oscillations determines the thickness of sheaths. A reduction in the amplitude of electron oscillations in the magnetic field therefore results in smaller sheaths and lower sheath voltage near the powered electrode. This leads to lower values of the auto-displacement, lower ion energies and lower voltage necessary to sustain the RF discharge.

4.4.7 Low-pressure non-thermal ICP RF discharges in cylindrical coil

Electromagnetic field in the RF-ICP discharges is induced by inductive coil (Figure 4.40) where the magnetic field is primary and the non-conservative electric field is low. The discharges are therefore generated at low pressures (usually less than 50 mTorr) to provide reduced electric field E/p sufficient for ionization. Coupling between the inductive coil and plasma can be interpreted as a transformer, where the coil is the primary multi-turn windings and plasma is the secondary single turn. The ICP discharge can be considered as a voltage-decreasing and current-increasing transformer, which allows it to reach a high current, high electric conductivity and high electron density at relatively low electric fields and voltages.

The low-pressure ICP discharges operate at electron densities of 10^{11}–10^{12} cm^{-3}, more than 10 times that of those for CCP; they are therefore also referred to as high-density plasma (HDP) discharges. The HDP discharges are widely used in electronics and other high-precision surface-treatment technologies, especially for treatment of polymers and tissue engineering. An important advantage of the discharges for high-precision surface treatment is the RF power coupling to plasma across a dielectric window or wall. Such 'non-capacitive' power transfer to plasma allows lower voltages to be operated across all sheaths at electrode and wall surfaces.

The DC plasma potential and energies of ions accelerated in the sheaths are typically 20–40 V, which is very good for the numerous surface-treatment applications. The ion energies can be independently controlled in this case by an additional capacitively coupled RF source called the *RF-bias*, driving the electrode on which the substrate for material treatment is placed. The ICP discharges are therefore able to provide independent control of the ion and radical fluxes by means of the main ICP source power and the ion-bombarding energies by means of power of the bias electrode.

Consider the inductive RF discharge in a long cylindrical tube placed inside a cylindrical coil. Physical properties of such a discharge are similar to those of the planar discharge, more convenient for applications. Electric field $E(r)$ induced in the discharge tube located inside the long coil is:

$$E(r) = -\frac{1}{2\pi r}\frac{d\Phi}{dt}, \qquad (4.77)$$

where Φ is magnetic flux crossing the loop of radius r perpendicular to the axis of the discharge tube and r is distance from the tube axis. Magnetic field is created in the system by electric current in the coil $I = I_c e^{i\omega t}$ and in the plasma. Assuming constant plasma conductivity σ and expressing plasma current as $j(r) = j_0(r)e^{i\omega t}$, current density distribution along the plasma radius is determined as:

$$\frac{\partial^2 j_0}{\partial r^2} + \frac{1}{r}\frac{\partial j_0}{\partial r} - \frac{1}{r^2}j_0 = i\frac{\sigma\omega}{\varepsilon_0 c^2}j_0. \qquad (4.78)$$

If pressure is very low and plasma can be considered collisionless, its conductivity is inductive:

$$\sigma = -i\varepsilon_0\frac{\omega_p^2}{\omega} \qquad (4.79)$$

where ω_p is plasma frequency. From Equations (4.78) and (4.79), the current density distribution in the ICP plasma is:

$$j_0(r) = j_b I_1\left(\frac{r}{\delta}\right), \qquad (4.80)$$

where $I_1(x)$ is the modified Bessel function and δ is the skin layer. Current density on the plasma boundary near the discharge tube j_b is determined by the non-perturbed electric field and plasma conductivity:

$$j_b = \sigma\frac{\omega a N}{2\varepsilon_0 c^2 l}I_c \qquad (4.81)$$

where a is radius of the discharge tube; N is number of turns in coil; l is length of the coil; c is speed of light; and I_c is amplitude of current in coil. Plasma conductivity grows with electric field and the skin layer is smaller than the discharge radius. Most of electric current is located in the relatively thin δ-layer on the discharge periphery. In this case, the Bessel function (Equation (4.80)) can be simplified as:

$$j_0(r) \approx j_b \exp\left(\frac{r-a}{\delta}\right). \tag{4.82}$$

Current in the coil I_c is determined by an external circuit and can be considered as a parameter. The electric field in plasma is related to the voltage on the plasma loop $E_p = U_p/2\pi a$, and logarithmically depends on other plasma parameters. Electric field in plasma at high currents is small with respect to electric field in the idle regime without plasma:

$$E = \frac{\omega M I_c}{2\pi a}$$

where M is the mutual inductance. The voltage drop U_p related to plasma can be neglected at high currents. The discharge current can therefore be expressed as:

$$I_d = -\frac{M}{L_d}I_c = -NI_c, \tag{4.83}$$

where $L_d = \mu_0 \pi a^2/l$ is geometrical inductance attributed to the plasma considered as a conducting cylinder. The discharge current flows in the opposite direction with respect to the inductor current, and the value of the plasma current significantly exceeds that of the current in the inductor. Electron density can then be expressed as:

$$n_e = j\frac{m\nu_{en}}{e^2 E_p} = \frac{NI_c}{l\delta}\frac{m\nu_{en}}{e^2 E_p}. \tag{4.84}$$

The thickness of the skin layer in the RF-ICP discharge can be calculated as:

$$\delta = \frac{2E_p l}{\omega \mu_0 N I_c} \propto \frac{1}{I_c}. \tag{4.85}$$

The skin layer is inversely proportional to the electric current in the inductor coil. Taking into account Equation (4.85), plasma density is proportional to the square of current in inductor coil:

$$n_e = \left(\frac{NI_c}{el E_p}\right)^2 \frac{\omega \mu_0 m \nu_{en}}{2} \propto I_c^2. \tag{4.86}$$

Similarly to the CCP discharges, ICP regimes are different at moderate and low pressures. In the *moderate-pressure regime*, energy relaxation length is less than the thickness of the skin layer. The heating of the electrons is determined by the local electric field and takes place in the skin layer. Ionization processes as well as plasma luminosity are therefore concentrated in the skin layer. The internal volume of the discharge tube, located closer to the tube axis with respect to the skin layer, is filled up with plasma only due to the radial inward plasma diffusion from the discharge periphery. If losses of charged particles in the internal volume due to recombination and diffusion along the axis of discharge tube are significant, plasma density can be lower in the central part of the discharge than in the periphery.

In the *low-pressure regime*, the energy relaxation length exceeds the skin layer. Heating of electrons takes place in the skin layer, but ionization processes are effective in the plasma volume where the electrons have maximum value of kinetic energy. Electron density on the discharge axis can significantly exceed that on the skin layer in this regime.

4.4.8   Planar-coil and other configurations of low-pressure non-thermal RF-ICP discharges

Figure 4.47 illustrates ICP discharges in the planar configuration, widely used for different surface-treatment applications (especially in electronics). It is similar geometrically to the conventional RF-CCP parallel plate reactor, but RF power is applied in this scheme to a flat spiral inductive coil which is separated from the plasma by quartz or another dielectric insulating plate. The RF currents in the spiral coil induce the image currents in the upper surface of the plasma corresponding to the skin layer. This discharge is therefore inductively coupled, and similar to the case of cylindrical geometry of the inductive coils.

The analytical relations derived above for the low-pressure RF-ICP discharge inside an inductive coil can be applied qualitatively for the planar coil configuration of the discharge. The planar ICP discharge

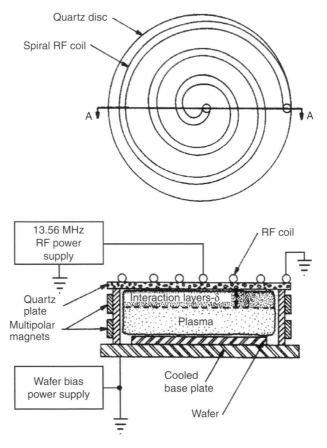

Figure 4.47   *ICP parallel-plate reactor.*

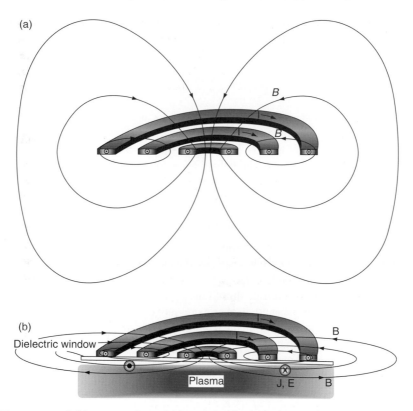

Figure 4.48 *RF magnetic field near a planar inductive coil: (a) without nearby plasma and (b) with nearby plasma.*

also includes two important elements: multi-polar permanent magnets and DC wafer bias. The multi-polar permanent magnets are located around the outer circumference of the plasma to improve plasma uniformity and confinement and increase plasma density. The DC wafer-bias power supply is used to control the energy of ions impinging on the wafer, which have ranged over 30–400 eV in the planar discharge system.

The structure of magnetic field lines in the planar coil configuration of the ICP discharges is more complicated than in the case of the cylindrical inductive coil. The RF magnetic field lines in the planar coil configuration in the absence of plasma are illustrated in Figure 4.48a. These magnetic field lines encircle the coil and are symmetric with respect to the plane of the coil. Deformation of the magnetic field in the presence of plasma formed below the coil is shown in Figure 4.48b. In this case, an azimuthal electric field and an associated current (in the direction opposite to that in the coil) are induced in the plasma skin layer. The total magnetic field is generated by both the multi-turn coil current and single-turn-induced plasma current. The dominant magnetic field component within the plasma is vertical near the axis of the planar coil, and horizontal away from the axis.

The *helical resonator discharge* is other special type of low pressure RF-ICP discharges. The helical resonator consists of an inductive coil (helix) located inside of cylindrical conductive screen, and can be considered as a coaxial line with an internal helical electrode. Electromagnetic wave propagates in such coaxial line with phase velocity much lower than the speed of light: $v_{ph} = \omega/k \ll c$, where k is wavelength and c is speed of light. This property allows the helical resonator to operate in the MHz frequency range and generate low-pressure plasma. The coaxial line of the helical discharge becomes resonant when an integral number of quarter-waves of the RF field fits between the two ends of the system, that is:

$$2\pi r_h N = \frac{\lambda}{4}, \tag{4.87}$$

where r_h is helix radius; N is number of turns in the coil; and λ is electromagnetic wavelength in vacuum. The helical resonator discharges effectively operate at frequencies of 3–30 MHz, and do not require a DC magnetic field. The resonators exhibit high Q values (600–1500 without plasma); electric fields are high, therefore facilitating the initial breakdown. The helical resonator discharges also have high impedance and can be operated without a matching network.

4.4.9 Non-thermal RF atmospheric pressure plasma jets as surface-treatment device

The RF atmospheric glow discharge or atmospheric-pressure plasma jet (APPJ) is one of the most developed atmospheric pressure glow (APG) systems. It has been used widely for the treatment of polymers, biopolymers and other fragile surfaces, as well as in microelectronics for the plasma-enhanced chemical vapor deposition (PECVD) of silicon dioxide and silicon nitride thin films.

Quasi-uniformity of the APPJ plasma at atmospheric pressure explains the importance of this discharge for the treatment of living tissues and for plasma-medical applications in general. The APPJ can be generated as a planar and co-axial system with a discharge gap of 1–1.6 mm and frequency in the MHz range (13.56 MHz). APPJ can be sustained in α- and γ-modes in contaminated helium and argon. The APPJ is an RF-CCP discharge that can operate uniformly at atmospheric pressure in noble gases, mostly in helium.

In most APPJ configurations, electrodes are placed inside the chamber and not covered by any dielectric (in contrast to DBD). The discharge in pure helium has limited applications; various reactive species such as oxygen, nitrogen, nitrogen trifluoride and so on are therefore added. To achieve higher efficiency and higher reaction rate, the concentration of the reactive species in the discharge has to be increased. If the concentration of the reactive species exceeds a certain level (which is different for different species, but in all cases is of the order a small percentage), the discharge becomes unstable.

The distance between electrodes in APPJ is usually about 1 mm, which is much smaller than the size of the electrodes (about 10 cm × 10 cm). The discharge can therefore be considered as one-dimensional (1D) and the effects of the boundaries on the discharge can be neglected.

Electric current in the discharge is sum of the current due to the drift of electrons and ions and the displacement current. Since mobility of the ions is usually 100 times smaller than the electron mobility, the current in the discharge is mostly due to electrons. Considering that the typical ionic drift velocity in APPJ discharge conditions is about 3×10^4 cm s^{-1}, the time needed for ions to cross the gap is about 3 μs which corresponds to a frequency of 0.3 MHz. The frequency of the electric field is much higher, meaning that ions in the discharge do not have enough time to move while electrons move from one electrode to another as the polarity of the applied voltage changes.

The typical space–time APPJ structure, showing the two sheaths and the positive column, is shown in Figure 4.49. The overall APPJ voltage consists of the voltage on the positive column V_p (plasma voltage) and the voltage on the sheath V_s. The voltage on the positive column V_p slightly decreases with an increase of the discharge current density. This happens because a reduced electric field E/N in plasma is almost constant and equal to $E/p \approx 2$ V (cm Torr)$^{-1}$ for helium discharge. If the density of neutral species is constant, the plasma voltage will be also be constant; as the electric current density decreases, the density of neutrals decreases slightly since high currents cause a gas temperature to rise. At higher gas temperatures, a lower voltage is needed to support the discharge and subsequently the plasma voltage decreases.

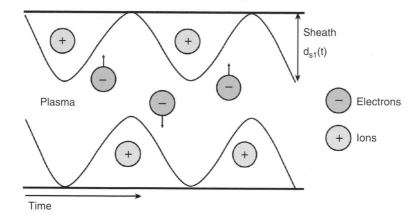

Figure 4.49 *APPJ: illustration of the space-time structure of RF-CCP discharge.*

Sheath thickness can be approximated from the amplitude of electron drift oscillations $d_s = 2\mu E/\omega \approx$ 0.3 mm, where μ is electron mobility, $\omega = 2\pi f$ is the frequency of the applied voltage and E is the electric field in plasma. Assuming the secondary emission coefficient $\gamma = 0.01$, the critical ion density $n_{p(crit)}$ in helium RF before the $\alpha - \gamma$ transition is about 3×10^{11} cm^{-3}. The corresponding critical sheath voltage is about 300 V. Typical power density for the APPJ helium discharge is of the order 10 W cm^{-2} (approximately 10 times higher than for the DBD discharges) including its uniform modifications.

The power density that can be achieved in the uniform RF discharge is limited by two major instability mechanisms: thermal instability and $\alpha - \gamma$ transition instability. Critical power density for the thermal instability in APPJ is about 3 W cm^{-2}. Stable APPJ can however be generated with a power density exceeding this threshold.

Suppression of the thermal instability in the APPJ conditions is due to the stabilizing effect of the sheath capacitance, which can be described by the R parameter: square of the ratio of the plasma voltage to the sheath voltage. The smaller value of R, the more stable is the discharge with respect to the thermal instability (see Figure 4.50). For example, if $R = 0.1$, the critical power density with respect to thermal instability is 190 W cm^{-2}. For the helium APPJ with $d_s = 0.3$ mm, $V_s = 300$ V and $d = 1.524$ mm, the parameter $R = (V_p/V_s)^2 = 0.36$, which corresponds to the critical discharge power density of 97 W cm^{-2}.

The APPJ discharge therefore remains thermally stable in a wide range of power densities as long as the sheath remains intact. Any major instability of the APPJ and loss of its uniformity is mostly determined by the $\alpha - \gamma$ transition or, in other words, by breakdown of the sheath. The $\alpha - \gamma$ transition in APPJ occurs because of the Townsend breakdown of the sheath, which occurs when ion density and sheath voltage exceed their critical values ($n_{p(crit)} = 3 \times 10^{11}$ cm^{-3}, $V_s = 300$ V).

It has been shown that the main mechanism of the discharge instability is the sheath breakdown, which eventually leads to thermal instability. More effective discharge cooling would not solve the stability problem, because it does not protect the discharge from the sheath breakdown. Nevertheless, the discharge cooling is important since the sheath breakdown depends on the reduced electric field that increased with temperature. Despite the fact that the thermal stability of helium discharge is better compared to the discharge with oxygen addition, a higher power is achieved with oxygen addition which prevents the sheath breakdown.

Summarizing, it is much easier to generate the uniform APG in helium and argon than in other gases (especially electronegative gases). The effect cannot be explained only by high thermal conductivity of helium. It is more important for uniformity that noble-gas-based discharges have a significantly lower voltage and therefore lower power density, helping to avoid the thermal instability. Pure nitrogen provides better

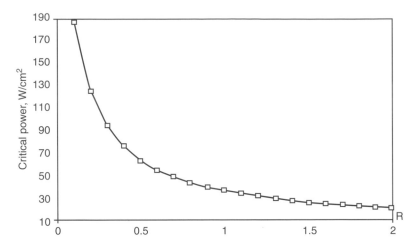

Figure 4.50 *Critical power density for APPJ discharge stability (area above the curve indicates unstable discharge).*

conditions for the uniformity than air. The presence of oxygen results in the electron attachment which causes higher voltage, higher power and finally leads to the thermal instability.

4.4.10   Atmospheric-pressure non-thermal RF plasma microdischarges: Plasma needle

In the RF range (13.56 MHz), the so-called *plasma needle* (Figure 4.51) developed by Eva Stoffels and her group attracted significant interest because of its medical applications. This discharge has a single-electrode configuration and operates in helium. It operates near the room temperature, allows the treatment of irregular

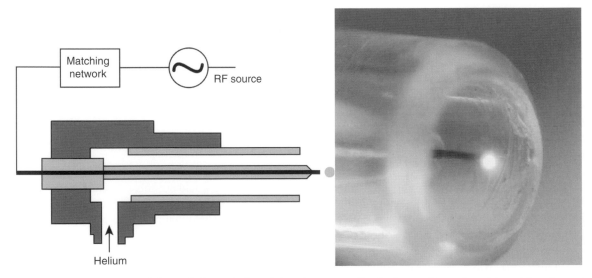

Figure 4.51 *Schematic (left) and photo (right) of the plasma-needle discharge. Reproduced from New Journal of Physics: Plasma medicine: an introductory review – Vol 11, 115012, 2009, with permission from IOP Publishing Ltd. OR Kieft, v.d. Laan & Stoffels, 2004 (see color plate).*

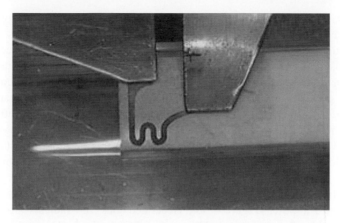

Figure 4.52   *Miniaturized ICP RF microplasma jet (Ichiki, Koidesawa and, Horiike, 2003, 2004).*

surfaces and has a small penetration depth. The plasma needle is capable of bacterial decontamination and localized cell removal without causing necrosis to the neighboring cells. Areas of detached cells can be made with a resolution of 0.1 mm. Radicals and ions from the plasma as well as UV radiation interact with the cell membranes and cell adhesion molecules, causing detachment of the cells. The plasma needle is confined in a plastic tube, through which helium flow is supplied (Figure 4.51). The discharge is entirely resistive with voltage 140–270 V_{rms}. The electron density is about 10^{11} cm^{-3}. Optical measurements show substantial UV emission in the range 300–400 nm; radicals O and OH have been detected. At low helium flow rates, densities of molecular species in the plasma are higher. Detailed modeling of the plasma-needle was performed by David Graves and Yukinori Sakiyama.

Conventional RF discharges, both ICP and CCP, have also been generated at microscale conditions at atmospheric pressure. These plasmas are non-equilibrium because of their small sizes and effective cooling. Reduction in size requires a reduction in wavelength (or increase in frequency).

A *miniaturized atmospheric-pressure ICP jet* has been developed for a portable liquid analysis system, also of interest for plasma medicine. The plasma device is a planar ICP source (Figure 4.52), consisting of a ceramic chip with an engraved discharge tube and a planar metallic antenna in a serpentine structure. The chip consists of two dielectric plates with an area of 15 mm × 30 mm. A discharge tube (1 mm × 1 mm × 30 mm) is mechanically engraved on one side of the dielectric plate. A planar antenna is fabricated on the other side of the plate. The atmospheric-pressure plasma jet with density of about 10^{15} cm^{-3} is produced using a compact very high frequency (VHF) transmitter at 144 MHz and power 50 W. The electronic excitation temperature in this microplasma system is 4000–4500 K.

4.4.11   Non-thermal low-pressure microwave and other wave-heated discharges

We consider three types of low-pressure wave-heated plasmas in this section: electron cyclotron resonance (ECR) discharges, helicon discharges and surface-wave discharges.

4.4.11.1   ECR discharges

In ECR discharges, a right-circularly polarized wave (usually at microwave frequencies, e.g. 2.45 GHz) propagates along the DC magnetic field (about 850G) in the ECR conditions, which provides the wave energy absorption by a collisionless heating.

The ECR-resonance between electromagnetic wave frequency ω and the electron cyclotron frequency $\omega_{Be} = eB/m$ allows the electron heating sufficient for ionization at lower electric fields. The electron cyclotron frequency can be calculated as $f_{Be} = 2.8B$. Electron heating in the ECR takes place because the gyrating electrons rotate in phase with the right-hand polarized wave, leading to a steady electric field over many gyro-orbits. Pressure in the ECR discharge is low for low electron-neutral collision frequency $\nu_{en} \ll \omega_{Be}$, and to provide the electron gyration long enough to obtain the energy necessary for ionization.

The ECR discharge with the microwave power injected along the axial magnetic field is depicted in Figure 4.53. The magnetic field profile is chosen to provide effective propagation of the electromagnetic wave from the quartz window to the zone of the ECR resonance without reflection at high plasma densities. A special magnetic field profile can provide multiple ECR resonance positions, as shown on the figure by the dashed line.

Low-pressure gas introduced into the discharge chamber forms plasma, which streams and diffuses along the magnetic field toward a wafer (Figure 4.53). Generated ions and radicals are able to effect surface treatment. A magnetic field coil at the wafer holder can be additionally used to modify the uniformity of etch or deposition. Typical ECR microwave discharge parameters are: pressure 0.5–50 mTorr, power 0.1–5 kW,

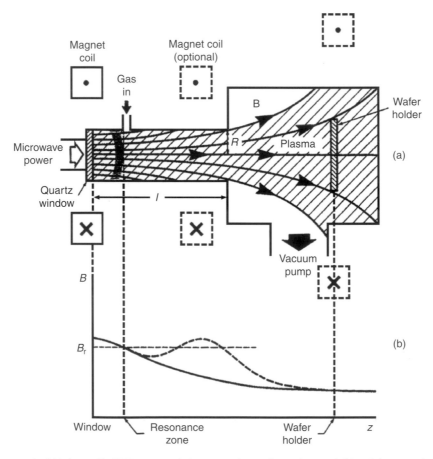

Figure 4.53   *A typical high-profile ECR system: (a) geometric configuration and (b) axial magnetic field variation, showing one or more resonance zones.*

microwave frequency 2.45 MHz, volume 2–50 L, magnetic field 1 kG, plasma density 10^{10}–10^{12} cm^{-3}, ionization degree 10^{-4}–10^{-1}, electron temperature 2–7 eV, ion acceleration energy 20–500 eV and source diameter 15 cm.

4.4.11.2   Helicon discharges

An antenna radiates the whistler wave which is subsequently absorbed in plasma. The helicon wave-heated discharges are excited at RF frequencies (13.56 MHz), and weak magnetic fields 20–200 G are required for the wave propagation and absorption. It is sustained by electromagnetic waves propagating in magnetized plasma in the helicon mode. The phase velocity of electromagnetic waves in magnetized plasma can be much lower than the speed of light, which allows operation in the wave propagation regime with wavelengths comparable to the system size even at radio frequencies. Magnetic field for helicon discharges applied for material processing are much lower than the level of magnetic field applied in ECR microwave discharges. Plasma density in these wave-heated discharges is 10^{11}–10^{12} cm^{-3}.

Excitation of the helicon wave is provided by an RF antenna that couples to the transverse mode structure across an insulating chamber wall. The electromagnetic wave mode propagates along the plasma column in the magnetic field, and is absorbed by plasma electrons. A helicon discharge is illustrated in Figure 4.54, depicting the material processing chamber which is located downstream from the plasma source. Plasma potentials in the helicon discharges are typically low (15–20 V).

Advantages of the helicon discharges with respect to ECR discharges are related to relatively low magnetic fields and applied frequency.

4.4.11.3   Surface-wave discharges

For surface-wave discharges, a wave propagates along the plasma surface and is absorbed by collisional heating of the plasma electrons near the surface. The heated electrons then diffuse from the surface into the bulk plasma. The surface-wave discharges can be excited by RF or microwave sources, and do not require a DC magnetic field. Plasma potential with respect to the walls is low (about $5T_e$, similarly to ICP). This results in effective generation of high-density plasmas at reasonable absorbed powers, which is attractive for intensive surface treatment.

The surface-wave discharges can generate the high-density plasma (HDP) with large diameter of about 15 cm. The absorption lengths of the electromagnetic wave surface modes are long in comparison to the ECR discharge. The surface wave discharge typically operates at frequencies in the microwave range of 1–10 GHz without an imposed axial magnetic field. It can be generated in planar rectangular configuration.

The electromagnetic surface wave, damping in both directions away from the surface, can be arranged in different configurations. One of them is a planar configuration on the plasma-dielectric interface. In another configuration, plasma is separated from a conducting plane by a dielectric slab. This planar system also admits propagation of a surface wave that decays into the plasma region. Although this electromagnetic wave does not decay into the dielectric, it is confined within the dielectric layer by the conducting plane. Finally, a surface wave is also able to propagate in the cylindrical discharge geometry. The surface wave propagates in this case on a non-magnetized plasma column confined by a thick dielectric tube.

4.4.12   Non-equilibrium microwave discharges of moderate and elevated pressures: Energy-efficient plasma source of chemically active species

The highest energy efficiency of generation of chemically active species (as well as plasma-chemical processes in general) can be achieved in non-equilibrium conditions with the contribution of vibrationally excited reagents. Such regimes require the generation of plasma with special parameters: electron temperature T_e

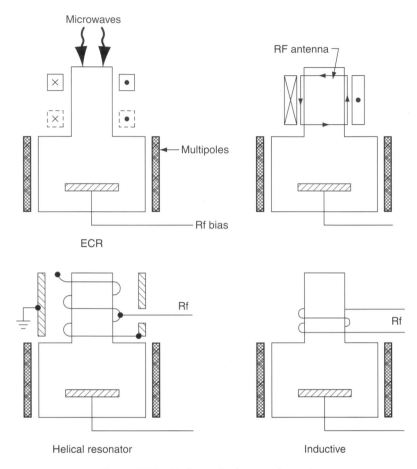

Figure 4.54 *Helicon discharge schematics.*

of about 1 eV and higher than translational temperature (≤ 1000 K) and ionization degree and specific energy input should be sufficiently high $n_e/n_0 \geq 10^{-6}$, $E_v \approx 1$ eV mol^{-1}. Simultaneous achievement of these parameters is difficult. For example, low-pressure non-thermal discharges have too high a specific energy input of 30–100 eV mol^{-1}, the streamer-based atmospheric pressure discharges have low specific power and average energy input and powerful steady-state atmospheric pressure discharges usually operate close to quasi-equilibrium.

Moderate-pressure microwave discharges are able to generate non-equilibrium plasma with the optimal parameters. Formation of an overheated filament within the plasma zone does not lead to the electric field decrease because of the skin effect. The electrodynamic structure permits microwave discharges to be sustained in non-equilibrium conditions $T_e > T_v \gg T_0$ at high specific energy inputs. For example, a steady-state microwave discharge can be sustained in CO_2 at frequency 2.4 GHz, power 1.5 kW, pressure 50–200 Torr and flow rate 0.15–2 sl s^{-1}. Specific energy input is 0.2–2 eV mol^{-1} and specific power is up to 500 W cm^{-3} (in conventional glow discharges it is about 0.2–3 W cm^{-3}). The vibrational temperature in the discharge can be of the order 3000–5000 K, and significantly exceeds rotational and translational temperatures which are about 1000 K.

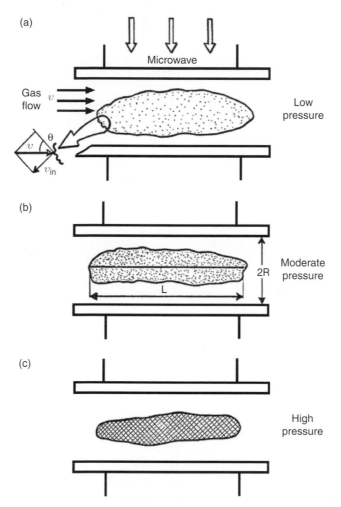

Figure 4.55 *Transition of (a) a diffusive microwave discharge into (c) a contracted one related to pressure increase.*

Moderate-pressure microwave discharges operate in three major regimes: diffusive, contracted and combined (Figure 4.55). They occur at different pressures p and electric fields E (Figure 4.56). The area above the curve 1–1 corresponds to breakdown conditions; microwave discharges are sustained below the curve. Curve 2–2 corresponds to $(E/p)_{max}$ sufficient to sustain ionization on the plasma front even at room temperature.

The *diffusive regime* (Figure 4.55a) takes place at $E/p < (E/p)_{max}$ and relatively low pressures 20–50 Torr). Ratio E/p decreases at higher pressures, and temperature on the discharge front should increase to provide the necessary ionization rate. Curve 3–3 determines the minimum reduced electric field $(E/p)_{min}$ when the microwave discharge is still non-thermal.

The *combined regime* (Figure 4.55b) occurs at intermediate pressures of 70–200 Torr and $(E/p)_{min} < E/p < (E/p)_{max}$. Curve 4–4 separates the lower-pressure regime of the homogeneous discharge and higher-pressure regime of the combined discharge. In the combined regime, a hot thin filament is formed inside a relatively large surrounding non-thermal plasma. The skin effect prevents penetration of the electromagnetic

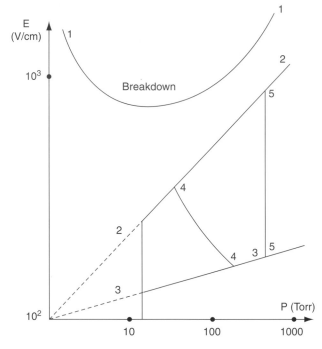

Figure 4.56 *Three regimes of moderate-pressure microwave discharges.*

wave into the filament, and most of the energy is absorbed in the strongly non-equilibrium plasma. The combined regime sustains the non-equilibrium plasma with high values of ionization degree and specific energy input.

This feature makes the regime interesting for plasma-chemical applications, where high energy efficiency is the most important factor. At high pressures (to the right of the curve 5–5), radiative heat transfer and the discharge front overheating become significant and the contracted discharge converts into the *thermal regime* (Figure 4.55c).

4.5 Coronas, DBDs, plasma jets, sparks and other non-thermal atmospheric-pressure streamer discharges

4.5.1 Corona and pulsed corona discharges

Of the several types of atmospheric-pressure cold discharges in air, *continuous corona* is probably the best known and the most widely applied. The continuous corona is a weakly luminous discharge which appears at atmospheric pressure near sharp points, edges or thin wires where the electric field is sufficiently large. Corona discharges are always non-uniform: strong electric field, ionization and luminosity are located in the vicinity of one electrode. Charged particles are dragged by the weak electric fields from one electrode to another to close the electric circuit. At the initial stages of the breakdown, the corona circuit is closed by a displacement current rather than by charged particle transport.

The polarity of an electrode where the high electric field is located distinguishes *negative corona* (around the cathode) and *positive corona* (around the anode). Ignition of the negative corona is based on the secondary emission from the cathode similarly to the Townsend breakdown:

$$\int_0^{x_{max}} [\alpha(x) - \beta(x)]\, dx = \ln\left(1 + \frac{1}{\gamma}\right), \tag{4.88}$$

where x_{max} is the distance from cathode where $\alpha(x_{max}) = \beta(x_{max})$. The distance $x = x_{max}$ can also be considered as the visible size of the corona. Ionization processes in positive corona are related to the cathode-directed streamers. Ignition can be described by the generalized Meek breakdown criterion:

$$\int_0^{x_{max}} [\alpha(x) - \beta(x)]\, dx \approx 18 - 20. \tag{4.89}$$

Taking into account that $\ln(1/\gamma) \approx 6-8$, ignition of the positive corona requires slightly higher electric fields. The igniting electric field for the coaxial electrodes in air, E_{cr}, can be calculated in units of kV cm^{-1} using the empirical *Peek formula*:

$$E_{cr} = 31\,\delta\left(1 + \frac{0.308}{\sqrt{\delta r}}\right), \tag{4.90}$$

where δ is the ratio of air density to the standard value and r is the radius of the internal electrode. The critical corona-initiating electric field in the case of two parallel wires in air can be calculated using a similar formula:

$$E_{cr} = 30\,\delta\left(1 + \frac{0.301}{\sqrt{\delta r}}\right). \tag{4.91}$$

Current-voltage characteristic $I(V)$ of a corona around a thin wire with radius r, length L and characteristic size R can be expressed:

$$I = \frac{4\pi \varepsilon_0 \mu V (V - V_{cr})L}{R^2 \ln(R/r)}, \tag{4.92}$$

where V_{cr} is the corona ignition voltage and μ is mobility of the charged particles providing conductivity outside the active corona volume. Noting that mobilities of positive and negative ions are nearly equal, the electric currents in positive and negative corona discharges are also close. Negative corona in gases without electron attachment (e.g. noble gases) provides much larger currents because electrons are able to rapidly leave the discharge gap without forming a significant space charge. Parabolic current-voltage characteristics such as Equation (4.92) are valid not only for thin wires, but also for other corona configurations:

$$I = \text{const} \times V(V - V_{cr}). \tag{4.93}$$

For example, a relevant expression for corona generated in atmospheric air between a sharp point cathode with radius $r = 3$–50 µm and a perpendicular flat anode of distance $d = 4$–16 mm is:

$$I = \frac{52}{d^2} V (V - V_{cr}).$$ (4.94)

The corona ignition voltage in this case $V_{cr} = 2.3$ kV does not depend on d. Based on the current-voltage characteristic described by Equation (4.93), power released in the corona discharge is:

$$P = \text{const} \times V^2 (V - V_{cr}).$$ (4.95)

For example, corona discharges generated in atmospheric-pressure air around the thin wire ($r = 0.1$ cm, $R = 10$ cm, $V_{cr} = 30$ kV) with voltage 40 kV releases relatively low power of about 0.2 W cm^{-1}. An increase of voltage and power of the continuous coronas results in a transition to sparks. Applications of the continuous coronas are limited by low current and power, which results in low rate of treatment of materials.

Increasing the corona power without transition to sparks becomes possible by using pulse-periodic voltages. The *pulsed corona* is one of the promising atmospheric-pressure non-thermal streamer discharges. The streamer velocity is about 10^8 cm s^{-1} and exceeds the typical electron drift velocity in an avalanche by a factor of 10. If the distance between electrodes is about 1–3 cm, the total time necessary for the development of avalanches, avalanche-to-streamer transition and streamer propagation between electrodes is about 100–300 ns. The voltage pulses of this duration are therefore able to sustain streamers and effective power transfer into non-thermal plasma without streamer transformations into sparks.

It should be mentioned that even continuous coronas are characterized by non-steady-state effects, including ignition delay, flashing and *Trichel pulses*. For the pulsed corona discharges, the key is development of the pulse power supplies generating sufficiently short voltage pulses with a steep front and short rise times. The nanosecond pulse power generates pulses with a duration 100–300 ns, sufficiently short to avoid the corona-to-spark transition. The power supply should provide a high voltage rise rate (0.5–3 kV ns^{-1}), which results in higher corona ignition voltage and higher power.

The high voltage rise rates also result in better efficiency of several plasma-chemical and plasma-medical processes requiring higher electron energies. In such processes, high values of the mean electron energy are necessary to decrease the fraction of the discharge power going to vibrational excitation of molecules, which stimulates ionization, electronic excitation and dissociation of molecules. The nanosecond pulse power supplies used for the pulsed coronas include Marx generators, simple and rotating spark gaps, thyratrons, thyristors with possible further magnetic compression of pulse and special transistors for the high-voltage pulse generation. The pulsed corona can be relatively powerful and sufficiently luminous. One of the most powerful pulsed corona discharges (10 kW in average power) operates in Philadelphia in the Drexel Plasma Institute (Figure 4.57).

4.5.2   Dielectric-barrier discharges (DBDs)

The corona-to-spark transition is prevented in pulsed corona by employing a nanosecond pulse power supply. Another approach to avoiding the spark formation in streamer channels is based on use of a dielectric barrier in the discharge gap that stops current and prevents spark formation. Such a discharge is called the dielectric barrier discharge (DBD). The presence of a dielectric barrier precludes DC operation of DBD, which usually operates at frequencies 0.05–500 kHz. Sometimes DBDs are called *silent discharges* due to the absence of sparks, which are accompanied by local overheating, generation of local shock waves and noise.

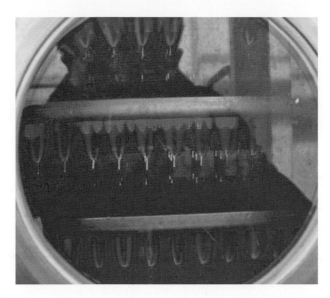

Figure 4.57   *10 kW pulsed corona discharge; view from one of six windows (see color plate).*

DBD has numerous applications because it operates under strongly non-equilibrium conditions at atmospheric pressure of different gases, including air, at reasonably high power levels and (in contrast to the pulsed corona) without using sophisticated pulse power supplies. DBD is widely applied for ozone generation, in UV-sources and excimer lamps, in polymer treatment (particularly to promote wettability, printability and adhesion), in plasma display panels (including plasma TV) and recently more and more for biological and medical applications. DBD can occur in a number of individual tiny breakdown channels, referred to as microdischarges, and has been intensively investigated regarding its relationship with streamers. The DBD gap includes one or more dielectric layers which are located in the current path between metal electrodes.

Two specific DBD configurations, planar and cylindrical, are illustrated in Figure 4.58. Typical clearance in the discharge gaps varies from 0.1 mm to several centimeters. Breakdown voltages of these gaps with dielectric barriers are practically the same as those between metal electrodes. If the DBD gap is a few millimeters, the required AC driving voltage with frequency 500 Hz–500 kHz is typically about 10 kV at atmospheric pressure. The dielectric barrier can be made from glass, quartz, ceramics or other materials of low dielectric loss and high breakdown strength.

Conventional DBD is not uniform and consists of numerous microdischarges distributed in the discharge gap, often moving between and interacting with each other. Formation of the DBD microdischarges, their plasma parameters and the interaction between them resulting in formation of the microdischarge patterns is an interesting and practically important subject. In some special cases, particularly in helium, DBD can be uniform without any streamers and microdischarges. Several special DBD modifications important for plasma-medical applications are discussed in the following sections.

4.5.3   Special modifications of DBD: Surface, asymmetric, packed bed and ferroelectric discharges

Closely related to DBDs are *surface discharges*, generated at dielectric surfaces embedded by metal electrodes and supplied by AC or pulsed voltage. The dielectric surface essentially decreases the breakdown voltage because of significant non-uniformities of electric field and local overvoltage. Effective decrease of the

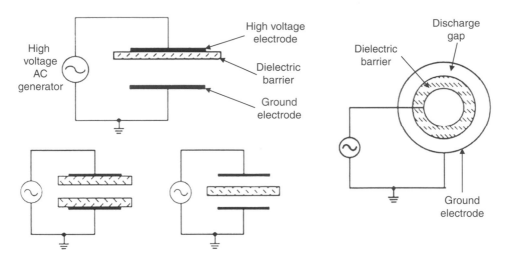

Figure 4.58   *Common DBD configurations.*

breakdown voltage can be reached in the surface discharge configuration (called the *sliding discharge*) with one electrode located on the dielectric plate and other electrode partially wrapped around (see Figure 4.59). A similar DBD configuration is called sometimes an *asymmetric DBD*. The sliding discharge can be quite uniform on large surfaces with linear sizes over 1 m at voltages not exceeding 20 kV. The component of the electric field normal to the dielectric surface plays an important role in generation of the pulse-periodic sliding discharge, which does not depend on the distance between electrodes along the dielectric.

   Two different modes of the surface discharges can be achieved by changing voltages: complete mode or the *sliding surface spark* and incomplete mode or the *sliding surface corona*. The sliding surface corona is ignited at voltages below the critical voltage and has a low current limited by charging the dielectric capacitance. Active volume and luminosity is localized near the igniting electrode and does not cover all the dielectric. The sliding surface spark occurs at voltages exceeding the critical voltage corresponding to breakdown. The plasma channels actually connect electrodes of the surface discharge gap. At low overvoltages, the breakdown delay is about 1 µs. The multi-step breakdown phenomenon starts with the propagation of a direct ionization wave, which is followed by a possibly more intense reverse wave related to the compensation of charges left on the dielectric surface. At higher overvoltages, the breakdown delay becomes shorter (nanoseconds). The complete mode takes place immediately after the direct ionization wave reaches the opposite electrode.

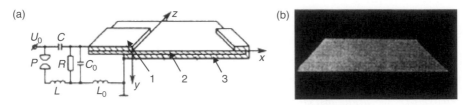

Figure 4.59   *(a) Illustration and (b) emission picture of a pulsed surface discharge (1: initiating electrode; 2: dielectric; 3: shielding electrode).*

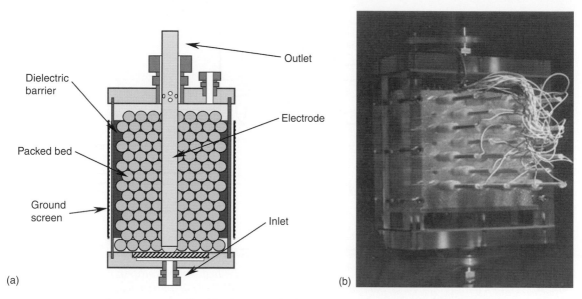

(a)   (b)

Figure 4.60   *Packed-bed corona discharge: (a) illustration and (b) photo.*

The surface discharge consists of many current channels in this regime. In general, the sliding spark surface discharge is able to generate the luminous current channels of very sophisticated shapes, usually referred to as the *Lichtenberg figures*.

The *packed-bed discharge* is a combination of DBD and sliding surface discharges. High AC voltage (about 15–30 kV) is applied to a packed bed of dielectric pellets and creates a non-equilibrium plasma in the void spaces between the pellets. The pellets refract the electric field, making it non-uniform and stronger than the externally applied field by a factor of 10–250 depending on the shape, porosity and dielectric constant of the pellets. A packed-bed discharge is depicted in Figure 4.60. The inner electrode is connected to a high-voltage AC power supply operated at 15–30 kV at a frequency of 60 Hz. The glass tube serves as a dielectric barrier to inhibit direct charge transfer between electrodes and as a plasma-chemical reaction vessel.

A special mode of DBD (called the *ferroelectric discharge*) uses the ferroelectric ceramics with high dielectric permittivity ($\varepsilon > 1000$) as the dielectric barriers. Ceramics based on $BaTiO_3$ are mostly employed for the discharges. The ferroelectric materials can have significant dipole moment in the absence of external electric field. An external AC voltage leads to overpolarization of the ferroelectric material and reveals strong local electric fields on the surface, which can exceed 10^6 V cm^{-1} and stimulates the discharge on the ferroelectric surfaces. Active volume of the ferroelectric discharge is located in the vicinity of the dielectric barrier. The discharge can be arranged, for example, using a packed bed of ferroelectric pellets. Non-equilibrium plasma is created in such a system in the void spaces between the pellets.

4.5.4   OAUGDP as quasi-homogeneous DBD (APG) modification

One-atmosphere uniform-glow discharge plasma (OAUGDP) was initially developed at the University of Tennessee by Roth and his colleagues. OAUGDP is similar to a traditional DBD, but it can be much more uniform which has been interpreted as being a result of the ion-trapping mechanism. The discharge transition from filamentary to diffuse mode in atmospheric air has been analyzed in the system depicted by Figure 4.61.

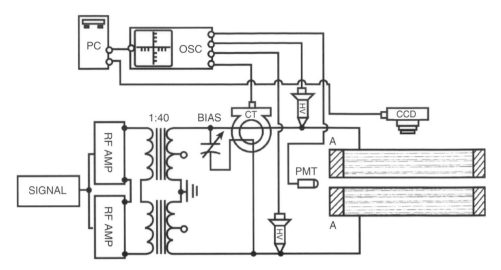

Figure 4.61 *OAUGDP experimental setup. A: water electrodes; HV: high-voltage probes; PMT: photomultiplier probe; CT: current transformer; BIAS: parasitic current elimination tool; C_V: variable capacitor; OSC: oscilloscope; PC: computer; SIGNAL: harmonic signal generator; RF AMP: radio frequency power amplifier; CCD: digital camera (Roth, 2006).*

The transition can lead not only to the diffuse mode but also to another non-homogeneous mode where the filaments are more numerous and less intense.

The stability of the filaments and transition to uniformity is related to the 'memory effect'. In particular, electrons deposited on the dielectric surface promote the formation of new streamers at the same place again and again by adding their own electric fields to the external electric field. The key feature of OAUGDP in promoting the transition to uniformity may be hidden in the properties of particular dielectrics that are not stable in plasma and probably become more conductive during a plasma treatment. In particular, plasma can further increase the conductivity of borosilicate glass used as a barrier, for example, by UV radiation. It transforms the discharge into the resistive barrier discharge (RBD) working at high frequency in the range of 1–15 kHz.

It is not only volumetric but also surface conductivity of the dielectric which can promote the DBD uniformity, if it is in an appropriate range. The memory effect can be suppressed by removing the negative charge spot formed by electrons of a streamer during the half period of voltage oscillation (i.e. before polarity changes), and the surface conductivity can help with this. On the other hand, when the surface conductivity is very high, the charge cannot accumulate on the surface during DBD current pulses of several nanoseconds and cannot stop the filament current.

4.5.5   Electronically stabilized DBD in APG discharge mode

DBD can be electronically stabilized in quasi-uniform APG mode. Uniform plasma has been generated, in particular, in argon DBD during the first cycles of voltage oscillations with relatively low amplitude (i.e. $\alpha d \sim 3$). Existence of the Townsend discharge at such low voltages requires an unusually high secondary electron emission coefficient (above 0.1). Such a high electron emission and the breakdown during the first low-voltage oscillations can probably be explained by taking into account the low surface conductivity of most polymers applied as barriers in the system. Surface charges occurring due to cosmic rays can be then

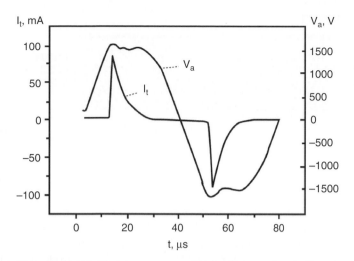

Figure 4.62   *Current and voltage wave-forms in electronically stabilized DBD in argon with active displacement current control (Aldea et al., 2005, 2006).*

easily detached by the applied electric field. A long induction time of the dark discharge is not required in this case, in contrast to the OAUGDP with glass electrodes.

Assuming that the major cause of the DBD filamentation is instability leading to the glow-to-arc transition, it has been suggested by Van De Sanden and his group that the glow mode is stabilized using an electronic feedback to fast current variations. Figure 4.62 shows the total current and voltage waveforms for DBD in argon with an active displacement current control achieved in the system. If the plasma is in series with a dielectric, an RC circuit is formed. The filaments are characterized by higher current densities and a smaller RC constant. Differences in RC constant can be used to 'filter' the filaments because they react differently to a drop in displacement current (displacement current pulse) for different frequency and amplitude. A simple LC circuit in which the inductance is saturated during the pulse generation has been used to generate the displacement current pulses.

This method of electronic uniformity stabilization has used for relatively high power densities (of the order 100 W cm^{-3}) and in a large variety of gases including Ar, N_2, O_2 and air.

4.5.6   Arrays of DBD-based microdischarges and kilohertz-frequency microdischarges

The power of a microdischarge is so small that individual microdischarges have limited applications. Some industrial applications therefore require the construction of microdischarge arrays or microplasma-integrated structures. The plasma TV is an example of such a complex structure. The simplest structure may consist of multiple identical microdischarges electrically connected in parallel.

For stable operation of such structures, each discharge should have a positive differential resistance. Most microdischarges have this property as a result of a significant increase in the power losses with a current increase. An example is the array consisting of microdischarges with inverted, square pyramidal cathodes (Figure 4.63). An optical micrograph of the 3×3 array of microdischarges of 50×50 μm each, separated (center-to-center) by 75 μm, has been made by Park *et al.* (2001). All of the microdischarges (700 Torr of Ne) have a common anode and cathode, that is, the devices are connected in parallel. Ignition voltage and current for the array are 218 V and 0.35 μA. The array is unable to operate at a high power loading (433 V and 21.4 μA) and emission from each discharge is spatially uniform.

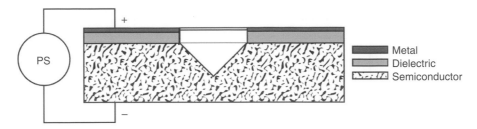

Figure 4.63   *Single DBD-based microdischarge from the array described by Park* et al. *(2001) and numerically simulated by Kushner (2004).*

Another example of a microdischarge arrays is the so-called *fused hollow cathode* (FHC) developed by Bárdos and Baránková (2001), based on the simultaneous RF generation of HCD plasmas in an integrated open structure with flowing gas. The resulting discharges are stable, homogeneous, luminous and volume filling without streamers. The power is of the order 1 W cm^{-2} of the electrode structure area. Experiments have been carried out with the system having a total discharge area 20 cm^2. The concept of the source is extremely suitable for scaling-up different gas throughputs.

AC microdischarges can be generated at different frequencies. Low- and medium-frequency AC microdischarges are related to DBD. A relatively simple large-area plasma source based on the micro-DBD approach has been developed by Tachibana and his group. An integrated structure called the coaxial-hollow micro dielectric-barrier discharge (CM-DBD) has been made by stacking two metal meshes covered with a dielectric layer of alumina with thickness about 150 μm. A test panel (diameter 50 mm) with hundreds of hollow structures (0.2 × 1.7 mm) was assembled. He or N$_2$ were used as the plasma gases at pressures 20–100 kPa and voltage below 2 kV, even at the maximum pressure. Bipolar square-wave voltage pulses were applied to one of the mesh electrodes. The pulse duration of both positive and negative voltages varied from 3 to 14 μsec with an intermittent time of 1 μs and repetition frequency 10 kHz. In each coaxial hole, the discharge occurs along the inner surface. The intensity of each microdischarge is uniform over the whole area. The extended glow with a length of some millimeters is observed in He but not in N$_2$. The electron density in He at 100 kPa is about 3 × 10^{11} cm^{-3}. The CM-DBD configuration therefore has a rather low operating voltage (typically 1–2 kV); the scaling parameter *pd* is several tens of Pa m, corresponding to the Paschen minimum. Plasma in the system is stable over a wide range of external parameters without filamentation or arcing.

Another kHz-range microdischarge is the capillary plasma electrode (CPE) discharge, proposed by Becker and his group. The CPE discharge uses an electrode design which employs dielectric capillaries that cover one or both electrodes. Although the CPE discharge looks similar to a conventional DBD, the CPE discharge exhibits a mode of operation that is not observed in DBD: the capillary jet mode. The capillaries with diameter 0.01–1 mm and a length-to-diameter (L/D) ratio from 10:1 to 1:1 serve as plasma sources and produce jets of high-intensity plasma at high pressure. The jets emerge from the end of the capillary and form a 'plasma electrode'.

The CPE discharge displays two distinct modes of operation when excited by pulsed DC or AC. When the frequency of the applied voltage pulse is increased above a few kHz, a diffuse mode similar to the diffuse DBD as described by Okazaki and Kogoma (1993) is first observed. When the frequency reaches a critical value (which depends on the L/D value and the plasma gas), the capillaries become 'turned on' and bright intense plasma jets emerge from the capillaries. When many capillaries are placed in close proximity to each other, the emerging plasma jets overlap and the discharge appears uniform. This capillary mode is the preferred mode of operation of the CPE discharge, and is somewhat similar to the fused hollow cathode (FHC). At high frequency even dielectric capillaries can work as hollow cathodes; for CCP-RF plasma in the gamma mode,

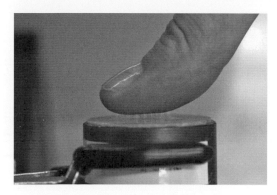

Figure 4.64   *Floating-electrode dielectric barrier discharge (FE-DBD) (see color plate).*

a dielectric surface is also a source of the secondary emitted electrons similarly to metal cathodes in glow or hollow-cathode discharges.

4.5.7   Floating-electrode dielectric barrier discharge (FE-DBD)

Developed at Drexel Plasma Institute, floating electrode dielectric barrier discharge (FE-DBD) is probably the most applied cold plasma source in plasma medicine today (see Figure 4.64).

The principal of operation of FE-DBD plasma can be explained with the help of a relatively simple model. Consider the insulated DBD electrode as a sphere of diameter D_{el} while the 'biological' object whose surface is being treated is modeled as a sphere of diameter D_{ob}. In the absence of the object, the electrode capacitance with respect to the far away (located at infinity) ground is $C_{el} = 2\pi\varepsilon_0 D_{el}$. If the biological object being treated has a relatively high dielectric constant (such as water), it effectively expels most of the electric field from its interior when it is brought close to the electrode. From that point of view this object behaves like a good conductor and its capacitance with respect to the far away ground can therefore also be modeled as $C_{ob} = 2\pi\varepsilon_0 D_{ob}$. The region between the object and the electrode can be modeled roughly as a parallel plate capacitor $C_{gap} = \pi\varepsilon_0 D_{el}^2/2g$ (where g is gap distance) if the gap is significantly smaller than the electrode diameter. Note that:

$$\frac{C_{gap}}{C_{ob}} = \frac{D_{el}^2}{4g D_{ob}} \ll 1 \tag{4.96}$$

for the typical choices of characteristic sizes. In the absence of any conduction current, the electrical models of the electrode by itself and the electrode near the treated object are well approximated by the circuits shown in Figure 4.65(a and b). When the electrode is removed from the ground, the magnitude of the applied voltage V is insufficient to create an electric field strong enough to cause the breakdown and discharge. However, when the object with a high dielectric constant is sufficiently close to the electrode, most of the applied voltage appears across the gap. This is because the capacitance of the object with respect to ground is much larger than the gap capacitance, and the voltage divides across these capacitors proportionally to the inverse of their size. This results in a strong electric field in the gap, which can then lead to breakdown and discharge.

The electrical circuit model can be further refined by taking into account non-linear resistance and the capacitance of the plasma created in the gap; the resulting circuit refinement is shown in Figure 4.65(c). The refined circuit does not change the main conclusion that most of the applied voltage appears across the plasma gap. At about 10 kHz, the following circuit parameters typical for our experiments can be estimated assuming

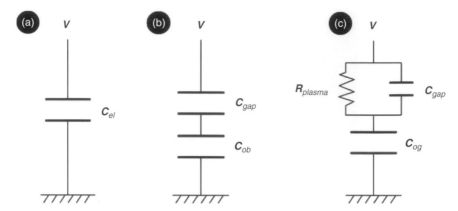

Figure 4.65   *Illustration of: (a) electrode itself, (b) electrode near the treated biological object and (c) plasma discharge on the treated object.*

that body diameter is roughly 1 m, the electrode diameter is about 25 mm and the gap is about 1 mm; gap capacitance is 4 pF; capacitive resistance (impedance) of the gap is 4.2 MΩ; plasma resistance is 5–10 MΩ; the biological object capacitance is 50 pF; and the capacitive resistance (impedance) of the object is 300 kΩ. Electrical safety of the biological object being directly treated by FE-DBD plasma is ensured because the current which the power supply delivers is less than 5 mA. Although such currents may cause some mild discomfort, they do not cause muscle or cardiovascular malfunction in a human and are therefore deemed safe by the US Occupational Safety and Health Administration.

The simplest FE-DBD system is based on a conventional dielectric barrier discharge and is basically a system driven by AC high voltage applied between two conductors where one or both are covered with a dielectric to limit the current and to prevent transition to an arc. Amplitude and waveform of the high-voltage signal are obviously quite important, which is discussed in the following section.

A simplified schematic of the plasma-medical treatment setup is shown in Figure 4.66. Here, the signal of any frequency, amplitude and waveform is current-amplified and voltage is then stepped up in the transformer,

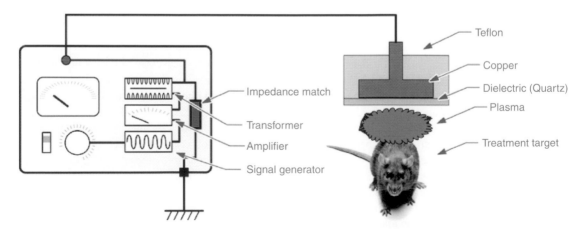

Figure 4.66   *General schematic of FE-DBD plasma-medical treatment of living tissue.*

which is then connected to the 'powered electrode'. The electrode is basically a conductor covered by a dielectric. Plasma is then generated between the surface of the dielectric and the treatment target.

4.5.8 Micro- and nanosecond pulsed uniform FE-DBD plasma

One of the key issues for the plasma-medical application of FE-DBD in air is its uniformity. The most promising method of uniform plasma generation in DBD is the application of very short pulses, shorter than the time required to create the DBD filaments. The way in which the voltage is delivered to the electrode can be quite important in FE-DBD; the main issue is the rise time of the voltage. In dielectric barrier discharges at atmospheric pressure, breakdown occurs following the spark breakdown mechanism. The initial electron avalanche transitions to streamer occur in approximately 20–60 ns. If the voltage is increased slowly (as is the case with sinusoidal excitation wave c. 1 V ns^{-1}), the number of streamers occurring during a single cycle can be quite high.

Since the initial streamer forms a pre-ionized channel, the probability of the next streamer striking in the same position is increased. In the case of a sinusoidal excitation wave, multiple streamers strike at the same position creating a rather energetic channel called a microfilament. The temperature in this channel can approach very high values (see Figure 4.67).

To solve this problem, it is helpful to employ pulses rather than a sinusoidal wave, allowing for a voltage rise time of c. 5 V ns^{-1} and a pulse duration of a few μs. The time between pulses is sufficient for the gas to return to the initial conditions so that, during the next pulse, streamers again strike randomly. These power supply systems operate on the basis of voltage amplification: a low voltage pulse or a sinusoidal wave is generated and the current is amplified and passed through a transformer with a high turns ratio. A high-voltage signal is then taken from the transformer's secondary coil and collected to the load.

DBD plasmas are generally classed as non-thermal as the temperature of ions and neutrals is significantly lower than that of electrons. However, especially in more filamentary regimes, the temperature in the filament can rise above room temperature and potentially damage the surface being treated, particularly in the case of sensitive surfaces such as polymer surface treatment or in biological applications. As can be seen from Figure 4.67, during a microsecond pulse the gas does not heat up in DBD. However, there is sufficient time for streamer formation and development into filaments. This way the discharge is more uniform than in the case of a sinusoidal excitation wave, but a non-uniformity is still observed as can be seen from the short-exposure photograph through the water electrode below (Figure 4.68). This way the treated surface is covered fairly uniformly in as little as a third of a second or less.

For some applications a more uniform plasma treatment might be needed, however. For this, a different power supply with a shorter pulse duration (high voltage pulses with short rise time of 5–20 ns) and a high pulse repetition frequency (up to 2000 Hz) has been applied. Figure 4.69 presents a Drexel Plasma Institute setup to capture a single FE-DBD nanopulse on photographic film. This setup allows photographic film to be quickly moved through plasma discharge. When the discharge is in pulsed mode and the time between pulses is kept sufficiently long (a few hundred Hertz), it is possible to obtain an image of a single pulse of plasma on the film plane. The result is a Lichtenberg figure of a completely uniform plasma field, shown in Figure 4.70. The nanosecond pulsed FE-DBD is therefore able to generate a completely uniform and filament-free plasma at atmospheric pressure and temperature in open air, very important for sterilization and general plasma-medical treatment of living tissues.

4.5.9 Spark discharges

When a streamer connects the electrodes in the absence of either a pulse power supply or a dielectric barrier which would prevent further growth of current, a spark may develop. The initial streamer channel is not very

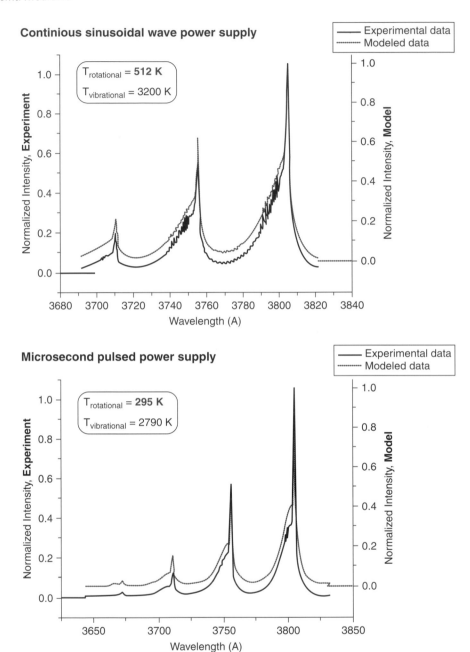

Figure 4.67   *Comparison of microsecond and nanosecond pulsed FE-DBD: rotational and vibrational (estimate) temperature measurements from second positive nitrogen system.*

(a)   (b)

Figure 4.68   *Photograph of microsecond pulsed FE-DBD plasma through water electrode: (a) 0.33 s shutter open time and (b) 0.0031 s shutter open time (see color plate).*

conductive and provides only a low current of about 10 mA. The fast ionization processes lead to a higher ionization degree, to higher current and to intensification of the spark. When the streamer approaches the cathode, its electric field grows and stimulates intensive formation of electrons in the cathode vicinity. New ionization waves much more intense than the original streamer start propagating along the streamer channel but in the opposite direction (from the cathode to anode) with velocities of up to 10^9 cm s^{-1}, which is referred

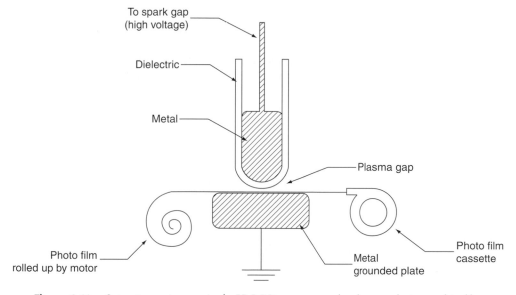

Figure 4.69   *Setup to capture a single FE-DBD nanosecond pulse on photographic film.*

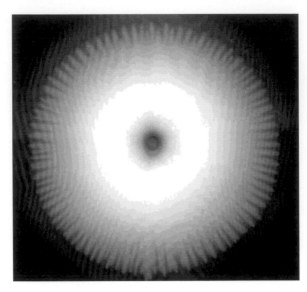

Figure 4.70 *Nanosecond-pulsed FE-DBD image produced on black and white photographic film (100 ISO sensitivity negative film).*

to as the *back ionization wave*. The high velocity of the back ionization wave is not directly the velocity of electron motion, but rather the phase velocity of the ionization wave. The back wave is accompanied by a front of intensive ionization and the formation of a plasma channel with sufficiently high conductivity to form a channel for the intensive spark. The radius of the powerful spark channel grows up to about 1 cm, which corresponds to a spark current of 10^4–10^5 A and current densities of 10^4 A cm^{-2}. The plasma conductivity becomes high and a cathode spot can be formed on the electrode surface. The voltage between the electrodes decreases, and the electric field becomes about 100 V cm^{-1}.

If the voltage is supplied by a capacitor, after reaching a maximum the spark current also decreases. The sparks can be modified by synergetic application of high voltages with laser pulses. The laser beams can direct spark discharges not only along straight lines but also along more complicated trajectories. Laser radiation is able to stabilize and direct the spark channel in space due to three major effects: local preheating, local photoionization and optical breakdown of the gas. At the laser radiation density of 30 J cm^{-2}, the breakdown voltage decreases by an order of magnitude. The length of the laser-supported spark in these experiments was up to 1.5 m.

Photoionization by laser radiation is able to stabilize and direct coronas by means of local pre-ionization of the discharge channel without significant change in the gas density. UV-laser radiation (for example, Nd-laser or KrF-laser) should be applied in this case. The most intensive laser effect on spark generation can be provided by the optical breakdown of the gases. The length of such a laser spark can exceed 10 m. The laser spark in pure air requires power density of Nd-laser ($\lambda = 1.06$ μm) exceeding 10^{11} W cm^{-2}.

Plasma generated by well-developed sparks can be qualified as thermal, and gas temperatures in these cases can be quite high and approach those of arc discharges. The application of spark discharges to biology and medicine therefore requires the use of special configurations to protect living tissue from the direct effect of high gas temperature, discussed in the following section.

Figure 1.1 *Non-thermal short-pulsed 40 kV FE-DBD plasma sustained directly between a dielectric-coated electrode and a human body.*

Figure 1.2 *In the 1950s, Stanley Miller of the University of Chicago synthesized amino acids in plasma from methane and inorganic compounds.*

Figure 1.13 *Professor Mounir Laroussi of Old Dominion University working with the non-thermal atmospheric-pressure plasma jet, a convenient tool for topical focused plasma treatment of living tissues.*

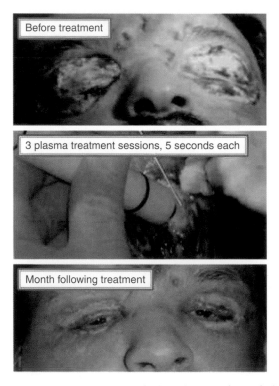

Figure 1.15 *Result of plasma treatment sessions (top: before; lower: after) of plasma treatment (middle) of a complicated ulcerous eyelid wound.*

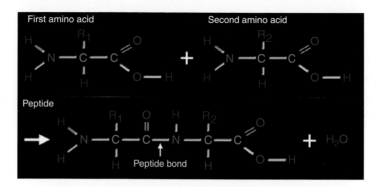

Figure 3.10 *Polypeptide formation and the peptide bond.*

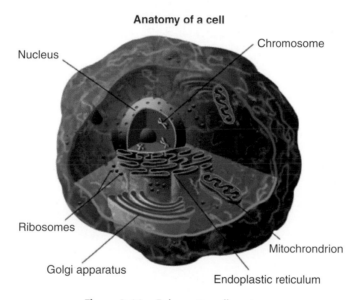

Figure 3.16 *Eukaryotic cell anatomy.*

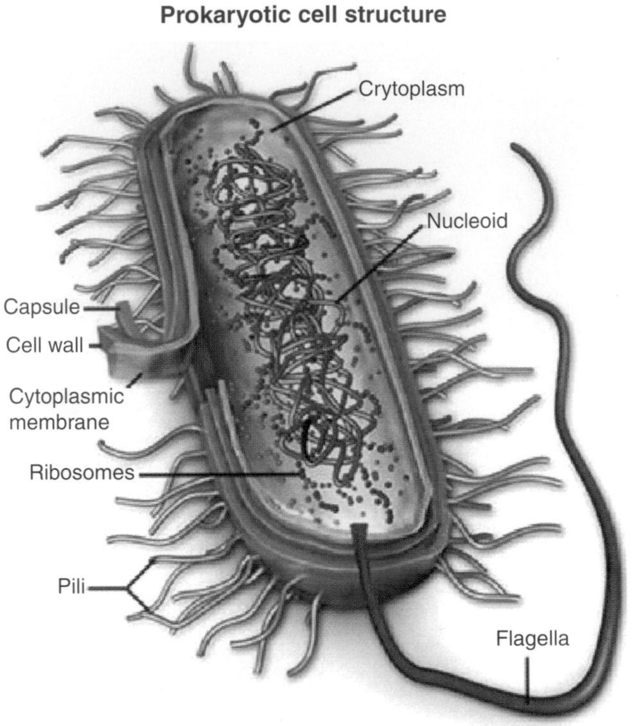

Prokaryotic cell structure

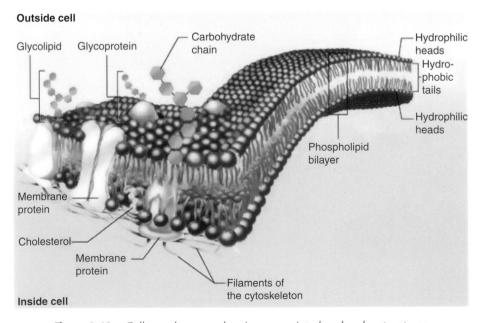

Crytoplasm

Nucleoid

Capsule

Cell wall

Cytoplasmic
membrane

Ribosomes

Pili

Flagella

Figure 3.17   *Anatomy of a bacterial cell.*

Outside cell

Glycolipid   Glycoprotein

Carbohydrate
chain

Hydrophilic
heads

Hydro-
phobic
tails

Hydrophilic
heads

Phospholipid
bilayer

Membrane
protein

Cholesterol

Membrane
protein

Filaments of
the cytoskeleton

Inside cell

Figure 3.18   *Cell membrane and various associated molecular structures.*

(a) A channel protein

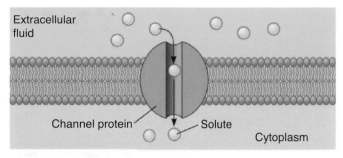

(b) A carrier protein

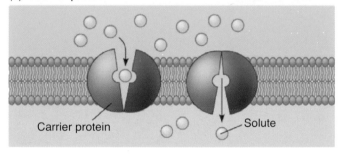

Figure 3.19 *Illustration of facilitated transport processes across lipid membranes.*

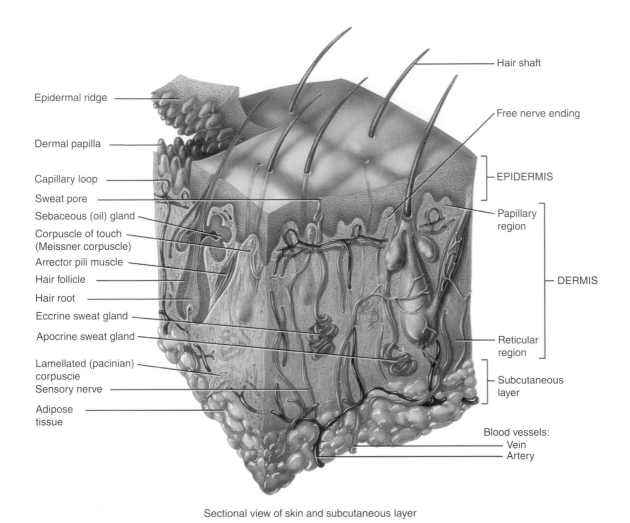

Epidermal ridge

Dermal papilla

Capillary loop

Sweat pore

Sebaceous (oil) gland

Corpuscle of touch
(Meissner corpuscle)

Arrector pili muscle

Hair follicle

Hair root

Eccrine sweat gland

Apocrine sweat gland

Lamellated (pacinian)
corpuscie

Sensory nerve

Adipose
tissue

Hair shaft

Free nerve ending

EPIDERMIS

Papillary
region

DERMIS

Reticular
region

Subcutaneous
layer

Blood vessels:
Vein
Artery

Sectional view of skin and subcutaneous layer

Figure 3.27 *Structure of skin and various processes originating from it.*

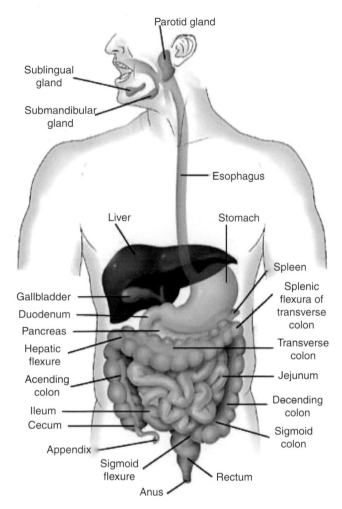

Figure 3.31 *Illustration of the digestive system.*

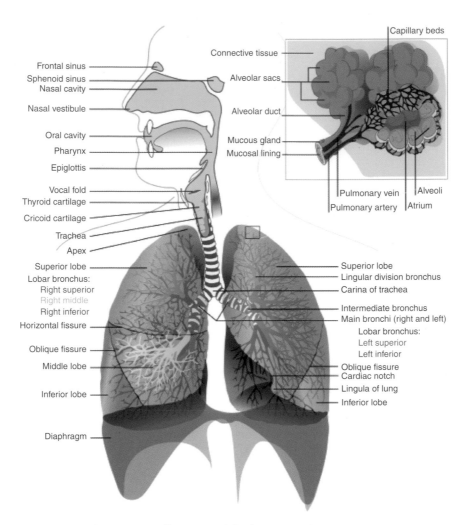

Figure 3.34 *Illustration of the human respiratory system.*

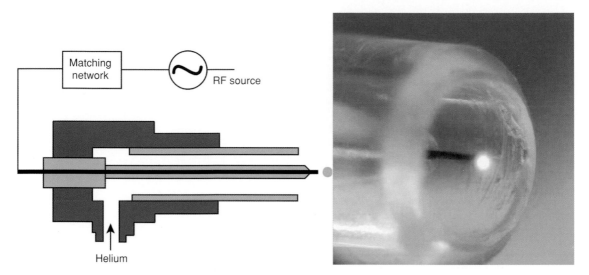

Figure 4.51 *Schematic (left) and photo (right) of the plasma-needle discharge. Reproduced from New Journal of Physics: Plasma medicine: an introductory review – Vol 11, 115012, 2009, with permission from IOP Publishing Ltd. OR Kieft, v.d. Laan & Stoffels, 2004.*

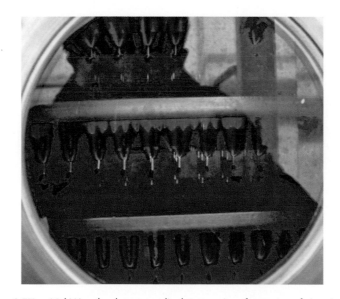

Figure 4.57 *10 kW pulsed corona discharge; view from one of six windows.*

Figure 4.64 *Floating-electrode dielectric barrier discharge (FE-DBD).*

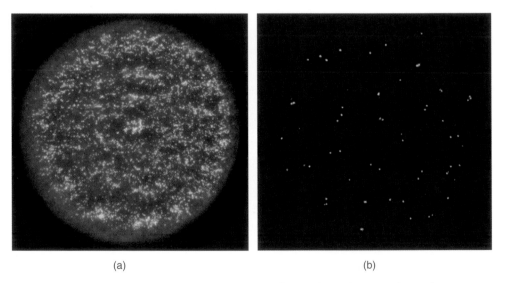

(a)   (b)

Figure 4.68 *Photograph of microsecond pulsed FE-DBD plasma through water electrode: (a) 0.33 s shutter open time and (b) 0.0031 s shutter open time.*

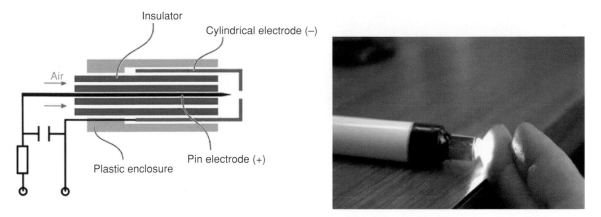

Figure 4.71 *PHD plasma system schematic (left) and in operation (photo, right) demonstrating that this plasma source is safe to the touch.*

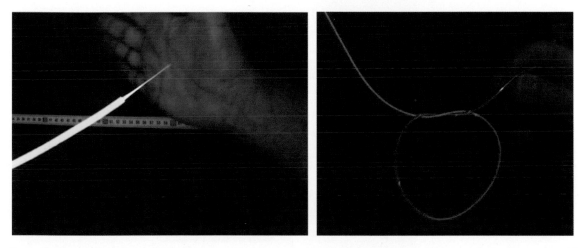

Figure 4.77 *Plasma gun shooting plasma 'bullets'. Left: 50 cm long, 4 mm in diameter dielectric guide with plume formation. Right: 50 cm long, 200 mm in diameter dielectric guide. The guide is twisted only for 'artistic' purposes (Robert et al., 2009).*

Figure 4.84 *Discharge emission dynamics; 1000 accumulations/image. The camera gate is 500 ps, λ = 250– 750 nm; image size 7 × 4 mm.*

Figure 4.88 Velocity of streamer propagation in water: 150 ps rise time; 1.5 MV ns^{-1}; growth rate vz = L/t ≈ 1 mm per 200 ps = 5000 km s^{-1}. Expansion rate vr = r/t ≈ 0.05 mm per 200 psec = 250 km s^{-1}.

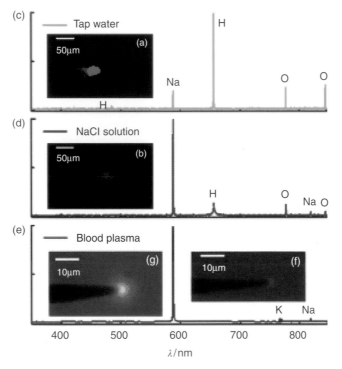

Figure 4.89 (a, b) OES and corona discharges produced using 5 kV stepped and pulsed voltage excitation in liquids. Demonstration of color changes as a function of the composition of the solution: (c) OES from tap water and (d) 0.5 M NaCl. (e) OES for 5 kV, 20 ns duration pulses in blood plasma; this spectrum does not show oxygen and hydrogen peaks and indicates that, for longer discharges, light emission is due to recombined ions and decomposition of the liquid (whereas in short discharges, light emission is due solely to recombined ions). Optical micrographs indicate that the (f) short-duration pulsed coronas are less than 3 μm in diameter and are significantly smaller than (g) discharges from microsecond-duration pulses, for which a larger discharge and bubble formation is seen.

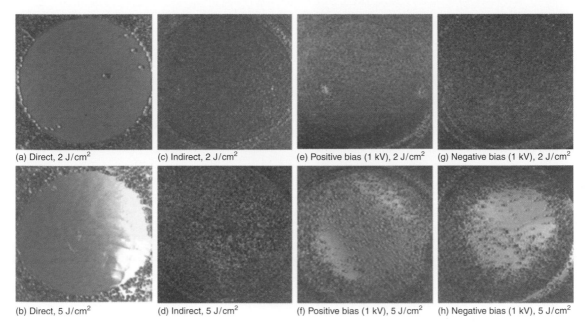

Figure 5.4 *Results of inactivation of* E. coli *on agar surface by (a, b) direct and (c–h) indirect plasma treatment. In the case of indirect treatment, the agar was either (c, d) grounded orDC-biased with (e, f) 1 kV (higher bias voltages were not employed due to occurrence of a corona discharge) positive polarity or (g, h) 1 kV negative polarity. For all cases, the plasma dose was kept at 2 J cm^{-2} (a, c, e, g) and 5 J cm^{-2} (b, d, f, h).*

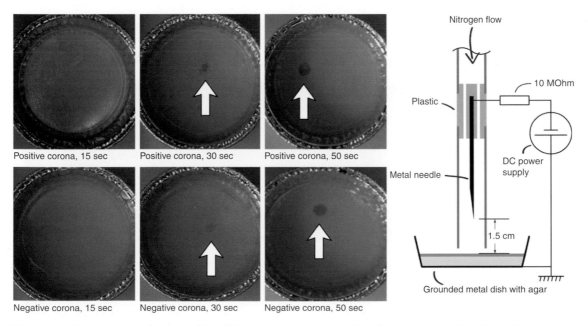

Figure 5.5 *Illustration and results of* E. coli *inactivation on agar surface by negative and positive polarity corona discharge in nitrogen.*

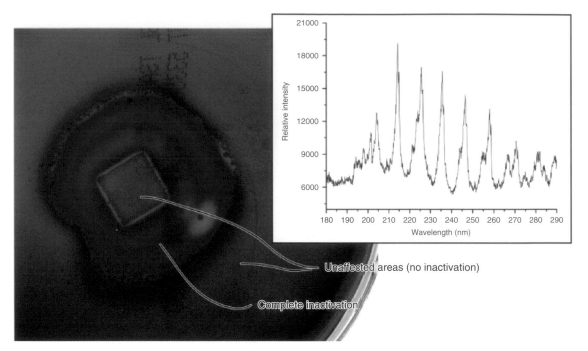

Figure 5.6 *Emission from DBD plasma over the cell surface in UV range (right) and results of inactivation of bacteria by direct plasma contact compared with only UV (left). No effect on bacteria protected from plasma by MgF$_2$ slide (10 mm square in the center) is observed. In this image the plasma surface power density was 0.6 W cm^{-2} and the treatment dose was 108 J cm^{-2}.*

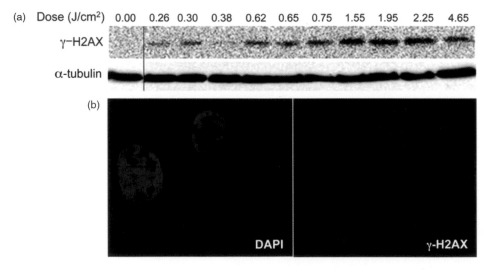

Figure 5.9 *Induction of DNA damage by DBD plasma. (a) MCF10A cells were treated with the indicated dose of plasma. After 1 hour of incubation, lysates were prepared and resolved by SDS-PAGE and representative immunoblots with antibody to γ-H2AX (top) or α-tubulin (bottom) are shown. (b) Indirect immunofluorescence was performed utilizing an antibody to γ-H2AX 1 hour after treatment of MCF10A cells with 1.55 J cm^2 DBD plasma.*

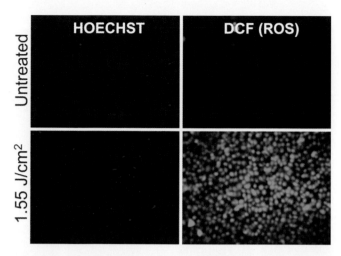

Figure 5.10 *CM-H$_2$DCFDA dye was preloaded into MCF10A cells for 30 min and the cells were then allowed to recover for another 30 min at 37°C. After recovery, the cells were treated with the indicated dose of DBD plasma (1.55 J cm^{-2}). One hour after the plasma treatment, intracellular ROS were detected using a fluorescence-enabled inverted microscope. The blue dye marks cell nuclei, while the green dye marks their fluorescence indicating ROS levels. Untreated cells clearly have much lower levels of ROS.*

Figure 5.11 *Confirming the effects of intracellular ROS by ROS scavenging. MCF10A cells were incubated for 2 hours with 4 mM NAC (+), followed by treatment with the indicated dose of DBD plasma. γ-H2AX (top) or α-tubulin (bottom) was detected by immunoblotting of cell lysates prepared 1 hour after plasma treatment.*

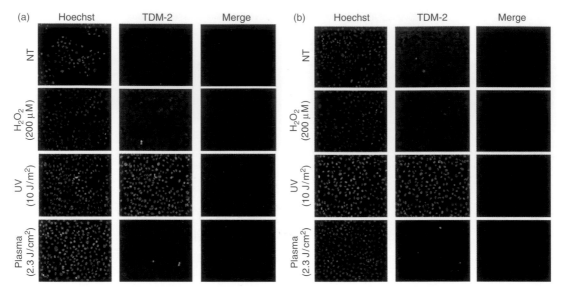

Figure 5.13 *Plasma treatment does not induce the formation of thymine dimers. MCF10A cells grown on glass coverslips were incubated (a) with or (b) without 4 mM NAC for 2 hours, followed by treatment with direct plasma (2.3 J cm^{-2}), UV (10 J m^{-2}), H$_2$O$_2$ (200 mM) or a no-treatment (NT) control. Cells were allowed to recover for 1 hour, then fixed and immunostained with TDM-2 primary antibody (kindly provided by Dr Toshio Mori at the Nara Medical University, Kashihara, Nara, Japan) and Alexa Fluor 594 anti-mouse secondary antibody to detect cyclobutane pyrimidine dimmers.*

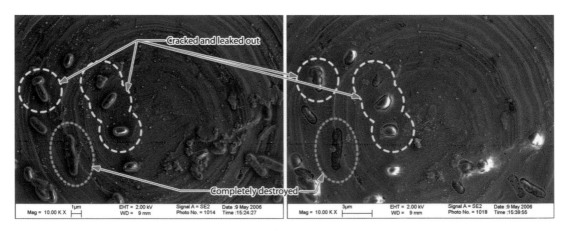

Figure 6.2 B. subtilis *before (left) and after (right) treatment with DBD plasma (120 s, 0.8 W cm^{-2}).*

Set screws

Powered electrode

Closed plastic
envelope containing
air-dried spores

Grounded stand

Figure 6.5   *Setup for DBD plasma treatment of air-dried spores contained inside a closed envelope.*

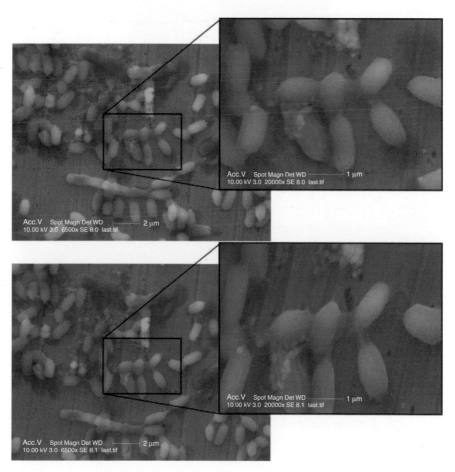

Figure 6.9   *SEM photomicrographs of* Bacillus cereus *spores on a stainless steel coupon (top) before and (bottom) after exposure to direct DBD plasma (same spot of sample in both photos).*

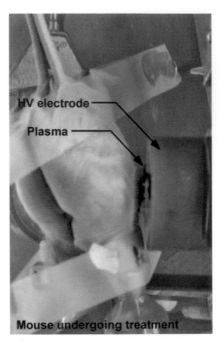

Figure 6.15 *FE-DBD direct treatment of living tissue: animal directly treated in plasma for up to 10 min remains healthy and no tissue damage is observed visually or microscopically immediately after and up to two weeks following the treatment.*

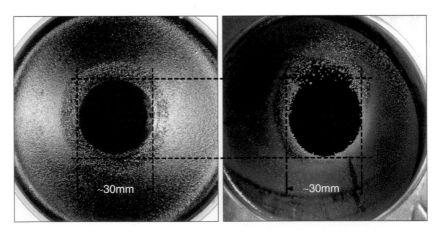

Figure 6.17 *Using a blower to shift sterilization region does not affect plasma and shows little effect of the afterglow: 15 seconds of treatment with (a) blower off and (b) blower on (air flows up).*

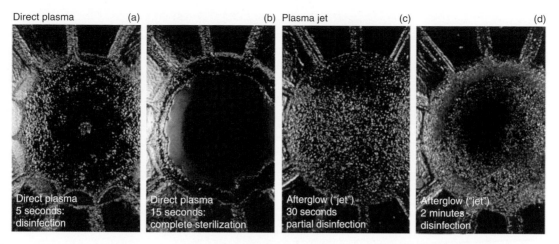

Figure 6.18 Direct application of plasma provides better sterilization efficiency than treatment by plasma afterglow: (a) 5 s and (b) 15 s of direct plasma compared with (c) 30 s and (d) 2 min of plasma jet.

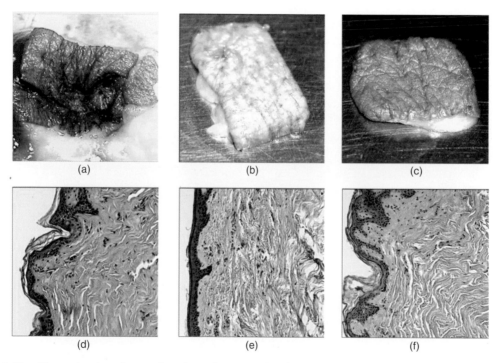

Figure 6.23 Photos (top) and tissue histology (bottom) of cadaver skin samples after FE-DBD treatment: (a, d) control; (b, e) after 15 s of treatment; and (c, f) after 5 min of treatment – no visible damage is detected.

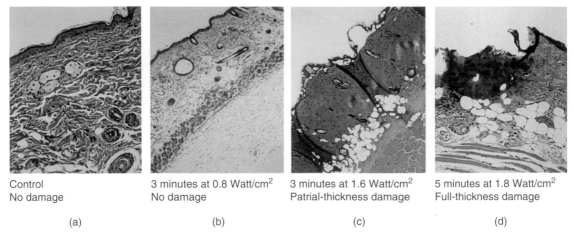

Control
No damage

3 minutes at 0.8 Watt/cm^2
No damage

3 minutes at 1.6 Watt/cm^2
Patrial-thickness damage

5 minutes at 1.8 Watt/cm^2
Full-thickness damage

(a)   (b)   (c)   (d)

Figure 6.25 *Histology of toxic and non-toxic to SKH1 skin plasma doses compared to untreated skin: (a) control; (b) 3 min at 0.8 W cm^{-2} (no damage); (c) 3 min at 1.6 W cm^{-2} (partial-thickness damage); and (d) 5 min at 1.8 W cm^{-2} (full-thickness damage).*

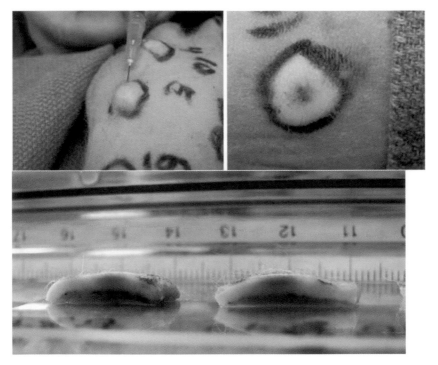

Figure 6.26 *Subcutaneous injection of the fluorescent dye (euthanized rat, top left), the rat skin after the plasma treatment (top right), and the skin sample cross-section just before measurement (bottom).*

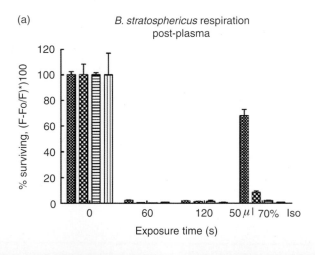

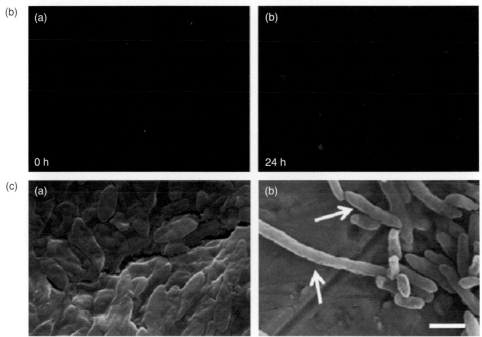

Figure 6.37 *(a) XTT assay was used to detect all culturable as well as viable but non-culturable (VBNC) cells of* Bacillus stratosphericus *after holding the cells (2, 6, 18 and 24 h) in nutrient medium upon plasma treatment over time. Each set had negative control (reagents without cells) and positive control (70% ethanol treated cells) for comparison. The experiments were repeated in triplicate. About 1 × 10⁸ CFU mL⁻¹ were starting cells (shown as 100%). (Table 1 demonstrates the percent deviations and can be read with this figure). Representative microphotographs of the smear of suspended cell pellet showing respiring* B. stratosphericus *from few initial survivors of plasma treatment for (b) 120 s after 2 h to (c) increased respiration after 24 h whichstill remained non-culturable. Scanning electron microscopic image showing heterogeneous morphology of* B. stratosphericus *(d) untreated and (e) after 120 s of plasma treatment under wet environment . Such cellular elongation is found associated with VBNC bacteria. Arrow points divided cells. Bar: 2 μm. XTT: 2,3-bis-(2-methoxy-4-nitro-5-sulfophenyl)-2H-tetrazolium-5-carboxanilide.*

Figure 7.1 *Pathogen Detection and Remediation Facility (PDRF), general view.*

Figure 7.3 *PDRF liquid impinger for bioaerosol sampling.*

Figure 7.4 *PDRF dielectric barrier grating discharge (DBGD) with air sterilization chamber.*

Figure 7.7 *The plasma unit with its multiple-electrode configuration (left). The same plasma unit, with the plasma discharge initiated (right). Note the screen of plasma covering the entire cross-section of air passage.*

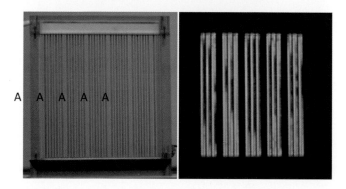

Figure 7.8 *The plasma unit with every fourth high-voltage électrode wire removed (left). The quartz tubes are retained in order to maintain the same airflow. This produces zones of indirect exposure (A). The image shown on the bottom is with the plasma.*

Figure 7.19 *Mobile Environmental Laboratory built in Drexel University carrying precise exhaust gas characterization facility as well as 10 kW wet (spray) pulsed corona discharge for treatment of large-volume low-concentration exhaust gases containing VOC.*

Figure 7.20 *Photo of the 10 kW pulsed corona discharge (generated in the water-spray configuration) through a window of the Mobile Environmental Laboratory.*

Figure 7.21 *Photo of the Mobile Environmental Laboratory demonstrating all 6 access windows to the 10 kW pulsed corona discharge system applied for treatment of large volume low concentration exhaust gases containing VOC.*

Figure 7.27 *Point-to-plane plasma discharge system for pulsed corona discharge and spark in water.*

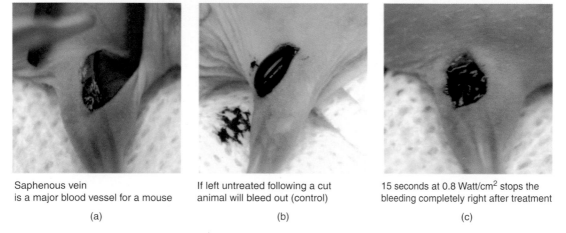

Saphenous vein
is a major blood vessel for a mouse

(a)

If left untreated following a cut
animal will bleed out (control)

(b)

15 seconds at 0.8 Watt/cm^2 stops the
bleeding completely right after treatment

(c)

Figure 8.3 *Blood coagulation of a live animal. (a) Saphenous vein is a major blood vessel for a mouse; (b) if a cut is left untreated, the animal will bleed out and (c) after 15 s at 0.8 W cm^{-2}, the bleeding is stopped.*

Figure 8.8 *Application of DBD to the blood plasma sample.*

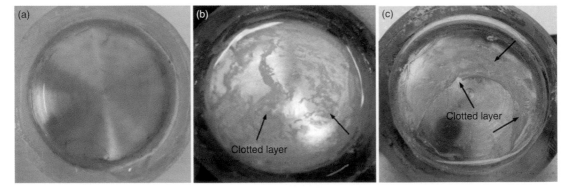

Figure 8.10 *Clotted formation of white layer in blood plasma sample with DBD treatment: (a) before DBD treatment, showing no coagulation; (b) after DBD treatment for 4 min, exhibiting a partially coagulated layer formation; and (c) after DBD treatment for 8 min, showing a white clotted layer on the overall surface of copper.*

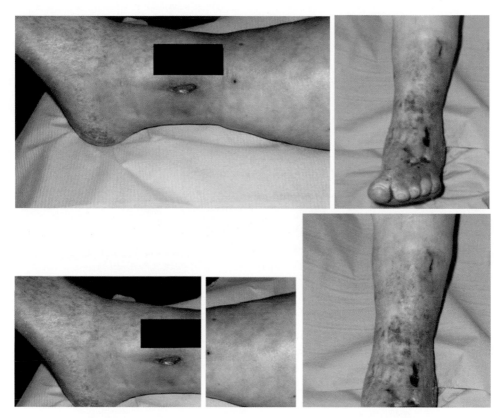

Figure 9.1 *Examples of different chronic wounds: top left is venous leg ulcer, top right is arterial leg ulcer, bottom left is ischemic diabetic foot ulcer, bottom right is grade 3 pressure ulcer.*

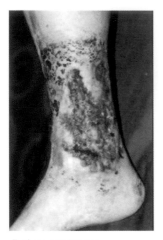 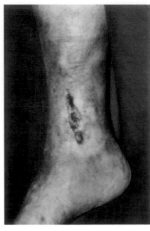 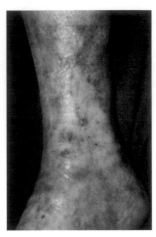

| Before treatment | 21st day of NO-therapy (10 seances) | After 2 months of NO-therapy |

Figure 9.4 *Reduction of ulcer size as a function of time during the Plazon-generator-based NO therapy. Photographs courtesy of A. Shekhter.*

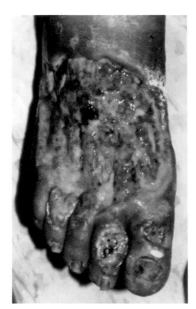 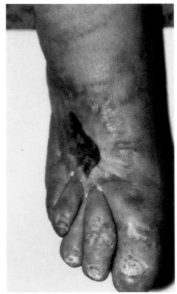

| Before treatment | After 4.5 months of NO-therapy (3 courses; 12 seances per course) |

Figure 9.7 *Example of a large ulcerated wound before and after 4.5 month duration of NO treatment. Photographs courtesy of A. Shekhter.*

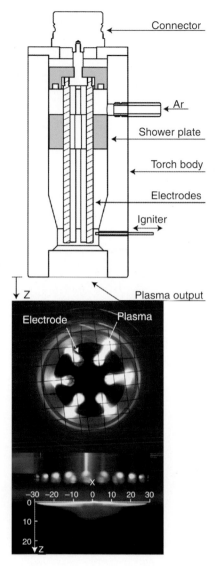

Figure 9.8 *Schematics and photograph of MicroPlaSter operation (courtesy of ADTEC Plasma Technology Co. Ltd., Hiroshima, Japan and London, UK).*

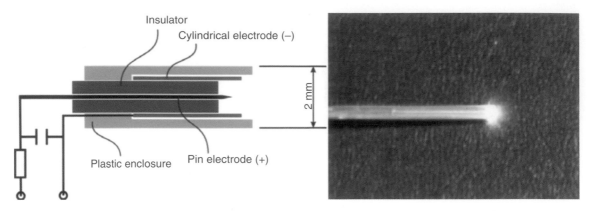

Figure 9.9 *Miniaturized pin-to-hole discharge system employed in IBS treatment demonstration on mice.*

Insulator
Cylindrical electrode (−)
2 mm
Plastic enclosure
Pin electrode (+)

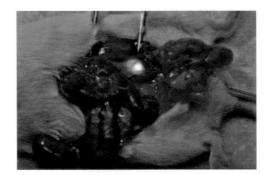

Figure 9.10 *The miniaturized pin-to-hole discharge in operation. The tip was inserted through the anal verge up to 4 cm into the mouse colon. The light due to plasma was observed through the intestinal wall when the abdomen was open.*

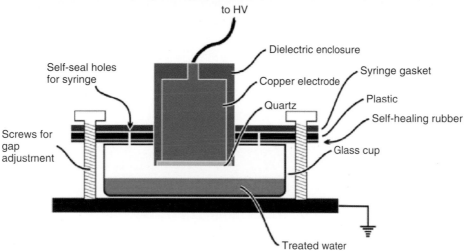

Figure 10.1 *(Lower) Schematic of the experimental setup; DBD electrode is placed in an enclosed glass chamber. The distance between the electrode and the liquid is 1.5 mm. (Upper) Photograph of the experimental setup with plasma.*

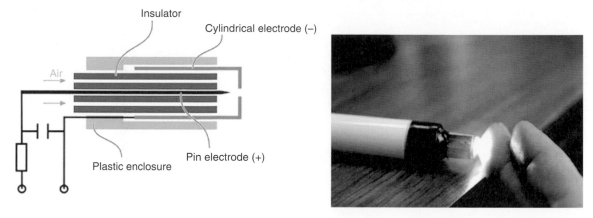

Figure 4.71   *PHD plasma system schematic (left) and in operation (photo, right) demonstrating that this plasma source is safe to the touch (see color plate).*

4.5.10   Pin-to-hole spark discharge (PHD), thermal microplasma source-generating ROS and NO for medical applications

A DC spark discharge plasma in a pin-to-hole spark discharge (PHD) configuration (Figure 4.71), developed in Drexel Plasma Institute, is of great interest for medical application because of its ability to generate locally a significant amount of reactive oxygen species (ROS) as well as reactive nitrogen species, especially NO.

A needle anode (1.5 mm diameter) is coaxially fixed in an insulator with gas inlet openings (c. 0.5 L min^{-1}) surrounded by an outer cylindrical cathode (7 mm diameter) with an axial opening (2 mm diameter) for plasma outlet. Both electrodes are made of stainless steel. The discharge is ignited by applying a 4 kV positive potential to the central electrode. To provide high discharge energy while keeping average gas temperature low, the electrode system is powered through a 0.33 µF capacitor. This forms a 35 µs dense energetic discharge with an average energy of c. 1.8 J pulse^{-1}. To prevent cell effects due to plasma discharge UV emission production, either straight or curved 50 mm Tygon tube extensions can be attached to the cathode.

The plasma discharge appears as a series of microdischarges, in which the first microdischarge is the most energetic with 1 µs duration and 0.6 J energy. Due to the low frequency and short duration of the plasma microdischarges, the average plasma gas temperature is relatively low (Figure 4.72). Gas temperature decreases from 75°C at 2 mm distance from the plasma device down to c. 37°C at 10 mm distance. By adding airflow to the device, the maximum plasma gas temperature becomes c. 50°C and a temperature of 37°C is reached 6 mm away from the device. Although the average gas temperature is slightly above room temperature, the plasma temperature itself is high enough to produce a significant NO quantity. A time-averaged plasma excitation temperature is 9030 ± 320 K.

The PHD plasma discharge radiates intensively in the UV range. The total plasma UV radiation with and without air flow is 90 and 140 µW cm^{-2}, respectively. DNA damage by UV radiation occurs at doses ranging from 0.5 to 50 mJ cm^{-2} for human cells and bacteria. Fatal DNA damage can be induced in cells after only 5 s of direct PHD plasma exposure. H_2O_2 concentration in PHD plasma-treated phosphate-buffered saline (PBS) (Figure 4.73) increases up to 60 µM with 30 s of direct plasma treatment (210 pulses). However, the PBS H_2O_2 concentration can be significantly reduced when either a straight or curved tube extension is used. In these configurations, H_2O_2 reaches a stable concentration of 1.5–2 µM in 5–10 s of treatment at 7 Hz (35–70 pulses), which is nearly 30 times lower than H_2O_2 measured in the direct treatment configuration.

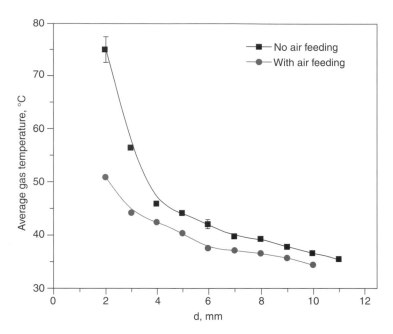

Figure 4.72   *Penetrating plasma afterglow average gas temperature decreases with distance from the discharge and with the addition of airflow.*

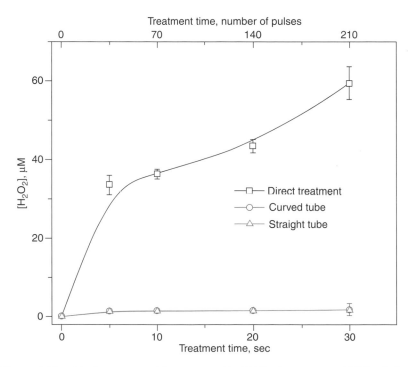

Figure 4.73   *Hydrogen peroxide generated in PHD plasma-treated PBS solution.*

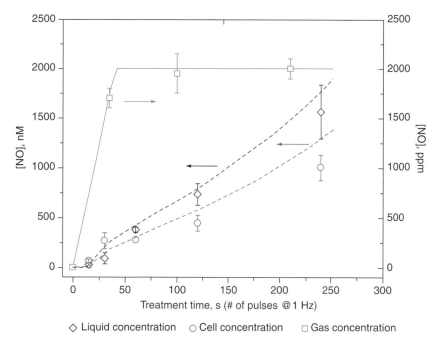

Figure 4.74 *NO concentration in gas, PBS and the endothelial cell monolayer following PHD plasma treatment with the straight tube configuration; dotted lines show numerically calculated NO concentration in PBS and endothelial cells.*

PHD plasma H_2O_2 production is of interest in wound-healing applications. In early wound healing, H_2O_2 activates cell surface tissue factor to restore hemostasis and modulates both leukocyte adhesion and motility in inflammation. ROS are also important in wound re-vascularization, since H_2O_2 induces vascular endothelial growth factor expression and is critical in its signaling. Finally, in late wound healing, H_2O_2 promotes re-epithelialization by activating matrix metalloproteinases and promoting epithelial cell motility. A plasma device that can deliver H_2O_2 and NO individually or in combination is therefore useful throughout the wound healing process.

NO concentrations in gas, liquid and cells following PHD plasma treatment and comparison of the experimental results with numerical simulations of NO concentrations in liquid and cells under diffusion-only conditions are presented in Figure 4.74. In gas, NO concentration rapidly increases with plasma treatment and reached a stable level of 1980 ± 50 ppm NO by around 50 s. In the liquid (PBS), NO concentration increases linearly up to 240 plasma pulses. Approximately 75% of NO from plasma-treated PBS diffuses through the cell membrane into the cytoplasm. A maximum of 1000 nM NO was detected in endothelial cells immediately following 240 plasma pulses (Figure 4.74). While gas NO concentration is saturated after 50 pulses, the NO level in liquid and cells continued to rise up to 240 pulses. This effect is likely related to diffusion time. NO diffusion through gas is relatively fast ($D = 0.3$ cm^2 s^{-1}); however, NO diffusion through water is significantly slower ($D = 2 \times 10^{-5}$ cm^2 s^{-1}). Plasma NO had to travel through a 50 mm tube filled with room air to reach the liquid surface, and additionally through 3 mm of PBS to reach the liquid bottom and the cells. Lower NO concentration in cells as compared to PBS may be due to the short NO half-life in liquid, since NO diffuses freely across cell membranes. Similarly, while a maximum NO concentration of 1400–1600 nM is detected in PBS with the straight tube configuration, c. 900 nM NO can be detected in PBS for the curved tube configuration for the same treatment duration.

4.5.11   Atmospheric-pressure cold helium microplasma jets and plasma bullets

DBD-based microdischarges, developed in particular by M. Kong, M. Laroussi and their groups (e.g. Laroussi *et al.*, 2012), have attracted much interest recently by offering a chamberless delivery of downstream reaction chemistry. This is ideal for polymeric surface modification, bacterial and biomolecule inactivation, wound healing and other biomedical applications.

A distinct characteristic of a jet configuration is its ability to generate a stable discharge in a region of inert gas and then to transport the plasma to a separate region of reactive gas for processing applications, thus providing chemical reactivity without compromising plasma stability. This spatial separation of the plasma generation and surface processing regions allows for flexibility in jet designs to vary and control both plasma dynamics and reaction chemistry. At frequencies below 100 kHz, sub-microsecond pulsed is found to effectively reduce gas temperature.

In the context of operational characteristics of atmospheric plasma jets, a major topic of current interest has been plasma bullets, a fast-moving train of highly luminous but discrete clusters along the plasma jet when imaged on a nanosecond scale. According to M. Kong and his group (e.g. Laroussi *et al.*, 2012), the chaotic mode of the jet is observed after gas breakdown. With increasing input power, the discharge enters the bullet mode; the continuous mode is observed when input power reaches a sufficient level. Transition from the chaotic mode through the bullet to the continuous mode is abrupt and distinct, with each mode having a unique set of operating characteristics.

A typical microplasma helium jet system is illustrated in Figure 4.75 (Walsh *et al.*, 2010). The device consists of a quartz tube measuring 5 cm in length with a 2 mm inner diameter and a 3 mm outer diameter.

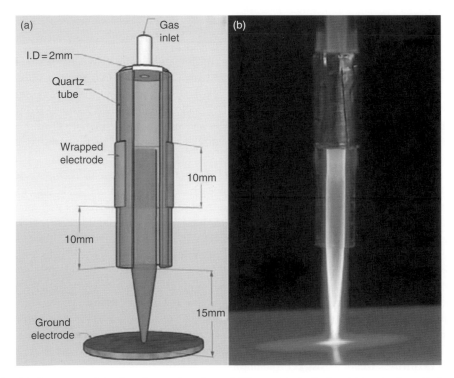

Figure 4.75   *(a) Structure of the jet electrode unit and (b) an atmospheric plasma jet in a 5 slm helium flow at 1 W input power. Reproduced from J L Walsh et al., J. Phys D; Appl. Phys. 43 075201, doi 10.1088/0022-3727/43/7/075201, 2010, with permission from IOP Publishing Ltd.*

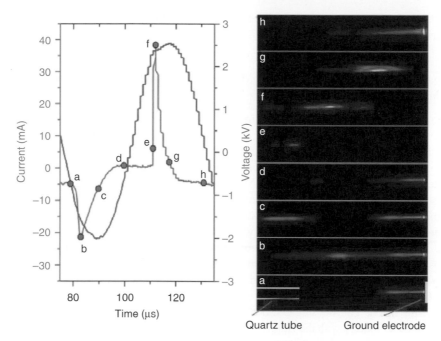

Figure 4.76 *Plasma jet in the bullet mode with the current and voltage traces (left panel) and 10 ns exposure images over one complete cycle of the applied voltage (right panel). In (a) of the right panel, the quartz tube is indicated with two horizontal lines on the left and the downstream electrode with a vertical line on the right. Reproduced from J L Walsh et al., J. Phys D; Appl. Phys. 43 075201, doi 10.1088/0022-3727/43/7/075201, 2010, with permission from IOP Publishing Ltd.*

A 1-cm-wide metallic electrode is wrapped tightly around the quartz tube at a distance of 1 cm from the tube nozzle. A second grounded metallic electrode, which can be dielectrically coated, is placed 1.5 cm downstream of the nozzle. A power source, capable of producing a variable-amplitude high-voltage sinusoid at a fixed frequency of 18 kHz, is connected to the wrapped electrode.

The chaotic mode is observed immediately after breakdown, and an increase in the input power eventually leads to the bullet mode and then to the continuous mode. A common feature of all three modes is that their current waveform has one distinct peak every positive half-cycle of the applied voltage. For the chaotic mode the negative half-cycles often have a current peak, but this is not always the case. For both the bullet mode and the continuous mode, there is one strong current peak every half-cycle which is typical of dielectric barrier discharges. These strong current peaks are due to the rapid increase in gas conductivity at plasma ignition followed by an accumulation of surface charge acting to reduce the gas voltage and extinguish the discharge. Figure 4.76 (Walsh *et al.*, 2010) shows a sequence of 10 ns exposure images taken throughout the positive and negative half-cycles of the applied voltage while the plasma jet was operating in the bullet mode.

Figure 4.76(a–d) were acquired during the negative half-cycle; consequently, the downstream ground electrode (on the right-hand side) can be considered as the instantaneous anode and the powered electrode wrapped around the dielectric tube as the instantaneous cathode (on the left-hand side). As the current pulse increases in magnitude, a faint bullet-like object can be observed leaving the grounded anode and traveling towards the cathode inside the quartz tube. It should be mentioned that plasma bullets tend to form in the positive half-cycle; in the case of Figure 4.76 the discharge in the negative half-cycle does not have the appearance of a plasma bullet. Note that plasma bullets are cathode-directed and are very similar to streamers in this regard. However, plasma bullets appear in a well-defined column without branching and their characteristics are similar to that of glow discharges.

4.5.12   Propagation of plasma bullets in long dielectric tubes and splitting and mixing of plasma bullets

The plasma bullets described in the previous section have the ability to propagate in long dielectric tubes. This has been widely investigated, in particular by J.M. Pouvesle, E. Robert and their group (e.g. Robert *et al.*, 2012). Such plasma bullets provide *in situ* plasma up to a few tens of centimeters away from the primary plasma-generation zone, which is especially interesting for various plasma-medical applications.

The so-called 'plasma gun', developed by J.M. Pouvesle, E. Robert and their group (e.g. Robert *et al.*, 2012), was used to generate a 'plasma bullet' in a glass-capillary channel composed of a circular ring of 15 cm in diameter between two straight sections (see Sarron *et al.*, 2011 for more details). The two straight portions define a symmetry axis for the assembly, with one being the inlet and the other the outlet for a rare gas flow. The DBD reactor where plasma generation occurs is 45 cm upstream from the ring center. This T-shaped glass reactor is flushed at 100 cm^3 min^{-1} with neon. It is equipped with an inner rod of 1 mm in diameter, constituting the cathode. The grounded anode is a 5-mm-wide annular ring positioned on the outer surface of the 6-mm-outer-diameter 1-mm-thick glass pipe. A −30 kV voltage pulse with a rise time of 40 ns is used to power the discharge from single-shot operation up to 500 Hz. A gas-tight connector allows for the connection of the DBD reactor with the ring.

This system was applied to demonstrate the plasma splitting at the ring inlet, the propagation of the two plasmas in the circular branches and these two plasmas mixing at the ring outlet. The intensified-charge-coupled-device imaging indicates that two 'colliding' plasmas present very similar properties and are synchronized with a jitter, induced by a slight variation of the propagation velocity (i.e. less than 5 ns).

Figure 4.77 (Robert *et al.*, 2009; Sarron *et al.*, 2011) presents the time-integrated photography captured by a conventional camera. The plasma propagation in the whole glass-capillary assembly is visualized through the neon red light emission. A continuous decrease in the plasma emission, which is associated with a regular

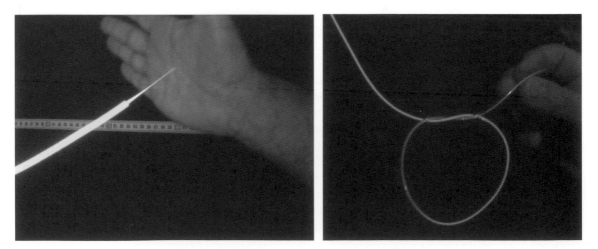

Figure 4.77   *Plasma gun shooting plasma 'bullets'. Left: 50 cm long, 4 mm in diameter dielectric guide with plume formation. Right: 50 cm long, 200 mm in diameter dielectric guide. The guide is twisted only for 'artistic' purposes (Robert* et al., *2009) (see color plate).*

diminution of the current pulse intensity, is observed along the propagation path. Another observation is the visualization of the plasma mixing and the successive generation of a fast-traveling plasma stream in the evacuation port.

Several 15 ns exposure snapshots are analyzed in Figure 4.77. The time origin corresponds to the first image, which presents the plasma stream traveling toward the circular portion. After 50 ns, the plasma split into the two branches of the circular ring. Such plasma separation was also observed in the structure, presenting up to ten bifurcations. For the three instants which are delayed by 250, 450 and 700 ns, the two-plasma propagation is observed. The good synchronization, the similar shape and the intensity in the two branches reveal a very balanced portioning of the incident stream energy. In connection with the ionization-wave dumping, the plasma volume decreases and the downstream tail evidenced at 250 ns disappears gradually. A continuous decrease in the propagation velocity is measured from the inner electrode tip, as in straight capillaries. A charge repulsion-like behavior is observed as the two facing plasmas influence each other. The mixing of these two plasmas results in the production of a plasma at the evacuation port which has an intensity greater than each of the plasma stream heads. This is an indication of an energy coupling, which may be associated with constructive ionization wave mixing. These features may be valuable for in vivo biomedical applications where structured, splitting and connecting paths are likely to exist. For additional information regarding these pulsed atmospheric-pressure plasma streams (PAPS), see Robert *et al.* (2012).

4.6   Discharges in liquids

4.6.1   General features of electrical discharges in liquids in relation to their biomedical applications

Electric breakdown of liquids is limited by their high-density short mean free path of electrons and therefore requires very high electric fields E/n_0 (see Paschen curves). Nevertheless, the breakdown of liquids cannot be performed at the extremely high electric fields required by Paschen curves, but only at those slightly exceeding breakdown fields in atmospheric-pressure molecular gases. The effect can be explained by different electrically induced mechanisms of formation of micro-voids and quasi-'cracks' in the liquids. Discharge can be sustained in water by pulsed high-voltage power supplies. The discharges in water usually start from sharp electrodes. If the discharge does not reach the second electrode it can be interpreted as pulsed corona; branches of such a discharge are referred to as streamers. If a streamer reaches the opposite electrode, a spark forms. If the current through the spark is high (above 1 kA), the spark is a pulsed arc.

Various electrode geometries have been used for plasma generation in water for the purpose of water treatment, and especially for water sterilization. Two simple geometries are a point-to-plane and a point-to-point. The former is often used for pulsed corona discharges, whereas the latter is often used for pulsed arc systems.

A concern over the use of pulsed discharges is the limitation posed by the electrical conductivity of water. In the case of a low electric conductivity of water (below 10 μS cm^{-1}), the range of the applied voltage that can produce a corona discharge without sparking is very narrow. In the case of a high electric conductivity of water (above 400 μS cm^{-1}), streamers become short, the efficiency of radical production decreases and a denser and cooler plasma is generated.

In general, the production of OH radical and O atoms is more efficient at water conductivity below 100 μS cm^{-1}. For the case of tap water, the bulk heating can be one of the problems in the use of corona discharges. At frequency of 213 Hz, the temperature of the treated water rises from 20°C to 55°C in 20 min, indicating a significant power loss.

Generally, plasma generated in water as well as above the water surface or around the water droplets can sterilize the liquid but also significantly change its composition for further applications in wound treatment and healing of different diseases. Plasma is also effectively applied to treat blood and other liquid constituents of living tissue and cells. The generation of plasma in such special bioliquids is discussed in particular by Yang, Cho, and Fridman (2012).

4.6.2   Mechanisms and characteristics of plasma discharges in water

The mechanisms of plasma discharges and breakdowns in liquids (specifically in water) can be classified as two groups: the first presents the breakdown in water as a sequence of a bubble process and an electronic process and the second divides the process into a partial discharge and a fully developed discharge such as an arc or spark.

In the first approach, the bubble process starts from a microbubble which is formed by the vaporization of liquid by local heating in the strong electric field region at the tips of electrodes. The bubble grows, and an electrical breakdown takes place within the bubble. In this case, the cavitation mechanism can explain the slow bush-like streamers. The appearance of bright spots is delayed from the onset of the voltage, and the delay time tends to be greater for smaller voltages. The time lag to water breakdown increases with increasing pressure, supporting the bubble mechanism in a sub-microsecond discharge formation in water. The time to form the bubbles is about 3–15 ns, depending on the electric field and pressure. The influence of the water electrical conductivity on this regime of the discharges is small. Bulk heating via ionic current does not contribute to the initiation of the breakdown. The power necessary to evaporate the water during the streamer propagation can be estimated using the streamer velocity, the size of the streamer and the heat of vaporization. Using a streamer radius of 30 μm, a power of 2 kW was estimated to be released into a single streamer to ensure its propagation in the form of vapor channels. In case of multiple streamers, the required power can be estimated by multiplying the number of visible streamers to the power calculated for a single streamer. In the electronic process, electron injection and drift in liquid take place at the cathode while hole injection through a resonance tunneling mechanism occurs at the anode. Breakdown occurs when an electron makes a suitable number of ionizing collisions in its transit across the breakdown gap.

For the second approach, the electrical discharges in water are divided into partial electrical discharges and arc and spark discharge, as described in Table 4.11. In the partial discharges, the current is mostly transferred by ions. For the case of high-electrical-conductivity water, a large discharge current flows which results in a shortening of the streamer length due to the faster compensation of the space-charge electric fields on the head of the streamer. Subsequently, a higher power density in the channel is obtained resulting in a higher plasma temperature, a higher UV radiation and the generation of acoustic waves. In the arc or spark discharges, the current is transferred by electrons. The high current heats a small volume of plasma in the gap between the two electrodes, generating quasi-thermal plasma. When a high-voltage/high-current discharge takes place between two submerged electrodes, a large part of the energy is consumed by the formation of a thermal plasma channel. This channel emits UV radiation, and its expansion against the surrounding water generates intense shock waves. The shock waves are weak or moderate for the corona discharge in water, but are strong for the pulsed arc or spark.

4.6.3   Physical kinetics of water breakdown: Thermal breakdown mechanism

The critical breakdown condition for a gas is described by the Paschen curve, from which the breakdown voltage for air can be calculated. A value of 30 kV cm^{-1} is a well-accepted breakdown voltage of air at 1 atm. When attempting to produce direct plasma discharges in water, it could be expected that a much greater breakdown voltage of the order 30 000 kV cm^{-1} might be needed due to the density difference between air and water. A large body of experimental data (e.g. Yang, Cho and Fridman, 2012) on the breakdown voltage in water however shows that, without special precautions, this voltage is of the same magnitude as for gases.

Table 4.11 *Pulsed discharges in water.*

Pulsed corona in water	Pulsed arc in water	Pulsed spark in water
Streamer-like channels	Current is transferred by electrons	Similar to pulsed arc, except for short pulse durations and lower temperature
Streamer-channels do not propagate across the entire electrode gap, i.e. partial electrical discharge.	Quasi-thermal plasma	Pulsed spark is faster than pulsed arc, i.e. strong shock waves are produced
Streamer length of order cm, channel width c. 10–20 μm	Arc generates strong shock waves within cavitation zone	Plasma temperatures in spark around a few thousand Kelvin
Electric current is transferred mostly by ions	High current filamentous channel bridges the electrode gap	
Non-thermal plasma	Gas in channel (bubble) is ionized	
Weak–moderate UV generation	High UV emission and radical density	
Relatively weak shock waves	Gap between two electrodes of <5 mm is needed	
Water treatment area is limited to a narrow region near the corona	Large discharges pulse energy (greater than 1 kJ per pulse)	
Pulse energy: a few joules per pulse, often less than 1 J per pulse.	Large current (about 100 A), peak current greater than 1000 A	
Frequency c. 100–1000 Hz	Electric field intensity at the tip of electrode 0.1–10 kV cm^{-1}	
Relatively low current, i.e. peak current < 100 A	Voltage rise time 1–10 μs	
Electric field intensity at the tip of electrode 10^2–10^4 kV cm^{-1}	Pulse duration c. 20 ms	
A fast-rising voltage of the order 1 ns, but less than 100 ns	Temperature >10 000 K	

This interesting and practically important effect can be explained by taking into account the fast formation of gas channels in the body of water under the influence of the applied high voltage (see previous section). When formed, the gas channels provide space for the gas breakdown inside the body of water. This explains why the voltage required for water breakdown is of the same magnitude as for gases.

The gas channels can be formed by the development and electric expansion of gas bubbles already existing in water, as well as by additional formation of the vapor channel through electrolysis or fast local heating and evaporation. We focus on the second mechanism, which is usually referred to as the *thermal breakdown*. When a voltage pulse is applied to water, it induces a current and the redistribution of the electric field. Due to the dielectric nature of water, an electric double layer is formed near the electrode which results in the localization of the applied electric field. This electric field can become high enough for the formation of a narrow conductive channel, which is heated up by electric current to temperatures of about 10 000 K. Thermal plasma generated in the channel is rapidly expanded and ejected from the narrow channel into water, forming a plasma bubble. The energy required to form and sustain the plasma bubble is provided by Joule heating in the narrow conductive channel in water.

The physical nature of thermal breakdown can be related to thermal instability of local leakage currents through water with respect to the Joule overheating. If the leakage current is slightly higher at one point, the Joule heating and hence temperature also grows there. The temperature increase results in a significant growth of local conductivity and the leakage current. Exponential temperature growth to several thousand degrees at a local point leads to formation of the narrow plasma channel in water, which determines the thermal breakdown. The thermal breakdown is a critical thermal-electric phenomenon taking place at the applied voltages exceeding a certain threshold value, when heat release in the conductive channel cannot be compensated by heat transfer losses to the surroundings. The thermal conditions of water are constant during the breakdown; water stays liquid away from the discharge with the thermal conductivity about 0.7 W mK^{-1}.

When the Joule heating between the two electrodes is larger than a threshold value the instability can occur, resulting in the instant evaporation and a subsequent thermal breakdown. When the Joule heating is smaller than a threshold value, electrolysis and the breakdown do not occur. The thermal breakdown instability is characterized by the instability increment describing the frequency of its development:

$$\Omega = \left(\frac{\sigma_0 E^2}{\rho C_p T_0} \right) \frac{E_a}{T_0} - D \frac{1}{R_0^2} \tag{4.97}$$

where σ_0 is the water conductivity; E_a is the Arrhenius activation energy for the water conductivity; E is the electric field; ρC_p is the specific heat per unit volume; T_0 is the temperature; R_0 is the radius of the breakdown channel; $D = 1.5 \times 10^{-7}$ m^2 s^{-1} is the thermal diffusivity of water. When the increment $\Omega > 0$, the perturbed temperature exponentially increases with time and results in thermal explosion. When $\Omega < 0$, the perturbed temperature exponentially decreases with time and results in the steady-state condition. For the plasma discharge in water, the breakdown voltage in the channel with length L can be estimated as:

$$V \geq \sqrt{\frac{DC_p \rho T_0^2}{\sigma_0 E_a}} \frac{L}{R_0}. \tag{4.98}$$

The breakdown voltage increases with L/R_0. Assuming $L/R_0 = 1000$, V ~ 30 kV.

4.6.4   Non-thermal short pulse electrostatic (electrostriction) water breakdown

The previous model of water breakdown has the nature of thermal breakdown and is valid only if breakdown voltage is applied long enough to evaporate water in a microchannel (at least hundreds of microseconds). The dormation of a microchannel void sufficient for breakdown can be achieved much faster electrostatically (electrostriction) without a corresponding increase in temperature.

Such an electrostatic model is depicted by Figure 4.78 (Yang, Cho, and Fridman, 2012). A thin needle electrode with a rounded tip was aligned in this model perpendicular to a ground plate electrode. High voltage

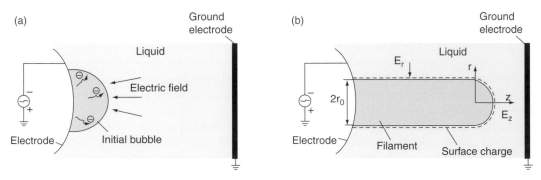

Figure 4.78   (a) Initiation of bubble formation; (b) schematic diagram of a cylindrical filament in water.

Φ_0 was applied to the needle electrode. Liquids become phase-unstable under a high electric field so that gas channels can be formed along electric field lines. The time required for breakdown ignition in the channels is estimated as $\tau_b = (k_I n_0)^{-1}$, where k_I is the direct ionization rate coefficient and n_0 is the molecule density. Under atmospheric pressure, n_0 is of the order 10^{19} cm^{-3} while k_I is of the order 10^{-10} to 10^{-9} cm^3 s^{-1} in the reduced electric field E/n_0 of 10^3 V cm^2; τ_b is therefore of the order 0.1 to 1 ns.

For negative discharges, due to the higher momentum transfer collision frequency and thus a low mobility in the liquid phase, electrons tend to deposit on the gas–liquid interface and charge it negatively. For positive discharges, the high mobility of electrons leaves the interface charged positively. Under both circumstances, it is possible that the charged interface can be pushed to displace the liquid under an external electric field by electrostatic force.

A simple calculation can be made to examine whether or not the electrostatic force is sufficient to overcome the resistance of water at the interface. The pressure due to the surface tension γ on a water interface of a spherical bubble with a radius of curvature r can be approximated by the Young-Laplace equation $p = 2\gamma/r$. With $r \sim 1$ µm and $\gamma = 72.8 \times 10^{-4}$ N m^{-1}, the surface tension pressure is c. 15 kPa. The ultimate strength of water of approximately 30 MPa must be exceeded to rupture the liquid. Considering the forces due to charged particles only and ignoring those due to field gradients and material property gradients, the electric force at the interface is simply the electrostatic force L. L is the product of charge density per unit area σ and the electric field E, i.e. $L = e\sigma E$ where e is the charge per electron. For $E = 10^8$ V cm^{-1}, σ should have a value of 10^{12} charges cm^{-2}. For electrons with an average energy of 1 eV, the electron thermal velocity can be estimated as 6×10^7 cm s^{-1}. A modest electron density of 10^{13} cm^{-3} will therefore provide the flux necessary to charge the surface to the breaking point within 1 ns.

Although these estimations for water rupturing also neglect both loss mechanisms and the energy requirements to overcome the hydrodynamic resistance, the electrostatic mechanism is still a likely candidate for streamer propagation and such forces may dominate at a timescale of nanoseconds. The growth of a plasma filament is determined by conservation equations of mass, momentum and energy. To quantify the breakdown process described above, the equations for the formation and propagation of the plasma-filled filaments are:

$$\frac{\partial \rho}{\partial t} + \nabla(\rho u) = \frac{2\lambda(T)T}{\Delta_v H r_0^2} \tag{4.99}$$

$$\frac{\partial u}{\partial t} + u\nabla u + \frac{1}{\rho}\nabla P = 0 \tag{4.100}$$

$$\frac{\partial}{\partial t}\left[\rho\left(Z + u^2\right)\right] + \nabla\left[\rho u\left(\frac{P}{\rho} + \frac{u^2}{2}\right)\right] = \kappa(T)E^2 \tag{4.101}$$

where t is time; ρ and P are the radial density and pressure inside the streamer, respectively; u is the velocity of streamer; T is the temperature; λ is the thermal conductivity; $\Delta_v H$ is the evaporation heat of water; r_0 is the radius of streamer; Z is the internal energy of ionized gas; E is the electric field strength; and κ is the electric conductivity.

For simplicity, the streamer can be assumed as a cylinder with a hemispherical tip as shown in Figure 4.78, with a reference frame fixed on the tip. The radius of the filament is r_0. Although it appears from photographic evidence that the filament is usually of a conical shape, the cylindrical approach is still a good approximation when the length of the filament is much greater than the radius. The electric conductivity κ inside the filament could be described as

$$\kappa = \frac{n_e e^2}{m v_{en}} \tag{4.102}$$

where m is the mass of electron and v_{en} is the frequency of electron-neutral collisions. Note that v_{en} is proportional to the gas number density and the value of v_{en}/p is usually of the order 10^9 s^{-1} Torr^{-1}. The density inside streamers during the initial phase of water breakdown is of the order of 10^{18} cm^{-3}.

With a room-temperature saturated water vapor pressure of 20 Torr, the electric conductivity inside the filament can be estimated to be of the order 10^7 S m^{-1}, a value comparable to that of metals. The filament can therefore be regarded as equipotential with the electrode and can be treated as an extension of the electrode throughout the expansion. The external fluid provides drag force and constant external pressure for the development of the filament. The electric field outside a slender jet can be described as if it were due to an effective linear charge density (incorporating the effects of both free charge and polarization charge) of charge density σ on the surface. Since the charge density in liquid can be ignored compared to that on the filament surface, the following equation for the space outside the filament can be obtained by applying the Laplace equation in the radial direction:

$$\frac{1}{r}\frac{\partial}{\partial r}\left(r\frac{\partial \Phi}{\partial r}\right) = 0 \tag{4.103}$$

with boundary conditions $\Phi|_{r=r_0} = \Phi_0$ and $\Phi|_{r=R} = 0$. R is the distance between the anode and cathode. Because the filament could be regarded as an extension of the electrode, R decreases as the streamer propagates through the gap. Solving the above equation with an assumption of negative discharge, the radial electric field E_r and local surface charge density σ_r can be written:

$$E_r = \frac{\partial \Phi}{\partial r} = -\frac{\Phi_0}{r_0 \ln{(R/r_0)}} \tag{4.104}$$

$$\sigma_r = \varepsilon E_{r0} = -\varepsilon_r \varepsilon_0 \frac{\Phi_0}{r_0 \ln{(R/r_0)}}. \tag{4.105}$$

There is no analytical solution for the electric field at the hemispherical tip of the filament. A frequently used approximation is $E_z \approx \Phi_0/r_0$. For the electric field at the tip of a needle in a needle-to-plane geometry:

$$E_z = -\frac{2\Phi_0}{r_0 \ln{(4R_0/r_0)}}. \tag{4.106}$$

Similarly, the local charge density at the tip is:

$$\sigma_z = \varepsilon E_z = -\varepsilon_r \varepsilon_0 \frac{2\Phi_0}{r_0 \ln{(4R/r_0)}}. \tag{4.107}$$

From Equations (4.104)–(4.107), it can be concludes that the radial direction electrostatic pressure $E\sigma$ exerted on the sidewall of the streamer was weaker than the axial direction electrostatic pressure on the tip. Note that both electrostatic pressures were roughly inversely proportional to r_0^2. This means that at the initial stage of the filament growth when r_0 is small, the electrostatic forces on both directions were strong and the filament grows both axially and radially.

A direct consequence of both the axial and radial expansions of the streamer channel is the launching of compression waves into adjacent liquids. At some critical point, the electrostatic force reaches a balance with hydrodynamic resistance acting on the surface in the radial direction first, while the filament continues to grow in the axial direction.

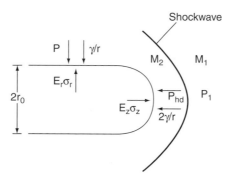

Figure 4.79   *Force balance for the present electrostatic model.*

Experimentally recorded propagation speeds of the filaments varies depending on the measurement techniques, ranging from a few to 100 km s^{-1}. Despite the discrepancy observed by different groups (e.g. Yang, Cho and Fridman, 2012), the propagation is clearly in the supersonic regime and the formation of shockwaves has to be taken into consideration (see Figure 4.79). The drag force on the tip of the streamer, which is a stagnation point, is equal to the force produced by the total hydrodynamic pressure:

$$P_{\text{hd}} = P_1 \left(\frac{2\alpha}{\alpha + 1} M_1^2 - \frac{\alpha - 1}{\alpha + 1} \right) + \frac{1}{2} \rho (C M_2)^2 \tag{4.108}$$

where P_1 is ambient pressure; the first term on the right-hand side is the pressure behind the shock front; α is the specific heat ratio of water; M_1 is the Mach number of the streamer; M_2 is the Mach number after the shock front; and C is the speed of sound in liquid. The relationship between M_1 and M_2 can be written:

$$M_2^2 = \frac{(\alpha - 1) M_1^2 + 2}{2\alpha M_1^2 + 1 - \alpha}. \tag{4.109}$$

Equating the hydrodynamic pressure to the sum of the electrostatic pressure and the pressure produced by surface tension at the tip yields the equation for streamer propagation:

$$4\varepsilon_r \varepsilon_0 \frac{\Phi_0^2}{r_0^2 \ln^2 (4R/r_0)} = P_1 \left(\frac{2\alpha}{\alpha + 1} M_1^2 - \frac{\alpha - 1}{\alpha + 1} \right) + \frac{1}{2} \rho (C M_2)^2 + \frac{2\gamma}{r_0}. \tag{4.110}$$

The balance between the electrostatic force and the force produced by the total hydrodynamic pressure in the radial direction can be given as

$$\varepsilon_r \varepsilon_0 \frac{\Phi_0^2}{r_0^2 \ln^2 (R/r_0)} = P_1 + \frac{1}{2} \rho (C M_2)^2 + \frac{\gamma}{r_0}. \tag{4.111}$$

Note that there are three unknowns (M_1, M_2 and r_0) in the above equations; it is therefore possible to solve Equations (4.109)–(4.111) simultaneously when the applied voltage Φ_0 and the interelectrode distance R are specified. The filament radius predicted by the model is shown in Figure 4.80. For a typical interelectrode distance of 1 cm, the filament radius increased from 3 μm to 50 μm as the applied voltage rose from 5 kV to 30 kV, in agreement with typical experimental values.

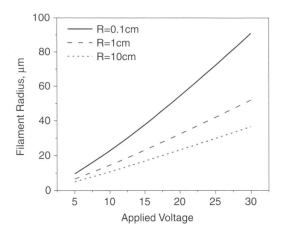

Figure 4.80 *Variations of filament radius as a function of applied voltage and interelectrode distance.*

Figure 4.81 shows the filament propagation speed as a function of Φ_0 and R. The calculated propagation speed from the model was around 15 km s^{-1}, which is also in agreement with relevant experiments. The Mach number increased moderately with the applied voltage, a phenomenon that is understandable from the point of view of energy conservation. For an applied voltage of 30 kV, the Mach number increased from 11.2 to 12.3 when the interelectrode distance decreased from 10 cm to 0.1 cm. This is consistent with the known property of negative streamers; the previous experiment showed that for a given voltage the propagation velocity was relatively constant as the streamer crossed the gap, and it increased as the streamer approached the plane electrode.

The electrostatic model is able to describe the water breakdown experiments with pulse duration much shorter than that required for a bubble formation due to electrolysis or for water evaporation. The pure electrostatic approach is however not sufficient to explain water breakdown and plasma generation in water by voltage pulses shorter than 10–30 ns, which correspond to the time required to move the liquid and create the sufficient void.

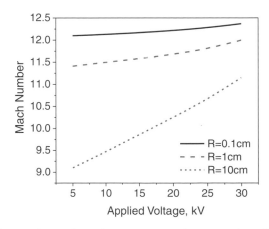

Figure 4.81 *Variations of the Mach number of streamer as a function of applied voltage and interelectrode distance.*

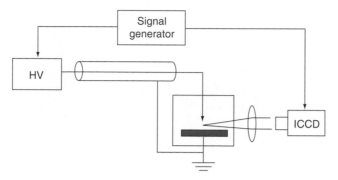

Figure 4.82 *Experimental setup for nanosecond pulsed plasma generation in water without voids and bubbles.*

4.6.5 Nanosecond pulse breakdown and plasma generation in liquid water without bubbles

The generation of plasma in water without any bubbles and voids has been demonstrated by Starikovskiy *et al.* (2011) using extremely short nanosecond pulsed discharge. To initiate the discharge in liquid water, two different pulsed power systems were used. The first power supply generated pulses with +16 kV pulse amplitude in 50 Ω coaxial cable (32 kV on the high-voltage electrode tip due to pulse reflection), 10 ns pulse duration (90% amplitude), 0.3 ns rise time and 3 ns fall time. Maximum pulse frequency was 5 kHz. The second system generated +112 kV pulses (224 kV on the gap) with 150 ps rise time and duration on the half-height of about 500 ps. Maximum frequency was 100 Hz. The discharge cell had a point-to-plate geometry with a high-voltage electrode diameter of 100 μm, and the high-voltage electrode was a cut-off platinum rod. The interelectrode distance was 4 mm, and the low-voltage electrode diameter was 18 mm (Figure 4.82). The water layer between the center of the discharge gap and window was 50 mm. Double-distilled and degased water was used without de-ionization or pH adjustment (pH ~ 6.4, conductivity ~5 μS cm^{-1}). The capacitance of the electrode system was below 1 pF and guaranteed the absence of RC effects for all signals with leading front rise time >50 ps.

Three elements were synchronized: the low-voltage pulsed power supply module, the high-voltage pulse switch and the intensifier of the ICCD camera. The delay between low-voltage and high-voltage modules of the power supply was 33 μs and required an accuracy of ±2 μs. The delay between the image intensifier and the high-voltage pulse switch was variable over the range 35–50 ns and required the jitter to be better than 100 ps to allow pulse-by-pulse signal accumulation without additional averaging of the emission signal. The pulse frequency in these experiments was fixed at $f = 1$ Hz.

The optical system used in this work allowed strong density perturbations (if any) of size larger than 2 μm to be measured and the absence of bubbles of those sizes to be confirmed. It was found that the discharge in liquid water develops on a timescale of picoseconds. The size of the excited region near the tip of the high-voltage electrode was c. 1 mm. The discharge had a complex multichannel structure (Figure 4.83) which changed from pulse to pulse. An image accumulation therefore leads to discharge structure elimination, but allows the emission dynamics to be traced with the 500 ps camera gate (Figure 4.84).

Figure 4.84 demonstrates the dynamics of emission from the discharge taken with 1000 accumulations. It is clearly seen that the discharge emission has a typical duration very close to the camera gate (500 ps) while the emission rise time has an even shorter timescale (c. 250 ps). The high-voltage rise time is about 150 ps and we can assume that the typical discharge development time is between 150 and 250 ps. The emitting region has a diameter of about 1 mm.

To analyze the spatial structure of the discharge, a longer camera gate (1 ns) without signal accumulation has been used. The temporal resolution of the image decreases and the observed emitting discharge phase

Figure 4.83 *Geometry of the discharge gap and ICCD camera field of view. Distilled water, U = 27 kV. The image was acquired 5 ns after the start of the pulse. The camera gate was 1 ns; field of view is 2.6 × 1.7 mm.*

becomes longer (Figure 4.85). In exchange, the spatial distribution of the emission can be observed, which looks like the gas-phase streamer discharge. The emitting channel's typical diameter was c. 50 μm and propagation length was 0.5–0.6 mm for $U = 27$ kV.

In the case of short rise time the discharge propagation velocity is up to 2000 km s^{-1} during the initial stage of the discharge, which corresponds to the voltage rise time (Figure 4.85). When the voltage reaches a maximum, the discharge propagation stops and a 'dark phase' appears ($t = 6$–9 ns in Figure 4.85). During this phase, the discharge cannot propagate because of space-charge formation and a decrease in the electric field.

Voltage decrease leads to the second stroke formation and the second emission phase (Figure 4.85, 10–13 ns). This means that the channels lose conductivity and the trailing edge of the nanosecond pulse generates significant electric field and excitation of the media (comparable to the excitation corresponding to the leading edge of the pulse). An ionization wave corresponding to the trailing edge of the high-voltage pulse was observed under different conditions.

The physics of the process in gas-phase discharges is rather simple. In condensed media, due to the very short plasma recombination time and solvation of electrons, the process becomes much more complex. The secondary ionization wave development definitely correlates with the pulse trailing edge. It should be emphasized that in the described experiments (Starikovskiy *et al.*, 2011) bubble formation was never observed, even after several hours of the discharge operation. The time between pulses was long enough to ensure complete heat dissipation. A single discharge energy was enough to increase the temperature by less than 50 K in the plasma channel but was not enough to evaporate the liquid. Experiments at elevated (up

Figure 4.84 *Discharge emission dynamics; 1000 accumulations/image. The camera gate is 500 ps, λ = 250–750 nm; image size 7 × 4 mm (see color plate).*

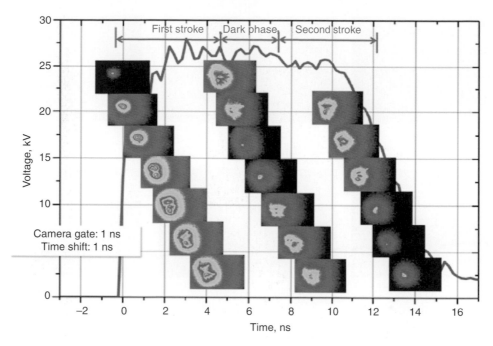

Figure 4.85 *Dynamics of the discharge emission and high-voltage potential on the electrode. Distilled water,*
U = 27 kV; camera gate is 1 ns; spectral response is 250–750 nm.

to 220 kV) voltage confirm the discharge propagation in the liquid phase. The experiments were performed using ultra-short pulses with typical duration of about 400 ps and rise time of 150 ps.

Figure 4.86 shows an integral image of the discharge (camera gate 100 ns) for different voltages. It is clear that the discharge formation depends on the voltage applied. At 92 kV, discharge formation takes longer than the pulse width and only the initial phase of the discharge formation can be observed (Figure 4.86a). Increasing the voltage elongates the plasma channels from 1–2 mm for 107 kV to 6–8 mm for 220 kV (Figure 4.86b–f). Typically, formation of 2–3 channels from the top and bottom edges of the electrode's tip is observed. The length of the channels decreases gradually with the voltage decrease, which for fixed pulse duration means a streamer velocity decrease. Below 90 kV the discharge cannot start until after 400 ps. As in the case with 10 ns pulses, no bubble formation was observed.

Figure 4.87 shows the dynamics of the discharge formation for $U = 224$ kV. The lambda-shaped pulse has no plateau and the dark phase of the discharge did not appear (see Figure 4.87). Discharge development takes an extremely short time (plasma channels reach a length of 0.5–0.8 mm and diameter c. 100 μm in 100–200 ps). The observed propagation velocity reaches 5 mm ns^{-1} (5000 km s^{-1}; c. 15% c). The typical channel diameter is $d = 50$–100 μm. The radial expansion velocity of the channel can therefore be estimated as $v_r = r/t \approx 0.05$ mm per 200 ps, which is equivalent to 250 km s^{-1} (see Figure 4.88).

4.6.6 Nanosecond-pulse uniform cold plasma in liquid water without bubbles: Analysis and perspectives for biomedical applications

While cold uniform plasma in liquid without bubbles and voids has great potential for biology and medicine, an understanding of this novel physical phenomenon is required for successful application in biomedical

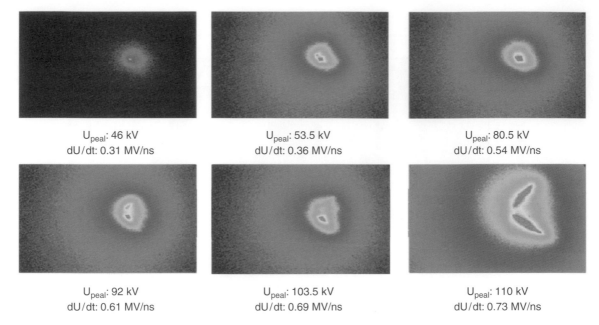

Figure 4.86 *Discharge development for different voltages. Pulse width 400 ps; ICCD camera gate 100 ns.*

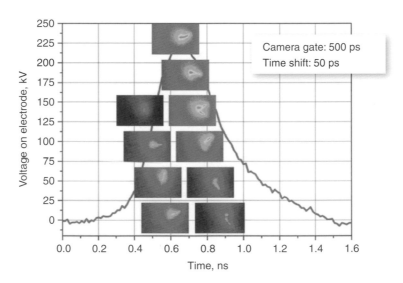

Figure 4.87 *Dynamics of discharge emission (distilled water). U = 220 kV; dU/dt = 1.46 MV ns^{-1}. The camera gate is 500 ps and the time shift between frames is 50 ps.*

Figure 4.88 *Velocity of streamer propagation in water: 150 ps rise time; 1.5 MV ns^{-1}; growth rate vz = L/t ≈ 1 mm per 200 ps = 5000 km s^{-1}. Expansion rate vr = r/t ≈ 0.05 mm per 200 psec = 250 km s^{-1} (see color plate).*

science. There is currently no complete adequate model of the non-equilibrium discharge development in dense media. The main problem is the completely different interaction dynamics as compared to gas-phase systems (in liquids we cannot neglect multi-particle interactions).

We can estimate the electrical field needed for water ionization using the 'dense-gas' approach (Starikovskiy *et al.*, 2011). The ionization rate becomes higher than the attachment rate for $E/n > 50$ Td. In condensed media, multibody attachments play a major role. Taking into account the strong dependence of the rate constant of ionization on E/n we can nevertheless assume that for $E/n > 100$ Td we can obtain fast ionization even in the condensed media. Under normal conditions this means that the electric field value should be close to or greater than $E_{crit} > 30$ MV cm^{-1}. Taking into account the geometrical parameters of the plasma channels (Figure 4.88), we can estimate the radius of the streamer's tip to be $r \sim 10$ μm and subsequent electric field as $E \sim 220$ MV cm^{-1} ($E/n \sim 670$ Td). The electric field near the discharge tip is therefore an order of magnitude higher than the critical electric field estimated using the dense-gas model. We can therefore assume that direct ionization by electron impact takes place in the streamer head even in liquid water, and that it is possible to generate liquid non-equilibrium plasma in different media.

It should be mentioned that the nanometer-scale non-uniformities of water and its dielectric properties (especially at nanosecond and sub-nanosecond timescales) can significantly facilitate uniform water break-down without bubbles and voids. Making a statement regarding the absence of bubbles and voids obviously requires detailed analysis of at micro- and sub-microscales. Water under laboratory conditions always contains trace impurities of dissolved gases.

There are several different mechanisms of vapor bubble (void) formation in the liquid under discharge conditions: electrostatic mechanism; evaporation due to conductive current and Joule heat release; and initial bubble expansion. In the described experiments, it was possible to control the bubbles of size above 5 μm (there was no trace of them, either before or after the discharge). Below this limit, we can assume the existence of bubbles. Nevertheless, the size of these initial bubbles should be very small in comparison to the plasma channel diameter (c. 100 μm). We can therefore analyze the possibility of initial bubble expansion from possible 'detection-limited size' to 'plasma channel size' during the discharge development time (c. 100 ps). It is clear that to form the void in the liquid we need to generate a strong shock wave to accelerate the liquid from the axis of the discharge channel. Simple estimations give us the required flow velocities of about 100–200 km s^{-1} in the radial direction.

Are these parameters achievable using a low-energy discharge in water? The analysis shows that liquid motion with a velocity of 30–35 km s^{-1} (compression ratio 3.8) is possible only in nuclear test explosion experiments. One such test has demonstrated shock wave propagation velocity $D = 43.95$ km s^{-1}, velocity of water behind shock wave front $U = 32.54$ km s^{-1}, compression ratio $s = r/r_0 = 3.852$ and a pressure behind shock wave front of $P = 1430$ GPa (c. 14 Mbar) at a distance of 2 m from the epicenter of the underground nuclear explosion. The total discharge energy in the described experiment can be estimated as the energy of a cylindrical capacitor with typical channel parameters. For $U = 32$ kV (16 kV in the cable), the estimation gives a maximum temperature increase of $\Delta T \sim 6$ K; for $U = 220$ kV (effective 55 kV in the cable at half-height of the bell-shape pulse) we have $\Delta T \sim 50$ K. The water overheat in the discharge channel in all cases is less than 50 K, and definitely not enough to generate an ultra-strong shock wave in order to create a significantly large void.

It is impossible to estimate the energy release in a potential 'gas bubble' because the voltage drop across such a 'bubble' is unknown. It can however be assumed that the voltage is proportional to the media density ('dense-gas' approximation, $E/n =$ constant to maintain the current value). In this case the energy release per molecule does not depend on the density and the temperature increase at all points of the channel (including the hypothetical 'gas bubble') is the same. An approximate estimation based on the charge transfer into the plasma channel (charging of the linear capacitor) gives a maximum temperature increase of about 50 K. For water vapor, this means a pressure increase of up to $P \sim 1.2$ atm.

The pressure increase in the liquid phase can be estimated using the relation $K_p \times \Delta P = \alpha_v \times \Delta T$, where α_v is a volumetric thermal expansion coefficient and K_p is a compressibility coefficient. This equation gives

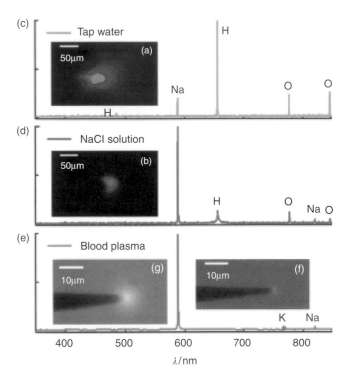

Figure 4.89   *(a, b) OES and corona discharges produced using 5 kV stepped and pulsed voltage excitation in liquids. Demonstration of color changes as a function of the composition of the solution: (c) OES from tap water and (d) 0.5 M NaCl. (e) OES for 5 kV, 20 ns duration pulses in blood plasma; this spectrum does not show oxygen and hydrogen peaks and indicates that, for longer discharges, light emission is due to recombined ions and decomposition of the liquid (whereas in short discharges, light emission is due solely to recombined ions). Optical micrographs indicate that the (f) short-duration pulsed coronas are less than 3 μm in diameter and are significantly smaller than (g) discharges from microsecond-duration pulses, for which a larger discharge and bubble formation is seen (see color plate).*

an estimation of $\Delta P = 180$ atm for $\Delta T = 50$ K due to the low compressibility of liquid water. This estimation explains the weak shock wave formation in the liquid even at relatively low discharge energy density, but is definitely not enough to create the 'voids' or 'low-density regions' over this extremely short timescale.

The same estimations for electrostatic forces give the maximal boundary propagation velocity c. 10 km s^{-1} and do not lead to 'voids' on the sub-nanosecond timescale. On the other hand, the reduced electric field value $E/n \sim 700$ Td explains the observed phenomena via the direct electron ionization.

Although a complete adequate model of the non-equilibrium discharge development in dense media is currently not available, the cold uniform plasma in liquid without bubbles and voids is still attracting interest for biomedical applications. For example, Staack *et al.* (2008) reported a non-thermal corona discharge in liquid without bubbles around electrodes with ultrasharp tips and around elongated nanoparticles (nanowires) for biomedical spectroscopy (see Figure 4.89).

Plasmas created with 50 nm probe tips or carbon nanotubes (CNTs) dispersed in solutions allow simultaneous chemical analysis of multiple dissolved elements within nanoseconds. The proposed optical emission spectroscopy (OES) method can be applied for ultrafast time-resolved multi-elemental analysis of liquid in microfluidic reactors, living biological systems or environmental sensors and for diagnostics of femtoliter volumes with one micrometer or better spatial resolution. Using this method, part-per-million concentrations of sodium, calcium and other elements in aqueous solutions and bioliquids have been measured.

5

Mechanisms of Plasma Interactions with Cells

The complexity of larger multicellular organisms such as the human body makes it more difficult to study their interaction with various plasmas. It is somewhat easier to study the interactions of cells with different plasmas experimentally. Initial steps in deciphering possible mechanisms of interactions between cells and plasma have been made over the last decade or so, but many parts of these mechanisms remain unclear and the entire subject remains a work in progress. The goal of this chapter is to summarize what is known on this subject at this time.

5.1   Main interaction stages and key players

Figure 5.1 illustrates what are believed to be the primary stages and the major players involved in all of the considered pathways of interaction between plasma and cells. Firstly, different active agents can be more or less important for different types of plasma treatment. When plasma is applied directly to the medium or to the substrate containing the cells, charges and local electric field can have important effects; this is in contrast to indirect plasma applications where only chemically active species and UV can participate in the process of interaction. Ozone is known to occur at higher concentrations at lower gas temperature in discharges, while concentrations of nitrogen oxides in air discharges tend to increase at higher gas temperatures. Different discharges produce different UV intensities over different wavelength ranges, which may be differently absorbed depending on the gas composition and the medium above the cells. The ability of the electric field to reach the cells and affect them in some way depends on the amount, conductivity and dielectric properties of the medium above the cells.

With some exceptions (such as sporulated bacteria), cells are rarely in a dry environment. Even if mammalian cells were capable of surviving for a short time without water around their cell membrane, it would be difficult to remove this water completely due to the strong hydrophilic nature of the outside surface of the mammalian membranes. The actual amount of water may however vary with circumstances. Plasma jets having their flow directed toward the surface of the cell medium may remove some of this medium from the cells, for example. The amount of medium left over the cells can be very roughly estimated by setting the surface tension forces of water equal to the shear-force-created tension due to the air velocity U at the surface

Plasma Medicine, First Edition. Alexander Fridman and Gary Friedman.
© 2013 John Wiley & Sons, Ltd. Published 2013 by John Wiley & Sons, Ltd.

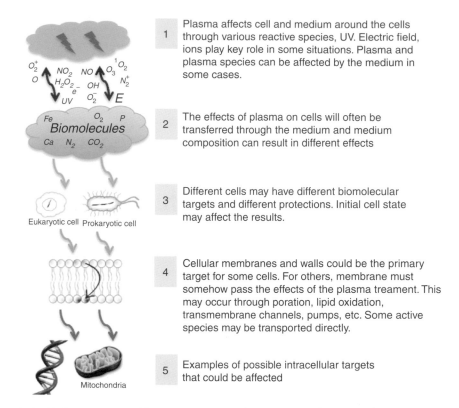

Figure 5.1 *Various stages and key participants in the interaction between plasma and living cells.*

of the water film. Assuming an airflow at the high velocity of 1 m s^{-1} is maintained roughly over the planar substrate length L of about 1 cm, the thickness of the water film on this substrate is approximately:

$$\delta \approx \frac{\mu U}{\gamma} L = \frac{1 \times 10^{-3} \times 1}{7 \times 10^2} 0.01 = 10 \tag{5.1}$$

where μ is the dynamic viscosity of water, γ is its surface tension and δ is the thickness of the water film (nm). In this situation a shear force (μN)

$$F \approx \mu U \frac{D_{cell}^2}{\delta} \approx 10 \tag{5.2}$$

is applied to an individual cell whose characteristic size is $D_{cell} = 10$ μm when this cell is attached to the substrate. This is a very large shear force for a single cell to handle, much larger than what an endothelial cell might experience due to blood flow for example. This illustrates the difficulty of removing water from the cells by mechanical means (such as gas flow). Similar difficulties would arise in other situations. It is therefore reasonable to assume that, in all cases of plasma treatment, there is a water layer greater than 10 nm remaining on the mammalian cells.

Given this water layer over the mammalian cells, we wonder to what extent a charge delivered through plasma to the surface of this water layer can influence processes at the cell membrane. The electrostatic influence can be estimated roughly by comparing the energy of electrostatic interactions of two electrons or ions (having a single electron charge) and the energy of thermal fluctuations. The effective distance at which two point charges in water have interaction energy equal to thermal fluctuation is known as the *Bjerrum length*:

$$\lambda_B = \frac{e^2}{4\pi \varepsilon_r \varepsilon_0 k_B T} \tag{5.3}$$

where e is the electron charge, k_B is the Boltzmann constant, T is the absolute temperature, ε_0 is the free space permittivity and ε_r is the relative dielectric constant of water, equal to about 80. The Bjerrum length is about 0.7 nm in water. The presence of free ions in water provides another mechanism for screening of cells from the direct electrostatic effects of charges that arrive at the water from plasma. The effective screening length in this case is the Debye length for electrolytes, given by:

$$\lambda_D = \sqrt{\frac{\varepsilon_r \varepsilon_0 k_B T}{2 N_A e^2 I}} \tag{5.4}$$

where I is the ionic strength of the solution and N_A is the Avogadro number. At typical intracellular salt concentrations (around 150–200 mM), the Debye length is $\lambda_D \approx 1$ nm. Given the estimate of about 10 nm of water on top of mammalian cells and that charges are effectively screened over the distance smaller than about 1 nm, we therefore conclude that charges arriving from plasma will have a negligible electrostatic influence on membranes of mammalian cells.

This does not rule out dynamic effects of charges and electric fields created by them. The dynamic effects may occur because the electrolyte solution surrounding the cells is not a perfect conductor. As a result, some time is required before complete screening of the plasma charges occurs. This time is given by:

$$t = \frac{\varepsilon_r \varepsilon_0}{\sigma} \tag{5.5}$$

where σ is the conductivity of the cell medium, which is typically around 1–2 S m^{-1}. For such conductivity, assuming the dielectric constant of water is still around 80 (the effective dielectric constant is actually reduced over the short timescale due to a finite rate of water polarization), the electric field screening time is found to be around 10 ns. Due to a reduced effective dielectric constant, in practice the time may be closer to few nanoseconds. In situations such as those typical for DBDs where plasma is generated in a pulsed manner, a strong electric field (such as the field at the head of a streamer) can therefore penetrate the medium for a few nanoseconds every time a pulse of electric field and charge deposition occurs on the surface of the cell medium. This may be sufficient to create biological effects, particularly given the repetitive nature of this process when cells are under a relatively thin film of aqueous solution.

Other effects on cells can occur due to chemical modification of the medium surrounding the cells. In this case, the exact chemical composition of the cell medium can play an important role. The nature of chemical species produced in plasma and their concentration is also critical in such mechanisms. Charges from plasma can play an important role in this chemistry in several ways. On the one hand, their presence can influence peroxidation and oxidation processes catalytically. On the other hand, they can change the medium acidity and, though this, various reactions and their rates.

Cellular membrane as the outermost barrier separating the cell interior from its environment is likely the first and most important biological object in the mechanism of interaction between plasma and cell. Various processes can occur at the cell membrane as a result of plasma treatment. The electric field that penetrates the cell medium for a short time can induce or enhance the formation of pores, a phenomenon known as electroporation. As well as the obvious possibility of membrane damage, this can influence the transport of ions and other biological molecules leading to a variety of signaling effects. Chemically active species created in the medium through its interaction with plasma can lead to oxidation or peroxidation of various molecules in the lipid membrane. Byproducts of lipid peroxidation, such as aldehydes, can end up within the cell. Transmembrane proteins can also be affected by the various chemically active species in the medium.

The interaction of plasma and cells does not end with the cell membrane. Through biological signaling, electroporation, lipid peroxidation or other mechanisms, the membrane can transfer the effects resulting from plasma treatment toward various intracellular targets. This transfer seems to occur through or be accompanied by intracellular oxidative stress. It is possible that this oxidative stress occurs through changes in the functionality of mitochondria. One of the targets of this oxidative stress appears to be intracellular DNA.

5.2   Role of plasma electrons and ions

5.2.1   Selection of biological targets and plasma generation methods

In studying the different roles of various plasma species, a convenient form of plasma, convenient biological targets such as different types of mammalian cells or bacteria and an appropriate medium for supporting the cell function all need to be chosen. With some exceptions, most of the studies that have investigated the role of different plasma species have so far been carried out with bacteria as the biological target, probably because this was the starting application (sterilization) for plasma in historical plasma medicine. Growing bacterial cultures can also be somewhat simpler experimentally than working with mammalian cells. Consequently, agar surfaces, bacterial medium and various solid substrates including glass and metal have been employed as supports for the bacterial sample.

Several considerations are involved in choosing the type of plasma to investigate the effect of charged species. Most importantly, it should be possible to apply the same type of plasma without substantial variation of its properties in different modes. In the direct application mode, charges from plasma should be able to directly enter the medium above the cells or supporting the cells. In the indirect application mode, there must be a way to filter out the charges without substantially affecting the plasma properties or impacting the generation and transport of neutral species. Corona discharges have been used to observe the effect of charged species by several researchers and are discussed in Section 5.2.4. However, most corona discharges generate too little power and do not produce the some of the neutral species (at sufficiently large concentrations) often considered in plasma medical applications.

Dielectric barrier discharges offer higher power and greater production of active species for studies of mechanisms of interaction between non-thermal plasma and cells, particularly when comparing direct and indirect plasma applications to determine the effects of charged species. In the direct plasma application mode, the cellular sample can be used as a second active electrode in the discharge formation; plasma is bound between the dielectric surface of the powered electrode and the surface of the cellular sample (usually surface of cell medium) being treated. In the indirect application mode, the plasma can be separated from the cellular sample by a grounded metal mesh and gas can blown through the discharge to carry active species outside the plasma generation region.

5.2.2   Comparison of direct DBD plasma treatment to indirect treatment with and without ion flux

The first investigations comparing the effects of direct and indirect modes of application were carried out at Drexel University (Fridman *et al.*, 2007b). These initial investigations focused on bacteria as the biological target and employed agar as the supporting surface. It was shown that direct application of plasma yields a faster rate of bacteria inactivation on agar by roughly two orders of magnitude as compared to indirect plasma application (Fridman *et al.*, 2007b). These investigations also demonstrated that the effects of active species without charges were negligible compared to the effects observed with the charges when bacteria were treated on agar. In short, using only UV radiation (placing a quartz window for UV transmission or a magnesium fluoride window for vacuum UV or VUV transmission) effectively removes the ability of DBD plasma to cause 6 log reduction (i.e. by a factor of 10^6) of bacterial population. Average gas temperature and the applied electric fields have also been shown to have a negligible effect on bacteria. The effects of neutral active species produced in DBD, on the other hand, could not be ignored. It has been shown that neutrals by themselves are able to sterilize as well as direct plasma treatment, although this takes an order of magnitude more time. The role of charges in killing bacteria on agar therefore appeared to be catalytic, in that charges increased the rate of bactericidal effects.

Later experiments by the same group went further in controlling the flux of charged particles (Dobrynin *et al.*, 2009); these experiments are described below. In case of direct plasma treatment, the discharge is ignited on the treated surface with 1.5 mm discharge gap. For indirect plasma treatment a grounded metal mesh is used as a second electrode (22 wires per centimetre, 0.1 mm wire diameter, 0.35 mm openings, 60% open area and weaved mesh). The gap between the mesh and the quartz dielectric insulator on the electrode is set to 1.5 mm, as for the direct treatment setup. To ensure that both DBD setups with and without the mesh produce the same amount of UV radiation, corrections were made for the UV transparency of the mesh using separate measurements. These demonstrated that the mesh cuts off only about 20–40% of UV light depending on UV wavelength (see Figure 5.2).

Agar placed into metal dishes was employed as the substrate on which bacteria were treated. These dishes are then biased with unipolar potential to extract charges from the discharge. Charge extraction was

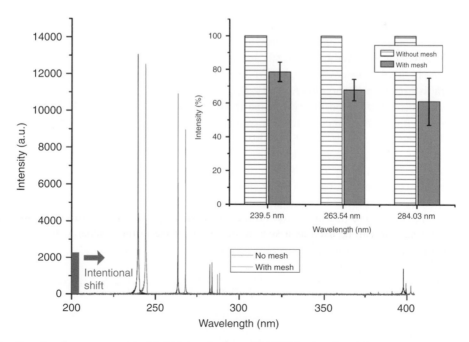

Figure 5.2 *Results of measurement of light intensity from FE-DBD in the ultraviolet spectrum measured at three peaks (239.5, 263.54 and 284.03 nm) without mesh (taken as 100% for each wavelength) and with mesh: a representative spectra and averaged data for the three peaks.*

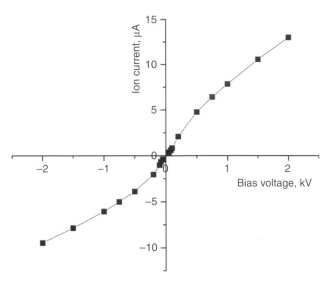

Figure 5.3 *Measurements of bias current during the plasma treatment between the metal dish with agar and the metal mesh separating the high-voltage DBD electrode from the surface of the agar. Note that agar is an electrically conducting medium.*

confirmed by current measurements in the dish–ground circuit (see Figure 5.3). The plasma was ignited in the same electrode-mesh configuration as for the indirect plasma treatment experiment. Distances between the powered electrode and the mesh, and between the mesh and the agar (note that both the dish and the agar are conductors), were both set to 1.5 mm.

It is difficult to establish the amount of fluid that could have accumulated over the bacteria in these experiments prior to or during the plasma treatment. Agar is a hydrogel and will elute water to its surface should water be somehow removed from the surface at some time. It can be expected that bacteria on agar are never completely dry and are covered by a minute amount of water, a condition to which the researchers referred to as 'moist' bacteria (Dobrynin *et al.*, 2009).

The results of direct, indirect and biased plasma treatment experiments are shown in Figure 5.4. When *E. coli* are treated with plasma directly they are exposed simultaneously to charged particles, UV and all active plasma components such as ozone (O_3), hydroxyl radicals (OH) and other excited molecular and atomic species; maximum inactivation effect is obtained in this case. The results of indirect plasma treatment show that it is significantly less effective, probably due to an absence of charged species in plasma afterglow as previously reported with skin flora (mix of staphylococci, streptococci and yeast) (Fridman *et al.*, 2007a). Applying bias potential to the agar leads to an increase in inactivation efficiency. These results show that the presence of charged species may lead to a significant increase of plasma treatment efficiency. Note that although Figure 5.4 appears to indicate that there is some observable difference between positive and negative bias, no statistically significant difference was observed in this work.

5.2.3 Effect of gas composition on antibacterial efficacy of direct DBD

Gas composition can certainly affect the concentration of various ions in plasma. In the presence of nitrogen, for example, N_4^+ ions readily form in the discharge due to the three-body reaction:

$$e^- + 2N_2 \rightarrow N_4^+ + 2e^-. \tag{5.6}$$

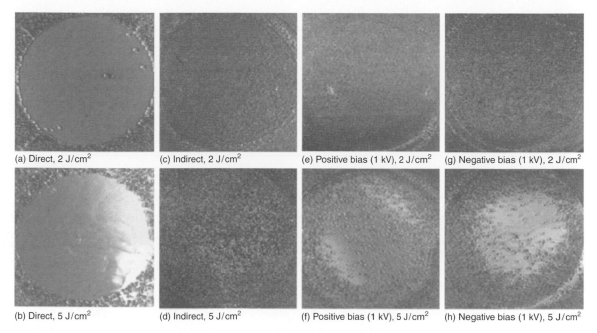

(a) Direct, 2 J/cm^2 (c) Indirect, 2 J/cm^2 (e) Positive bias (1 kV), 2 J/cm^2 (g) Negative bias (1 kV), 2 J/cm^2

(b) Direct, 5 J/cm^2 (d) Indirect, 5 J/cm^2 (f) Positive bias (1 kV), 5 J/cm^2 (h) Negative bias (1 kV), 5 J/cm^2

Figure 5.4   *Results of inactivation of* E. coli *on agar surface by (a, b) direct and (c–h) indirect plasma treatment. In the case of indirect treatment, the agar was either (c, d) grounded orDC-biased with (e, f) 1 kV (higher bias voltages were not employed due to occurrence of a corona discharge) positive polarity or (g, h) 1 kV negative polarity. For all cases, the plasma dose was kept at 2 J cm^{-2} (a, c, e, g) and 5 J cm^{-2} (b, d, f, h) (see color plate).*

Similarly, positive oxygen-based ions can form due to:

$$e^- + 2O_2 \rightarrow O_4^+ + 2e^-. \tag{5.7}$$

At the same time, in the presence of oxygen the DBD discharge produces superoxide anions O_2^-. In the hope of clarifying the effect of different ions in DBD, Drexel University researchers (Fridman *et al.*, 2007b) used different gases during application of DBD to bacteria seeded on agar. The effect of treatment was observed as before on *E. coli*, skin flora (mix of streptococci, staphylococci and yeast obtained from human patient samples) and *B. subtilis* spores treated by the DBD plasma on agar in various gases followed by a 24-hour incubation and colony counts. Gases tested included air, O_2, N_2, Ar, He and N_2/NO mixture (700 ppm NO). Complete inactivation (>7 log reduction in colony-forming units) of *E. coli* was achieved in direct plasma treatment in air and oxygen at 2 J cm^{-2}. In other gases, a tested dose of over 12 J cm^{-2} was required to achieve any visible effect and much higher doses were needed to achieve complete inactivation (compared to O_2 and air, no significant effect was observed in Ar or He at all, even at >600 J cm^{-2}). The results are summarized in Table 5.1 (Dobrynin *et al.*, 2009).

Although it can be concluded that the presence of oxygen significantly enhances bacterial inactivation with DBD, no clear conclusion could be drawn regarding which specific ions play key roles. This is particularly the case given that some bacterial inactivation could be observed using DBD in nitrogen.

Table 5.1   *Effect of gas composition on inactivation efficiency.*

Gas	Inactivation
O_2	Sterile at 2 J cm^{-2}
Air	Sterile at 2 J cm^{-2}
N_2	Some visible disinfection at ≥ 12 J cm^{-2}
Ar	
He	
N_2/NO (700 ppm)	

5.2.4   Effect of positive and negative ions in nitrogen corona discharge

In DBD treatment it is difficult to apply charged species while holding back the neutral species which are usually produced in such discharges at significant concentrations. DC corona discharge in a flow of dry nitrogen at 0.5 SLPM (standard liters per minute) was used to study the effects of ions without significant flux of chemically active neutral species (Dobrynin *et al.*, 2009). A stainless steel needle electrode is placed inside a ceramic tube with an inner diameter of 5 mm. It is powered through a 10 MΩ resistor to produce ion flow with an average current of approximately 20 μA for both positive and negative polarities; voltage is then varied around 1 kV to produce the same current for both polarities. A grounded metal dish of 60 mm diameter with agar and *E. coli* is then used as a second grounded electrode, completing the circuit. There is also a 2 mm gap between the end of the tube and agar surface to provide gas output from the system. Nitrogen (99.999% pure) is used to ensure that the concentration of reactive oxygen species is reduced in the discharge.

The results show that ions of both polarities are able to inactivate bacteria. The effect becomes visible after $\sim$25 s of treatment for both cases and positive ions show slightly higher efficiency; however, the difference between positive and negative ions is no more than $\sim$10–15% (Figure 5.5). No water evaporation was observed for treatment times below 1 hour and the effect of gas flow alone was analyzed and shown to have no effect on bacteria (results not shown, but see Fridman *et al.* 2007b for a description of a similar experiment). This experiment serves as a second indication of the potential importance of ions in the inactivation of bacteria. It may not completely eliminate the role of UV and ROS however, since minute amounts of water and oxygen are present even in high-purity nitrogen.

5.3   Role of UV, hydrogen peroxide, ozone and water

5.3.1   Effect of UV in DBD treatment

The mechanism of UV-based sterilization is widely studied, and indeed in many cases it plays a major role (Moisan, 2002; Boudam *et al.*, 2006). Experiments analyzing the role of UV radiation produced by direct plasma have been conducted (Fridman *et al.*, 2007b). One way to protect bacteria from everything that is generated in plasma except for the UV photons is to place a quartz glass on top of the treated surface. Experiments described by Fridman *et al.* (2007b) used quartz which is transparent to UV photons of >200 nm wavelengths and MgF_2, which is transparent to VUV photons of >140 nm wavelengths (Baydarovtsev *et al.*, 1985). As can be seen in Figure 5.6, bacteria that are protected from direct discharge by a 10 mm square MgF_2 slide are unaffected (highest dose used was 600 J cm^{-2} with no observed difference between untreated bacteria and MgF_2-protected bacteria); we can therefore conclude that the action of ultraviolet radiation for our case can be neglected. However, even though we observe no visible effect on bacteria by UV/VUV photons, it should not be discounted completely as UV (especially VUV) is known for its synergetic effect in

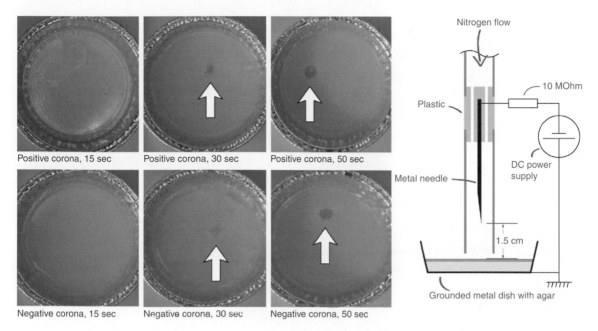

Figure 5.5 *Illustration and results of* E. coli *inactivation on agar surface by negative and positive polarity corona discharge in nitrogen (see color plate).*

interaction and destruction of model polymers (Vasilets *et al.*, 1981; Baydarovtsev *et al.*, 1985; Ponomarev *et al.*, 1989).

5.3.2 Effect of hydrogen peroxide

In the presence of water molecules in room-temperature-atmospheric-pressure air, a certain amount of OH molecules are produced in DBD plasma. This is a highly reactive radical and, as such, will quickly react with the organic molecules present in the bacteria surroundings ; this probably limits its effectiveness to directly affect bacteria. However, the product of the recombination of two polar OH molecules on charge centers, a hydrogen peroxide molecule, is much less reactive and may pass through the cell membrane relatively easily causing a variety of downstream lethal effects (for example, fatal DNA damage; Imlay *et al.*, 1988; Henle and Linn, 1997). The ability of hydrogen peroxide to sterilize is widely known and has been well studied. Experiments were carried out to analyze the role of hydrogen peroxide specifically in DBD-plasma-based inactivation of bacteria. Because measurement of H_2O_2 concentrations at the surface of agar is a challenging problem, this concentration was approximated from the measured amount of H_2O_2 produced in liquid. The concentration of H_2O_2 was measured in distilled water and phosphate-buffered saline (PBS) with H_2O_2 specific test strips. The dependence of peroxide concentration on the treated volume is almost linear, and increases by a factor of 3 when the amount of treated liquid is decreased by a factor of 10 (see Figure 5.7). The estimated amount of liquid under the electrode (surface area of $\sim$5 cm^2) in the case of 'moist' agar is a few microliters; we can therefore expect the concentration of peroxide to be of the order a few tens of mmol L^{-1} for a plasma dose of several J cm^{-2}.

To determine the concentration of H_2O_2 that causes the same sterilization effect on agar in direct plasma treatment, 50% by volume of water solution of H_2O_2 was used and further diluted with distilled de-ionized

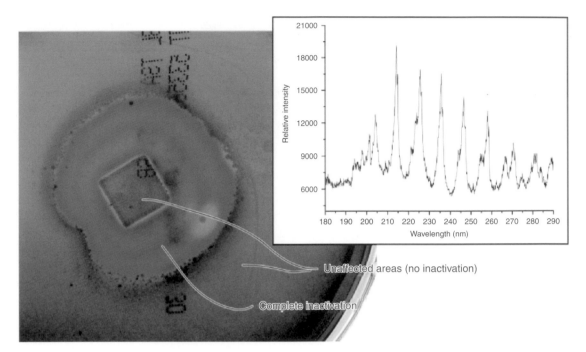

Figure 5.6   *Emission from DBD plasma over the cell surface in UV range (right) and results of inactivation of bacteria by direct plasma contact compared with only UV (left). No effect on bacteria protected from plasma by MgF$_2$ slide (10 mm square in the center) is observed. In this image the plasma surface power density was 0.6 W cm^{-2} and the treatment dose was 108 J cm^{-2} (see color plate).*

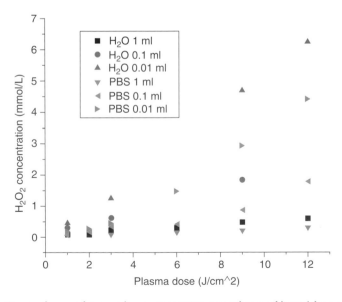

Figure 5.7   *Dependence of peroxide concentration on volume of liquid for water and PBS.*

Table 5.2   *Effect of hydrogen peroxide on inactivation of* E. coli.

H_2O_2 volume %	H_2O_2 mmol L^{-1}	Inactivation result
50	20 000	Complete inactivation
5	2 000	
0.5	200	Some visible disinfection
5×10^{-2}	20	No visible result
5×10^{-3}	2	
5×10^{-4}	0.2	

water. The 0.1 mL droplet of the solution is poured onto the bacteria that are placed on agar and spread over the whole agar surface. The results show that concentration of H_2O_2 that corresponds to about 0.5 J cm^{-2} of direct DBD plasma is >200 mmol L^{-1} (Table 5.2). As can be seen from Figure 5.8, direct DBD plasma produces at best about 6.5 mmol L^{-1} H_2O_2 at a plasma dose of more than an order of magnitude higher than that required for inactivation by plasma. It can therefore be concluded that while hydrogen peroxide may have some effect, it is not the key mechanism by which direct plasma inactivation occurs.

5.3.3   Effect of ozone

The bactericidal effect of ozone is well known and has already been utilized in industry for some time (Broadwater 1973; Ohmi *et al.*, 1992). DBD generates ozone in room-temperature-atmospheric-pressure air, which may be partly responsible for the observed bacterial inactivation. The antimicrobial effect of ozone generated by the DBD has been evaluated in two ways: (1) by comparing the results to those found with an ozone generator and (2) by scavenging ozone in DBD. The DBD in room-temperature-atmospheric-pressure

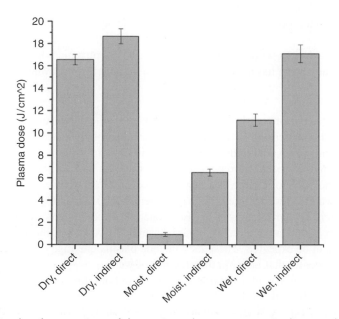

Figure 5.8   *Results of comparison of dry, moist and wet treatment in direct and indirect setup.*

air at ~60% relative humidity produces about 28 ppm of ozone as measured outside of the discharge zone (recorded by ozone-specific spectrophotometer MedOzon 254/5). An ozone generator (~500 ppm maximum output, Quinta Inc) was used to produce the same concentration of ozone in room-temperature-atmospheric-pressure air without plasma (Gallagher *et al.*, 2004, 2007; Gutsol *et al.*, 2007). No inactivation effect was observed on *E. coli* cultures on agar and on skin flora (mix of streptococci, staphylococci and yeast) in as much as 30 minutes of treatment. However, it is likely that ozone concentration within the DBD discharge is significantly higher. To test for the possibility that ozone within the discharge has a much stronger effect, nitric oxide was employed to scavenge the DBD produced ozone (Clyne *et al.*, 1964; Clough and Thrush, 1967; Fontijn *et al.*, 1970) as described by the reactions:

$$NO + O_3 \rightarrow NO_2^* + O_2 \tag{5.8}$$

$$NO_2^* \rightarrow NO_2 + h\nu. \tag{5.9}$$

DBD plasma in gas produced by diluting a stock 700 ppm NO in N_2 with oxygen leads to the same inactivation efficiency as when pure N_2 with O_2 mixture is used, while the measured ozone concentration is 0 in the first case and 28 ppm in the second. It can therefore be concluded that, as for UV and H_2O_2, ozone does not play a major role in the inactivation of bacteria.

5.3.4   Effects of water and its amount

Water is present in all the plasma treatment cases discussed above. It can be present in different amounts however and this can, in principle, affect the results and possible mechanism of interaction between bacteria and plasma. To analyze this possibility, Drexel University researchers separated DBD treatment conditions into three groups:

1. dry treatment, when a droplet with bacteria is placed on a glass slide and then dried (~1 hour);
2. moist treatment, when a droplet with bacteria is placed on an agar surface and left until the agar appears dry (~1 hour); and
3. wet treatment, when a droplet with bacteria is placed in a cavity on a glass hanging drop slide and treated immediately before any water has a chance to evaporate (although some evaporation likely occurs during the treatment).

DBD plasma treatment was performed directly or indirectly under the conditions described above. A significant difference that could be described as qualitative, not just quantitative, was found to exist between these three cases. While bacteria are covered with an extremely thin (the exact amount is unknown, hypothesized to be anywhere from few nanometers to microns) layer of free water on agar, in solution bacteria are submerged in water and therefore the effect of charged particles may be significantly diluted. In addition, if in the case of bacteria on agar the water layer is of the order nanometers, bacteria may experience short but very strong electric field pulses associated with streamers. In case of bacteria dried on glass slides when there may be only a few molecules of water bound to bacteria, gas-phase plasma effects such as those observed in the treatment of polymers by plasma can be expected. These may include plasma chemistry and ion bombardment, rather than typical oxidative processes that occur in the presence of water.

The results of comparison of dry, moist and wet treatment in both direct and indirect setups are shown in Figure 5.8. For every case, 0.2 mL of 10^8 CFU per ml *E. coli* solution was used and treated with plasma in ~1 J cm^{-2} increments. It is clear that, as was previously shown by Fridman *et al.* (2007b), direct plasma treatment achieves inactivation at significantly lower doses than indirect in the case of 'moist' treatment on agar. The ratio of plasma doses in direct and indirect treatment decreases significantly as the amount of water around the bacteria increases in the 'wet' treatment case. Overall this suggests that the importance of charges directly in contact with the treated sample diminishes as the amount of water increases, while neutral species from plasma continue to play a significant role.

While the difference in dose required for inactivation of bacteria in moist and wet conditions may be attributed to the amount of water protecting the organism, the case of dried bacteria is not as clear. The only clear conclusions that can drawn from this experiment are: (1) that the presence of water is required for effective interaction of plasma with bacteria and (2) that charges play a much more important role when there is a relatively small amount of water, while the relative importance of neutral species in the interaction of plasma with bacteria increases as the amount of water increases.

5.4   Biological mechanisms of plasma interaction for mammalian cells

While the effects of different active species as initiators of interaction between plasma and living systems was investigated primarily using bacteria as the biological targets, downstream biological mechanisms due to plasma activity have also been studied for mammalian cells. This is partly due to the desire to begin applying plasma in real medical applications where tissues are mammalian. In addition, some mechanisms are easier to investigate using mammalian cells as targets.

5.4.1   Intracellular ROS as key mediators of plasma interaction with mammalian cells

Several researchers and research groups have speculated that biological effects of plasma treatment are somehow mediated by reactive oxygen species (ROS). Researchers from Drexel University (Kalghatgi *et al.*, 2011) not only confirmed this hypothesis, but also established that somehow it is the intracellular ROS which are the essential mediators of plasma interaction with mammalian cells. They employed breast epithelial cells (MCF10A) that adhered to the surface of glass slides and were covered by a relatively small amount of cell medium (100 µL, about 0.3 mm fluid layer) during DBD treatment (doses ranged 0.13–7.8 J cm^{-2}).

DNA damage as detected by phosphorylation of a histone protein called H2AX was used in this work as the primary readout of the plasma interaction with the cells. Using DNA as the readout of the effects of interaction is convenient for several reasons. On the one hand, knowing what happens to DNA is important in determining long- and short-term toxic effects of plasma treatment. On the other hand, DNA damage, repair and regulation can result from a large number of intracellular processes.

The phosphorylated form of H2AX, γ-H2AX, can be detected by its attachment to a highly specific fluorescently labeled antibody. This can be done directly in the treated cells. Figure 5.9 provides an example of such immunofluorescence detected DNA damage in cellular nuclei. The same figure also shows quantitatively the amount of γ-H2AX as a function of plasma exposure dose detected through the immunoblotting process, where material from the treated cells is dragged by electric field through gel to separate the specific protein being sought. Figure 5.9 essentially indicates that phosphorylation of H2AX (and therefore DNA damage) does increase with the increasing dose of the DBD treatment. This is the first important conclusion that is not entirely intuitive, since DBD plasma does not generate any significant amount of penetrating radiation or charged particles that can penetrate through cell medium, through the cell membrane, through nuclear membrane and into the nuclei in order to affect the DNA. This result clearly suggests that some kind of biological chain reaction/diffusion process takes place through the cell medium and across the cell membrane in order to create the observed effects.

Following DBD plasma treatment, ROS were detected inside cells by using specific fluorescent indicators such as 5-(and- 6)-chloromethyl-2′, 7′-dichlorodihydrofluorescein diacetate, acetyl ester (CM-H$_2$DCFDA,

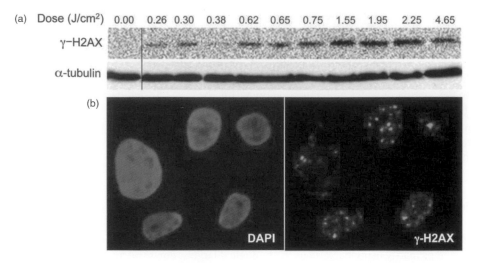

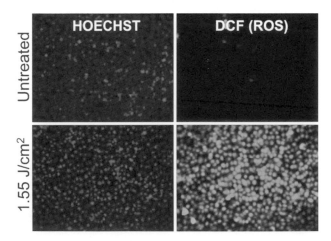

Figure 5.9 *Induction of DNA damage by DBD plasma. (a) MCF10A cells were treated with the indicated dose of plasma. After 1 hour of incubation, lysates were prepared and resolved by SDS-PAGE and representative immunoblots with antibody to γ-H2AX (top) or α-tubulin (bottom) are shown. (b) Indirect immunofluorescence was performed utilizing an antibody to γ-H2AX 1 hour after treatment of MCF10A cells with 1.55 J cm² DBD plasma (see color plate).*

Molecular Probes). An example illustrating that plasma-treated cells have much higher levels of intracellular ROS is illustrated in Figure 5.10. To test the hypothesis that the intracellular ROS are directly responsible for the observed DNA damage, the cells were incubated with a well-known modification of cysteine protein called N-acetyl cysteine (NAC). NAC effectively acts as a scavenger of intracellular ROS by increasing levels of intracellular glutathione. The resulting levels of DNA damage were compared at different plasma doses for cells with and without NAC using immunoblotting-based measurements of γ-H2AX. Figure 5.11

Figure 5.10 *CM-H₂DCFDA dye was preloaded into MCF10A cells for 30 min and the cells were then allowed to recover for another 30 min at 37°C. After recovery, the cells were treated with the indicated dose of DBD plasma (1.55 J cm⁻²). One hour after the plasma treatment, intracellular ROS were detected using a fluorescence-enabled inverted microscope. The blue dye marks cell nuclei, while the green dye marks their fluorescence indicating ROS levels. Untreated cells clearly have much lower levels of ROS (see color plate).*

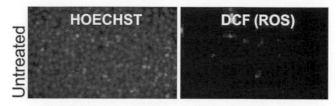

Figure 5.11 *Confirming the effects of intracellular ROS by ROS scavenging. MCF10A cells were incubated for 2 hours with 4 mM NAC (+), followed by treatment with the indicated dose of DBD plasma. γ-H2AX (top) or α-tubulin (bottom) was detected by immunoblotting of cell lysates prepared 1 hour after plasma treatment (see color plate).*

clearly demonstrates that cells in which NAC scavenged intracellular ROS showed no detectable level of DNA damage in contrast to cells that were not incubated with NAC. This is probably the first definitive confirmation of the hypothesis that intracellular ROS are directly responsible for mediating interactions between cells and plasma, or at least any cellular responses that involve DNA.

5.4.2 DNA damage and repair as a consequence of DBD plasma treatment

DNA damage is a relatively common occurrence in cells, and can occur in several different forms due to various causes. Intracellular ROS that occur as a byproduct of cellular respiration is one cause of occasional DNA damage. UV radiation from the environment is another typical cause. Medical therapies such as ionizing radiation (a common cancer therapy) and more recently photodynamic therapy cause substantial DNA damage primarily though generation of OH radicals and singlet oxygen, respectively, directly within cells.

The probability of any given type of DNA damage depends on the underlying cause. The most common type of DNA damage due to UV, for example, is formation of thymine dimers in the DNA molecule. Cellular-respiration-produced ROS most often cause single-stranded DNA breaks. These breaks are usually repaired by cellular processes that employ the second undamaged DNA strand as the source of information for the appropriate DNA sequence. Ionizing radiation creates one of the most difficult to repair DNA lesions called double-strand breaks (DSB), where both complementary DNA strands are broken. These types of DNA breaks are more difficult to repair when the reliable source of information (the second DNA strand that acts as a repair blueprint) is missing. As a result, ionizing radiation is more likely to cause DNA mutations. Hydrogen peroxide creates DNA damage similar to the ionizing radiation and is often employed to simulate the effects of ionizing radiation.

It is important to note that DSBs can also occur in the process of other type of DNA damage repair. These DSBs do not increase the likelihood of mutation because a reliable repair blueprint exists. Phosphorylation of H2AX used to detect DNA as described above does occur in the process of repair that involves DSB. However, this may be the type of DSB that occurs in the process of repair, rather than a DSB created directly by an external influence. It would therefore not be correct to conclude that plasma treatment causes double-stranded DNA breaks simply because the damage is detected using γ-H2AX.

Different types of DNA damage are associated with different repair pathways. All these pathways proceed through phosphorylation of different proteins and involve different kinases (recall that kinases are enzymes that help phosphorylate proteins). Three most common kinases involved in DNA repair are called ATM (Ataxia Telangiectasia Mutated), DNA-PK (DNA-dependent protein kinase) and ATR (Ataxia Telangiectasia and Rad3-related). Roughly speaking, ATM and DNA-PK kinases are activated as a result of double-stranded breaks caused by some external agent, while ATR is typically activated in association with repair of single-stranded DNA damage or UV damage. Identifying the type of kinase that is activated following the plasma

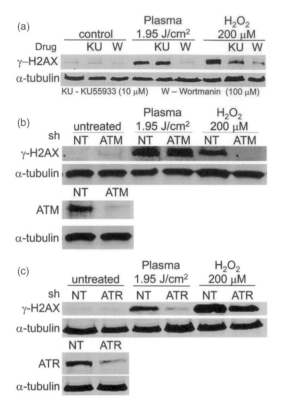

Figure 5.12 *ATR dependence of non-thermal plasma induced phosphorylation of H2AX. (a) Immunoblot of γ-H2AX (top) and α-tubulin (bottom) from MCF10A cells exposed to the DBD plasma at a dose of 1.95 J cm^{-2} or 200 mM H$_2$O$_2$ in the presence (+) or absence (−) of 100 mmol L^{-1} Wortmannin (W) or 10 mmol L^{-1} KU55933 (KU). (b) MCF10As were depleted of endogenous ATM by shRNA for 72 hours (bottom, immunoblot of ATM after ATM or non-targeting shRNA). Cells were then plated on glass cover slips, covered by 100 µL of medium and exposed to DBD plasma at a dose of 1.95 J cm^{-2} or 200 mM H$_2$O$_2$. (c) MCF10As depleted of endogenous ATR by shRNA for 72 hours (bottom, immunoblot of ATR after ATR or non-targeting shRNA). (b, c) After knockdown (shRNA introduction), cells were plated on glass cover slips for 24 hours followed by exposure to the DBD plasma at a dose of 1.95 J cm^{-2} or 200 mM H$_2$O$_2$. After 1 hour incubation, lysates were prepared and resolved by SDS-PAGE and representative immunoblots with antibody to γ-H2AX (top) or α-tubulin (bottom) are shown.*

treatment therefore provides important information on the type of DNA damage that occurs in cells, the line of investigation pursued by Kalghatgi *et al.* (2011).

Researchers (Kalghati *et al.*, 2011) observed that phosphorylation of H2AX in response to DBD plasma treatment (as well as H$_2$O$_2$) was markedly reduced in cells pretreated with 100 mM of Wortmannin (Figure 5.12a), which is known to inhibit ATM, ATR and DNA-PK (Abraham, 2004). In contrast, 1 hour pretreatment with 10 mM KU55933, an ATM-specific inhibitor (Hickson *et al.*, 2004), did not significantly reduce the phosphorylation of H2AX in response to plasma treatment, whereas it significantly reduced it in response to H$_2$O$_2$ (Figure 5.12a). These findings indicate that ATR and/or DNA-PK is required for the phosphorylation of H2AX in response to DBD plasma, although they do not rule out that other kinases may be activated.

To more directly assess the role of ATM and/or ATR, shRNAs were utilized. As a brief reminder, RNA are molecules which translate DNA code into proteins and regulate protein expression from the DNA. Short hairpin RNA or shRNA is an RNA molecule that has a sharp hairpin tight turn and can be used to silence gene expression via a phenomenon called RNA interference. shRNA is now a fairly common tool in the arsenal of molecular biology methods. This molecule can be introduced into cells using a virus as a delivery mechanism (vector); its presence and activity (gene silencing) within cells is inherited through cell division.

It was found (Kalghatgi *et al.*, 2011) that ATM shRNA effectively reduced levels of ATM and blocked phosphorylation of H2AX in response to hydrogen peroxide, but did not significantly affect the phosphorylation of H2AX induced by the DBD plasma treatment (1.95 J cm^{-2}) as compared to non-targeting shRNA (Figure 5.12b). These findings confirm the results with inhibitors and indicate that ATM is not the primary mediator of H2AX phosphorylation in response to plasma treatment. Depletion of ATR by shRNA, on the other hand, reduced phosphorylation of H2AX in response to plasma treatment by 92% relative to non-targeting shRNA and by 40% in response to hydrogen peroxide (Figure 5.12c). Taken together, these findings clearly demonstrate that non-thermal plasma treatment activates ATR kinase.

Activation of ATR by UV is through the formation of thymine dimers resulting in the replication fork collapse (Ward *et al.*, 2004) (replication fork is involved in DNA replication). To determine whether plasma treatment resulted in the formation of thymine dimers, immunofluorescence with an antibody that detects cyclobutane pyrimidine dimers (TDM-2) was performed on cells after treatment with UV or DBD plasma, alternatively. As shown in Figure 5.13a, all of the cells treated with UV showed the presence of cyclobutane pyrimidine dimmers; however, there was no evidence of thymine dimers in cells treated with the DBD plasma. Moreover, pretreatment of cells with NAC did not prevent the formation of bulky adducts/thymine dimers in UV-treated cells (Figure 5.13b), further supporting that the effects of the DBD plasma are different from UV.

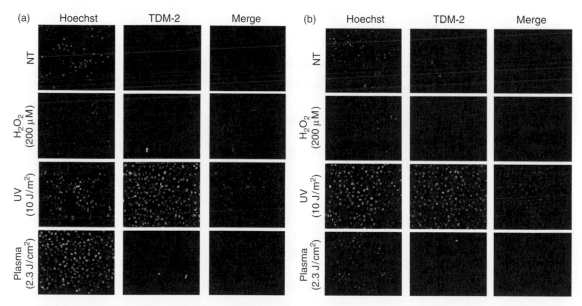

Figure 5.13 *Plasma treatment does not induce the formation of thymine dimers. MCF10A cells grown on glass coverslips were incubated (a) with or (b) without 4 mM NAC for 2 hours, followed by treatment with direct plasma (2.3 J cm^{-2}), UV (10 J m^{-2}), H$_2$O$_2$ (200 mM) or a no-treatment (NT) control. Cells were allowed to recover for 1 hour, then fixed and immunostained with TDM-2 primary antibody (kindly provided by Dr Toshio Mori at the Nara Medical University, Kashihara, Nara, Japan) and Alexa Fluor 594 anti-mouse secondary antibody to detect cyclobutane pyrimidine dimmers (see color plate).*

In summary, this investigation confirmed that the intracellular ROS generation following the DBD plasma treatment results in DNA damage; this DNA damage very different from the types of DNA damage that typically occur in response to ionizing radiation, hydrogen peroxide or UV radiation. Not only does this suggest that plasma is a novel tool that differs from ionizing radiation, hydrogen peroxide and UV in its mechanisms of interaction with cells, but also that this novel tool is probably less dangerous than tools such as ionizing radiation (which is more likely to lead to dangerous mutations).

5.4.3   Effect of the cell medium in plasma interaction with mammalian cells

As already noted in the above discussion related to the treatment of bacteria with DBD plasma, the direct application of plasma produces a much stronger effect when the cells have only a little water around them compared to the effect of indirect treatment where neutral species play the dominant role. Direct and indirect treatments do not differ in the inactivation of bacteria as much when the bacteria are surrounded by a greater amount of water. Similar conclusions can be made for the treatment of mammalian cells. Figure 5.14 illustrates that direct and indirect (with a grounded mesh placed between the cell sample and the insulated high-voltage DBD electrode) treatments produce similar results as measured by phosphorylation of H2AX, with only a slightly stronger effect due to the direct treatment. This clearly supports the idea that plasma charges (and possibly electric fields generated by them) play a critical role when the amount of water around the cells is smaller; their role is however diminished (at least at the doses considered) when the amount of water around the cells is much larger.

Consistent with the result that intracellular ROS mediated the DBD-plasma-induced DNA damage, Kalghatgi *et al.* (2011) also found that longer incubation in the original 100 µL of medium in which

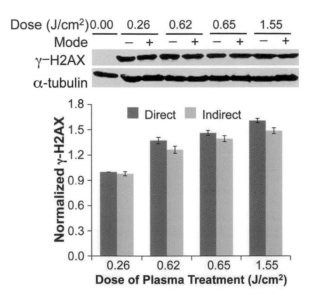

Figure 5.14 *Comparison of direct and indirect treatments. Cells were subjected to plasma directly (D) or indirectly where a grounded mesh that filters charged particles was placed between the electrode and the medium (I). Representative immunoblots with γ-H2AX (upper panel) or α-tubulin (lower panel) are shown. The graphs below the immunoblots show quantification from three independent experiments using the Odyssey Infrared Imaging System (LI-COR Biosciences, Lincoln, NE, USA). The γ-H2AX signal was normalized to the amount of α-tubulin and data are expressed relative to the lowest dose (set at 1.0).*

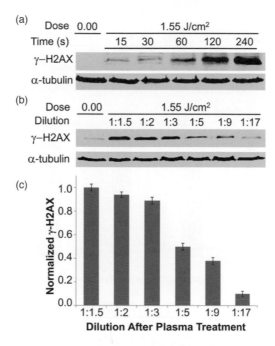

Figure 5.15 *The effects of plasma on cells are dependent on ROS concentration. (a) Cells on cover slips overlaid with 100 μL cell culture media were treated with plasma (1.55 J cm⁻²), followed by dilution in 2 mL of media at the indicated holding time after treatment. Cell lysates were collected 1 hour after dilution in 2 mL of media. Immunoblots with γ-H2AX or α-tubulin are shown. (b) Medium was separately treated with plasma and then diluted immediately after treatment as indicated. Cells were then exposed to the treated and diluted medium for 1 min, followed by 1-hour incubation in 2 mL of fresh medium. Immunoblots with γ-H2AX or α-tubulin are shown. (c) Dependence of the γ-H2AX normalized to α-tubulin as a measure of DNA damage on the dilution of the plasma-treated medium by untreated medium.*

cells were treated (before dilution into 2 mL of medium) resulted in higher levels of γ-H2AX (Figure 5.15a). Additionally, when the treated medium was subjected to different dilutions 1 min after treatment, the amount of damage correlated with the dilution; that is, damage was greater at the lowest dilution (Figure 5.15b and c). These data suggest that the generation of intracellular ROS and the induction of DNA damage are the results of the plasma interaction with the extracellular medium. The results also suggest that the effects of plasma depend strongly on the concentration of active species in the medium as well as the length of exposure of cells to these active species.

To further establish that the effects of DBD plasma are due to modification of the cell medium by the plasma treatment as opposed to a direct effect of the cells, medium was treated separately and added to cells. The term 'separated treatment' was used to describe the form of treatment where medium (100 μL) on a coverslip (without cells) was treated with the DBD plasma and then transferred to a fresh coverslip with cells already on it. The effect of medium separately treated with the DBD plasma and added to cells was not significantly different from the effect of direct treatment of cells overlaid with the same type of medium (Figure 5.16a). Effectively, this implies that chemical phenomena occurring in the medium as a result of its treatment by the DBD plasma play a critical role in transferring the effects of the plasma treatment to cells. Moreover, the treated medium is capable of 'storing' the effect of plasma treatment, at least for some time (this is discussed in more detail in Chapter 10). To test the stability of this storage, separately treated medium

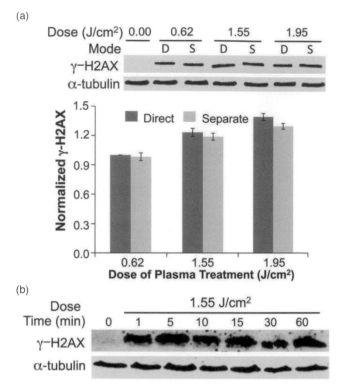

Figure 5.16   *Effects of DBD plasma are stored in the medium. (a) Cells were subjected to direct treatment with plasma (D) or to medium (100 µL) that was exposed to plasma and then transferred to the cells (separated, S). (A, B) Representative immunoblots with γ-H2AX (top) or α-tubulin (bottom) are shown. The graphs below the immunoblots show quantification using Odyssey. The γ-H2AX signal was normalized to the amount of α-tubulin. Data are expressed relative to lowest dose, which was set at 1.0. (b) Medium (100 µL separated treatment) was subjected to plasma and transferred to cells after holding for 1–60 min. After 1-min incubation with cells, cover slips with treated medium and cells were transferred to a dish with 2 mL of medium.*

was held for increasing times before being added to cells. Induction of DNA damage by the treated medium was not reduced by holding the medium up to 1 hour prior to adding it to cells (Figure 5.16b), suggesting that the active species formed in the medium are relatively stable.

The question of which components of the medium are responsible for the storage and which are responsible for the transfer of the plasma treatment arises. Is the organic content of the medium important or can the observed behavior be attributed to water and other inorganic medium content? The second question is answered by comparing the effects of separated-medium treatment to separated treatment of medium reduced only to its inorganic content (phosphate buffered saline or PBS). Little or no DNA damage was observed in cells exposed to the separately treated PBS (Figure 5.17a), whereas separately treated medium induced the DNA damage as anticipated. This suggests that stable organic components in the medium, such as organic peroxides, are probably the main solution mediators of the plasma treatment.

Cell culture medium is composed of amino acids, glucose, vitamins, growth factors and inorganic salts, as well as serum. It has been shown (Gebicki and Gebicki, 1993) that γ-radiation (ionizing radiation) induces the formation of amino acid and protein hydroperoxides in aqueous solutions containing bovine serum albumin

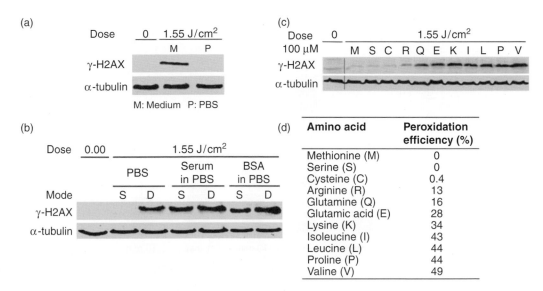

Figure 5.17 *Amino acid hydroperoxides produced by plasma treatment of organic medium induce DNA damage. (a) Cells were treated with 100 μL of medium or PBS that was separately exposed to 1.55 J cm^{-2} plasma. (b) 100 μL of PBS, medium without serum, or PBS with 100 mg mL^{-1} BSA were treated with plasma (1.55 J cm^{-2}) and immediately added to cells on a coverslip (separated, S). Cells overlaid with 100 mL of the indicated solution were treated with plasma (1.55 J cm^{-2}) (direct, D). (c) Solutions containing the indicated amino acid (100 μL) were separately treated with plasma and then added to MCF10A cells. (A, B, C) After 1-minute incubation, cells on cover slips were diluted in 2 mL medium, followed by lysis and Western blot for γ-H2AX and α-tubulin. (d) Peroxidation efficiency of various amino acid components of cell culture medium when treated with ionizing radiation. For each amino acid, the amount of DNA damage induced is proportional to the peroxidation efficiency.*

(BSA) or individual amino acids (Gebicki and Gebicki, 1993). Equivalent levels of H2AX phosphorylation were induced in cells subjected to separately treated serum-containing medium, serum-free medium or PBS with BSA, but not PBS alone (Figure 5.17b), suggesting that amino acid peroxidation may be involved. Peroxidation efficiency is widely variable among different amino acids (Gebicki and Gebicki, 1993).

To determine whether the observed results were related to the peroxidation efficiency of organic components in cell culture medium, 11 different amino acids with a range of peroxidation efficiencies were dissolved individually in PBS, separately treated with plasma and then added to cells. As shown in Figure 5.17c, phosphorylation of H2AX was directly proportional to the peroxidation efficiency of the amino acids. Valine produced the most significant level of damage and serine and methionine produced no detectable DNA damage (Figure 5.17d). There is a direct correlation between the peroxidation efficiency of 11 different amino acids and the level of DNA damage, providing strong support for the hypothesis that organic peroxides are produced in the plasma-treated medium and are responsible for the observed effects on DNA.

5.4.4 Crossing the cell membrane

The above sections described various experiments demonstrating that non-thermal plasma generated outside the cell medium is capable of initiating a sequence of events that induce intracellular ROS and various other associated biological effects, including repairable DNA damage. It was found that chemical modifications of organic molecules in the cell medium play a key role in transferring the observed effects from the plasma

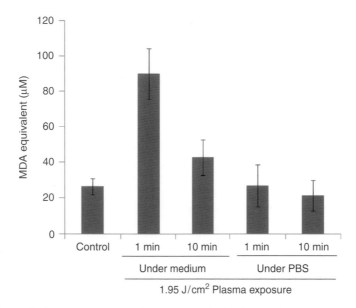

Figure 5.18 *Non-thermal plasma treatment of MCF10A cells leads to release of malondialdehyde (MDA), a commonly used marker for measuring lipid peroxidation in mammalian cells. Plot shows MDA equivalent for untreated cells and cells treated at indicated dose. After treatment, the cells are held in the treatment fluid for a minute and subsequently 2 mL of medium is placed onto the cells. After that, a quencher of lipid peroxidation is added to the medium at the specified times (1 min or 10 min).*

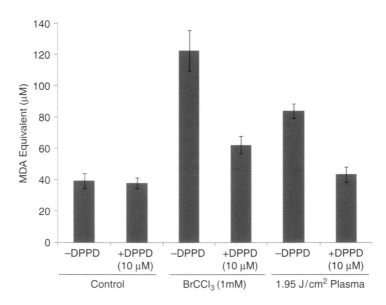

Figure 5.19 *Protecting cells against oxidation. MCF10A cells were incubated for 1 hour with 10 mM DPPD (+DPPD) or cell culture medium (-DPPD), followed by treatment at the indicated dose of DBD plasma or with 1 mM bromotrichloromethane (BrCCl$_3$). Lipid peroxidation was measured via release of MDA. Plot shows MDA equivalent for untreated cells (control) and cells treated at indicated plasma dose or with 1 mM BrCCl$_3$.*

phase into the cell. Biological mechanisms by which the effects initiated by plasma pass across the cell membrane remain unclear.

One of the simplest such mechanisms could be non-specific lipid peroxidation. Lipid peroxidation was discussed in Chapter 3 and is a relatively common outcome of the presence of oxidizing agents in the extracellular space. Various byproducts of lipid peroxidation, such as malondialdehyde (MDA), have been known to create bulky adducts on DNA which is a form of damage requiring repair. MDA is one of the most abundant carbonyl products of lipid peroxidation. It reacts with DNA to form adducts to deoxyguanosine and deoxyadenosine. One of the major adducts to DNA is a pyrimidopurinone called M_1G. Site-specific mutagenesis experiments indicate that M_1G is mutagenic in bacteria and is repaired by the nucleotide excision repair pathway (Marnett, 1999; Niedernhofer *et al.*, 2003; Jeong and Swenberg, 2005).

The fact that MDA is produced as a result of liquid peroxidation due to DBD treatment of breast epithelial (MCF10A) cells under the conditions described above has been confirmed by Kalghatgi (2012). Figure 5.18 shows also that MDA production due to the DBD treatment depends on the presence of organic components of the medium and does not occur in PBS treatment. It is also interesting to note that observed MDA is reduced at later time points when the cells are treated under the medium. If the MDA were produced due to peroxidation by the intracellular ROS, this type of behavior would not be expected since the intracellular ROS continue to increase for several minutes after the treated medium is diluted with untreated medium. A

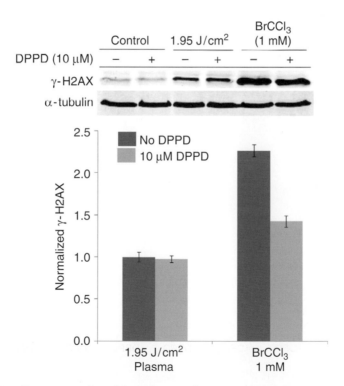

Figure 5.20   *MCF10A cells were incubated for 1 hour with 10 mM DPPD (+DPPD) or cell culture medium (−DPPD), followed by treatment at the indicated dose of DBD plasma or with 1 mM bromotrichloromethane (BrCCl₃). Representative immunoblot with γ-H2AX (upper panel) or α-tubulin (lower panel) are shown. The graph below the immunoblot provides quantification of triplicate samples in three separate experiments. The γ-H2AX signal was normalized to the amount of α-tubulin. Data (±standard deviation) are expressed relative to plasma treatment in the absence of DPPD, which was set to 1.0.*

decrease of MDA within the first 10 minutes after the dilution therefore indicates that MDA is produced due to peroxidation by the active species in the medium, rather than the intracellular ROS.

Cells can be protected from peroxidation of the lipid membrane by various lipophilic molecules that scavenge oxidizing agents. Lipid peroxidation is a form of chain reaction (discussed in Chapter 3) and these antioxidants essentially act to terminate the chains in these reactions. One type of molecule that has been employed for this purpose is N-diphenyl-phenyl-enediamine (DPPD). The ability of DPPD to block lipid peroxidation due the DBD plasma treatment is illustrated in Figure 5.19 (Kalghatgi, 2012), which also compares the effects of the plasma treatment against bromotrichloromethane ($BrCCl_3$), a known inducer of lipid peroxidation.

Having proven that DPPD provides effective protection against lipid peroxidation, it is not possible to test the hypothesis that the DNA damage associated with intracellular ROS is the result of lipid peroxidation. Figure 5.20 shows (Kalghatgi, 2012) the results of the experiment where cells are incubated with the DPPD antioxidant and treated by plasma in order to observe DNA damage (H2AX phosphorylation). The clear conclusion emerging from this experiment is that DNA damage observed in the plasma treatment is not the result of lipid peroxidation. Therefore, lipid peroxidation cannot explain the mechanism by which intracellular ROS species are generated as a consequence of the DBD plasma treatment, and the exact mechanism remains a mystery at the time of writing (2012).

6

Plasma Sterilization of Different Surfaces and Living Tissues

Sterilization is probably the first, most-investigated and best-known biological application of cold plasmas. Non-thermal plasma is an effective source of active species and factors such as radicals, ions, excited atoms and molecules, UV radiation and so on which are able to deactivate, kill or even completely disintegrate bacteria, viruses and other microorganisms without any significant temperature effects. Plasma is also of interest for applications in sterilization and disinfection of different surfaces (especially subtle and temperature-sensitive) such as living tissues (including skin, wounds, ulcers), liquids (including water, blood and other bio-liquids) and air streams. Disinfection usually implies a reduction of the population of microorganisms by a couple orders of magnitude, while sterilization usually requires a reduction in the number of microorganisms by at least a factor of 10^4–10^5.

6.1   Non-thermal plasma surface sterilization at low pressures

This section focuses on some examples of plasma sterilization and disinfection of different 'not-alive' surfaces (e.g. medical instruments and equipment) in low-pressure plasma discharges. The sterilization effect can be achieved in this case either through direct plasma interaction with microorganisms or application of plasma afterglow.

6.1.1   Direct application of low-pressure plasma for biological sterilization

Earlier non-thermal plasma sterilization experiments have been carried out mostly at low gas pressures. It should also be mentioned that initial low-pressure plasma sterilization methods implied the application of gas mixtures containing components with germicidal properties such as H_2O_2 and aldehydes (Boucher, 1980; Jacobs and Lin, 1987). The advantages of plasma sterilization are highlighted in that gases (such as air, He/air, He/O_2, N_2/O_2) without germicidal properties of their own can be used; they become biocidal only when the plasma is ignited.

Such studies have been motivated by the required decontamination of interplanetary space probes and sterilization of medical tools, and have been performed using RF and microwave low-pressure discharges in

Plasma Medicine, First Edition. Alexander Fridman and Gary Friedman.
© 2013 John Wiley & Sons, Ltd. Published 2013 by John Wiley & Sons, Ltd.

oxygen and O_2-N_2 mixtures (Moreau *et al.*, 2000; Bol'shakov *et al.*, 2004). The effect of the low-pressure RF oxygen plasma on bacteria has been investigated by Bol'shakov *et al.* (2004) in both inductively coupled (ICP) and capacitively coupled (CCP) plasma discharges. The ICP provided better efficiency in destroying biological matter due to higher electron and ion densities in this mode. High densities of atomic and electronically excited oxygen in synergy with UV photons induced chemical degradation of the biological materials followed by volatilization of the decomposition products (CO_2, CO, etc.). DNA degradation was evaluated for both ICP and CCP modes of the low-pressure non-thermal RF plasma. It was found that at the same power the ICP discharge destroyed over 70% of supercoiled DNA in 5 s, while only 50% of DNA was destroyed under the same conditions in the CCP discharge. Analysis of the direct effect of low-pressure plasma has contributed significantly to basic understanding of the contribution of different plasma-generated species to sterilization.

6.1.2   Effect of low-pressure plasma afterglow on bacteria deactivation

Effect on bacteria of the low-pressure N_2/O_2 afterglow plasma generated using a surfatron source driven by microwave power with frequencies 0.915 and 2.450 GHz has been investigated by Moreau *et al.* (2000) and Moisan *et al.* (2001). The *survival curves* (colony-forming units or CFUs versus treatment time) in the experiments with *Bacillus subtilis* spores exhibited three inactivation phases, as depicted in Figure 6.1 (Philip *et al.*, 2002).

The first phase presented in this figure, which exhibited the shortest *D*-value (decimal value of time required to reduce an original concentration of microorganisms by 90%, one $\log_{10}$ reduction), corresponds to the action of UV radiation on isolated spores or on the first layer of stacked spores. The second phase, which

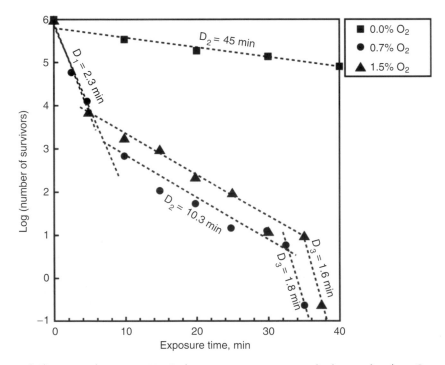

Figure 6.1   B. subtilis *survival curves in N_2-O_2 low-pressure microwave discharge afterglow. Gas pressure in the sterilization chamber is 5 Torr (Philip* et al., *2002).*

is characterized by the slowest kinetics, can be attributed to a slow erosion process by active species (such as atomic oxygen, O). The third phase starts after the outer spore debris are cleared during the second phase, hence allowing UV to hit the genetic material of the still living spores. The D-value of this phase is close to that of the first phase.

The survival curves of the low-pressure plasma inactivation of *B. subtilis* spores have only two phases (Moisan *et al.*, 2001). The radical debris removal stage is not required, and UV radiation can dominate both phases. The first phase is similar to that described in the previous paragraph (see Figure 6.1). The second phase represents spores that are shielded by others and require longer irradiation time to accumulate a lethal UV dose. A key feature of the experiments is that at low UV intensity there is a lag time before inactivation. Minimum UV dose should be achieved before irreversible damage to the DNA strands occurs (Moisan *et al.*, 2001).

Practical applications of the low-pressure non-thermal plasmas for surface sterilization include the treatment of medical instruments or equipment without use of high temperature and aggressive chemicals, as well as disinfection in low-pressure extraterrestrial conditions (e.g. on Mars or other planets).

6.2 Surface microorganism inactivation by non-equilibrium high-pressure plasma

6.2.1 Features of atmospheric-pressure air plasma sterilization

Non-thermal atmospheric-pressure discharges, particularly in air, are very effective and convenient in the deactivation of microorganisms. Atmospheric-pressure non-thermal discharges in air are characterized by a high density of strongly reactive oxidizing species and are not only able to kill microorganisms without any notable heating of the substrate, but they also destroy and decompose the microorganism. This ability of complete but cold destruction of microorganisms in atmospheric-pressure plasmas is attractive for complete sterilization of spacecrafts in the framework of the planetary protection program.

Atmospheric-pressure cold dielectric-barrier-discharge (DBD) plasma has been demonstrated as an effective tool in the destruction of resistive microorganisms such as *Bacillus subtilis* (spores), *Bacillus anthracis* (anthrax spores), and *Deinococcus radiodurans* (microorganisms surviving strong radiation of nuclear materials). Effective sterilization (5-log reduction) and significant destruction (cell wall fracture and leakage as well as complete disintegration) of *Bacillus subtilis* spores in DBD air plasma (120 s treatment, D-value 24 s) is illustrated in Figure 6.2.

The range of applications of plasma sterilization at atmospheric pressure is wide, from medical instruments and spacecrafts to different food products. For a review of these applications, see Fridman (2003).

6.2.2 Kinetics of atmospheric-pressure plasma sterilization

The mechanisms and kinetics of the sterilization processes are more sophisticated for the high-pressure strongly non-equilibrium plasmas because of the significant contribution of collisional gas-phase processes; a wider variety of active species are therefore involved in the sterilization kinetics. In particular, the germicidal effect of non-thermal atmospheric-pressure plasma is characterized by different shapes of the survival curves (see previous section) depending on the type of microorganism, the type of the medium supporting the microorganisms and the method of exposure (Laroussi, 2005).

'Single-slope' survival curves (1-line curves) for the inactivation of the bacteria strains have been observed in atmospheric-pressure plasma sterilization by Herrmann *et al.* (1999), Laroussi *et al.* (1999) and Yamamoto, Nishioka and Sadakata (2001), illustrated in Figure 6.3. The D-value ranges in this case from 4 s to 5 min.

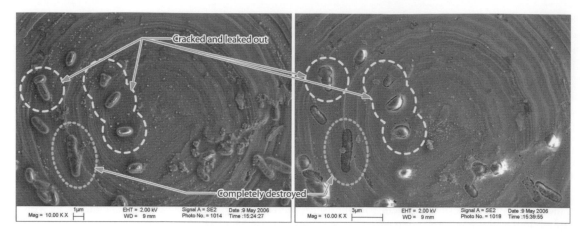

Figure 6.2   B. subtilis *before (left) and after (right) treatment with DBD plasma (120 s, 0.8 W cm⁻²) (see color plate).*

Two-slope survival curves (2 consecutive lines with different slopes) have been observed in atmospheric-pressure plasma by Kelly-Wintenberg *et al.* (1998), Laroussi, Alexeff and Kang (2000) and Montie, Kelly-Wintenberg and Roth (2000). The D-value of the second line (D_2) is smaller (shorter time) in these systems than the D-value of the first line (D_1). The D_1 value is dependent on the species being treated and the D_2-value is dependent on the type of surface supporting the microorganisms. The two-slope survival curve can be explained by taking into account that the active plasma species during the first phase react with the outer membrane of the cells, inducing damaging alterations. After this process is sufficiently advanced, the reactive species can quickly cause cell death which results in a rapid second phase.

Multi-slope survivor curves (3 kinetic phases or more) have also been observed in sterilization by non-thermal atmospheric-pressure plasmas. This type of kinetics can be illustrated by deactivation of *Pseudomonas aeuroginosa* on filter exposed to He/air DBD plasma; see Figure 6.4 (Laroussi, Alexeff and Kang, 2000).

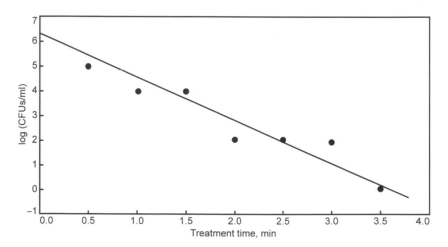

Figure 6.3   *Single phase surviving curve:* E. coli *on Luria-Bertani broth exposed to He-air plasma generated by DBD at atmospheric pressure.*

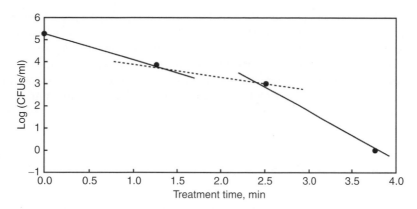

Figure 6.4 *Multi-phase surviving curve:* Pseudomonas aeruginosa *on filter exposed to He-air plasma generated by DBD at atmospheric pressure (Laroussi, Alexeff and Kang, 2000).*

Interpretation of the multi-slope kinetic effect is similar in this case to that described in the previous section regarding sterilization by low-pressure plasma (see Figure 6.1).

The bacteria inactivation kinetics therefore reveals the complexity of mechanisms of sterilization by non-equilibrium high-pressure plasmas. Several factors can impact the killing process: type of bacteria, type of medium in/on which the cells are seeded, number of cell layers, the type of exposure, the contribution of UV, operating gas mixture and so on. If UV plays an important or dominant role, the survival curves generally tend to exhibit a first rapid phase (small *D*-value) followed by a second slower phase (outer layer of bacteria is deactivated by UV during the fast first phase; then slower erosion of the bacteria proceeds during the longer second phase). When UV does not play a significant role, such as in the case of air plasma (where contribution of active oxidizing species is essential), the single-phase survival curves are mainly observed. Plasma sterilization kinetics at high pressures are generally quite sophisticated and depend on the mechanisms of plasma sterilization in specific atmospheric-pressure discharges.

6.2.3 Cold plasma inactivation of spores: Bacillus cereus and Bacillus anthracis (anthrax)

Bacillus species, which are ubiquitous in the environment, are aerobic or facultative anaerobic gram-positive bacteria. The genus Bacillus is divided into three broad groups depending, among other characteristics, on the morphology of the spore: *Bacillus cereus*; *Bacillus anthracis* (anthrax) and *Bacillus thuringiensis*. Morphological and chromosomal similarities between these species have prompted the view that *Bacillus anthracis*, *Bacillus thuringiensis* and *Bacillus cereus* are all varieties of a single species.

Bacilli can produce a dormant cell type called a spore in response to nutrient-poor conditions. Bacterial spores have little or no metabolic activity and can withstand a wide range of environmental assault including heat, UV and solvents. To kill or inactivate Bacillus spores, 0.88 mol L^{-1} hydrogen peroxide at a pH of 5.0 should be applied for 3 hours to sterilize a spore suspension of 10^6 spores mL^{-1}. Alternatively, 10^6 rad of gamma irradiation will sterilize 10^6 spores mL^{-1}.

Bacillus anthracis spores, as opposed to vegetative cells, are the infectious form and cause anthrax. The spores of *Bacillus anthracis* represent a noteworthy bioterrorism agent. They can be easily distributed in dry form in parcels and letters via the postal service (which occurred in 2001 when anthrax-contaminated letters sent through the US postal service killed 5 people and made 23 others ill), in aerosols or in contaminated water. In response to these possibilities, an effective low-energy cost-effective method of spore inactivation or sterilization is required. An attractive method of spore inactivation is plasma treatment.

In the following sections, we discuss the inactivation of Bacillus spores (in dry form and suspended in water) with the use of atmospheric-pressure DBD plasma on surfaces as well as inside closed volumes (e.g. envelopes). Inactivation of bacteria in spore form, both in liquid or air-dried on a surface, requires higher doses of DBD plasma treatment, and up to 5-log reduction can be achieved within a minute of exposure to plasma. Note that that the mechanisms by which direct plasma inactivates spores and vegetative bacteria may be quite different.

6.2.4   Atmospheric-pressure air DBD plasma inactivation of Bacillus cereus and Bacillus anthracis spores

In the experiments of Dobrynin *et al.* (2010), the Bacillus spores were treated in DBD plasma at room temperature in air. The powered electrode was made of a 2.5-cm-diameter solid copper disk covered by a 3.5-cm-diameter 1-mm-thick quartz dielectric. The discharge gap was kept at 1.5 mm. The microsecond-pulsed DBD discharge was ignited by applying an AC-pulsed high voltage of 30 kV magnitude (peak-to-peak), 1.3 kHz frequency, between the electrodes. Current peak duration was 1.2 μs and the corresponding plasma surface power density was 0.3 W cm^{-2}. The spores were treated in either dry form or in aqueous suspension on glass slides. The slides were placed on top of grounded metal, and the plasma was ignited directly between the powered electrode and the treated spores.

Another set of experiments were carried out using indirect plasma treatment. In this case, a grounded metal mesh (22 wires per cm, 0.1 mm wire diameter, 0.35 mm openings, 60% open area and weaved mesh) was used as a second electrode. The gap between the mesh and the second electrode was kept at 1.5 mm, the same as that for the direct treatment setup. The distance between the mesh and the treated surface was also kept at 1.5 mm. Ten microliters of *Bacillus cereus* or *Bacillus anthracis* spores at concentrations of 10^7, 10^6 and 10^5 spores mL^{-1} in distilled water were placed in hanging-drop glass-slide wells and treated for 5–45 s with direct DBD plasma. After treatment, bacteria were appropriately diluted and plated. Spores were also placed inside a plastic chamber, covered with a glass cover slide and dried for c. 30 min with a constant air flow of c. 0.1 L min^{-1} (control experiments with only gas flow through the chamber showed no loss of collected spores). Ten microliters of *Bacillus cereus* at 10^8, 10^7 and 10^6 spores mL^{-1} in distilled water were used in another experiment where dried spores were treated with direct DBD plasma for 5–45 s and then washed out of the chamber with 30 mL of distilled water, appropriately diluted and plated.

Similar experiments were conducted using *Bacillus anthracis* spores, except that 10 μL of Bacillus anthracis spores were placed inside either plastic or paper envelopes, dried in room-termpature air for 1 hour and treated with DBD plasma. In these experiments, the discharge was ignited in the volume of either the plastic chamber or the paper or plastic envelope (envelopes were slightly inflated to ensure that walls were in contact with powered and grounded electrodes; see Figure 6.5).

In the first series of experiments conducted by Dobrynin *et al.* (2010), the spores of *Bacillus cereus* and *Bacillus anthracis* were treated in water droplets on the surface of glass slides using atmospheric-pressure DBD plasma in room-termpature air. The inactivation kinetics of these Bacillus spores, at three different concentrations, after DBD plasma treatment are shown in Figure 6.6. The number of colonies after treatment is plotted on a logarithmic scale as a function of the treatment dose and time, which are directly related.

Although the spores of both species were effectively inactivated (up to 5-log reduction in less than a minute of treatment), log N (i.e. killing kinetics) was quite different for *Bacillus cereus* (2-step) versus *Bacillus anthracis* (linear) spores. In addition, anthrax spores appeared to be slightly more resistant to plasma treatment (see Figure 6.6). These species have similar morphology, metabolism and physiology, and a similar behavior of inactivation kinetics.

Figure 6.5 *Setup for DBD plasma treatment of air-dried spores contained inside a closed envelope (see color plate).*

In contrast to the inactivation of spores suspended in distilled water, the inactivations by the plasma treatment of air-dried *Bacillus cereus* and *Bacillus anthracis* spores both appear to be linear. Figure 6.7 shows the results of the plasma inactivation of three different concentrations of *Bacillus cereus* spores dried inside a plastic chamber. The inactivation of 10^7 per mL anthrax spores inside a closed paper envelope is shown in Figure 6.8

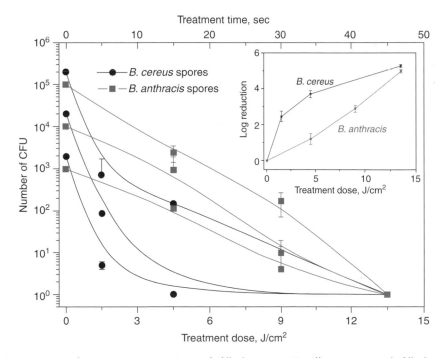

Figure 6.6 *Inactivation of various concentrations of (filled squares)* Bacillus cereus *and (filled circles)* Bacillus anthracis *spores in water on glass slides using direct DBD plasma treatment.*

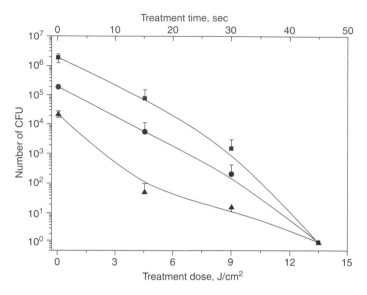

Figure 6.7   *Inactivation by direct DBD plasma of various concentrations of dry* Bacillus cereus *spores contained inside a closed plastic chamber.*

These results are interesting, as they show the high efficiency of DBD-plasma-based systems to sterilize within temperature-sensitive materials. This is important in view of the 2001 bioterrorism attacks when anthrax spores in envelopes were distributed through the US postal service, resulting in illness and death.

To determine whether morphological differences have occurred, spore morphology was analyzed before and after DBD plasma treatment. *Bacillus cereus* spores were deposited onto stainless steel coupons and were observed by scanning electron microscopy (SEM). The samples were then exposed to DBD plasma

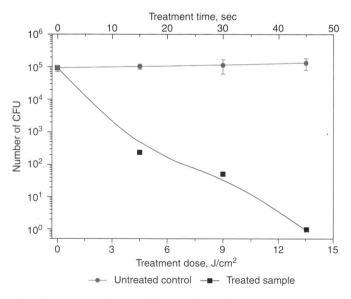

Figure 6.8   *Inactivation by direct DBD plasma of dry* Bacillus anthracis *spores contained inside a closed paper envelope.*

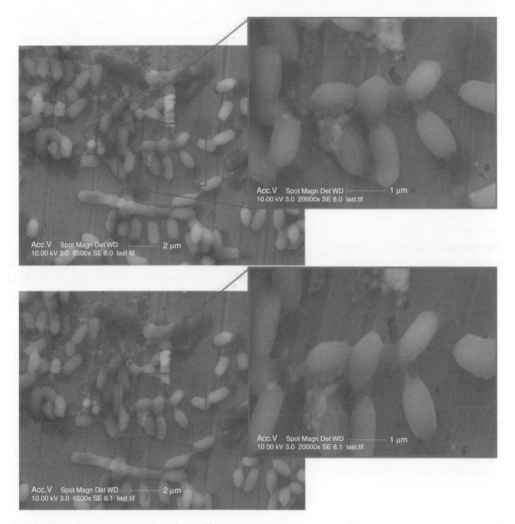

Figure 6.9   *SEM photomicrographs of* Bacillus cereus *spores on a stainless steel coupon (top) before and (bottom) after exposure to direct DBD plasma (same spot of sample in both photos) (see color plate).*

for 1 min (corresponding to a dose of 18 J cm^{-2}), and SEM photomicrographs of the same spores were acquired again. A comparison of the photomicrographs of spores before and after treatment (Figure 6.9) shows that the inactivation mechanism is not the erosion of the spores, since their protective coats appear to be undamaged. We therefore believe that the diffusion of chemically active oxygen species into spores followed by the damage of internal macromolecules or molecular systems may be the primary mechanism of spore inactivation.

6.2.5   Decontamination of surfaces from extremophile organisms using non-thermal atmospheric-pressure plasma

Non-thermal plasma has the ability to kill any microorganisms including those which are extremely resistant to other sterilization methods, the so-called extremophiles. As an example, the experiments of Cooper *et al.*

(2009) showed that non-thermal DBD plasma compromises the integrity of the cell membrane of *Deinococcus radiodurans*, an extremophile organism. In samples of *D. radiodurans*, which were dried in a laminar flow hood, it was observed that DBD plasma exposure resulted in a 6-log reduction in CFU count after 30 min of treatment. When the *Deinococcus radiodurans* cells were suspended in distilled water and treated, it took only 15 s to achieve a 4-log reduction of CFU count.

Such investigations are especially interesting in the planetary-protection field of research, where the proliferation of terrestrial bacteria beyond Earth is known as *forward contamination*. When sending spacecraft and probes to other planets and moons, it is extremely important to prevent the contamination of the environment, particularly when searching for native life. If bacteria from the location in which a sample was collected returns to Earth and proliferates, then *reverse contamination* has taken place. The goal with this regard is to achieve surface sterilization of spacecraft materials with complete disintegration of spores and bacteria. This effect was successfully demonstrated by cold ambient-air plasma, which leads to sterilization, lysing and disintegration of microorganisms (Cooper *et al.*, 2009).

D. radiodurans has been chosen for such plasma sterilization experiments since if we can show that cold plasma can destroy one of the toughest microorganisms found on Earth, then we can claim that we can probably destroy less-robust organisms. *D. radiodurans* is very resistive to radiation, temperature change, reactive oxygenated species and vacuum. It can withstand an instantaneous radiation dose of 5000 Gy with no loss of viability (60 Gy sterilizes a culture of *E. coli*), an instantaneous dose of up to 15 000 Gy with 37% viability loss and exposure to space vacuum (c. 10^{-6} Pa) for three days with decreased cell survival by four orders of magnitude.

Figure 6.10 shows the test results obtained with *D. radiodurans* after DBD plasma treatment. At $t = 0$, the number of viable *D. radiodurans* was approximately 10^6 CFU. After 30 min of DBD treatment, there was a 6-log reduction in the number of *D. radiodurans* cells. Temperature measurements show that during plasma treatment the average temperature was 26°C. Imaging techniques were applied to visualize the effect of treatment with cold plasma on the morphology of microorganisms. The aforementioned viability measurements were supported by SEM images taken before and after 30 min of DBD treatment of *D. radiodurans* on blue steel at 1 W cm^{-2} (Figure 6.11). A comparison of Figure 6.11a (before DBD plasma treatment) and Figure 6.11b (after 30 min of DBD plasma treatment) clearly shows that DBD plasma causes significant morphological changes and the 'physical' destruction of *D. radiodurans* on the sterilization

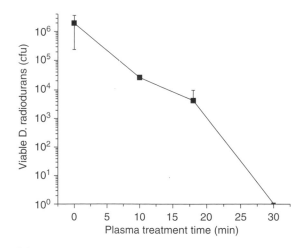

Figure 6.10   *Viability measurements of dry* D. radiodurans *after DBD plasma treatment.*

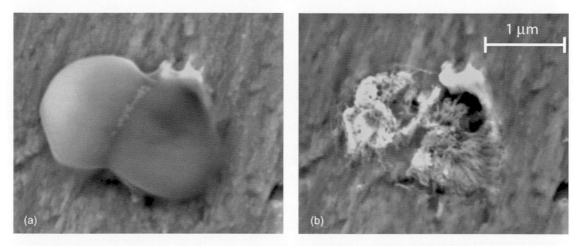

Figure 6.11 *SEM images of* Deinococcus radiodurans *on blue steel (a) before and (b) after 30 min of DBD plasma treatment.*

surface at 30°C. Although the cell was considered to be dead, its remnants continue to inhabit the surface. SEM images were also taken of *D. radiodurans* on surgical-grade stainless steel, where the bacteria were treated for 20 min (Figure 6.12).

6.2.6   Plasma sterilization of contaminated surgical instruments: Prion proteins

Another critical sterilization challenge which can be resolved by application of non-thermal plasma stems from the fact that the majority of medical contamination is not only by microorganisms but also by biomolecules such as proteins, DNA and lipids. The obvious example in this regard is the contamination by misfolded prion proteins, widely regarded as the etiologic agent of spongiform neurodegenerative pathologies such as bovine spongiform encephalopathy (BSE), scrapie and Creutzfeldt-Jakob diseases (CJD). Prion proteins are

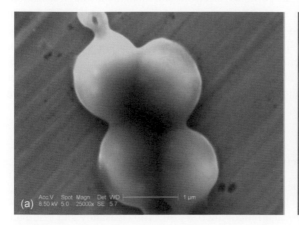
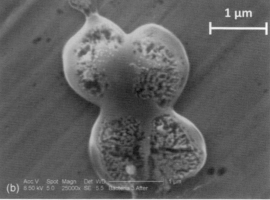

Figure 6.12   *SEM images of D. radiodurans on surgical-grade stainless steel (a) before and (b) after 20 min of DBD plasma treatment.*

resistant to all current conventional sterilization strategies including autoclaving, ionizing radiation, ethylene oxide and formaldehyde, and as a result have forced the unsustainably expensive option of single-use surgical instruments.

Recent studies demonstrate that non-thermal plasmas, and in particular microplasma jets developed by Michael Kong and his group (e.g. Iza *et al.*, 2008), are capable of prion protein destruction. A helium-oxygen atmospheric-pressure plasma jet has been successfully applied by this group to inactivate proteins deposited on stainless-steel surfaces. Using a laser-induced fluorescence technique for surface protein measurement, a maximum protein reduction of 4.5-log has been achieved by varying the amount of the oxygen admixture into the background helium gas. This corresponds to a minimum surface protein of 0.36×10^{-15} mole mm^{-2} or 0.36 femtomole mm^{-2}. Plasma reduction of surface-borne protein is through protein destruction and degradation, and it's typically biphasic reduction kinetics are largely influenced by the thickness profile of the surface protein.

By interplaying the protein inactivation kinetics with optical emission spectroscopy, it has been shown in this research that the main protein-destroying agents are excited atomic oxygen (via the 777 and 844 nm emission channels) and excited nitride oxide (via the 226, 236 and 246 nm emission channels). It has also been demonstrated that the most effective protein reduction is achieved through a synergistic effect of atomic oxygen and nitride oxide. The studies of Michael Kong and his group represent an important step towards confirmation of the efficacy of non-thermal plasma as a sterilization technology for surgical instruments contaminated by prion proteins.

6.3   Plasma species and factors active for sterilization

6.3.1   Direct effect of charged particles in plasma sterilization

Plasma sterilization is a fairly complicated process determined by multiple plasma species and factors including charged and excited species, reactive neutrals, UV radiation and so on. The contribution of these factors varies among different types and regimes of non-equilibrium plasma discharges; it also differs from the point of view of the induced biological pathways. The synergistic nature of interaction between the plasma factors is important in sterilization, similarly to the case of plasma treatment of polymers.

We first consider the general contribution of the charged particles (electrons and ions) in plasma sterilization. The charged particles play a major role in sterilization, especially in the case of direct plasma treatment when plasma is in direct contact with a microorganism and electric current can cross the sterilization zone. The advantages and greater effectiveness of direct (versus indirect) plasma sterilization and the key contribution of charged particles in this case is discussed in Section 6.4. The contribution of the charged particles in sterilization can be divided into three effects: direct biochemical effect of electrons and ions; direct effect of ion bombardment; and effect of electric field induced by the charged particles. These three effects are discussed in the following sections.

6.3.2   Biochemical effect of plasma-generated electrons in plasma sterilization

Electrons make a significant contribution to plasma-chemical modification of polymers and especially biopolymers (Fridman, 2008). The effectiveness of electrons in sterilization is related in particular to the significant depth of their penetration into the polymers and biopolymers $x_0 \approx 2$ μm (Rusanov and Fridman, 1978a, b). The characteristic depth of inactivation of bioorganisms (sterilization) provided by non-thermal plasma electrons generated by DBD air plasma has been investigated and simulated by Rusanov and Fridman (1978a, b) as:

$$x^2 \approx x_0^2 (\ln \sigma_{i0} N - \ln \ln K) \qquad (6.1)$$

where $\sigma_{i0} \approx 10^{-9}$ cm^2 is the electron-impact bio-deactivation cross-section; N (cm^2) is the total time-integrated flux of electrons to the surface of a bioorganism; and $K \gg 1$ is destruction degree. Assume for example that $N = 10^{13}$ and $K = e$; the characteristic bio-deactivation depth provided by the plasma electrons is about 6 μm.

The greater significance of electrons with respect to other charged particles (ions) in sterilization is due to their ability to provide a relatively high electron flux to the treated surface, especially in the case of direct plasma treatment. Charged particles compete in sterilization effectiveness with reactive neutral species (OH, ozone, etc.) characterized by c. 10^5 times the number density in non-thermal plasma. Electrons having very high drift velocities (10^7 cm s^{-1} and above) are nevertheless able to provide fluxes higher than that of the reactive neutral species. For electronegative gases (in particular, air), the application of discharges (e.g. DBD) operating in the 'breakdown regimes' (i.e. when ionization rates exceed those of electron attachment, preventing significant formation of negative ions O_2^-), is required.

Plasma sterilization is characterized by a significant depth of penetration, which means that deactivation of bioorganisms occur under a layer of water and bio-solution. From this point of view (and also remembering that the microorganisms themselves contain a lot of water), it is interesting to analyze the behavior of the plasma electrons when they reach and penetrate the water surface, becoming the hydrated (aqueous) electrons e_{aq}. *Hydrated electrons* are surrounded by polar water molecules and remain quite stable, providing the deep penetration of the electrons in water. They react with different admixtures in water, in particular with biomolecules and peroxide (converting it to OH and OH$^-$, which is the Fenton reaction). In most important processes, a hydrated electron e_{aq} is able to convert oxygen dissolved in water into *superoxide*, the highly reactive O_2^- ion-radical, as follows:

$$e_{aq} + O_2(H_2O) \rightarrow O_2^-(H_2O). \qquad (6.2)$$

The superoxide is a precursor to other strong oxidizing agents, including singlet oxygen and peroxynitrite. Note that certain cells in the human body produce superoxide as an antibiotic 'weapon' to kill invading microorganisms. The superoxide also acts as a signaling molecule in the regulation of cellular processes. In biological systems with a sufficient level of acidity, the superoxide is converted into hydrogen peroxide and oxygen in a reaction called the *superoxide dismutation*:

$$2O_2^- + 2H^+ \rightarrow H_2O_2 + O_2. \qquad (6.3)$$

The superoxide dismutation can be spontaneous or can be catalyzed and therefore significantly accelerated by the enzyme superoxide dismutase (SOD). The superoxide not only generates hydrogen peroxide (H$_2$O$_2$), but also stimulates its conversion into OH-radicals which are actually extremely strong oxidizers (very effective in sterilization through a chain oxidation mechanism). The H$_2$O$_2$ conversion into OH, known as the *Fenton reaction*, proceeds as a redox process provided by oxidation of a metal ion (e.g. Fe^{2+}):

$$H_2O_2 + Fe^{2+} \rightarrow OH + OH^- + Fe^{3+}. \qquad (6.4)$$

Restoration of the Fe^{2+} ions takes place in the Fe^{3+} reduction process provided by the superoxide ion-radicals O_2^-:

$$Fe^{3+} + O_2^- \rightarrow Fe^{2+} + O_2. \tag{6.5}$$

Superoxide can also react in water with nitric oxide (NO) producing peroxynitrate (OONO$^-$), another highly reactive oxidizing ion-radical, as follows:

$$O_2^- + NO \rightarrow OONO^-. \tag{6.6}$$

Summarizing, electrons are able to provide significant sterilization effect even deep under the surface of a liquid, especially bio-liquid. In water (and water solutions), plasma electrons are able to be kinetically and quickly converted into hydrated electrons e_{aq} and the superoxide O_2^- ion-radicals (Equation (6.2)). The further sequence of conversion leads to hydrogen peroxide H_2O_2 (by dismutation; Equation (6.3)) and such extremely strong oxidants as hydroxyl OH and peroxynitrate (OONO$^-$). These oxidants are very effective in deactivation of microorganisms. Electrons are therefore not directly responsible for bio-inactivation, but their effect is observed through a long sequence of plasma-chemical and biochemical conversions often related to reactive oxygen species (ROS).

6.3.3 Bio-chemical effect of plasma-generated negative and positive ions

Taking into account the balance of charges on the treated surfaces, the effect of ambipolar diffusion and acceleration of ions in the sheaths, fluxes of positive ions to the treated surface can be significant (although usually lower than fluxes of reactive neutral species and drift-related pulsed fluxes of electrons in the breakdown discharge regimes, see Section 6.3.2). Direct plasma fluxes of negative ions are not as high as those of positive ions (but obviously can be the same) because of the typically negative charge of surfaces and the weaker fields in the anode sheaths. The major negative ion in non-thermal air plasma is O_2^-. When these ions are immersed in water (or water-based bio-solution), they become superoxides O_2^- and form hydrogen peroxide H_2O_2 and OH according to the above-described mechanisms (Equations (6.3)–(6.5)) of ROS conversion. This effect is similar to the case of sterilization provided by chemical effect of plasma electrons (see Section 6.3.2). Fluxes of the negative ions from plasma can be relatively low, which can somewhat decrease their contribution to sterilization. The combined electron/negative ion effect is mostly in the generation of ROS, which play an extremely important role in plasma-stimulated biochemistry. This is discussed in more detail in Chapters 5, 9 and 10.

The primary biochemical effect of positive plasma ions M^+ (e.g. N_2^+) on microorganisms can be illustrated by the effect of the positive plasma ions M^+ on a water surface (either a biological solution where the microorganisms are located, or the microorganisms themselves). Interaction of the plasma ions M^+ with water molecules starts with fast charge transfer processes to the water molecules, characterized by relatively low ionization potential:

$$M^+ + H_2O \rightarrow M + H_2O^+. \tag{6.7}$$

Water ions then react rapidly with water molecules producing H_3O^+ (which makes the water acidic) and OH radicals, playing an important role in sterilization:

$$H_2O^+ + H_2O \rightarrow H_3O^+ + OH. \tag{6.8}$$

The positive plasma ions make the water acidic. The effect of hydrated electrons is to stabilize the plasma-stimulated decrease of pH (Equations (6.2)–(6.5)), which requires special conditions. Non-thermal plasma interaction with water conventionally decreases pH by making the water slightly acidic. The contribution of both electrons and ions (positive and negative) in one way or another leads to effective generation of OH and other highly reactive oxidants, resulting in the intensive oxidation of biomaterials and a strong sterilization effect.

6.3.4 Sterilization effect of ion bombardment

The deactivation of microorganisms can be due to destruction of the lipid layer of their membranes or other membrane damage by ion bombardment from plasma. This effect is especially important in the case of direct plasma treatment of the microorganisms. Although ion bombardment makes a bigger contribution in low-pressure plasmas when the ion energies are quite high, the effect of ion bombardment in atmospheric-pressure discharges with relatively low ion energies can also be significant. The average energy of the ions $\langle \varepsilon_i \rangle$ in the ion bombardment process in the high-pressure non-equilibrium ($T_e \gg T_0$) plasmas can be estimated as energy received by the ions in electric field E during their displacement on the mean free path $eE\lambda$ (Fridman and Kennedy, 2004, 2011):

$$\langle \varepsilon_i \rangle \approx T_0 + T_e \sqrt{\delta} \tag{6.9}$$

where T_e, T_0 are electron and gas temperatures in plasma and δ is the average fraction of energy lost by electrons in the electron-neutral collisions. Taking into account that $\delta \approx 0.1$ in molecular gases, average energies of the ion bombardment (even in high-pressure plasma systems) can reach 0.3–0.7 eV. Although these energies are not sufficient to break strong chemical bonds, they are high enough to break the hydrogen bonds responsible for integrity of lipid layers forming cellular membranes.

In addition to chemical effects, ions are therefore able to provide deactivation of microorganisms through mechanical effects of bombardment (and possibly related osmotic pressure effects). The effect of ion bombardment in plasma medicine is generally significantly limited by the contribution of the liquid medium usually protecting cells and living tissues, discussed in Chapter 9.

6.3.5 Sterilization effect of electric fields related to charged plasma particles

The direct effect of externally applied electric fields on sterilization and deactivation of microorganisms (electroporation) in non-thermal plasma systems (as well as thermal) can usually be neglected (Neumann, Sowers and Jordan, 2001). At the same time, the effects of electric fields related to collective motion and deposition of charged particles can be significant. According to Mendis, Rosenberg and Azam (2000) and Laroussi, Mendis and Rosenberg (2003), deposited charged particles can play a significant role in rupture of the outer membrane of bacterial cells.

It has been shown that electrostatic forces caused by charge accumulation on the outer surface of the cell membrane can overcome the tensile strength of the membrane and cause its rupture. Charged bacterial cells with micrometer sizes experience large electrostatic repulsive forces proportional to the square of the charging potential Φ and inversely proportional to the square of the radius of the cells curvature r. The charging (floating) potential Φ is negative and depends on the ratio of the ion mass to the electron mass. The condition for disruption of the bacterial cell membrane can be expressed in this case as (Mendis, Rosenberg and Azam, 2000):

$$|\Phi| > 0.2\sqrt{r\Delta F_t}, \tag{6.10}$$

where Δ is the thickness of the membrane and F_t is its tensile strength. This mechanism can be relevant for gram-negative bacteria with irregular surface membranes. These irregularities with small radii of curvatures can result in localized high electrostatic forces (Laroussi, Mendis and Rosenberg, 2003).

The high strengths of electric fields and electroporation effect are achieved not only because of the charge deposition on irregular surfaces of gram-negative bacteria, but also due to the collective and strongly localized motion of the charged particles, especially in streamers of DBDs. The strengths of electric fields in the DBD streamers significantly exceed those of the externally applied electric fields. While the externally applied electric fields in the atmospheric-pressure discharges are not sufficient for electroporation and sterilization, electric fields in the streamers are able (in principle) to make a contribution in plasma deactivation of microorganisms. The biological effect of very high electric fields of streamer heads, especially of those in short-pulsed DBD plasma systems, is not limited to electroporation; it can also be related to triggering of the intracellular signaling, which is especially important in the case of plasma interaction with cells.

6.3.6 Effect of plasma-generated active neutrals: ROS and RNS

Reactive neutral species generated in non-thermal plasmas make a significant contribution to sterilization and treatment of biomaterials, especially at high pressures. Strongly non-equilibrium air plasmas, for example, are excellent sources of reactive oxygen-based and nitrogen-based species (ROS and RNS) such as O, O_2 ($^1\Delta_g$), O_3, OH, NO and NO_2. Some of the neutrals (such as active atoms and radicals O, OH) as well as excited species O_2 ($^1\Delta_g$) are extremely reactive but have a very short lifetime; they are therefore only effective for sterilization inside or within close vicinity of plasma. Other neutrals (such as saturated molecules O_3 and passive radicals NO, NO_2) are not so extremely reactive, but have a relatively long lifetime and can be effective for treatment of biomaterials far from the plasma zone where they have been generated.

The kinetics of the air plasma generation of major reactive neutrals such as O_3, O, OH, NO and NO_2 has been analyzed in detail in the special plasma-chemistry-focused book by Fridman (2008) regarding plasma synthesis of ozone and nitrogen oxides. The concentrations of O_3 and NO_2 in DBD air plasma for sterilization applications has been measured by Laroussi and Leipold (2004) and Minayeva and Laroussi (2004). The concentrations of O_3 and NO_2 are presented in Figures 6.13 and 6.14 as functions of the discharge power and air flow rate. While plasma specialists mostly relate the non-thermal plasma sterilization effect to ROS, RNS (especially nitrogen oxides NO and NO_2 as well as peroxinitrite O_2NO) can also make a significant contribution to the total sterilization effect. The contribution of RNS is especially important in the case of quasi-thermal and transitional (warm) air discharges, where RNS generation is significant and exceeds that of ROS.

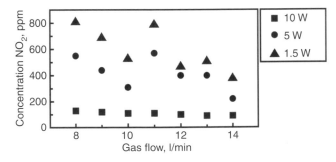

Figure 6.13 *Concentration of NO_2 generated in air DBD as a function of flow rate at three power levels from 1.5 W to 10 W.*

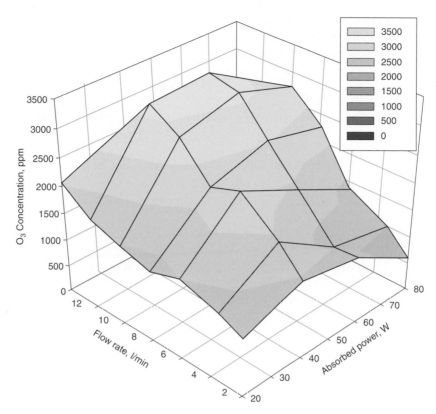

Figure 6.14 *Concentration of ozone O_3 generated in air DBD as a function of flow rate and dissipated discharge power.*

6.3.7 Effect of plasma-generated active neutrals: OH-radicals and ozone

The germicidal effects of ozone are caused by its interference with the cellular respiration system. Generally, the oxygen-based and nitrogen-based plasma-generated reactive neutral species have strong oxidative effects on the outer structures of cells. Cell membranes are made of lipid bilayers, an important component of which is unsaturated fatty acids. The unsaturated fatty acids give the membrane a gel-like nature, which allows transport of the biochemical agents across the membrane. Since unsaturated fatty acids are susceptible to OH attacks (Laroussi and Leipold, 2004), the presence of the hydroxyl radicals can compromise the function of the membrane lipids, whose role is to act as a barrier against the transport of ions and polar compounds in and out of the cells (Bettleheiem and March, 1995).

Protein molecules, which are basically linear chains of amino acids, are susceptible to oxidation by atomic oxygen, OH or metastable electronically excited oxygen molecules. Proteins play the role of gateways that control the passage of various macromolecules in and out of cells.

Where bacteria are of the gram-positive type, they are able to form spores which are highly resistive states of cells. Spores are made of several coats surrounding a genetic core. These coats are also made of proteins susceptible to chemical attack of reactive neutrals.

The reactive neutral species generated by air plasmas therefore greatly compromise the integrity of the walls, coats and membranes of the cells of microorganisms. Extremely high reactivity in biochemical reactions is provided by hydroxyl radicals OH generated in plasma. The OH radicals easily pass through membranes and cannot be kept out of cells. Hydroxyl radical damage is so fast that it is only limited by the diffusion rate. Hydroxyl attaches to an organic substrate R (e.g. a fatty acid) forming a hydroxylated adduct-radical ROH:

$$OH + R \rightarrow ROH \text{ (radical)}. \tag{6.11}$$

The adduct-radical ROH can be further oxidized by metal ions or oxygen, resulting in an oxidized but stable product:

$$ROH \text{ (radical)} + Fe^{3+} \rightarrow HOR + Fe^{2+} + H^+, \tag{6.12}$$

$$ROH \text{ (radical)} + O_2 \rightarrow HOR + O_2^- + H^+. \tag{6.13}$$

Two adduct-radicals ROH can also react with each other, forming a stable cross-linked-but-oxidized product:

$$ROH \text{ (radical)} H + ROH \text{ (radical)} H \rightarrow R - R + 2H_2O. \tag{6.14}$$

A significant part of OH oxidation of bioorganic materials is due to initial dehydrogenization of the substrate organic molecules, RH:

$$OH + RH \rightarrow R + H_2O. \tag{6.15}$$

The resulting oxidized substrate bioorganic compound R is a radical, and reacts with molecules in chain processes. It could react with oxygen to produce a peroxyl radical R-OO-:

$$R + O_2 \rightarrow R - OO - . \tag{6.16}$$

The peroxyl radical is highly reactive, and can react with another organic substrate RH propagating the chain oxidation mechanism:

$$(R - OO-) + RH \rightarrow ROOH + R. \tag{6.17}$$

This type of biochemical chain reaction of OH is common in the oxidative damage of fatty acids and other lipids, and demonstrates why radicals such as the hydroxyl radical can cause so much more damage than might otherwise be expected. Plasma-generated ozone is at least partially dissociated when immersed in water, liquid-based or liquid-like biomaterial:

$$(O_3)_{aq} \rightarrow (O_2)_{aq} + O_{aq}. \tag{6.18}$$

Atomic oxygen created in the process intensively reacts with the bioorganic molecules RH producing R-radicals and OH-radicals:

$$O + RH \rightarrow R + OH, \tag{6.19}$$

which then stimulate the same biochemical chain oxidation mechanism as described by Equations (6.15)–(6.17). Ozone O_3 can therefore be added to the intensive OH-based biochemical oxidation.

Similar damage caused by hydroxyl radicals and other reactive oxygen species can occur in proteins and with nucleic acids (mainly DNA). Proteins are highly susceptible to oxidative damage, particularly at sites where sulfur-containing amino acids are found. DNA can be oxidatively damaged at both the nucleic bases (the individual molecules that make up the genetic code) and at the sugars that link the bases. Oxidative damage of DNA results in degradation of the bases, breaking of the DNA strands by oxidation of the sugar linkages or cross-linking of DNA to protein (a form of damage particularly difficult for the cell to repair). Although all cells have some ability to repair oxidative damage to proteins and DNA, excess damage can cause mutations or cell death (Christophersen *et al.*, 1991).

Non-thermal atmospheric-pressure plasma-generated reactive neutral species, particularly OH and O_3, are therefore very effective in the deactivation of microorganisms and in treatment of biomaterials. The reactive neutrals, especially OH, are not only generated in plasma and transported to the surface of biomaterials, but also produced in the biomaterials or tissues from plasma-generated charged particles. In the case of direct plasma treatment with sufficiently high electric fields, the flux of charged particles can dominate the production of reactive neutrals in the biomaterial. Otherwise, the reactive neutrals are mostly generated in plasma and then transported to the surface of biomaterials and tissues. Numerical kinetic data on the effect of O_3 and OH on deactivation of different microorganisms are summarized in the following section in connection with plasma sterilization of air streams.

6.3.8   Effects of plasma-generated active neutrals: Hydrogen peroxide (H_2O_2)

A significant and very special cold plasma sterilization effect can be attributed to hydrogen peroxide H_2O_2. Often the effect of plasma-generated hydrogen peroxide is combined with the effect of plasma-generated OH. Hydrogen peroxide is a relatively stable ROS with respect to OH, and can often be considered as a 'carrier' of OH to be released using the Fenton mechanism (see above). The key active particle in this is actually OH, which is initially produced in plasma. This plasma-generated OH is not so destructive however because it is too chemically active and immediately reacts with the first available organic molecule. Hydrogen peroxide generated by OH recombination can therefore be used as the OH 'carrier', which later releases OH through the Fenton mechanism.

Such an ROS scheme is typical not only for sterilization but for several plasma-medical methods discussed in Chapter 9. An especially strong effect is due to plasma-stimulated generation of hydrogen peroxide in water or water solution, especially in phosphate-buffered saline (PBS) or a biological medium. Conventional DBD in water produces more than 10–100 micromoles of H_2O_2 which are able to penetrate cells and stimulate double-strand breaks of DNA after just seconds of treatments. It is important to note that the sterilization effect of hydrogen peroxide itself is not so significant, although its effect in acidic solution can be extremely strong. Recalling that cold plasma makes water acidic, the combined plasma generation of peroxides and acids is useful for cold-plasma sterilization of bioorganisms submerged in water or water solutions. This subject is discussed in Chapter 7 on the biological effect of plasma-treated water, the effect of a medium in plasma medicine and plasma pharmacology.

6.3.9   Contribution of plasma-generated heat and temperature to plasma sterilization

When discussing sterilization, the first factor which usually comes to mind is the thermal effect. Many conventional sterilization methods are based on the use of either moist heat or dry heat with a typical treatment time of c. 1 hour. Moist heat sterilization is usually performed in autoclaves at atmospheric pressure and at a typical temperature of 121°C. Dry heat sterilization requires temperatures close to 170°C. Most non-thermal plasma discharges operate at lower temperatures (from room temperature to c. 70°C), which does not provide thermal sterilization (Laroussi and Leipold, 2004).

Neglecting the effect of heat in non-thermal plasma sterilization, we should keep in mind that even strongly non-equilibrium discharges can be characterized by elevated temperatures in some localized intervals in space or time. For example, non-equilibrium gliding arc discharges are on average cold, but they can provide a significant short-term local increase of gas temperature exceeding 1000°C. Such local but intensive heating is able to make a significant contribution to sterilization, not only killing microorganisms also completely disintegrating them.

When analyzing sterilization in DBD, which is traditionally considered as a non-thermal discharge, the thermal effect should also be taken into account. The thermal effect is related to the DBD microdischarge channels, where temperature can locally reach several hundred degrees. Overheating of the microdischarge channels occurs if they are relatively thick due to the non-uniformity of electrodes or particular conditions of the electrodes (wetness, dirt, surface conductivity, etc.). These problems are especially important when living tissue is used as one of the DBD electrodes, discussed in Chapter 9. Although the thermal sterilization effect of non-thermal plasma discharges is not significant by itself, the strong contribution of radicals or charged species discussed in the previous sections can be enhanced by even a small temperature increase.

Generally, plasma sterilization is a complicated process where even a minor factor can make an important synergetic contribution in combination with the major plasma sterilization factors. Local overheating of plasma discharges can lead to the formation of shockwaves and to the *shockwave sterilization effect*. This effect can be significant for discharges in water and related water sterilization (see Chapter 7). Summarizing experimental results, it can be definitely concluded that the heat effect is not a major contributor to the sterilization effect in non-thermal and strongly non-equilibrium plasmas.

6.3.10 Effect of UV radiation

Plasma is a source of UV radiation at different wavelengths, and can be effective in sterilization. Four relevant ranges of the UV wavelengths can be highlighted:

1. vacuum ultraviolet (VUV) of wavelength 10–100 nm;
2. UV-C of wavelength 100–280 nm;
3. UV-B of wavelength 280–315 nm; and
4. UV-A of wavelength 315–400 nm.

VUV-photons have high energy sufficient for breaking chemical bonds. Their efficiency in sterilization is limited, however, by a very short penetration depth. The efficiency of UV-A/B photons in sterilization is limited by the low energy of the photons. UV-C photons have energy sufficient for significant reconstruction of organic molecules, and are characterized by a sufficiently large penetration depth; UV-C photons are therefore the most effective in direct sterilization processes.

UV radiation in the 200–300 nm wavelength range with doses of several mJ cm^{-2} causes lethal damage to cells. There are several biological mechanisms of UV deactivation of bioorganisms, including dimerization of thymine bases in the bacterial DNA strands which inhibits the ability of the bacteria to replicate properly.

The contribution of UV radiation to the total plasma sterilization effect depends on the effectiveness of the specific plasma to radiate in the UV-C range. Low-pressure plasma discharges are able to provide significant UV radiation in the range of wavelengths effective for sterilization. Non-thermal atmospheric-pressure plasmas are typically not very effective sources of high-energy UV radiation.

According to Laroussi and Leipold (2004), no significant UV emission occurs in non-thermal atmospheric-pressure plasmas below 285 nm, which reduces the contribution of UV to sterilization in such plasmas. Even when the direct UV contribution to sterilization is negligible, it can be important in synergy with other factors such as radicals and charged particles. Such an effect has been discussed above in relation to non-thermal plasma treatment of polymers. Kinetic data on the effect of UV on deactivation of different microorganisms are summarized in Chapter 7 in connection with plasma sterilization of air streams.

6.4 Physical and biochemical effects of atmospheric-pressure air plasma on microorganisms

6.4.1 Direct and indirect effects of non-thermal plasma on bacteria

Non-thermal plasma is effective in the killing of parasites, bacteria, fungi and viruses (Fridman *et al.*, 2006, 2007b; Lu and Laroussi, 2006; Raizer and Zenker, 2006; Sakiyama and Graves, 2006; Sladek, Baede and Stoffels, 2006; Stoffels, 2006). The sterilization of different surfaces including living tissues can be divided into two approaches: indirect and direct. *Indirect treatment* uses a jet or flow of products (plasma afterglow) generated in remotely located plasma discharge. *Direct treatment*, by contrast, uses the tissue itself as an electrode that participates in creating the plasma discharge, as illustrated in Figure 6.15. Plasma in this case is contained between the quartz surface of the powered high-voltage electrode and the surface of bacteria or tissue being treated.

The distinction between direct and indirect treatment is not only related to the proximity of the plasma and the tissue; direct contact with plasma brings charged energetic particles to the plasma–tissue interface. By contrast, no charged particles are usually taken out of the plasma region by a jet even if the plasma region is located only a fraction of a millimeter away. The difference between direct and indirect plasma treatment is especially crucial in the case of plasma treatment of living tissues, that is, in plasma medicine. Fridman *et al.*

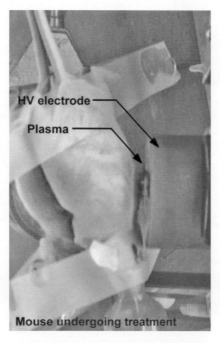

Figure 6.15 *FE-DBD direct treatment of living tissue: animal directly treated in plasma for up to 10 min remains healthy and no tissue damage is observed visually or microscopically immediately after and up to two weeks following the treatment (see color plate).*

(2007b) have demonstrated that direct treatment can achieve sterilization much faster without any thermal effects. The effect of charged particles in sterilization can be significantly stronger than that of the long-living active neutrals and longer-wavelength UV radiation (VUV is absorbed by air at atmospheric pressure within a few microns).

Significant advantages of the direct plasma treatment (Fridman *et al.*, 2007b) have been demonstrated by comparison of two types of experiments with DBD discharges. In one set of experiments a surface covered by bacteria was employed as a DBD electrode; a second smaller-area DBD electrode was placed over the first. Areas of sterilization with and without air flow parallel to the surface were compared. In another set of experiments, a surface covered by bacteria was employed as a DBD electrode. The rate of sterilization in the system was compared to when the surface was separated from the DBD discharge by a grounded mesh electrode (see Figure 6.16).

Bacterial samples for the experiments were collected from de-identified skin samples from human cadavers. Bacteria from the skin samples consisted of staphylococci, streptococci and Candida species of yeast. They were transferred onto the blood agar plate, cultured and diluted in $10 \times$ PBS to approximately 10^9 CFU mL^{-1}. To quantify sterilization efficiency, 20 μL drops of 10^9 CFU mL^{-1} bacteria were placed on agar surface, left

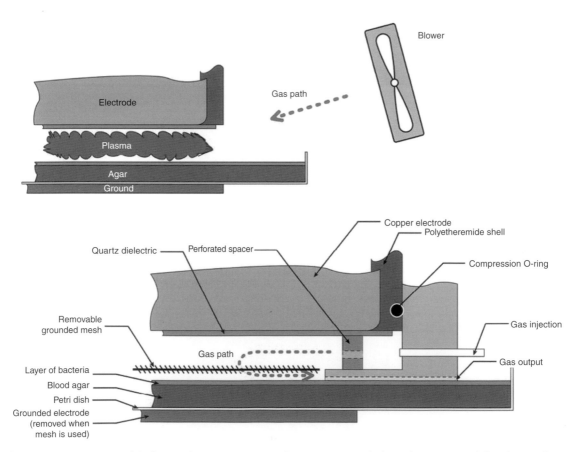

Figure 6.16 *Direct and indirect plasma experimental setups: grounded mesh is removed for direct plasma treatment and replaced for indirect treatment (above) or air is blown through plasma to carry out long-lived species (below).*

Table 6.1 *FE-DBD plasma sterilization of bacteria (in CFU mL^{-1}); comparison of direct and indirect plasma treatment.*

Original bacteria concentration	Bacteria concentration after 5 s of DBD treatment	Bacteria concentration after 10 s of DBD treatment	Bacteria concentration after 15 s of DBD treatment
10^9	850 ± 183	9 ± 3	4 ± 4
10^8	22 ± 5	5 ± 5	0 ± 0
10^7	6 ± 6	0 ± 0	0 ± 0

to dry for 5 minutes and treated with plasma. This volume was selected as it spread to c. 1 cm^2 over the agar surface; the area covered by the bacterial sample drop was therefore entirely within the area covered by the insulated plasma electrode. Following the treatment, the drop was spread over the entire agar surface and incubated in air at 37°C for 24 hours. Bacterial colonies (CFUs) were then counted and the results are listed in Table 6.1.

There is a clear difference in agar appearance between untreated partially sterilized and completely sterilized agar. Based on this difference, it is reasonable to classify the appearance of the bacterial surface into five categories:

1. untreated (10^9 CFU mL^{-1});
2. partially disinfected (10^9–10^7 CFU mL^{-1});
3. disinfected (10^7–10^4 CFU mL^{-1});
4. partially sterilized (10^3–10 CFU mL^{-1}); and
5. completely sterilized (0 CFU mL^{-1}).

To quantify the extent of sterilization, 1 mL of 10^9 CFU mL^{-1} was poured over the entire agar surface. These samples were left to dry for 3 hours and then treated by plasma. Here plasma only covered a portion of the Petri dish, while bacteria covered the entire dish. Following the 24 hour incubation period, the extent of sterilization was clearly visible. Areas where bacteria were killed looked like uncontaminated agar, while areas that received no treatment changed color and appearance significantly as bacteria grew. The complete sterilization area is easiest to identify: it corresponds to the agar area, that is, completely clear from bacteria as seen in Figure 6.17. Partial sterilization is also relatively easy to define since the number of CFUs is relatively small and can be counted. Disinfection is more difficult to assess because of the difficulty of counting the large number of CFUs. The complete sterilization gradually fades into untreated areas, forming a 'grayscale' fade that gradually increases from 0 CFU mL^{-1} in the sterile zone to 10^9 CFU mL^{-1} in the untreated zone.

6.4.2 FE-DBD experiments demonstrating higher effectiveness of direct plasma treatment

The experimental system developed to compare the effectiveness of direct and indirect plasma treatment of microorganisms (Fridman *et al.*, 2007b) is illustrated in Figure 6.16. An agar-covered Petri dish with bacteria spread on its surface from the earlier prepared samples of 10^9 CFU mL^{-1} concentration was used to test the sterilization. Specifically, one milliliter of such bacterial sample was transferred onto the blood agar, spread over the entire plate and dried for 3 hours prior to the plasma treatment. After the plasma treatment, the samples were incubated in air at 37°C for 24 hours to assess the ability of the DBD plasma to sterilize.

Figure 6.17   *Using a blower to shift sterilization region does not affect plasma and shows little effect of the afterglow: 15 seconds of treatment with (a) blower off and (b) blower on (air flows up) (see color plate).*

Control experiments were performed with the gas flow only. Bacteria were completely unaffected by flowing room air in up to 30 minutes. Placed on a metal substrate, the agar dish with bacteria on its surface acted as one of the DBD electrodes; bacteria are at the plasma–agar interface. The other smaller-area DBD electrode consists of 2.5-cm-diameter solid copper disk separated from the plasma by a 1-mm-thick quartz dielectric. Plasma gap, or the distance between quartz-plasma and agar-plasma interfaces, was set to 1.5 mm. Alternating polarity sinusoidal voltage of 35 kV magnitude (peak-to-peak) and 12 kHz frequency was applied between the copper electrode and the aluminum substrate under the agar Petri dish. Flow speed for this experiment was measured between the electrode and the substrate, directly outside of the electrode, to be 0.8 ± 0.3 m s^{-1}. Electrical and calorimetric measurements indicated the DBD power of 0.8 ± 0.2 W cm^{-2}.

When there is no air flow parallel to the surface, the area directly under the second smaller electrode is exposed simultaneously to charged particles, UV and the plasma afterglow containing active neutrals such as ozone (O_3), nitric oxide (NO), hydroxyl radicals (OH) and other excited molecular and atomic species. In the presence of the airflow, the area under the second smaller electrode experiences practically the same conditions as without the flow. However, in the presence of flow, the plasma afterglow will also impinge on the areas of the substrate that are not directly under the second smaller electrode.

Results comparing the sterilization with and without flow are illustrated in Figure 6.17. Complete sterilization by direct plasma treatment under the insulated electrode is achieved in 15 s. If air is passed parallel to the substrate through the plasma region, the disinfection area shifts by several millimeters but a 'tail' of disinfection appears. Even though complete sterilization is not observed within this 'tail', different degrees of disinfection are evident.

Locations closer to the plasma region have a greater degree of disinfection. This suggests that plasma 'afterglow', or the gas carried through and outside the plasma, contains active species and radicals that have bactericidal effects (although direct plasma is substantially more potent as a sterilizing agent). Complete sterilization, however, did not extend to the region reachable by the plasma afterglow in as much as 5 min of treatment. This clearly demonstrates that plasma-generated active neutrals are much less effective than direct plasma where charged particles participate in the process.

The first experiments (Fridman *et al.*, 2007b) separated the effects of plasma afterglow from those of charged particles and UV. In the second type of experiments (described below), the plasma afterglow together with UV is compared to the direct plasma, also involving charged particles.

Direct plasma   (a)   (b) Plasma jet   (c)   (d)

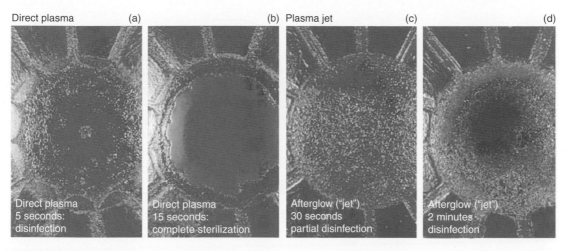

Figure 6.18   *Direct application of plasma provides better sterilization efficiency than treatment by plasma afterglow: (a) 5 s and (b) 15 s of direct plasma compared with (c) 30 s and (d) 2 min of plasma jet (see color plate).*

The direct treatment is obtained similarly to the first type of experiments. However, instead of creating airflow parallel to the substrate, provisions were made to allow the introduction of air into the discharge gap (Figure 6.16). To obtain indirect treatment including UV effects, an agar surface covered by bacteria was once again employed as a DBD electrode; however, this surface was separated from the DBD discharge by a mesh electrode as illustrated in Figure 6.16 (right). With a grounded mesh in place, bacteria are no longer at the plasma interface as plasma is bounded by the grounded mesh. The mesh is placed 1 mm above the surface of the agar which permits only the passage of UV, radicals and other longer-living particles created in the plasma gap between the mesh and the quartz-insulated electrode above it. The second electrode used with the mesh was the same electrode that was employed in the direct treatment setup illustrated in Figure 6.16 (left). The setup used to introduce air into the discharge gap is illustrated in Figure 6.16 on the right. The air was introduced into the discharge gap to enhance extraction of the plasma afterglow through the mesh.

Comparison of direct (without the mesh) and indirect (with the mesh) plasma sterilization is shown in Figure 6.18. It is clear that plasma that comes in direct contact with bacteria is able to sterilize significantly faster than afterglow or jet. Direct plasma treatment for as little as 5 s results in the appearance of a sterilization region (spot near the center). Complete sterilization occurs within 15 s when direct treatment is employed, while only partial disinfection can be achieved within the same time frame with indirect treatment. Over 5 min of indirect treatment is required to achieve sterilization results similar to direct treatment obtained within 15 s. Therefore, even with substantial UV radiation, indirect plasma treatment is substantially weaker than direct treatment provided by charged particles.

6.4.3   Surface versus penetrative plasma sterilization

Effective plasma treatment of bacteria, cells and other biomaterials does not require direct contact with plasma but is possible even when they are separated from plasma and protected by a layer of intermediate substance, in particular when submerged within a culture medium (Kieft *et al.*, 2004, 2005; Fridman *et al.*, 2007b). The studies described in this section are focused mostly on plasma treatment of cells. The objective of the

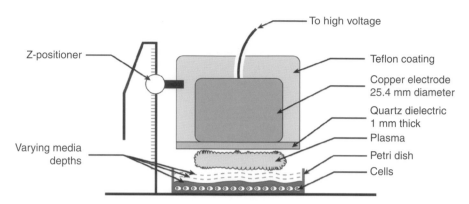

Figure 6.19   *Schematic of FE-DBD treatment of melanoma cells.*

discussion is clarification of depth of penetration of plasma effect, which is actually similar for the cases of treatment of bacteria and treatment of cells.

Detachment of mammalian vascular cells was observed after indirect treatment by the non-thermal atmospheric-pressure 'plasma needle' when the thickness of the liquid layer covering the cells was c. 0.1 mm (Kieft *et al.*, 2005). Effective direct treatment of cells that are covered by up to a half a millimeter of cell culture fluid was demonstrated by Fridman *et al.* (2007b), discussed in Chapter 10. Experiments were carried out using a DBD discharge with submerged melanoma cell culture as illustrated in Figure 6.19. The discharge gap between the bottom of the 1-mm-thick quartz plate covering the copper electrode and top surface of the fluid was set to 1.5 mm. AC voltage of 35 kV and 12 kHz was applied to the copper electrode, while the aluminum Petri dish remained at a floating potential to mimic the conditions that exist when the discharge is used to treat living tissue. Surface power density of the DBD was 0.8 ± 0.2 W cm^{-2}.

Melanoma cells were cultured on the surface of an aluminum dish. Cells were allowed to grow while the dish was kept within an incubator. The original growth medium was then removed and a measured volume of fresh culture medium was added while the cells remain attached to the bottom surface. The volume of the culture medium was measured to obtain the desired depth which was varied over the range 0–0.33 mm (Figure 6.19). The submerged cells were placed into the DBD treatment setup and treated by plasma for periods of time ranging 0–30 s. Immediately after the plasma treatment, the culture medium used during the treatment was replaced with a fresh medium. The cells were then analyzed by Trypan Blue exclusion test to determine their viability through the cellular membrane integrity. Trypan Blue only enters cells through permeabilized cytoplasmic membranes.

Figure 6.20 shows the percentages of affected cells and treatment times. No effect on the cells is observed in control experiments where plasma treatment does not occur. Without the medium protecting the cells, cell wall fracture is achieved within 10 s of treatment and all cells uptake the ink (Trypan Blue), indicating that the cell wall has been compromised and the cell is dead. Even though the DBD plasma is not uniform, all the cells are affected by the treatment in as little as 10 s. When a layer of media is introduced which covers these cells, the number of compromised cells decreases but it is clear that the plasma treatment is permeabilizing these cells through a layer of liquid cell growth medium. The DBD plasma treatment leads to notable changes in the pH of the culture medium (see Figure 6.21). Deep penetration of the plasma sterilization effect would appear to be the cause, but this is not the case.

Special control experiments were carried out to understand the role of the changing chemistry of the culture medium on the melanoma cells. In these experiments, grown cell culture was transferred into the culture medium pre-treated by the DBD discharge in a manner identical to that used to treat the submerged

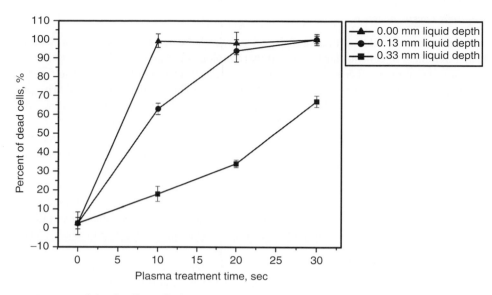

Figure 6.20 *Percent of dead cells (cells that uptake Trypan Blue) after treatment at different liquid depths. The thickness of the cell layer at the bottom of the dish is c. 30 μm.*

cell culture. Changes in the pH of the culture medium were identical to those found in experiments with the submerged cells. As before, the cells were placed into fresh culture medium after remaining in the pre-treated culture medium for the amount of time equal to the plasma pre-treatment time. Similarly to the control (untreated) cells, cells which were transferred into the culture medium pre-treated by the plasma in this fashion and then placed into the fresh medium did not uptake Trypan Blue. The results of placing cells in acidified media temporarily indicate that the DBD plasma does not affect the cells simply through changes in the chemical composition of the culture medium.

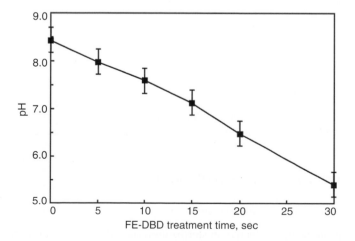

Figure 6.21 *pH change of the buffer medium under FE-DBD treatment.*

6.4.4   Apoptosis versus necrosis

Non-thermal plasma can stimulate both apoptosis and necrosis in treated cells. As in the previous section, we discuss plasma treatment of cells to clarify the variety of biological sequences of plasma treatment.

Apoptosis, or programmed cell death, is a complex biochemical process of controlled self-destruction of a cell in a multicellular organism (Johnstone, Ruefli and Lowe, 2002; Silverthorn *et al.*, 2004). This process plays an important role in maintaining tissue homeostasis, fetal development, immune cell 'education', development and aging. Examples of apoptosis that occur during normal body processes include the formation of the outer layer of skin, the inner mucosal lining of the intestine and the endometrial lining of the uterus, which is sloughed off during menstruation. During apoptosis, cellular macromolecules are digested into smaller fragments in a controlled fashion. The cell ultimately collapses into smaller intact fragments that can be removed by phagocytosis without damaging the surrounding cells or causing inflammation.

In contrast, during necrosis (also termed 'accidental cell death') the cell bursts and the cellular contents spill out into the extracellular space, which can cause inflammation. Necrosis is induced by cellular injury, for example, extreme changes in osmotic pressure or heat that lead to adenosine tri-phosphate (ATP) depletion of the cell.

Apoptosis and necrosis effects of plasma treatment on mammalian cells have been studied by Stoffels (2003). Chinese hamster ovarian cells were used as a model. The cells were exposed to a RF small-volume cold plasma generated around the tip of a needle-shaped electrode (*plasma needle*). Necrosis (cell death due to catastrophic injury) occurs in the plasma needle treatment for powers above 0.2 W and exposure times longer than 10 s. The cell membranes are damaged and the cytoplasm is released. Lower doses of exposure lead to apoptosis. If the power level and exposure time were reduced significantly to c. 50 mW and 1 s, the Chinese hamster ovarian cells partly detach from the sample, take a more rounded shape and do not undergo apoptosis (Stoffels, 2003).

The sub-lethal plasma effect on bacterial cells has been demonstrated by Laroussi, Richardson and Dobbs (2002). When a sub-lethal plasma exposure is administered to bacterial cells, a change in their metabolic behavior can occur. To demonstrate these effects, Laroussi, Richardson and Dobbs (2002) addressed the biochemical impacts of plasma on *Escherichia coli* with sole carbon substrate utilization (SCSU) experiments using Biolog (Hayward, CA) GN2TM 96-well microtiter plates. The purpose of the SCSU experiments was to determine if exposure to plasma altered the heterotrophic pathways of the bacteria. It was assumed that any changes in metabolism would be indicative of plasma-induced changes in cell function. The Biolog GN2 plate comprised a control well and 95 other wells, each containing a different carbon substrate. Color development of a redox dye present in each well indicated utilization of that particular substrate by the inoculated bacteria. The 95 substrates were dominated by amino acids, carbohydrates and carboxylic acids.

The plasma exposure caused an increase of utilization of some substrates and a decrease of utilization in others, indicating noticeable changes in the corresponding enzyme activities, without causing any lethal impact on the cells. Plasma-stimulated apoptotic behavior in melanoma skin cancer cell lines has been demonstrated by Fridman *et al.* (2007b). This effect has important potential for cancer treatment because the cancer cells frequently acquire the ability to block apoptosis and are therefore more resistant to chemotherapeutic drugs. This effect is discussed in Chapter 9 in the medical treatment of melanoma skin cancer.

6.5   Animal and human living tissue sterilization

6.5.1   Direct FE-DBD for living tissue treatment

Direct plasma treatment can be very effective in sterilization and other aspects of tissue treatment. The direct plasma treatment implies that living tissue itself is used as one of the electrodes and directly participates in the active plasma discharge processes. Figure 6.15 illustrates direct plasma treatment (sterilization) of skin of a live mouse. DBD plasma is generated in this case between the quartz-covered high-voltage electrode and the mouse as a second electrode.

Direct application of the high-voltage (10–40 kV) non-thermal plasma discharges in atmospheric air to treat live animals and people requires a high level of safety precautions. Discharge current should be limited to below the values permitted for treatment of living tissue. Moreover, discharge itself should be homogeneous enough to avoid local damage and discomfort. The creation of special atmospheric discharges to effectively solve these problems is an important challenge for plasma medicine.

Fridman *et al.* (2006) developed the floating-electrode dielectric barrier discharge (FE-DBD) for this purpose. One of the electrodes is a dielectric-protected powered electrode and the second active electrode is human or animal skin or organ – without human or animal skin or tissue surface present, the discharge does not ignite. In the FE-DBD setup, the second electrode (e.g. a human) is not grounded and remains at a floating potential. Discharge ignites when the powered electrode approaches the surface to be treated at a distance (discharge gap) of less than c. 3 mm, depending on the form, duration and polarity of the driving voltage. See Chapter 4 for a discussion of the physics and engineering aspects of the FE-DBD plasma source. A simple illustration of the FE-DBD discharge (Fridman *et al.*, 2006) is illustrated in Figure 6.19.

The typical value of plasma power in initial experiments was kept at c. 3–5 W and power density 0.5–1 W cm^{-2}. Further development of the FE-DBD discharge is related to optimization of shape of the applied voltage to minimize the DBD non-uniformities and related possible damaging effects. The best results have been achieved by generation of the FE-DBD in the pulsed mode with pulse duration <30–100 ns (Ayan *et al.*, 2007). This results in the no-streamer discharge regime, providing sufficient uniformity of the non-damaging direct plasma treatment even when the second electrode is a living tissue (and therefore wet and essentially non-uniform). As soon as the atmospheric discharge is safe, it can be applied directly to the human body as is illustrated in Figure 6.22. Highly intensive non-thermal plasma devices can therefore be directly applied to living animal or human tissue for different medical and cosmetic treatment.

6.5.2   Direct FE-DBD plasma source for living tissue sterilization

Sterilization of living animal or human tissue with minimal or no damage is of importance in a hospital setting. Chemical sterilization does not always offer a solution. For example, transporting chemicals for

Figure 6.22   *FE-DBD: one electrode is protected by a dielectric barrier, another electrode is living tissue to be treated.*

Table 6.2   *Human skin flora sterilization using FE-DBD.*

Original concentration	5 s of FE-DBD	10 s of FE-DBD	15 s of FE-DBD
10^9	850 ± 183	9 ± 3	4 ± 4
10^8	22 ± 5	5 ± 5	0 ± 0
10^7	6 ± 6	0 ± 0	0 ± 0

sterilization becomes a major logistics problem in a military setting, while use of chemicals for sterilization of open wounds, ulcers or burns is not possible due to the extent of damage they cause to punctured tissues and organs. Non-thermal atmospheric-pressure plasma is non-damaging to the animal and human skin but a potent disinfecting and sterilizing agent.

Human tissue sterilization has been investigated by Fridman *et al.* (2006). Bacteria in this case were 'skin flora' (a mix of bacteria collected from cadaver skin containing Staphylococcus, Streptococcus and yeast). Direct FE-DBD plasma sterilization leads roughly to a 6-log reduction in bacterial load in 5 s of treatment, as illustrated in Table 6.2.

A similar level of the skin flora sterilization using the indirect DBD approach requires 120 s and longer of plasma treatment at the same level of the discharge power. Sterilization of the mix of bacteria collected from cadaver skin generally occurred in the experiments after 4 s of treatment in most cases and 6 s in a few cases.

Non-thermal atmospheric plasma, especially when applied directly, is therefore an effective tool for sterilization of living tissue. It opens interesting possibilities for pre-surgical patient treatment, sterilization of catheters (at points of contact with human body), sterilization of wounds and burns as well as treatment of internal organs in gastroenterology. Specific medical examples of plasma sterilization in relation to wounds and surgical conditions are discussed in Chapters 9 and 10 focused on specific plasma-medical applications.

6.5.3 Toxicity (non-damaging) analysis of direct plasma treatment of living tissue

Plasma is an excellent sterilization tool for different surfaces. The key aim of direct plasma skin sterilization is for the skin to remain intact after the treatment; the problem of non-damage is the key issue of the entire plasma-medical techniques. In this regard, the magnitude of the energy or dose of direct plasma treatment is very significant. Comparing the direct DBD plasma effect with radiation therapy (or radiation biology) it can be concluded that, kinetically, they are quite similar, including possible double-strand breaks in DNA. However, typical DBD plasma doses after seconds of treatment are c. 10^6 Gy (J kg^{-1}). Similar treatment effects in the case of gamma-radiolysis require only 0.5–1 Gy. Typical direct non-thermal plasma is c. 10^6 times more 'energetic' than penetrating radiation. Alternatively, the plasma effect is less 'penetrating' than the ionizing radiation by a factor of 10^6.

The importance of non-damage to healthy tissue is the key issue for successful application of non-thermal plasma in medicine. A topical treatment which damages the tissue surface would not be acceptable to the medical community, and so cadaver tissue was first tested followed by escalating skin toxicity trials on SKH1 hairless mice in the FE-DBD experiments of Fridman *et al.* (2006). Cadaver tissue was treated by FE-DBD plasma for up to 5 min without any visible or microscopic change in the tissue, as verified with tissue sectioning and staining via the hematoxylin and eosin (H&E) procedure illustrated in Figure 6.23.

Based on the knowledge that FE-DBD plasma has non-damaging regimes, an animal model was tested by Fridman *et al.* (2007b). In an SKH1 mouse model, the skin treatment was carried out at varying doses to locate damaging power/time (dose) combination and skin damage was analyzed in two stages. First, the animal was

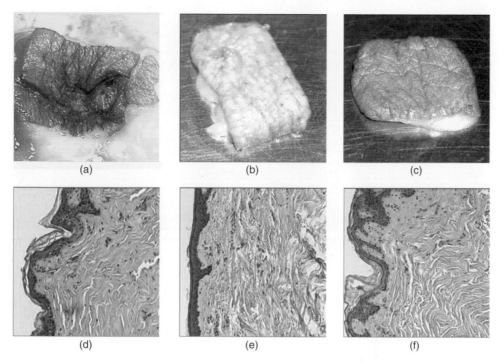

Figure 6.23 *Photos (top) and tissue histology (bottom) of cadaver skin samples after FE-DBD treatment: (a, d) control; (b, e) after 15 s of treatment; and (c, f) after 5 min of treatment – no visible damage is detected (see color plate).*

treated at what was deemed to be a toxic (damaging) dose based on trials with cadaver skin tissue. Once the dose where the damage was visible was determined, a new animal was treated at a lower dose. If no damage was observed at that dose, two more animals were treated. If no damage was observed in all three, the dose was deemed 'maximum acceptable dose'. Once the maximum dose was located, three animals were treated at that dose and left alive under close observation for two weeks. Based on the experimental matrix, a dose at 0.6 W cm^{-2} for 10 min was deemed maximum acceptable prolonged treatment and a dose of 2.3 W cm^{-2} for 40 s was deemed maximum acceptable high-power treatment.

Histological (microscopic) comparison of control SKH1 skin sample with toxic and non-toxic plasma doses shows regions where the plasma dose is fairly high while the animal remains unaffected (animal after the treatment: Figure 6.24; histological samples: Figure 6.25). Note that sterilization was achieved at 2–4 s at high-power treatment of 0.8 ± 0.2 W cm^{-2} and at 10 ± 4 s at half that power. The variation in time for sterilization is attributed to the initial contamination level of the animal (same for cadaver tissue); in other words, some skin samples are simply cleaner than others. Following the investigation on mice, an investigation on pigs was carried out achieving the same results.

The toxicity due to the FE-DBD treatment of living tissue not only depends on the treatment dose (discharge power and treatment duration), but also strongly depends on the shape of voltage applied to the discharge. Pulsing of the DBD discharges can significantly decrease its damaging ability. Application of nanosecond pulses completely prevents the formation of streamers and therefore DBD microdischarges, which helps to reduce the toxicity of the direct plasma-medical treatment of living tissue (Ayan *et al.*, 2007).

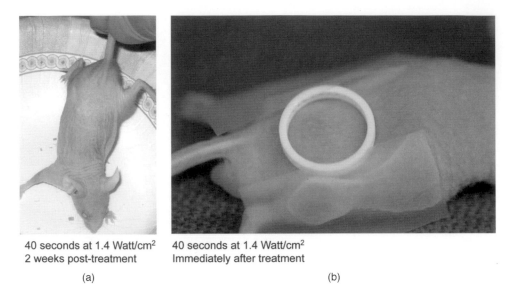

40 seconds at 1.4 Watt/cm²   40 seconds at 1.4 Watt/cm²
2 weeks post-treatment   Immediately after treatment

(a)   (b)

Figure 6.24   *Animal remains after a reasonably high FE-DBD plasma dose (40 s at 1.4 W cm⁻²; more than 10 times higher than needed for skin sterilization): (a) immediately after treatment and (b) 2 weeks after treatment.*

6.6   Generated active species and plasma sterilization of living tissues

6.6.1   Physico-chemical in vitro tissue model: Production and delivery in tissue of active species generated in plasma

As discussed in Chapter 5, one of the major mechanisms of direct plasma-medical treatment of tissues is related to the plasma-catalysis of the reduction/oxidation reactions occurring in the biological system. The

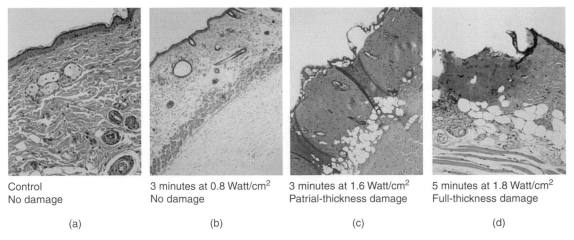

Control   3 minutes at 0.8 Watt/cm²   3 minutes at 1.6 Watt/cm²   5 minutes at 1.8 Watt/cm²
No damage   No damage   Patrial-thickness damage   Full-thickness damage

(a)   (b)   (c)   (d)

Figure 6.25   *Histology of toxic and non-toxic to SKH1 skin plasma doses compared to untreated skin: (a) control; (b) 3 min at 0.8 W cm⁻² (no damage); (c) 3 min at 1.6 W cm⁻² (partial-thickness damage); and (d) 5 min at 1.8 W cm⁻² (full-thickness damage) (see color plate).*

action of specific charged or neutral active species or radiation is frequently associated with the corresponding specific effect (e.g., anti-inflammatory effect of NO, highly oxidative hydroxyl radical and other ROS). With the vast amount of recent research it is becoming clear that plasmas are indeed able to cause some, perhaps positive, clinical effects in patients. It is therefore important to analyze plasma itself, not only its effects on the organism.

Clearly, plasma in contact with dirty oily bloody tissue is not the same as that generated between two perfect electrodes. For this reason, Dobrynin *et al.* (2012) have developed a simple in vitro model where the second electrode for plasma generation is an agarose gel. In the following section, we compare measurements of plasma-generated species on agarose gel to those on a store-bought chicken meat and to those on a live rat wound. Surprisingly, the findings suggest that properly prepared agarose gel may actually serve as a very good model of penetration of plasma-generated species into the tissue. This provides an opportunity to analyze plasma (perform spectroscopic, microwave and other measurements) with a second electrode of a simple agarose gel model and obtain results which are a sufficiently accurate representation of the real environment. These experiments are generally focused on the deep penetration of the plasma-generated active species of the living tissues.

6.6.2   FE-DBD plasma system for analysis of deep tissue penetration of plasma-generated active species

FE-DBD plasma was generated by applying alternating polarity pulsed (1 kHz) voltage of c. 20 kV magnitude (peak-to-peak) and a rise time of 5 V ns^{-1} between the insulated high-voltage electrode and the sample undergoing treatment (Dobrynin *et al.*, 2012). The powered electrode comprised a 1.5-cm-diameter solid copper disk covered by a 1.9-cm-diameter 1-mm-thick quartz dielectric. The discharge gap was kept at 1.5 mm. Current peak duration was 1.2 μs, and the corresponding plasma surface power density was 0.3 W cm^{-2}.

In the case of ex vivo measurement in rat tissue, a special pen-size electrode was used. A 1-mm-thick polished clear fused quartz (Technical Glass Products, Painesville, OH) was used as an insulating dielectric barrier and a handheld pen-like device with the quartz tip was used for treatment. In this case, the average power density for the active area of the high-voltage electrode was kept at a level of c. 0.74 W for 6 mm electrode diameter. Agarose gels were prepared using the standard procedure with pure agar powder (Fisher) in either distilled water or PBS (Fisher).

In order to determine the best concentration of agarose gels which would closely represent tissue, we have used agar at concentrations of 0.6%, 1.5% and 3% weight percentage. Agarose gels at 0.6% concentration were reported to closely resemble in vivo brain tissue with respect to several physical characteristics. 4% agar phantoms are widely used as a tissue models for radiology studies. The 1.5% concentration of agar was chosen as a median point which is often used as a microbiological substrate.

Measurements of H_2O_2 and pH penetration into agarose gels (0.6%, 1.5% and 3% wt) and tissues were made using Amplex UltraRed reagent (Invitrogen, ex/em: 530/590 nm) and Fluorescein (Sigma Aldrich, ex/em: 490/514 nm) fluorescent dyes, respectively. In the case of H_2O_2, 75 μL PBS containing 100 μM Amplex UltraRed with 200 U μL^{-1} horseradish peroxidase (MP Biomedicals) were placed between 1-mm-thick 4 × 4 cm agar slices and incubated for c. 15 min before the treatment, in order to provide the presence of the dye in the agar volume. For the pH measurement, the agarose gels were prepared by adding fluorescein dye before it solidified. In order to measure the H_2O_2 and pH in tissue, dyes were injected using a syringe into a 1-cm-thick 4 × 4 cm skinless chicken breast tissue samples at various points up to a depth of 1 cm.

Ex vivo measurements were made in rat tissue using an animal (hairless Sprague-Dawley male rat) which had been euthanized just before the procedure. A 200 μL dye solution (Amplex UltraRed) was injected subcutaneously using a sterile syringe, and the animal skin was treated with FE-DBD plasma at various time

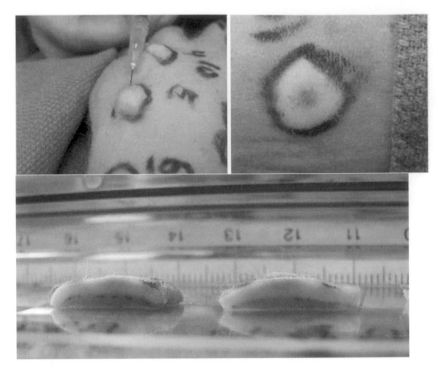

Figure 6.26 *Subcutaneous injection of the fluorescent dye (euthanized rat, top left), the rat skin after the plasma treatment (top right), and the skin sample cross-section just before measurement (bottom) (see color plate).*

intervals after 5 min incubation period (Figure 6.26). Immediately after the treatment, skin tissue samples were extracted and analyzed as follows. Treated samples were sliced in a vertical direction with thickness of 1 mm, and fluorescence was measured using an LS55 (Perkin Elmer) fluorescent spectrometer equipped with XY reader accessory (Figures 6.27 and 6.28). To obtain calibration curves for hydrogen peroxide in the plasma-treated samples, a standard stabilized 3% H_2O_2 (Fisher) water solution properly diluted to obtain various concentrations was used.

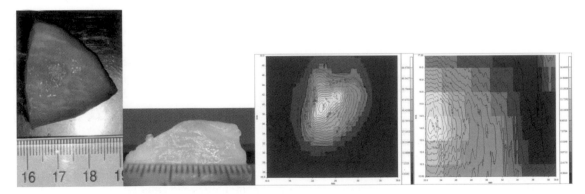

Figure 6.27 *Chicken breast after plasma treatment with H_2O_2 fluorescent dye: photograph and fluorescent images from the top and side of the sample (arbitrary units).*

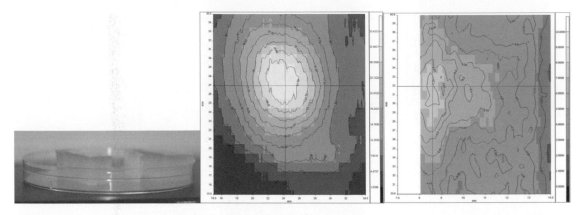

Figure 6.28 *Agarose gel after plasma treatment with H_2O_2 fluorescent dye: photograph and fluorescent images from the top and side of the sample (arbitrary units).*

6.6.3 Deep tissue penetration of plasma-generated active species

The results of the H_2O_2 measurements made of the FE-DBD system described in the previous section in dead tissue are depicted on Figure 6.29. The figure highlights that, with longer treatment time, the depth of penetration as well as concentration of hydrogen peroxide increases. In general, several millimoles of H_2O_2 are produced in tissue after plasma treatment, while it diffuses 1.5–3.5 mm deep. The same tendency is observed in the case of tissue acidity change, shown on Figure 6.30 (florescence intensity of fluorescein decreases with lower pH, and the data are presented in arbitrary units). However, the effect of pH lowering leads to deeper penetration of up to 4.5–5 mm.

In order to develop a simple realistic in vitro model of tissue that has simple physicochemical characteristics, in order to investigate depth of reactive species penetration, three concentrations of agarose media were used.

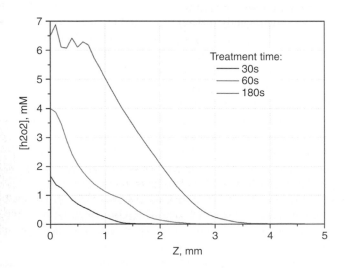

Figure 6.29 *Profiles of H_2O_2 in tissue after plasma treatment.*

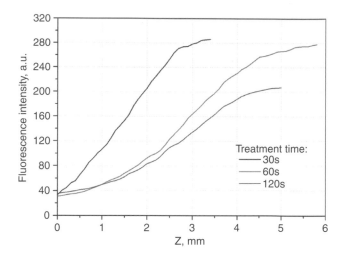

Figure 6.30   *Profiles of pH in dead tissue after plasma treatment.*

The measurement results for H_2O_2 produced by plasma treatment in agar gels together with the results for dead tissues for the same treatment doses are shown in Figure 6.31.

Hydrogen peroxide concentration on the agar gel surface was different for different agar densities: 0.5 mM for 0.6%, 0.7 mM for 3% and 1.9 mM for 1.5% gels after 1 min of plasma treatment. The results of hydrogen peroxide penetration measurements in skin tissues of a euthanized rat are shown in Figure 6.32. Compared to the dead chicken breast tissue, the depth of penetration appears to be very similar: up to 4 mm after 2 min treatment (although concentration is almost an order of magnitude higher due to corresponding higher power of the discharge by a factor of 8).

Surprisingly, the depth of H_2O_2 penetration for all types of agarose media was about the same. Figure 6.33 shows the depths at which the same level of 0.05 mM of H_2O_2 was detected in agar gels and tissue for different plasma exposure times. In contrast to hydrogen peroxide, the dynamics of acidity change inside the agar gels was significantly different for different agar media compositions (Figure 6.34). Tissue appeared to have better buffer properties compared to non-buffered agarose gels. However, the addition of PBS to agar produces required similarities in terms of pH changes.

The results show that, in the case of real tissue, active species produced by plasma on the surface may travel in tissue volume to depths of up to several millimeters. The particular depth of penetration is determined by diffusion and reaction rates, which depend highly on the type of tissue and its biochemical characteristics. Measurement of these parameters is an extremely complicated task, but the creation of a simple model that closely represents certain tissue is possible experimentally.

Here the agarose gels of various concentrations have been used in order to mimic tissue physical properties. For the case of hydrogen peroxide, as shown on Figure 6.31 and Figure 6.33, depth of diffusion is about the same for all types of agar. The concentration of H_2O_2 varies however, being the closest for the 1.5% agar weight percent case and c. 3 times lower for both 0.6% and 3% wt agar gels. This behavior may possibly be explained by both diffusion properties and reaction rates of H_2O_2 in agar. Acidity of tissue was measured to be consistently increased (lowering of pH) as a result of exposure to the discharge. The tissue compared to agarose gels prepared in distilled water acts as a significantly better buffer, and the depth of pH changes inside such gels is much greater. In fact, a significant drop in the whole volume of a phantom was observed – up to 1 cm thickness of agar. This problem may be addressed simply by adding a buffer into the agarose media as

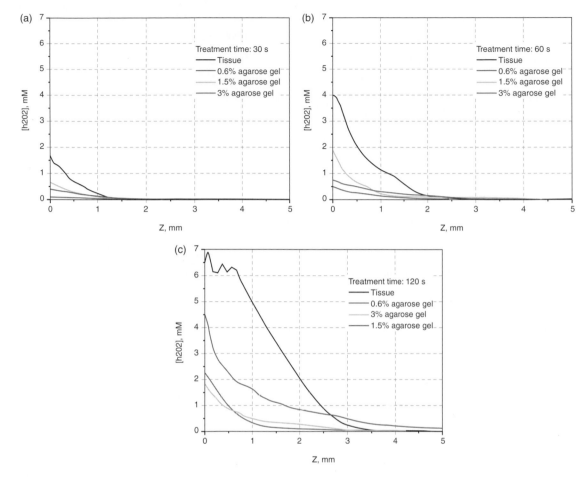

Figure 6.31 *The profiles of H_2O_2 in non-buffered agarose gels and dead tissues after plasma treatment during: (a) 30 s; (b) 60 s; and (c) 120 s.*

shown on Figure 6.34 in the form of 1.5% wt agar prepared in 1x PBS. It is interesting that a simple agar gel model may express similar physicochemical properties as a real tissue, resulting in comparable penetration effects of the plasma-generated active species.

6.7 Deactivation/destruction of microorganisms due to plasma sterilization: Are they dead or just scared to death?

6.7.1 Biological responses of *Bacillus stratosphericus* to FE-DBD plasma treatment

Plasma physicists involved in plasma sterilization research often ask the tricky question: we see that the microorganisms are not culturable (not proliferating, not multiplying) after plasma treatment, but does this mean that they are dead? Consider the analogy: I have a good friend who has no children, but he is very much alive (he is not a bacterium though). Physicists are also curious about the possibility of 'resurrection' after the bacterial death during the plasma sterilization. Although physicists are very excited

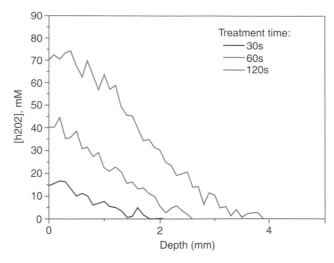

Figure 6.32   *The profiles of H_2O_2 in euthanized rat tissue after plasma treatment.*

about this question, microbiologists usually keep quiet knowing that the concepts of 'death' and 'resurrection' can be very sophisticated for microorganisms. A special attempt to answer these questions was made by Cooper *et al.* (2010), who focused on the biological responses of one of the robust organism, *Bacillus stratosphericus.*

Application of such a robust microorganism, especially interesting for NASA, allows the variety of biological responses to non-thermal plasma to be observed. DBD plasma was applied over various durations to *B. stratosphericus* either surface-dried or suspension in de-ionized water and viability, culturability and viable

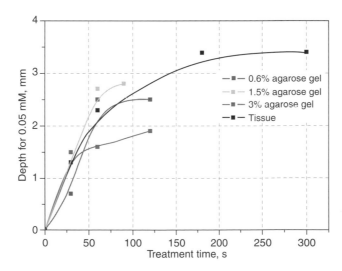

Figure 6.33   *Depth of H_2O_2 penetration at concentration of 0.05 mM: comparison between agarose gels and dead tissue.*

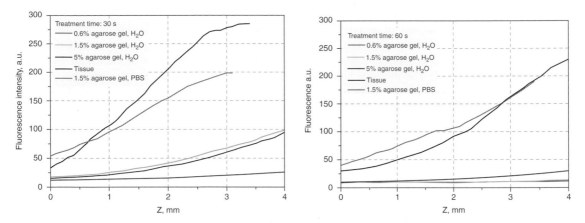

Figure 6.34 *The profiles of pH in agarose gels and dead tissues after the plasma treatment.*

but non-culturability (VBNC) were assayed using standard techniques. Depending upon the exposure of *B. stratosphericus* to DBD plasma, three viability states were obtained:

1. *viable and culturable* at low plasma doses;
2. *viable-but-not-culturable* at higher plasma doses; or
3. *disintegrated bacteria* at higher plasma doses.

In the case of VBNC, although the respiration levels of the organism were at relatively low levels immediately after plasma treatment, over the course of 24 hours respiratory activity increased eight times; it was found still non-culturable during colony assays, however. The loss of culturability in the case of VBNC is hypothesized to be induced as one of the responses to oxidative stress; it remains unclear if the response is temporary or indefinite. Appropriate plasma powers should obviously be used to avoid the VBNC-like status. It was also shown that *Bacillus stratosphericus* exhibits similar phenomena of etching and disintegration of the membrane and damage to DNA with plasma treatment in dry air; furthermore, when treated in liquid, the oxygenated species are able to inactivate bacteria to cause lethal and sub-lethal damages. Oxidative stress is able to induce a viable but non-culturable (VBNC) state in bacteria, which remains consistent with bacteria whose membrane is not peroxidized beyond the repairable limit by plasma.

Note that *Bacillus stratosphericus* tolerates 17% NaCl and is resistant to UV as well as selected antibiotics such as penicillin, vancomycin and erythromycin. Its resistance characteristics make it an interesting choice for understanding the effect of plasma exposure. The organism is multi-resistant and originally isolated from high altitude where radiation, UV and other rays are likely affecting it. The studies of Cooper *et al.* (2010) were focused on comparing culturability, membrane integrity, bacterial morphology and respiration capabilities to analyze the effects of FE-DBD exposure at doses which are lethal to other bacteria, and to show that robust bacteria are able to adapt and survive these treatments.

6.7.2 FE-DBD plasma treatment of *Bacillus stratosphericus*

To understand how DBD plasma influences the *B. stratosphericus* cell morphology, viability and culturability, Cooper *et al.* (2010) carried out a series of experiments using various plasma exposure times, both under dry and wet (cell suspension) environmental conditions. They looked for the cell wall and membrane-associated physical changes using SEM; genomic DNA-associated changes using PCR amplification of remnants of DNA (post-plasma treatment); viability of bacteria using standard colony count assays; cell membrane integrity using live/dead fluorescence assays; culturability and dormancy using standard respiratory metabolic activity with XTT assays; and morphological features peculiar to non-culturability under oxidative stress-induced cell elongation using SEM. The distance between the plasma probe and biological samples can be finely adjusted, and exact distance can be calculated. The function of quartz dielectric cover lowers the temperature generated across the biological samples being exposed to plasma.

6.7.2.1   Destruction of Bacillus stratosphericus *through etching phenomenon when present on dry surface*

To prove the efficacy of the plasma device system a highly resistant phenotype of *B. stratosphericus*, originally isolated during astrobiological studies, was selected. This spore-bearing bacterium is multi-drug resistant and demonstrates extreme tolerance to salinity and UV rays, making it the ideal challenge organism.

The physical damage to the cell envelope was first analyzed. Figure 6.35 indicates surprisingly clear evidence of punch-out porosity to bacterial cells when the dried samples are treated with plasma on stainless steel surface. Figure 6.35 shows the SEM images acquired at three different times and the observed graded changes. The images of untreated (a), 60 s plasma-treated (b) and 120 s plasma-treated (c and d) specimens are shown for comparison. Typical porosities through which cytoplasmic contents can freely leak out or, conversely, plasma effects can penetrate inside cells in fraction of seconds are exhibited. Similarly, the bacterial cells on dry surface were treated and harvested cells were tested for their DNA amplifications (Cooper *et al.*, 2010).

Figure 6.35e is the findings of PCR products run on agarose gel electrophoresis. The gel shows amplified products of untreated samples in normal phosphate buffered-saline (150 mmol L^{-1}) sodium chloride and 150 mmol L^{-1} sodium phosphate, pH 7.2 at 25°C (positive control) and plasma-treated samples. A plasma exposure of 60 s was sufficient to disintegrate DNA into possibly minute fragments, affecting the given genetic regions of DNA which the specified primers failed to amplify. Figure 6.35e therefore also suggests extensive DNA damage. The concurrent colony count assay did not reveal any growth (zero colonies).

6.7.2.2   Inactivation of Bacillus stratosphericus *when present in fluid (cell suspension)*

Contrary to the plasma treatments of bacteria present on dry surfaces, wet treatment took a relatively longer time to inactivate bacilli but demonstrated a similar pattern. Figure 6.36a shows the responses of *B. stratosphericus* to plasma when present in fluid medium (cell suspension). It appears from Figure 6.36a that the bacteria were viable from 60 s to 120 s of 24 hour post-treatment, and hundreds of microorganisms were live even after 60 s of treatment. The relationship between plasma exposure time and the decrease in number of viable cells was almost linear (Figure 6.36a).

6.7.2.3   Plasma treatment compromises cell membrane integrity over time

Bacterial viability was also assessed by fluorescent dye-labeling of the bacterial cells. The cells whose membrane integrity has been compromised have their genomic DNA stained predominantly red by propidium iodide (otherwise impermeable to cells); healthy cells are predominantly stained green upon taking up a cell permeable dye, SYTO9 (Molecular Probes, Invitrogen). The ratio of live to dead cells can be calculated. Alternatively, a fluorescence microplate reader can be used to capture red and green fluorescence signals to generate real-time live /dead graphs. We did not see any significant difference between these methods; and findings were close to each other with a deviation of $\pm$ 2–5% (data not shown).

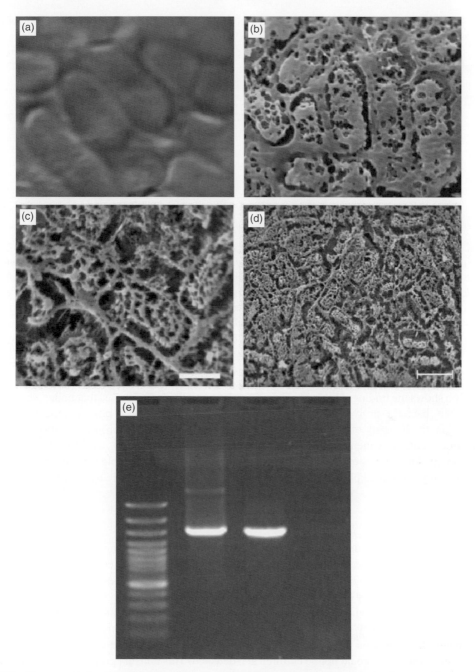

Figure 6.35 *Representative SEM images of plasma treatment of* Bacillus stratosphericus *on dry surface shows etching of bacterial cell envelopes. Such types of changes were visible from 60 s plasma treatment and onwards. (a) Untreated; (b) 60 s plasma-treated and (c, d) 120 s plasma-treated samples are seen. Bar in (c) 1 μm and (d) 2 μm. (e) Agarose gel showing PCR amplified products of DNA isolated from B. stratosphericus. A drastic reduction in the amount of DNA by plasma treatment is observed when surface-dried* B. stratosphericus *is exposed to dielectric barrier discharge plasma for 60 s onwards and compared with untreated (0 s) sample as mentioned under materials and methods.*

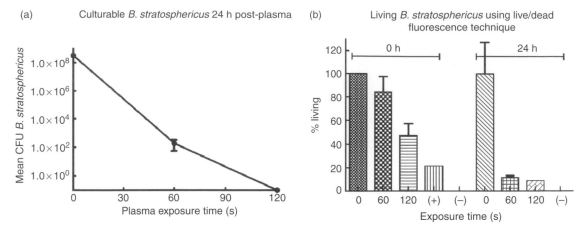

Figure 6.36   *(a) A representative colony count assay showing viable and culturable* Bacillus stratosphericus *as colony-forming units (CFU) 60 s post-plasma treatment under wet environment (suspended cells), with c. 6 log reduction. No colonies were noted at 120 s post-plasma treatment. Bar, standard error mean. (b) A representative live/dead fluorescence assay demonstrating a loss of cell membrane potential and compromise of membrane integrity upon application of plasma over time under wet environment (cell suspension treatments). Although viability of* B. stratosphericus *is reduced in plasma-dose-dependent manner, there are undamaged (viable) cells seen even by the end of 120 s of treatment which could not be detected by culture (colony assay). Cells were either immediately assayed (0 h) or 24 h post-treatment.*

Figure 6.36b shows the findings of live /dead assay and essentially demonstrates that under wet environment (fluid) treatments, bacteria survives even up to 120 s when detected immediately post-treatment. The assay also demonstrates that the integrity of cell membrane of these organisms was compromised substantially and that at 24 h post-treatment, these membrane-associated changes were significant ($p < 0.05$) when compared with 0 h post-treatment. These types of changes indicate that the cells remained viable for longer time (e.g. 120 s), even after relatively prolonged plasma treatments. Colony assay did not reveal any growth (Figure 6.36a).

6.7.2.4   Cells exist in VBNC state

A series of experiments were undertaken to determine whether the increasing amount of plasma-induced stress leads to the death of bacterial cells and subsequent sterilization or cells remain in VBNC state. Figure 6.37 shows the responses of *B. stratosphericus* over plasma exposure times. This XTT assay determines the respiring bacterial cells by measuring an orange-colored metabolic product of XTT; the amount is proportional to the number of viable cells. From Figure 6.37a it appears that, after 60 s of plasma treatment, cells probably undergo latency stage (dormancy) as early as 2 hours post-treatment through >24 hours (variable percent, depending on the amount of plasma energy applied). These small fractions of cells were viable, but upon colony assay could not be cultured and detected. This unusual property is possessed by a very few organisms. This newly reported *B. stratosphericus* species has shown VBNC state upon plasma-induced stress. The 70% ethanol treatment of *B. stratosphericus* cells are seen as built-in control, for comparison. Table 6.3 shows the percent surviving cells and an analysis of a minimum of three sets of experiments in triplicate.

Looking at the extreme narrow range of standard error measurement, the data are convincing. Figure 6.37b and c shows the low-power fluorescence microscopic images, taken concurrently at 2- and 24-hour holdings

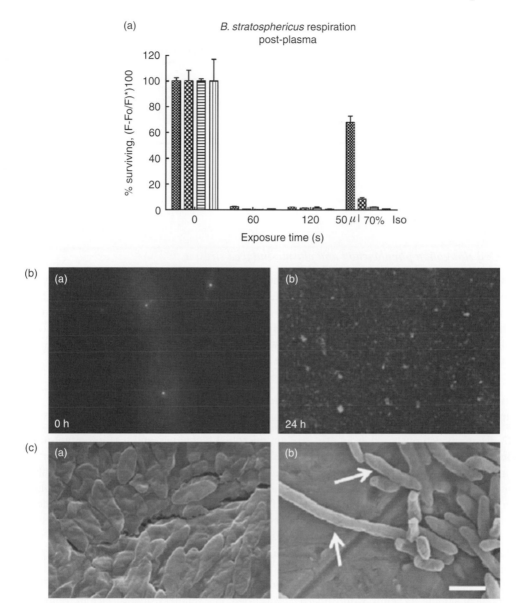

Figure 6.37 *(a) XTT assay was used to detect all culturable as well as viable but non-culturable (VBNC) cells of Bacillus stratosphericus after holding the cells (2, 6, 18 and 24 h) in nutrient medium upon plasma treatment over time. Each set had negative control (reagents without cells) and positive control (70% ethanol treated cells) for comparison. The experiments were repeated in triplicate. About 1 × 10⁸ CFU mL⁻¹ were starting cells (shown as 100%). (Table 1 demonstrates the percent deviations and can be read with this figure). Representative microphotographs of the smear of suspended cell pellet showing respiring B. stratosphericus from few initial survivors of plasma treatment for (b) 120 s after 2 h to (c) increased respiration after 24 h whichstill remained non-culturable. Scanning electron microscopic image showing heterogeneous morphology of B. stratosphericus (d) untreated and (e) after 120 s of plasma treatment under wet environment . Such cellular elongation is found associated with VBNC bacteria. Arrow points divided cells. Bar: 2 μm. XTT: 2,3-bis-(2-methoxy-4-nitro-5-sulfophenyl)-2H-tetrazolium-5-carboxanilide (see color plate).*

Table 6.3   *XTT assay showing respiratory status of the cells of* Bacillus stratosphericus *post-plasma treatment.*

Treatment time (s)	Observation time(post-plasma)			
	2 hour	6 hour	18 hour	24 hour
0	100.0 ± 3.19	100.0 ± 2.65	100.0 ± 11.04	100.0 ± 1.5
60	2.31 ± 0.50	0.45 ± 0.06	0.31 ± 0.04	0.65 ± 0.12
120	0.08 ± 0.13	0.69 ± 0.08	0.67 ± 0.19	0.65 ± 0.13

in nutrient broth (NB) medium, 120 s post-plasma treatment. The image of 24 hour panel (c) exhibited a relatively higher XTT activity, but colony assay did not show any growth of bacteria (plates were incubated and observed for 7 days).

6.7.2.5   *Cell elongation phenomenon and plasma-induced stress*

Various types of cellular stress and variable morphogenesis have been reported in other bacterial species. Since we observed a shift from viable to VBNC status of Bacillus during plasma treatment, we looked at the changes in morphologic features of these bacteria. After 60–120 s plasma treatment under wet environment conditions, we observed that *B. stratosphericus* cells undergo cellular elongation. Figure 6.37d and e is a representative SEM image of such a finding, suggesting that Bacillus also undergoes cellular elongation like that of many other gram-positive and gram-negative organisms during VBNC stage.

6.7.3   Analysis of deactivation/destruction of *Bacillus stratosphericus* due to non-thermal plasma sterilization

FE-DBD plasma is demonstrated as an effective antimicrobial technique, and it is important to understand the bacterial response to its exposure. As described Section 6.7.1, Cooper *et al.* (2010) observed that the DBD normal atmospheric non-thermal plasma treatment of *B. stratosphericus* results in three viability states: viable and culturable at low plasma doses, VBNC bacteria and disintegrated bacteria at higher plasma doses.

Bacteria in the VBNC state retain the ability to perform functions such as respiratory activity, metabolism of incorporation of radio-labeled substrates and cellular elongation. The detection of bacteria in this state requires assays which are independent of culturability. Assessment is therefore made by comparing the responses of bacteria under plasma-induced stress due to culturability, membrane integrity, bacterial morphology and respiration.

It is known that plasma-mediated inactivation mechanisms differ according to the amount of fluid present in the biological sample. When *B. stratosphericus* was dried on a stainless steel surface in room-temperature air and then exposed to DBD plasma, an etching of the cell membrane by charged particles was observed. A flux of a cocktail of neutral or charged particles or photons within the plasma interacts directly with the cell envelope and penetrates the bacteria at numerous locations, thus exposing internal components directly to plasma. In 120 s of plasma treatment, the physical disintegration of bacteria can be clearly observed (Figure 6.35a).

Similar pore formation and disintegration of cell envelope have been observed in the form of a release of cellular components including genomic DNA, resulting in loss of viability of *E. coli* and *Bacillus subtilis* (Hong *et al.* 2009). The resulting physicochemical reaction was so strong that DNA was completely destroyed. In 60 s of plasma treatment, nearly all DNA was disintegrated to the extent that PCR amplification of the remnants was not successful (Figure 6.35b). Under dry environmental conditions, longer treatment of plasma probably led to severe ionic etching of the membrane of *B. stratosphericus*. This resulted in pore formation and may be responsible for leakage of cytosol into the surroundings. Such observations were made earlier (Hong *et al.*, 2009).

In the studies discussed above, all biomaterial was collected from a defined area under treatment. Even if DNA is released from such fenestrated membrane, PCR amplification would not have missed it. Furthermore, in experiments with agarose gel electrophoresis, there was no DNA ladder of fragmented DNA (which was surprising) and PCR amplification was therefore set up. One possibility could be that the power applied to generate plasma was too high and would therefore destroy the DNA within in fraction of seconds. The lack of fragmentation may be explained as a result of DNA destruction to the extent that remaining pieces are too small to be amplified by the PCR technique. The lack of fragmentation may also be partly explained by a linearization of the chromosomal DNA by plasma treatment and the subsequent cleavage of polynucleotides by exonuclease enzymes, thus rapidly degrading the remaining DNA.

Nonetheless, the DNA digestion by either plasma and /or enzymes indicates that DNA is destroyed extensively where it drastically reduces the probability of bacterial viability. The subsequent colony count assay did not show any growth on trypticase soy agar (TSA) plates. It was observed that the multidrug-resistant, high–salinity- and high-temperature-tolerating *B. stratosphericus* loses its culturability in <120 s of plasma treatment (Figure 6. 36a) when treated under wet environment conditions (cell suspension). At this point, it is not uncommon to assume complete sterilization has taken place; with further analysis however, it can be seen that viable *B. stratosphericus* remains.

ROS and other oxidants are generated by plasma, and their lethal effect on bacteria is known. Researchers (Hong *et al.* 2009) demonstrated that *E. coli* and *B. subtilis* spores are inactivated by ROS, and the effect was proportional to the oxygen radical species generated by the atmospheric plasma. ROS generation has been shown to induce the VBNC state via oxidative stress (Gourmelon *et al.* 1994).

In the studies by Cooper *et al.* (2010), the response to hydrogen peroxide treatment (a positive control (+) in Figure 6.36b) and the generated ROS produced by DBD plasma were measured. *B. stratosphericus* was shown to inhibit its ability to culture while maintaining its membrane integrity in a small percent of the cells (Figure 6.36b) and a baseline respiration (Figure 6.37a and Table 6.3). In other studies, inactivation kinetics is found to be plasma-exposure-time-dependent, as well as bacterial-species specific (e.g. *Salmonella*, *Staph. aureus*, *E. coli*, *Bacillus atrophaeus*, *Clostridium botulinum*).

Bacterial species which require a higher exposure of plasma treatment are likely to be exposed to sub-lethal doses, which may lead to a stage of incomplete inactivation. Such exposures may be responsible for the genetic switch from viable to VBNC state. Recent studies show that this genetic mechanism is not present in all microorganisms, but is largely bacterial-species specific. It is likely that *B. stratosphericus* is a highly resistant organism and therefore requires prolonged exposure to plasma treatment; it is also possible that this bacterium possess a genetic switch for VBNC stage under adverse conditions. Complete inactivation and destruction of *B. stratosphericus* cells was observed in 5 min of plasma treatment. A reduction in respiration rate >98% is logical for bacteria which enter a dormant or quiescent lifestyle change. Transfer to growth media is typical to increase respiration in such microorganisms and is also observed in our case.

Further observations revealed that the bacteria producing the baseline signal (residual) after plasma treatment represent a small fraction of surviving bacteria when compared to 24 hour post-treatment condition, which showed about 8 times more bacteria were respiring (Table 6.3). The later were viable but remained non-culturable. This increase in respiration gives rise to the question of whether this viability is reversible or irreversible. It can be assumed that this change in the level of respiration may have resulted from incubation in lysogeny broth (LB) medium (i.e. in nutrients) for 24 hours post-plasma treatment, as it is known to enhance respiration of bacteria that has experienced an external stressor. Correlation of the culturability with membrane integrity and respiration activity is indicative of the VBNC state. The observation in cellular morphology shows that the *B. stratosphericus* vegetative cells are also able to elongate (Figure 6.37d and e), an additional indication of VBNC bacteria. Upon plasma treatment, *Bacillus stratosphericus* revealed a heterogeneous population comprising predominantly elongated larger cells and normal cells. The cells were overall more flattened with negligible visible binary fission. Such features were found to be associated with the VBNC stage of both Gram-positive and Gram-negative bacteria. This may also suggest that a small portion of cells may be able to undergo cell division in later life.

It is evident that such VBNC bacteria are able to maintain their antibiotic resistance markers and, during restoration of classical progeny division, they continue to express the resistance trait and therefore represent an additional risk to human health. In either case, a complete inactivation of such resistant bacteria using appropriately higher and optimized plasma doses is advisable to minimize potential future threat.

6.7.4   Bottom line for plasma physicists: Plasma sterilization can lead to VBNC state of microorganisms

An exposure to DBD plasma at otherwise 'lethal' doses induces a VBNC state in *B. stratosphericus* cells which are not completely inactivated /disintegrated by such treatment (Cooper *et al.*, 2010). It is possible that such doses might lead to the activation of a genetic switch from viable to VBNC state. The presence of VBNC cells poses a major public health hazard. These cells cannot be detected by traditional culture methods, and the cells may remain potentially pathogenic under favorable conditions. It can be hypothesized that the ROS produced by plasma may be inducing this state. The mechanisms of inactivation in *B. stratosphericus* may also be different under dry (directly ionic interactions with cell envelop) and wet environment conditions (probably via ROS). To ensure the death of such bacteria, a relatively longer plasma treatment time is advisable.

When plasma physicists involved in plasma sterilization research ask the 'tricky' question – when microorganisms are not culturable (not proliferating, not multiplying) after sufficient plasma treatment, is that possible that they are not dead but only 'not culturable'? – the answer is probably 'yes'. After insufficient plasma treatment, organisms can remain in the VBNC state.

7

Plasma Decontamination of Water and Air Streams

Plasma treatment of water and air attracts significant interest focused both on biological sterilization of gas and liquid flows as well as on their chemical cleaning, which is also closely related to health issues. The recent outbreaks of food-borne pathogens, environmental protection challenges as well as the growing threat of bioterrorism have highlighted the need for effective methods to decontaminate and sterilize not only different surfaces but also gases (especially air) and liquids (especially water including potable, waste as well as sea water). These problems are discussed in this chapter, beginning with decontamination and sterilization of atmospheric air.

7.1 Non-thermal plasma sterilization of air streams

7.1.1 Direct sterilization versus application of filters

In contrast to plasma-based surface sterilization (Chapter 6) and water sterilization (Section 7.5), only a few plasma researchers have been focused on air decontamination using non-thermal plasma systems. Most of them have only been successful in the case of coupling the plasma technology with high-efficiency particulate air (HEPA) filters to both trap and kill microorganisms (Jaisinghani, 1999; Gadri, 2000; Kelly-Wintenberg, 2000). Usually such coupling of filtration and plasma systems means trapping the microorganisms with the filter, followed by sterilization of the filter using non-thermal plasma discharges. The downside of relying on HEPA filters is that they have limited efficiency at trapping submicron-sized airborne microorganisms and they also cause significant pressure losses in heating, ventilation and air conditioning (HVAC) systems, giving rise to higher energy and maintenance costs.

A relatively large-scale experimental facility, named the Pathogen Detection and Remediation Facility (PDRF; Figure 7.1), was designed by Gallagher *et al.* (2005, 2007) and built in the Drexel Plasma Institute to perform air decontamination experiments using a dielectric barrier grating discharge (DBGD). The PDRF combines a plasma device with a laboratory-scale ventilation system with bioaerosol sampling capabilities. Experiments with PDRF show that direct contact of the bioaerosol and the DBGD plasma for a very short duration can cause an approximate 2-log reduction (97%) in culturable *Escherichia coli* with a c. 5-log reduction (99.999%) measured in the 2 min following exposure. Fast treatment times within plasma are due

Plasma Medicine, First Edition. Alexander Fridman and Gary Friedman.
© 2013 John Wiley & Sons, Ltd. Published 2013 by John Wiley & Sons, Ltd.

Figure 7.1   *Pathogen Detection and Remediation Facility (PDRF), general view (see color plate).*

to a high airflow rate, the high velocity of the bioaerosol particles in flight and a small discharge length of the DBGD device, which results in a residence time of treatment of c. 1 ms. Direct plasma treatment therefore leads to extremely fast and effective deactivation of the airborne bacteria. Analysis of rapid mechanisms and kinetics of the fast sterilization processes in air are discussed in Section 7.1.6.

7.1.2   Pathogen detection and remediation facility

The PDRF system is a bioaerosol treatment facility which provides a recirculating airflow environment for the DBD air decontamination experiments (Gallagher *et al.*, 2005, 2007). The bioaerosol can be treated with repeated passes through the same plasma discharge. The sealed recirculating system allows for complete control over relative humidity (RH), which is important because even small fluctuations in RH have been shown to significantly decrease the survivability of airborne bacteria. The PDRF is a plug flow reactor where air flow is turbulent so that radial variation of the bacterial concentration is minimized. An illustration of the PDRF is provided by Figure 7.2.

The PDRF system has a large total volume of 250 L and operates at high airflow rates ($\geq$25 L s^{-1}), typical of indoor ventilation systems. The system has an inlet with attached collison nebulizer for bioaerosol generation and two air-sampling ports connected to a vacuum air-sampling system. The system also has a large-volume barrel that contains a series of aluminum baffle plates and a variable-speed centrifugal blower motor that drives the air through the DBGD treatment chamber. The PDRF system residence time, that is, the time for one bioaerosol particle to make one complete revolution through the whole system, is c. 10 s.

An important element of the PDRF is the *air sampling system*. To separate the decontamination effect of direct exposure to DBD plasma from the remote exposure of ozone and other long-lived chemical species that can interact with bioaerosols downstream of the discharge, a sampling method was devised by Gallagher *et al.* (2005, 2007) so that air samples are taken just before and after bioaerosol passes through the discharge

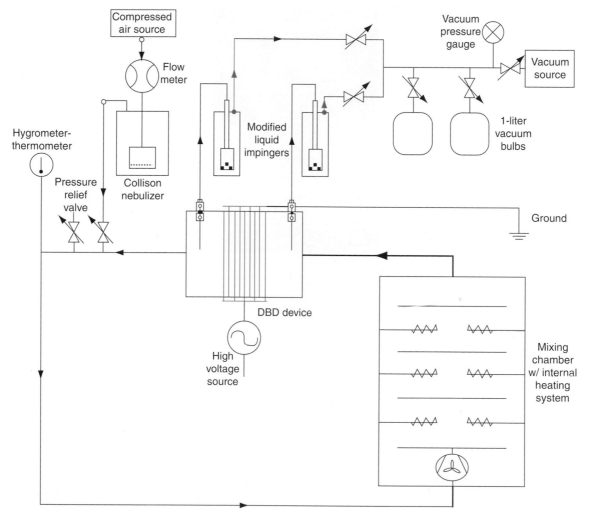

Figure 7.2 *General schematic of the PDRF.*

area. As there are only two sample ports, each set of two air samples measures the change in viability of bacteria on a 'per pass' basis through the discharge. For each of the subsequent sets of air samples, the sample taken 'before plasma' can give a measurement of the change in viability due to the effect of ozone from the previous 'after plasma' sample.

Liquid impingement was the chosen air-sampling method because it minimizes desiccation stress on the bacteria by directly depositing them into a buffered saline solution. Liquid impingers operate by drawing a sample of air through an inlet tube submerged in a solution, thereby causing the air stream to strike the liquid bed, trapping aerosols in the solution through forces of inertia. The Ace Glass Inc. AGI-30 is the most commonly used liquid impinger which contains a critical orifice that limits the maximum air sampling rate to 12.5 L min^{-1}. The PDRF vacuum air-sampling system was designed to sample as large a volume of air as possible (c. 1 L) in the shortest period of time (c. 1 s) in order to avoid disturbing the flow inside the system.

Figure 7.3   *PDRF liquid impinger for bioaerosol sampling (see color plate).*

To accommodate this high air-sampling rate, the AGI-30 impinger was modified by replacing the standard critical orifice with a hollow tip with several jet ports. Figure 7.3 shows an image of the modified air samplers.

The overall efficiency of this type of sampler is in the range $6 \pm 3\%$. To be injected into the PDRF, the bacterial culture was placed into a BGI 24-jet collison nebulizer operating at 40 psi for a period of 45 s (nebulizing rate: 1.1 mL min^{-1}). The DBGD device was then switched on for a period of 10 s and during this time the first two air samples were taken in sequential order: before and after passing through plasma. The discharge time of 10 s is used because this time is just enough to take two samples and is equal to the time of the bioaerosol to make one complete revolution in the PDRF system. Measurement of the decontamination effectiveness of the DBGD is therefore made on a per-pass basis, ensuring that each bioaerosol particle has been treated once by the discharge.

The next set of air samples are taken c. 2 min later; this is the amount of time required to remove and replace the air samplers with the next set of pre-sterilized samplers. This process is repeated until the typical number of air samples, usually 6, is achieved. Each of the pre-sterilized air samplers is initially filled with 30 mL of sterile PBS solution and after sampling, each solution is transferred to a sterile 50 mL centrifuge tube for assaying. The concentration of *E. coli* in phosphate-buffered saline (PBS) solution used for nebulization is on average 10^8 bacteria mL^{-1}.

A detailed balance of the plasma-treated airborne bacteria was achieved in the experiments of Galagher *et al.* (2005, 2007) using the flow cytometry. The flow-cytometric measurements were made using fluorescence-activated cell sorting (FACS) calibur flow cytometer with 488 nm excitation from an argon ion laser at 15 mW. Fluorochromes with a high affinity for nucleic acid SYBR Green I and propidium iodide, PI (Molecular Probes) were used for flow cytometry. The SYBR Green I, a green fluorescent nucleic acid stain, has been shown to stain living and dead Gram-positive and Gram-negative bacteria (Porter *et al.*, 1993). PI is a red fluorescent dye that intercalates with dsDNA and only enters permeabilized disintegrated cyto-plasmic membranes (Gregory *et al.*, 2001).

Figure 7.4   *PDRF dielectric barrier grating discharge (DBGD) with air sterilization chamber (see color plate).*

7.1.3   The dielectric barrier grating discharge (DBGD) applied in the PDRF

A special DBD configuration referred to as the dielectric barrier grating discharge or DBGD has been applied in the PDRF to provide effective direct sterilization of air flow. The DBGD consists of a thin plane of wires with equally spaced air gaps of 1.5 mm. The high-voltage electrodes are 1-mm-diameter copper wire shielded with a quartz capillary dielectric that has an approximate wall thickness of 0.5 mm. The total area of the DBGD discharge including electrodes is 214.5 cm^2 and without electrodes is 91.5 cm^2.

Figure 7.4 shows an image of the DBGD device which has two air sample ports located at a distance of 10 cm from each side of the discharge area so that bioaerosol can be sampled immediately before and after it enters the plasma discharge. When the PDRF system is operated at a flow rate of 25 L s^{-1}, the air velocity inside the DBGD discharge chamber is 2.74 m s^{-1} and the residence time of treatment, that is, the duration of one bioaerosol particle (containing one *E. coli* bacterium) passing through the DBGD device is approximately 0.73 ms.

The DBGD device is operated using a quasi-pulsed power supply that delivers a distinct sinusoidal current-voltage waveform with a very fast rise time that nearly simulates a true square-wave pulse. The period between pulses is approximately 600 µs, peak-to-peak voltage is 28 kV and pulsed current is nearly 50 A (peak-to-peak value). The average power of the discharge is approximately 330 W and, considering the discharge area of 91 cm^2, the power density is 3.6 W cm^{-2}. The majority of power is discharged in the very short duration of the pulse itself, which has a period of 77 µs and average pulse power of 2600 W. Since the residence time of a bioaerosol particle passing through the discharge area is 0.73 ms and the period between pulses is 0.6 ms, each bioaerosol particle that passes through the DBGD device experiences about 1 pulse of DBD discharge power. The typical concentration of bioaerosol in an experiment is approximately 5 × 10^5 bacteria per liter of air, which translates to approximately 9 × 10^3 bacteria within the cross-section of discharge area at any given time (in each 2 mm wide cross-section of flow passing through the DBGD).

7.1.4   Rapid and direct plasma deactivation of airborne bacteria in the PDRF

The culture test results from four replicate trials of DBGD-treated *E. coli* bioaerosol in the PDRF system are shown in Figure 7.5 (Gallagher *et al.*, 2005, 2007). Control experiments show a small but negligible change in the surviving fraction of *E. coli* over the total experimental period of 6 min. In the DBGD-treated trials, an approximate 1.5-log reduction (97%) in the surviving fraction of *E. coli* was measured between samples 1

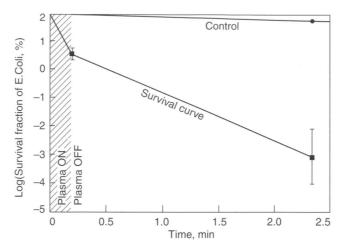

Figure 7.5 *Culture test results of DBD-treated* E. coli *in PDRF system in four replicate trials. The gray shaded areas indicate the time when the DBD discharge was ignited. Control experiments are also shown.*

and 2 (before and after plasma treatment). It is important to note that the plasma was only ignited in a short period during each set of samples, which is noted by the gray shaded areas on Figure 7.5.

The second decrease in the surviving fraction of *E. coli* is shown between samples 2 and 3, which occurred during the time between plasma treatments. In this second decline, the rest of culturable bacteria (3% from initial number) decreased by an additional 99.95% (3.5 logs) when the plasma discharge was switched off. Additionally, samples 4, 5 and 6 taken after second (sample 4) before and after third 'pass' (samples 5, 6) through plasma shown no culturable *E. coli* in the probes (results not shown). This means that, after the second and third passes through plasma, the number of culturable *E. coli* was less than the detection limit ($10^{-4}\%$).

Flow cytometry has been employed to detect the presence *E. coli* in each of the 6 air samples taken during experiments. Where colony-counting techniques are limited to detecting only culturable (i.e. visibly growing) bacteria, flow cytometry is capable of detecting the physical presence of bacteria in a sample regardless of culturability. Flow cytometry analysis of air samples taken during the trials indicates that the total number of bacteria (both active and inactive) remains almost constant; the DBGD device is therefore not acting as an electrostatic precipitator and the concentration of bioaerosol particles remains undisturbed for the duration of each experiment. Flow cytometry results also showed that bacterial outer membranes of *E. coli* were not disintegrated after up to three passes of direct exposure in the DBGD plasma device. However, culture test results demonstrated a 97% reduction in culturable *E. coli* with a millisecond exposure time in DBGD plasma (one pass through discharge) and a subsequent 3.5-log reduction in the 2 min following treatment.

The direct plasma exposure time of 0.73 ms (per pass) allows enough time for bacteria to be attacked by all chemically active components of plasma: charged particles, UV radiation, OH radicals, atomic oxygen and ozone, which is one explanation for the initial 97% reduction in culturability. Subsequent remote exposure to ozone in the 2 min following direct plasma treatment, as well as the high sensitivity of airborne bacteria to slight atmospheric changes, may account for the remaining 3.5-log reduction.

Direct plasma treatment of airflow therefore permits its rapid sterilization (much faster than corresponding indirect plasma treatment), similarly to the case of surface sterilization. This conclusion has been supported by special experiments with the PDRF where a small number of the DBGD wires were eliminated (see Section 7.2.2). Some relatively small fraction of the air stream was not treated directly in plasma in this case but only indirectly, which results in a much higher number of bacteria surviving the treatment or, in other words, a significant decrease in the sterilization efficiency.

7.1.5 Phenomenological kinetic model of non-thermal plasma sterilization of air streams

Sterilization of airborne microorganisms by plasma-generated UV and active chemical species (atoms, radicals, excited and charged particles as well as active stable molecules such as O_3 etc.) can be described in terms of conventional gas-phase chemical kinetics. The sterilization process is considered in this case as an elementary reaction, which occurs with some probability during a 'collision' of the microorganisms and relevant plasma-generated active specie. Rate and effectiveness of the 'elementary sterilization reaction' can be described in such a chemical kinetic approach in terms of conventional reaction rate coefficients.

Such an approach to plasma sterilization kinetics, proposed by Gangoli *et al.* (2005), is obviously a phenomenological one because it cannot describe detailed biochemical pathways of very complicated processes of deactivation of microorganisms. However, the phenomenological plasma sterilization kinetics permit generalization of the experimental kinetic data obtained in different discharge systems (with different concentrations of different relevant active plasma components) applied to deactivation of different types of microorganisms. Relevant reaction rate coefficients for plasma sterilization (initially focused only on ozone, OH-radicals and UV) were calculated by Gangoli *et al.* (2005) through analysis of a large amount of empirical data (available in the food and water decontamination literature). These data resulted from many experiments in which ozone and UV radiation were used to destroy a variety of bacteria, viruses and spores in various liquids and surfaces (Broadwater, 1973; Burelson, 1975; Khadre, Yousef, and Kim *et al.*, 2001). The reaction rate coefficients k were calculated assuming second kinetic order of the microorganism deactivation:

$$\frac{d[M]}{dt} = -k[M][A] \tag{7.1}$$

where $d[M]/dt$ is the change of concentration of viable microorganisms with time, $[M]$ is the concentration of microorganisms and $[A]$ is the concentration of active chemical species A. The reaction rate coefficient for interaction of A with the microorganism, k_A, is defined in this case as:

$$k_A = \frac{\ln\left(\frac{1}{S}\right)}{[A]t}, \tag{7.2}$$

where S is the surviving fraction of microorganism M and t is treatment time. Using the same approach, the model accounts for the destructive effects of the plasma-generated UV on each individual microorganism by assigning the reaction rate constants k_{UV}. Taking into account that the density of active species $[A]$ in Equations (7.1) and (7.2) are replaced in the case of UV radiation to energy flux $[J]$ ($\mu J\ s^{-1}\ cm^{-2}$), in the case of UV radiation the sterilization reaction rate coefficient is not measured in $cm^3\ s^{-1}$ (which is conventional for the second-order reactions) but in $cm^2\ \mu J^{-1}$ (see Table 7.1 which lists all these rate coefficients).

Table 7.1 *Phenomenological reaction rate coefficients for the non-equilibrium plasma sterilization of bacteria* (E. coli) *using ozone* O_3, *hydroxyl radicals OH and UV radiation.*

Ozone (O_3)	1.5×10^{-16} ($cm^3\ s^{-1}$)
Hydroxyl radical (OH)	3.6×10^{-13} ($cm^3\ s^{-1}$)
UV radiation	3.8×10^{-3} ($cm^2\ \mu J^{-1}$)

Regarding hydroxyl radicals, Von Gunten (2003) showed that the inactivation rate of pathogens by OH radicals in their system is c. 10^6–10^9 times faster than that of O_3. The interaction between OH radicals and *E. coli* were recently studied by Cho *et al.* (2004). It was found that the *CT* value (product of characteristic concentration *C* and time of exposure *T*) corresponding to a 2-log reduction of *E. coli* inactivation by OH is approximately 0.8×10^{-5} mg min L^{-1}. In comparison, the O_3 *E. coli* interaction *CT* value (4×10^{-2} mg min L^{-1}) is higher. Summarizing experimental data, for the phenomenological kinetic analysis it can be taken that the inactivation ability of OH exceeds that of O_3 by a factor of 10^3–10^4 (Gangoli *et al.*, 2005). Table 7.1 summarizes the phenomenological reaction rate coefficients for the inactivation of *E. coli* by ozone O_3, hydroxyl OH and UV radiation. Application of the phenomenological kinetic model to analysis of air sterilization PDRF is considered in the following section.

7.1.6   Kinetics and mechanisms of rapid plasma deactivation of airborne bacteria at the PDRF

In the framework of the above-introduced kinetic model of the plasma sterilization, the following formula describes the rate of inactivation of airborne microorganisms inside a DBD discharge from the combination of effects of ozone, hydroxyl and UV radiation:

$$\frac{\mathrm{d}[M]}{\mathrm{d}t} = -k_{O_3}[M][O_3] - k_{UV}[M][I] - k_{OH}[M][OH] \tag{7.3}$$

where $[M]$ is the concentration of microorganisms; $[O_3]$ and $[OH]$ are concentrations of the active chemical species; $[I]$ is energy flux of UV radiation; and k_{O3}, k_{OH} and k_{UV} are corresponding reaction rate coefficients (see Table 7.1).

Since the DBD is operated in air, the typical concentration of ozone is of the order 100 ppm, which corresponds to c. 2.7×10^{15} cm^{-3} at standard conditions. The G-factor for OH formation (number of OH radicals produced per 100 eV energy of the discharge) is c. 0.5 (Penetrante *et al.*, 1997), which corresponds to OH concentration of c. 15 ppm (4×10^{14} cm^{-3}) under PDRF conditions (DBD power 330 W, 25 L s^{-1} flow rate). The UV radiation intensity in the wavelength range 200–300 nm is below 50 μW cm^{-2}. To be consistent with the transient nature of the PDRF experiments, the changes in colony-forming unit (CFU) numbers of *E. coli* for a given volume (cm^3) are tracked as polluted air travels through the experimental system.

Three zones are considered based on differential interaction modes of *E. coli* (shown in Figure 7.6): (1) direct plasma interaction (DPI); (2) post-plasma interaction (PPI); and (3) interaction with only O_3 (IO). During DPI, the concentrations of O_3, OH and the UV intensity are assumed to be at a constant steady-state value. DPI only lasts for c. 1 ms due to the HVAC conditions of operation (25 L s^{-1}). The *E. coli* then goes through PPI, where it interacts with simultaneously recombining O_3 and OH species and decreasing UV intensity (absorption of UV radiation by ozone, cross-section c. 10^{-17} cm^2; Fix *et al.*, 2002). Ozone and OH radical deactivation reactions considered in the PPI simulation are limited to the following:

$$\mathrm{OH} + \mathrm{O_3} \rightarrow \mathrm{O_2} + \mathrm{HO_2}, \quad \mathrm{OH} + \mathrm{OH} + M \rightarrow \mathrm{H_2O_2} + M, \quad \mathrm{O_3} + M \rightarrow \mathrm{O_2} + \mathrm{O} + M. \tag{7.4}$$

According to the phenomenological kinetic modeling, OH radical interaction time with pathogens is c. 600 μs outside plasma which is reasonable when compared to Ono and Oda (2001). The IO part of the model is used to describe the situation when plasma is turned 'off' in experiments. Since there is no production of new OH or UV during IO, the *E. coli* only interacts with the remnant ozone for the duration of 2 min (the time between sets of air samples taken during experiments).

Keeping in mind the above considerations, the predicted variation in concentration of active species outside plasma with length from the discharge zone is shown in Figure 7.6. The simulation results depicted in

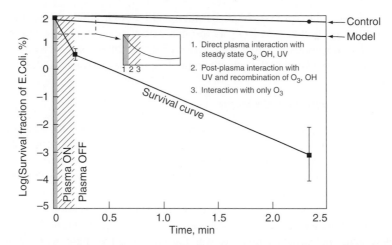

Figure 7.6   *Discrepancy between the modeling and experimental results due to consideration of only a few active agents from plasma. Modeling regions are divided into 1: in plasma, where* E. coli *encounters steady-state O_3, OH and UV concentrations and intensities; 2: post-plasma interaction, where* E. coli *interacts with the above three species in the downstream flow as they recombine (O_3 and OH) or go through extinction (UV); 3:* E. coli *interacts only with long-living O_3.*

Figure 7.6 predict a 14% decrease in viable *E. coli* during one pass of direct contact time with DBGD plasma. However, during further interaction with ozone (outside the DBGD plasma zone) through the re-circulating airflow system, a total of 36% inactivation is predicted for a treatment time of c. 10 s (DPI: 1 ms; PPI: 10 s; shaded area in Figure 7.6).

This result is obviously not in agreement with the PDRF experiments, where a 97% decrease was achieved during the same times and modes of treatment (see above). The disagreement is due to the fact that the model only considers the effect of OH, UV radiation and O_3 as the primary active components of plasma, when it is known that there are additional active species present (especially charged particles which make a significant contribution to direct plasma treatment of microorganisms). Considering only UV radiation and the most aggressive plasma neutral species OH and O_3 cannot explain the initial 1.5-log reduction during 1 ms treatment in DBGD plasma and subsequent 4-log reduction after 2 min of treatment by remnant species, as was observed experimentally (Gangoli *et al.*, 2005). Active species, such as charged species (ions and electrons) and other radicals (atomic oxygen and nitric oxides), are contributing to the inactivation process and may account for the remainder of the observed bactericidal activity of DBGD plasma.

7.2   Direct and indirect effects in non-thermal plasma deactivation of airborne bacteria

7.2.1   Major sterilization factors

7.2.1.1   *Reactive oxygen species*

Reactive oxygen species (ROS) are a well-known sterilization agent (Fang, 2004) produced in abundance in air plasma (Zheng *et al.*, 2009). ROS, such as hydroxyl radical, molecular and atomic oxygen ions (e.g. superoxide anion and singlet oxygen) and ozone, have significant bactericidal properties (Beckman and Koppenol, 1996). The sterilization effect produced by plasma has been discussed in detail and theories have been put forward to explain the cause and the mechanism of sterilization (Kelly-Wintenberg *et al.*, 1998;

Lerouge, 2001; Lu, Leipold, and Laroussi, 2003; Laroussi and Leipold, 2004). Most studies that have used DBD discharge point in the direction of ROS as the main sterilizing agent. Plasma discharge produces a mixture of species that are highly reactive with various biomolecules as well as able to produce other bio-reactive species due to interaction with each other (Lu, Leipold and Laroussi, 2003).

7.2.1.2   *Role of ozone in inactivation of bacteria*

Ozone has been widely used commercially as a sterilizing agent (Karlson, 1989). The only method used to generate ozone is plasma discharge (usually DBD) (Kogelschatz, 2003a, b). Ozone is long-living species and is predominantly present in the atmosphere after plasma discharge. The sterilization effect by ozone is a complex phenomenon. Ozone reacts with almost every major component of the bacterial cell and its effects have been observed on the cell membrane, DNA, and RNA (Komanapalli and Lau, 1996). Scanning electron microscope (SEM) images of *E. coli* treated with ozone concentration of 0.167 mg min^{-1} L^{-1} for 30–60 min have shown major damage to surface morphology and membrane lysis. Ozone was shown to react with nucleic acids *in vitro* and thymine was the most reactive base with ozone. Ozone acts as a general protoplasmic oxidant. Kowalski, Bahnfleth and Whittam (1998) investigated the effects of high concentration of airborne ozone on *E. coli* and *Staphylococcus aureus*. In their study, 30–1500 ppm of ozone was pumped into a chamber containing Petri dishes with bacteria on their surface. It was observed that for high concentrations of ozone there was a 4-log reduction in viable bacteria counts after 480 s of exposure. This would suggest that the action of ozone on bacteria is slow in comparison with direct plasma treatment (Fridman *et al.*, 2007a–d; Gallagher, 2007).

7.2.1.3   *Role of direct exposure to plasma on inactivation*

Microorganisms may be treated by plasma in two ways: (1) where the organism comes in direct contact with the whole contents of plasma (charged species, UV and VUV, long and short lived species, etc.); or (2) where the organism is separated from plasma and only the long-lived plasma-generated species and/or UV and VUV light are used for treatment. We refer to method (1) as *direct* treatment and method (2) as *indirect* treatment. For the case of air sterilization by plasma, direct exposure would mean that bacteria in air are flown through the plasma discharge. In the case of indirect air sterilization, the bacteria are not flown through plasma. The reactive species that have a shorter lifetime recombine with each other or may be deactivated before they can affect the bacteria (Laroussi, 2005). Fridman *et al.* (2007a–d) showed that direct exposure is more effective in bacteria inactivation during surface treatment. Investigating the effects of direct exposure will help us determine if the short-lived species, such as ions, electrons and ROS, are responsible for the inactivation.

7.2.2   PDRF: experimental procedure

The flow inside the PDRF system is not interrupted during the plasma treatment or the sampling procedure. The flow rate is maintained at 25 L s^{-1} so the entire volume is circulated within 10 s; for this reason, the plasma treatment procedure consisted of turning the plasma discharge on for 10 s to treat all air in the chamber. The sampling time points were kept constant over all experiments. All experiments were performed with the same initial conditions and the only parameter that was changed was the type of treatment:

1. direct plasma exposure as described before;
2. 75% direct exposure (where 75% of bacteria pass through plasma and 25% do not);
3. indirect plasma exposure: treatment by ozone (injection of the same amount of ozone as produced by plasma).

Figure 7.7   *The plasma unit with its multiple-electrode configuration (left). The same plasma unit, with the plasma discharge initiated (right). Note the screen of plasma covering the entire cross-section of air passage (see color plate).*

7.2.2.1   *Direct exposure*

The plasma discharge setup consists of 21 high-voltage wire electrodes insulated by quartz and 22 grounded wires. When the discharge is initiated, DBD plasma is produced in the air gaps between the electrodes and grounded wires. This creates a screen of plasma that bacteria have to pass through. This arrangement of the plasma discharge is shown in Figure 7.7.

7.2.2.2   *Indirect (75% direct) exposure*

The two methods of investigating the effect of direct exposure to plasma are as follows.

1. Introducing a barrier between the plasma and the sample to be treated, thus removing the influence of the ions and reactive oxygen species produced by plasma. This method is not possible for these experiments as the flow of air is perpendicular to the electrodes and any obstruction will lead to changes in airflow and a pressure drop will be introduced inside the system.
2. Reducing the total area of direct exposure and letting a certain percentage of the sample (in this case, the bioaerosol) be treated indirectly by the long-living species such as ozone. In the DBGD setup this can be achieved by reducing the number of active electrodes, creating gaps in the screen produced by plasma. This way, a certain percentage of bacteria are prevented from coming in contact with the plasma and are treated by plasma indirectly.

To understand the influence of direct and indirect plasma exposure of bacteria, every fourth high-voltage electrode in the DBGD discharge was removed and the plasma discharge initiated. This discharge, shown in Figure 7.8, occurs across the air gaps between the wires. Since the area of the discharge is 75% of the total cross-sectional area, this can be termed as a 75% direct and 25% indirect exposure. In this case, as the plasma discharge is away from the path of 25% of the bacteria flow, it can be considered indirect treatment for the area where there is no discharge. The selection of the electrodes to be removed was made in such a way that the indirect treatment is distributed across the cross-section of the airflow. The pulsed voltage input to the plasma discharge is 28 kV and the current was measured to be 50 A (peak-to-peak value). The total power dissipated in the plasma was 100 W.

Figure 7.8   *The plasma unit with every fourth high-voltage electrode wire removed (left). The quartz tubes are retained in order to maintain the same airflow. This produces zones of indirect exposure (A). The image shown on the bottom is with the plasma (see color plate).*

7.2.2.3   Indirect exposure (ozone treatment)

The major long-living species created by DBD plasma in volume is ozone. To isolate the effect of plasma-generated ozone on the airborne bacteria, we produced ozone elsewhere and injected it into the chamber. For this, ozone concentration generated by DBGD inside the system was measured, total ozone production was calculated and then the ozone generator (MedOzone, Russia) was adjusted to produce the same amount of ozone. The bacteria were introduced into the system through the process of nebulization as usual. A pre-treatment air sample was taken and the ozone generator was switched on for 10 s, the same time as plasma treatment. The ozone was allowed to pass through the experimental system and a post-ozone-treatment sample was taken. The time of sampling was kept the same as that for the plasma experiment. Further samples were taken before and after ozone exposure.

7.2.2.4   Growth and preparation of bacterial strains for nebulization

The microorganism used in our studies was *Escherichia coli* K12 substr. MG1655. Strains were prepared as frozen stocks. The frozen stock was transferred to 10 mL culture tube containing Luria Bertani (LB) media. The culture was grown overnight in an incubator shaker at 37°C. The culture was then transferred to centrifuge tubes and spun at 3500 rpm for 1 minute. The supernatant was removed and the pellet was again washed with deionized water. The final solution was prepared by adding the bacterial pellet to 30 mL of deionized water.

7.2.2.5   Injection of bacteria and air sampling

A 24 jet nebulizer (BGI Inc., Waltham MA) was connected to the system. Deionized water was added to the nebulizer and the nebulizer was operated at 40 psi input pressure. This injection was performed to increase the humidity inside the system. The humidity was increased to 70% RH and the nebulizer was disconnected. The bacterial solution was then added to the nebulizer and it was connected back to the system. The nebulizer was run again at 40 psi for 45 s. The nebulizer was then disconnected and removed. Sampling of the air inside the system was performed using specially modified AGI impingers. A negative air pressure system was used to acquire the samples from the uninterrupted circular flow system.

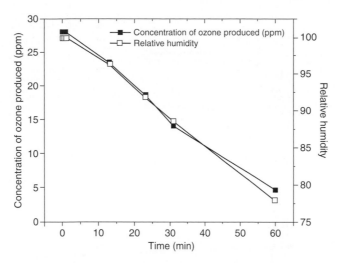

Figure 7.9   *Evolution of ozone generated by DBGD discharge and relative humidity inside the PDRF system.*

7.2.2.6   *Analysis of surviving cells*

The samples were taken in 1X PBS solution. Each sample was then diluted using the serial dilution method. The dilutions were then plated on brain-heart infusion (BHI) broth agar plates. These plates were incubated overnight at 37°C inside an aerobic incubator. The number of colonies growing on the plates was counted the next day to determine the number of bacteria present in the sample.

7.2.3   PDRF: experimental results

The concentration of ozone produced by plasma was measured using an ozone meter. The plasma discharge was initiated and run for 10 s and the concentration of generated ozone was observed to be 28 ppm. A separate ozone generator was employed for producing ozone. It has an intake for air and outlet for the ozone generated. It was observed that at 0.5 SLPM, the amount of ozone generated inside the system by the generator is the same as generated by DBD plasma for 10 s, that is, 28 ppm (Figure 7.9).

Figure 7.10 shows the summary of the experimental results regarding the direct and indirect exposure. 100% direct exposure to plasma leads to the greatest degree of inactivation. A 25% drop in the direct exposure leads to inactivation to drop to 29% from the 97% observed for 100% direct exposure experiments. Inactivation experiments with pure ozone produced the least degree of inactivation. The 10 s exposure to ozone resulted in only 10% inactivation of airborne *E. coli*. Pure ozone failed to produce complete inactivation by the time the next sample was taken. The third sample represents the pre-treatment sample for the second 10 s treatment (second pass through plasma). No viable *E. coli* were detected in either plasma treatments by this time, signifying the clear superiority of plasma exposure over pure ozone.

Two major factors affecting the sterilization of air using plasma were investigated: direct plasma treatment and ozone treatment. On the one hand, it is known that ozone is a relatively slow sterilization agent. On the other hand, experiments shows significant sterilization effect during the time when plasma is off, and ozone is probably the only active agent. If ozone was indeed the major inactivating agent in plasma, the same dosage of pure ozone would produce the same inactivation as seen with plasma discharge. The experiments with ozone injection have confirmed that ozone alone has slow damaging action. To determine the effect of direct exposure and the ions and charges associated with it, the total direct exposure was reduced to 75%

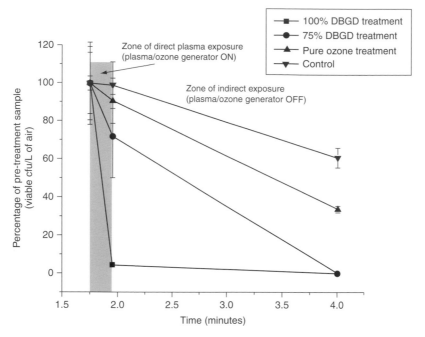

Figure 7.10   *Experiment results. The dark shaded region denotes the first 10-s-treatment with 100% plasma (■), 75% plasma (●)/ozone (▲) as compared to the control runs with no plasma/ozone (▼). The error bars represent the standard deviation of mean of 3 trials. Timescale is from the instance of injection of bioaerosol. The third sample is a pre-treatment sample for the second treatment.*

and sterilization experiments were performed. The results indicate that the inactivation dropped from 97% to 29% percent. In other words, a small reduction in the direct exposure results in a much larger reduction in the inactivation immediately after plasma exposure, so the effect of direct exposure has significant non-linearity.

The simplest explanation of this non-linearity is synergism between ozone and direct plasma exposure in bacteria inactivation. In addition, our assumption that 25% reduction in plasma area means that 25% of bacteria are not subject to direct plasma exposure is probably an oversimplification. Since all bacteria were inactive with 75% treatment after 4 min, this probably meant that all bacteria received a dosage of direct treatment, for example, in the form of UV radiation.

The direct exposure to plasma disturbs the bacteria membrane and the charges stick to the membrane. While complete membrane breakdown requires a field of several kV cm^{-1} and longer time periods (Sugar and Neumann, 1984), we know that charge absorption leads to pores opening much faster (in millisecond and tens of microseconds time range; Schoenbach *et al.*, 2000). Follow-up 2-minute ozone action on bacteria with disturbed membranes provides complete sterilization. Investigation of the influence of direct exposure shows that there is 3.5-log reduction during the much longer post-plasma exposure. This means that after the initial 97% reduction, remaining bacteria keep flowing through the system when ozone enters the bacteria and further reacts with the membrane to inactivate them.

Fan *et al.* (2004) observed that there is a synergism between negative air ions produced by plasma and ozone on bacterial cell death. The bactericidal effect of negative air ions in addition to ozone was found to be far greater than ozone by itself. In Fan *et al.*'s experiments, the viability of *Escherichia coli* was reduced to 40% of the first sample after 11 hours of negative air ion (NAI) treatment, as compared to 70% for treatment with ozone alone.

The humidity inside the system plays a role in the inactivation. The bacteria in our experiments are in the form of a bioaerosol, which consists of the bacteria enclosed in a fine droplet of water. As the bioaerosol travels inside the system, the droplet shrinks. The shrinkage of the water droplet depends on the relative humidity; at 50% RH, the droplets shrink to one-tenth of their size in 4 ms. Our experiments were performed at higher RH. As the liquid can act as a protective shield around the bacteria, shrinking of the droplet causes the bacteria to be more vulnerable to the charges and ROS produced by DBGD plasma.

Alhough Muranyi *et al.* (2008) have recently demonstrated that the fastest inactivation of plasma-treated *Aspergilus Niger* spores occurs at high relative humidity (70%), their experiments consisted of treating bacteria placed on surfaces and not airborne bacteria.

Due to the humid air inside our system, OH radical is expected to be formed. It is known that one main path for the generation of the hydroxyl radical in a DBD system is the photo-dissociation of ozone into atomic singlet oxygen and the reaction of this radical with water molecule (Falkenstein, 1997). This can be another synergetic mechanism that explains non-linearity in direct plasma treatment. With this knowledge and our experimental results, we conclude that the main cause of inactivation is the synergetic action of short-living plasma agents (charges, radiation and radicals such as OH) that disturb the membrane and ozone. This synergy creates a toxic environment for the bacteria, ultimately resulting in inactivation. The results of air sterilization experiments lead us to believe that the main reason for fast in-flight bacteria inactivation is the synergetic action of short-living plasma agents (direct plasma treatment) and ozone. The direct exposure to plasma inactivates part of bacteria and disturbs membranes of others. Follow-up ozone treatment of direct plasma-treated bacteria provides complete sterilization.

7.3   Non-thermal plasma in air-decontamination: Air cleaning from SO_2 and NO_x

7.3.1   Plasma cleaning of industrial SO_2 emissions

Industrial SO_2 emissions cause acid rain and result in serious environmental and medical disasters. The world's most severe acid rain regions are today Northern America, Europe and China. Major sources of the SO_2 emissions are coal-burning power plants (e.g. in China, 26 million tons of the hazardous SO_2 is produced, almost 40% of the total SO_2 emissions), steel works, non-ferrous metallurgical plants as well as oil refineries and natural gas purification plants. SO_2 emissions in air are usually high-volume and low-concentration: the SO_2 fraction is usually of the order hundreds of ppm, while total flow of polluted air in one system can reach a million m^3 per hour (Pu and Woskov, 1996). Oxidation of SO_2 in air to SO_3 results in the rapid formation of sulfuric acid, H_2SO_4. The kinetics of this process is limited by very low rates of natural oxidation of SO_2 in low-temperature air conditions.

The formation of SO_3 and sulfuric acid from the emitted SO_2 does not occur in the stacks of the industrial or power plants (burning sulfur-containing fuel) but later on in clouds, which ultimately results in acid rain. Non-thermal plasma can be used to stimulate oxidation of SO_2 into SO_3 inside the stack. This allows the sulfur oxides to be collected in the form of sulfates, for example fertilizer $(NH_4)_2SO_4$, if ammonia NH_3 is admixed to the plasma-assisted oxidation products (Pu and Woskov, 1996). An important advantage of the non-thermal plasma method is the possibility of simultaneous exhaust gas cleaning from SO_2 and NO_x. Taking into account the very large exhaust gas volumes to be cleaned from SO_2, the non-thermal atmospheric-pressure plasma systems mostly applied for the purpose are based on electron beams and pulsed corona discharges (Baranchicov *et al.*, 1990, 1992; Chae, Desiaterik, and Amirov, 1996; Jiandong *et al.*, 1996; Mattachini, Sani, and Trebbi, 1996).

The crucial question to be addressed is what is the energy cost of the plasma cleaning process, which strongly depends on the treatment dose rate (current density in the case of electron beams), air humidity, temperature and so on. As mentioned above, non-thermal plasma simply stimulates exothermic oxidation of SO_2 to SO_3, which is then removed chemically. Nevertheless, the process can be quite costly in terms of energy. At low current densities ($j < 10^{-5}$ A cm^{-2}) conventional for many experimental systems, the energy cost of the de-SO_2 process is usually high: of the order 10 eV per SO_2 molecule. In this case, cleaning the exhaust of a 300 MW power plant containing 0.1% (1000 ppm) of SO_2 requires c. 12 MW of plasma power. Even when such an energy cost of the plasma exhaust cleaning is acceptable, the use of such powerful electron beams or pulsed coronas is questionable. Decreasing the energy cost requires use of very specific plasma parameters, particularly dose rates and current densities in the case of electron beams (Baranchicov *et al.*, 1990, 1992).

7.3.2 SO_2 oxidation to SO_3 using relativistic electron beams

Application of relativistic electron beams (electron energies $\geq$ 300–500 keV) is attractive for exhaust gas cleaning in large-size ducts, taking into account that propagation length of the high-energy electrons in atmospheric air can be estimated as 1 m per 1 MeV (Fridman and Kennedy, 2004, 2011). The high-energy electrons are also characterized by the lowest energy cost of formation of charged particles (c. 30 eV per electron-ion pair), which reduces the energy cost of SO_2 oxidation in air stimulated by the chain ion-molecular reactions. Electron-beam-generated plasmas provide effective SO_2 oxidation to SO_3, which is removed afterwards using conventional chemical methods (in particular, admixture of NH_3, and filtration of $(NH_4)_2SO_4$).

For example, the electron-beam experiments of Baranchicov *et al.* (1990, 1992) were carried out in SO_2-air mixture with molar composition SO_2: 1%, H_2O: 3.4%, O_2: 19.6%, N_2: 76% at total pressure 740 Torr and temperature 298 K. A cylindrical reactor in the experiments had volume 800 mL and diameter 80 mm. Experiments were carried out with pulse-periodic beams and single-pulse beams. In the case of pulse-periodic beams (a pulse duration 100 ns, repetition frequency 0.1 kHz) with electron energy 300 keV, current densities varied over the range 1–10 A cm^{-2}. In the case of single-pulse beams, the current density varied over the range 10^{-3}–10 A cm^{-2}, corresponding to specific energy input in gas in the range 10^{-4}–3×10^{-2} J cm^{-3}. Energy costs of SO_2 oxidation into SO_3 in the electron-beam experiments of Baranchicov *et al.* (1990, 1992) are presented in Figure 7.11 as a function of the electron-beam current density. In the high current density range ($j_{eb} \geq$ 0.1 A cm^{-2}), the energy cost of SO_2 oxidation is proportional to the square root of the relativistic electron-beam current density:

$$A_{SO_2} \propto \sqrt{j_{eb}}. \tag{7.5}$$

Without taking into account oxidation in droplets, the minimum achieved value of the energy cost A_{SO_2} is 2–3 eV mol^{-1} SO_2. Including oxidation in the droplets, the minimum achieved oxidation energy cost is 0.8 eV mol^{-1} SO_2. The low oxidation energy costs are due to plasma-stimulation of chain oxidation mechanisms at relatively high values of the electron-beam current density. Electron beams with lower current densities ($\leq 10^{-4}$ A cm^{-2}) are unable to initiate chain oxidation, which results in an oxidation energy cost of c. 10 eV mol^{-1}; see Figure 7.11.

7.3.3 SO_2 oxidation to SO_3 using continuous and pulsed corona discharges

Corona discharges, and especially pulsed corona discharges, are also effective in SO_2 oxidation to SO_3 and therefore in the exhaust gas cleaning of sulfur oxides (Baranchicov *et al.*, 1992; Chae, Desiaterik, and Amirov, 1996; Jiandong *et al.*, 1996; Mattachini, Sani, and Trebbi, 1996). Baranchicov *et al.* (1992) investigated SO_2 oxidation in air using both continuous and pulsed corona discharges in a system similar to that described in

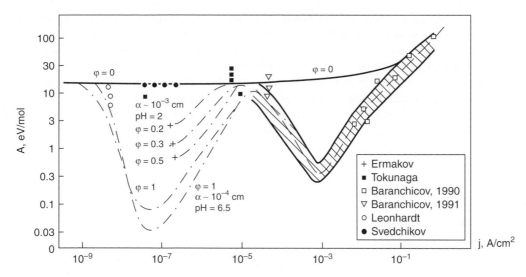

Figure 7.11   *Energy cost of SO₂ oxidation as a function of electron beam current density. Dashed line shows energy cost decrease by the cluster effect; dotted line shows energy cost decrease by the effect of droplets.*

the previous section. The continuous corona was generated by applying a constant 20 kV voltage to the central hexahedron electrode. The pulsed corona was sustained by negative or positive voltage pulses (30–70 kV) with duration 50–200 nseand repetition frequency 50 Hz.

Energy cost of the SO_2 oxidation process in the continuous and pulsed corona discharges (Baranchicov *et al.*, 1992) is also presented in Figure 7.11 as a function of current density (dose rate). The minimum energy cost for the pulsed coronas is 3–5 eV mol^{-1} SO_2 and for the continuous coronas is 10 eV mol^{-1} SO_2. The minimum SO_2 oxidation energy costs are therefore similar for the pulsed corona discharges and electron beams. The SO_2 oxidation energy cost dependence on the current density (dose rate) has similar behavior for electron beams, continuous and pulsed corona discharges. At low current densities ($\leq 10^{-4}$ A cm^{-2}), the energy cost of SO_2 oxidation to SO_3 remains almost constant at around 10 eV mol^{-1} SO_2. At high current densities (≥ 0.1 A cm^{-2}), the energy cost follows Equation (7.5) with an expected minimum at 10–100 mA cm^{-2}.

The low values of the SO_2 oxidation energy cost at current densities in the range 10–100 mA cm^{-2} are explained by plasma stimulation of chain mechanism of SO_2 oxidation, discussed in the following section. There is no chain mechanism for SO_2 oxidation to SO_3 in air in conventional gas-phase chemistry, which results in the above-mentioned relatively high energy cost of the oxidation. Even ozone is not effective in the gas-phase SO_2 oxidation. The chain oxidation of SO_2 is however possible in the liquid phase and even in small clusters, which can be stimulated in non-thermal plasma.

7.3.4   Plasma-stimulated liquid-phase chain oxidation of SO₂ in droplets

Atomic oxygen and other reactive oxygen species generated in non-thermal air plasma with energy cost c. 10 eV per particle are able to oxidize SO_2 to SO_3 with the same energy cost, which probably explains experiments with low current densities (see Figure 7.11). The SO_2 oxidation energy costs of 1–3 eV mol^{-1}, observed in plasma at higher current densities and discussed above, can be explained only by plasma stimulation of chain oxidation processes. There is no chain-mechanism of SO_2 oxidation to SO_3 in conventional gas-phase chemistry (without a catalyst).

Plasma is able to stimulate the chain oxidation of SO_2 either in liquid droplets or ionic clusters generated in non-thermal discharges in humid air. Plasma-stimulated liquid-phase chain oxidation of SO_2 into SO_3 occurs in droplets formed in non-thermal discharges by water condensation around H_2SO_4, ions and other active species. When the cold plasma is generated in atmospheric air containing sulfur compounds, the active species and especially sulfuric acid generated in the system immediately lead to water condensation and formation of mist. The presence of the water droplets allows the further chain process of SO_2 oxidation to SO_3 and sulfuric acid in liquid phase. Liquid-phase chain oxidation of SO_2 into SO_3 and sulfuric acid has been investigated by Deminsky *et al.* (1990). The oxidation mechanism in droplets depends on their acidity, which can be presented through the following kinetic schemes.

7.3.4.1   *Acidic droplets (pH < 6.5)*

The major ion formed in this case in water solution by gaseous SO_2 is the HSO_3^- ion:

$$SO_2(g) + H_2O \rightarrow HSO_3^- + H^+ \tag{7.6}$$

The product of oxidation in this case is the sulfuric acid ion HSO_4^-. The HSO_3^- oxidation in the acidic solution into sulfuric acid HSO_4^- is the chain process starting from the chain initiation through the HSO_3^- decomposition in collision with an active particle M (in particular, OH-radical):

$$M + HSO_3^- \rightarrow MH + SO_3^-. \tag{7.7}$$

Chain propagation in the acidic solution starts with attachment of oxygen and proceeds through active sulfur ion-radicals SO_3^-, SO_4^- and SO_5^-:

$$SO_3^- + O_2 \rightarrow SO_5^-, \quad k = 2.5 \times 10^{-12} \tag{7.8}$$

$$SO_5^- + HSO_3^- \rightarrow SO_4^- + HSO_4^-, \quad k = 1.7 \times 10^{-16} \tag{7.9}$$

$$SO_5^- + SO_5^- \rightarrow 2SO_4^- + O_2, \quad k = 10^{-12} \tag{7.10}$$

$$SO_4^- + HSO_3^- \rightarrow HSO_4^- + SO_3^-, \quad k = 3.3 \times 10^{-12} \tag{7.11}$$

$$SO_5^- + HSO_3^- \rightarrow SO_3^- + HSO_5^-, \quad k = 4.2 \times 10^{-17} \tag{7.12}$$

$$HSO_5^- + HSO_3^- \rightarrow 2SO_4^{2-} + 2H^+, \quad k = 2.0 \times 10^{-14} \tag{7.13}$$

Termination of the chain SO_2 oxidation is due to recombination and destruction of the active sulfur ion-radicals SO_3^-, SO_4^- and SO_5^-:

$$SO_3^- + SO_3^- \rightarrow S_2O_6^{2-}, \quad k = 1.7 \times 10^{-12} \tag{7.14}$$

$$SO_5^- + SO_5^- \rightarrow S_2O_8^{2-} + O_2, \quad k = 3.3 \times 10^{-13} \tag{7.15}$$

$$SO_3^-, \ SO_4^-, \ SO_5 + \text{admixture} \rightarrow \text{destruction}. \tag{7.16}$$

7.3.4.2 *Neutral and basic droplets (pH > 6.5)*

The major ion formed in this case in water solution by gaseous SO_2 is the SO_3^{2-} ion:

$$SO_2(g) + H_2O \rightarrow SO_3^{2-} + 2H^+ \qquad (7.17)$$

The product of SO_2 oxidation in this case is the sulfuric acid ion SO_4^{2-}. The SO_3^{2-} oxidation in the neutral and basic solutions into the sulfuric acid ion SO_4^{2-} is also a chain process. It starts from the chain initiation (through the SO_3^{2-} decomposition in collision with an active particle M, in particular, OH radical) and formation of an active SO_3^- radical similarly to Equation (7.7):

$$M + SO_3^{2-} \rightarrow MH^- + SO_3^-. \qquad (7.18)$$

Chain propagation in the neutral and basic solutions also starts with attachment of oxygen as in Equation (7.8) and then also proceeds (similarly to acidic solutions) through the active sulfur ion-radicals SO_3^-, SO_4^- and SO_5^-:

$$SO_3^- + O_2 \rightarrow SO_5^-, \quad k = 2.5 \times 10^{-12} \qquad (7.19)$$

$$SO_5^- + SO_3^{2-} \rightarrow SO_4^{2-} + SO_4^-, \quad k = 5 \times 10^{-14} \qquad (7.20)$$

$$SO_4^- + SO_3^{2-} \rightarrow SO_4^{2-} + SO_3^-, \quad k = 3.3 \times 10^{-12} \qquad (7.21)$$

$$SO_5^- + SO_3^{2-} \rightarrow SO_5^{2-} + SO_3^-, \quad k = 1.7 \times 10^{-14} \qquad (7.22)$$

$$SO_5^{2-} + SO_3^{2-} \rightarrow 2SO_4^{2-}, \quad k = 2 \times 10^{-14} \qquad (7.23)$$

The chain termination reactions in the case of neutral and basic droplets are mostly the same as for the case of acidic droplets described by Equations (7.14)–(7.16). The SO_3^- ion-radical dominates the chain propagation in both cases of low and high acidity. The SO_3^- ion-radicals can be significantly replaced by SO_5^- ion-radicals in propagation of the oxidation chain reaction, when oxygen (O_2) is in excess in the liquid droplets.

Radiation yield of the SO_2 liquid-phase oxidation (number of the oxidized SO_2 molecules per 100 eV of the plasma energy) is presented in Figure 7.12 as a function of the current density and acidity (Potapkin *et al.*, 1993, 1995). The radiation yield is very high (up to 10^6, which corresponds to the chain length 10^5) for the case of basic droplets and relatively low (c. 100, which corresponds to the chain length 10) for the

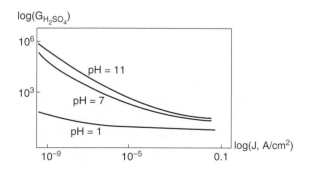

Figure 7.12 *Dependence of the radiation yield of SO_2 liquid phase oxidation on the current density and acidity.*

case of acidic droplets. This can be explained by taking into account the much faster kinetics of $HSO_3{}^-$ chain oxidation with respect to that of $SO_3{}^{2-}$.

Figure 7.12 also illustrates decrease of the radiation yield of the SO_2 liquid-phase oxidation with current density (dose rate of the plasma treatment), which is due to acceleration of the chain termination rate as the square of density of charged particles (Equations (7.14), (7.15)). The SO_2 oxidation chain length can be very high at low dose rates (up to 10^5). The energy-efficient experiments with SO_2 oxidation by relatively high current electron beams and pulsed corona discharges (Baranchicov *et al.*, 1990, 1992) cannot be explained by the liquid-phase chain mechanism (compare Figures 7.11 and 7.12). The high current and high dose rate plasma-chemical systems can provide the SO_2 chain oxidation in clusters, which is discussed in the following section.

7.3.5   Plasma-catalytic chain oxidation of SO_2 in clusters

The plasma-catalytic chain oxidation of SO_2 in clusters is not limited by dose rate and acidity, which therefore explains the high energy efficiency of the process in the high-current electron beam systems and pulsed corona discharges (Baranchicov *et al.*, 1990, 1992). The long-chain oxidation process is provided in this case by the complex ions (clusters), which catalyze SO_2 oxidation similar to the general plasma catalytic mechanism described in Section 2.3.5. The plasma-stimulated cluster mechanism of the chain SO_2 oxidation in air (Potapkin *et al.*, 1993, 1995) starts with electron-beam or pulsed corona generation of negative oxygen ions by the electron attachment to molecular oxygen (which is usually faster in the system than attachment to SO_2 and water clusters):

$$e + O_2 + M \rightarrow O_2^- + M, \quad k = 3 \times 10^{-31}. \tag{7.24}$$

The simple oxygen ions then attach to water molecules creating the negative complex ions (ion clusters) in the step-wise clusterization:

$$O_2^-(H_2O)_n + H_2O + M \rightarrow O_2^-(H_2O)_{n+1} + M, \quad k = 3 - 6 \times 10^{-28}. \tag{7.25}$$

Attachment of a SO_2 molecule leads to reconstruction of the ion cluster and formation of the most thermodynamically stable cluster core:

$$O_2^-(H_2O)_n + SO_2 \rightarrow O_2^-(H_2O)_n SO_2 \rightarrow O_2^- SO_2(H_2O)_n \rightarrow SO_4^- O_2(H_2O)_n. \tag{7.26}$$

The core ion of the cluster (SO_4^-) is characterized by the peroxide structure and is a strong oxidant. Attachment of a second SO_2 molecule to the cluster can lead to a strongly exothermic oxidation with formation of sulfuric acid ($\Delta H \approx -2$), and subsequent detachment of an electron from the acidic cluster:

$$SO_2^- O_2(H_2O)_n + SO_2 \rightarrow SO_2^- O_2 SO_2(H_2O)_n \rightarrow H_2SO_4 H_2SO_4(H_2O)_n + e. \tag{7.27}$$

Release of the electron during formation of the sulfuric acid results in the chain propagation of the SO_2 oxidation chain reaction mechanism described by Equations (7.14)–(7.17). We should note that the SO_4^- ion, playing the key role in the chain propagation, exists in two configurations: a high-energy active peroxide isomer and a low-energy non-active cyclic isomer. Plasma-excited species in air promote O_2 penetration into the core of the cluster and stimulate formation of the chemically active isomer of the SO_4^- ion. Contribution of the excited species to formation of the chemically active isomer of the SO_4^- ion explains suppression of the chain oxidation mechanism at low current densities and low dose rates (see Figure 7.11), when concentration of the excited species is low. An increase of the chain oxidation energy cost at higher currents (Figure 7.11) is due, as mentioned, to domination in this case of recombination and chain termination processes (proportional to the square of the concentration of active charged particles) over the chain propagation reactions (linear with respect to the concentration of active charged particles). The two effects explain the optimal value of the current density (or dose rate) of about 10^{-3} A cm^{-2}, resulting in an energy cost of SO_2 oxidation close to 1 eV mol^{-1} SO_2.

7.3.6 Simplified mechanism and energy balance of the plasma-catalytic chain oxidation of SO_2 in clusters

To analyze the energy balance and efficiency of plasma cleaning of SO_2 in exhaust gases, it is convenient to simplify the chain oxidation mechanism considered above to the kinetics of simple ions, neutrals and excited molecules (Potapkin, Rusanov and Fridman, 1989; Baranchikov *et al.*, 1992). The plasma-assisted SO_2 chain oxidation mechanism starts with the formation of negative ions O_2^- through the three-body electron attachment process (7.24). The SO_2^- ions are then formed due to a fast exothermic charge exchange process:

$$O_2^- + SO_2 \rightarrow O_2 + SO_2^-. \tag{7.28}$$

The SO_2^- ions are then converted into the chemically active peroxide configuration of SO_4^- ions in collisions with excited oxygen molecules:

$$SO_2^- + O_2^* \rightarrow SO_4^-. \tag{7.29}$$

Formation of the oxidation product SO_3 takes place in the reactions of the active SO_4^- ions with the excited oxygen molecules:

$$SO_4^- + O_2^* \rightarrow SO_3 + O_3^-. \tag{7.30}$$

Restoration of electrons and chain propagation is provided by ion-molecular reactions and associative detachment producing additional sulfuric acid:

$$O_3^- + SO_2 \rightarrow SO_3^- + O_2, \tag{7.31}$$

$$SO_3^- + H_2O \rightarrow H_2SO_4 + e. \tag{7.32}$$

With an excess of water vapor, the ion-molecular processes (7.30) and (7.32) are limiting reactions of the chain oxidation mechanism, which allows the energy cost of the SO_2 oxidation in the steady-state conditions to be expressed as (Baranchikov *et al.*, 1992):

$$A = \frac{W_i}{\frac{4G_v}{G_i\beta} + \delta}, \tag{7.33}$$

$$\beta = \left(1 + \frac{n_1}{n_i} + \frac{n_i}{n_2}\right) \left\{ 1 + \left[1 - \frac{4n_i/n_2}{\left(1 + \frac{n_1}{n_i} + \frac{n_i}{n_2}\right)^2} \right]^{-1} \right\} \tag{7.34}$$

where n_i is the steady-state ion density generated in atmospheric air (density n_0) by, for example, a relativistic electron beam (current density j; electron velocity c; electron charge e), which can be calculated as:

$$n_i = \sqrt{\frac{k_{ib} j n_0 G_i}{k_r^{ii} ec}};$$ (7.35)

where characteristic ion concentrations n_1 and n_2 are determined by the expressions:

$$n_1 = k_{VT} n_0 / k_{(11.82)},$$ (7.36)

$$n_2 = k_{(11.84)} n_{SO_2} \frac{k_r^{ii} G_v}{G_i};$$ (7.37)

G_i and G_v are the radiative yields of ions and vibrationally excited molecules; k_{ib}, k_r^{ii}, k_{VT} are rate coefficients of relativistic electron beam ionization, ion-ion recombination and VT relaxation; and W_i is the energy cost of generation of an ion.

The factor δ characterizes non-chain mechanisms of the plasma-stimulated SO_2 oxidation in air, including:

$$SO_2 + O \rightarrow SO_3, \quad SO_3 + H_2O \rightarrow H_2SO_4,$$ (7.38)

$$SO_2 + OH \rightarrow HSO_3, \quad HSO_3 + OH \rightarrow H_2SO_4.$$ (7.39)

The factor δ can be calculated using the degradation cascade method (Fridman and Kennedy, 2004, 2011), and makes no significant contribution to the total oxidation energy cost as defined by Equation (7.33). As can be seen from Equation (7.33), when $n_1 < n_2$ (relevant to the experimental conditions corresponding to SO_2 oxidation stimulated by the relativistic electron beams and pulsed coronas; Baranchikov *et al.*, 1990, 1992), the SO_2 oxidation energy cost can be significantly lower than the energy cost of an ion (W_i) because of the long oxidation chain length. The minimum value of the SO_2 oxidation energy cost can be calculated in this case as:

$$A^{min} = \frac{G_i W_i}{2G_v} = \frac{100}{G_v}$$ (7.40)

which is achieved at the optimal electron beam current densities (or the corresponding dose rates) resulting in optimal value of the ion density:

$$n_i = \sqrt{n_1 n_2}.$$ (7.41)

Taking into account that the G_v factor can be in the range of 100–300, the simplified kinetic model considered quite satisfactory describes experimental dependence $A(j)$ as shown in Figure 7.11. Specifically, the model explains minimal values of the energy cost of the plasma-stimulated SO_2 oxidation to SO_3 in air as 0.3–1 eV mol^{-1} SO_2 (Baranchicov *et al.*, 1992).

7.3.7   Plasma-stimulated combined oxidation of NO_x and SO_2 in air; simultaneous industrial exhaust gas cleaning from nitrogen and sulfur oxides

An important advantage of plasma-assisted exhaust gas cleaning, especially in the case of power plant exhaust, is the possibility of simultaneous oxidation of SO_2 and NO_x to sulfuric and nitric acids. The products can then be collected in the form of non-soluble sulfates and nitrates, for example, in the form of fertilizers $(NH_4)_2SO_4$ and $(NH_4)NO_3$ if ammonia NH_3 is admixed to the plasma-assisted oxidation products (Pu and Woskov, 1996). Mechanisms of simultaneous NO_x and SO_2 oxidation stimulated by electron beams have been investigated in numerous experiments and simulations (Paur, 1991, 1999; Potapkin *et al.*, 1995). It is interesting that the simultaneous NO and SO_2 oxidation appears to be more effective than their plasma-assisted individual oxidation in air. In the presence of SO_2, the radiative yield of NO oxidation to NO_2 and nitric acid can exceed 30 (energy cost below 3 eV mol^{-1} NO). The low-energy-cost effect of the simultaneous NO and SO_2 oxidation can be explained by the chain co-oxidation mechanism, propagating through the radicals OH and HO_2:

$$HO_2 + NO \rightarrow OH + NO_2 \tag{7.42}$$

$$OH + SO_2 \rightarrow SO_3 + H, \quad H + O_2 + M \rightarrow HO_2 + M. \tag{7.43}$$

It should be mentioned that the gas-phase chain process described by Equations (7.42) and (7.43) is limited by the formation of certain concentration of NO_2, when the plasma-generated atomic nitrogen causes significant destruction of the product in the reverse reactions:

$$N + NO_2 \rightarrow 2NO, \quad N + NO_2 \rightarrow N_2O + O. \tag{7.44}$$

To suppress the reverse reactions, Matzing (1989) proposed multi-stage plasma treatment with intermediate removal of the oxidation products. More interesting quenching effects occur in the plasma-generated droplets. In such droplets at certain pH conditions, it is not only possible to stabilize NO_2 by formation of nitric acid in the solution, but also to reduce NO_2 to molecular nitrogen N_2 with simultaneous oxidation of sulfur from S(IV) to S(VI):

$$NO_2 + 2HSO_3^- \rightarrow 1/2\ N_2 + 2HSO_4^-. \tag{7.45}$$

More kinetic details regarding the simultaneous plasma-chemical cleaning of industrial exhaust gases from sulfur and nitrogen oxides using relativistic electron beams can be found, in particular, in Potapkin *et al.* (1993, 1995).

While discussing the non-thermal plasma treatment of large-volume exhausts, we should point out the interesting concept of the *radical shower*, applied in particular to deNO$_x$ flow processing (Yan *et al.*, 2001; Chang, 2003; Wu *et al.*, 2005). The radical shower approach implies that plasma only directly treats a portion of the total flow (or even separate gas) producing active species (in particular, radicals), which then treat the total gas flow as a 'shower'. Although the approach is strongly limited by the mixing efficiency of the plasma-treated and plasma-untreated gases, the radical shower gas cleaning can be very energy efficient, especially in the case of very high flow rate values of the exhaust gas streams. Because the radical shower approach is often executed using corona discharges, it is sometimes referred to as the 'corona radical shower'.

7.4 Non-thermal plasma decontamination of air from volatile organic compound (VOC) emissions

7.4.1 General features of non-thermal plasma treatment of VOC emissions in air

Volatile organic compounds (VOC) form a class of air pollutants that has been addressed by environmental regulations in the past few decades due to their toxicity and contribution to global warming. Cleaning of the VOC-contaminated air streams has a significant impact on human health, and is an important medical issue.

The control of air pollution from dilute large-volume sources, such as paint spray booths, paper mills, pharmaceutical, food and wood processing plants, is a challenging problem. Conventional technologies widely used in industry such as carbon adsorption/solvent recovery or catalytic/thermal oxidation, regenerative thermal incineration and especially regenerative thermal oxidation (RTO) systems are prohibitively expensive and are not economical for large gas flow rates (50 000–250 000 scfm) and low VOC concentrations (<100 ppm).

Among the emerging low-temperature VOC treatment technologies are low-temperature catalysis, biofiltration and non-thermal plasma, considered in this section. Catalysts easily suffer from plugging, fouling or poisoning by particulates and non-VOC materials in the exhaust stream, which results in high maintenance cost. The major disadvantage of biofilters is their large specific footprint, typically 100–400 square feet per 1000 scfm of treated gas. Biofilter systems and filter materials may also require costly maintenance and replacement.

Application of the non-thermal plasma method (particularly pulsed coronas, DBD and electron beams) solves most of the problems typical of the alternative VOC treatment methods. Application of non-thermal plasma however requires that energy cost is reduced, as for plasma $deSO_2$ and $deNO_x$ processes. Plasma approaches usually becomes more energy efficient and competitive with other cleaning methods when stream volumes are large and the concentration of pollutants is small. Non-thermal atmospheric-pressure plasma systems that have been investigated for VOC abatement include pulsed corona (Yamamoto *et al.*, 1992, 1993; Hsiao *et al.*, 1995; Sobacchi *et al.*, 2003); DBD (Evans *et al.*, 1993; Rosocha *et al.*, 1993; Neely *et al.*, 1993); surface discharge (Oda *et al.*, 1991; Masuda, 1993); gliding arc discharge (Czernichowski, 1994; Mutaf-Yardimci *et al.*, 1998; Fridman *et al.*, 1999); microwave discharge (Bailin *et al.*, 1975); and electron beam plasma (Penetrante *et al.*, 1996a, b).

The highest energy efficiency for cleaning large-volume air exhaust from low-concentration VOC has been achieved in the plasma systems with the highest electron energies (such as electron beams and pulsed coronas), where energy cost of generation of oxidation active species (such as OH) is lowest. Non-thermal plasma systems have been successfully applied for cleaning high-volume low-concentration (HVLC) VOC exhausts, particularly from volatile hydrocarbons (acetone, methanol, pinene, etc.), sulfur-containing compounds (dimethyl sulfide, H_2S etc.) and chlorine-containing compounds (vinyl chloride, trichloroethylene, trichloroethane, carbon tetrachloride etc.). Some of the processes are discussed in the following sections.

7.4.2   Mechanisms and energy balance of treatment of VOC exhaust gases from paper mills and wood processing plants

Electric energy cost is one of the crucial factors characterizing plasma treatment of HVLC VOC exhausts. To analyze energy cost of the VOC treatment in non-thermal plasma, we should point out that the mechanism of the plasma oxidation of hydrocarbons can be generally interpreted as the low-temperature burning out of the VOC in air to preferentially CO_2 and H_2O using different plasma-generated active oxidizers (OH, atomic oxygen, electronically excited oxygen, ozone etc.), but especially OH-radicals. The non-thermal plasma oxidation (burning out) of a small admixture of the volatile hydrocarbons in air stream can be described by the following illustrative kinetic mechanism. Of the different plasma-generated active oxidizers responsible for the cold burning-out of VOC, OH radicals play a major role. The formation of OH radicals in air plasma is due to air humidity and is provided by numerous channels, starting with the charge exchange from any positive air ion M^+ to water ion H_2O^+:

$$M^+ + H_2O \rightarrow M + H_2O^+. \tag{7.46}$$

The H_2O^+ ions formed through the charge exchange then react with water molecules producing the active OH radicals in the fast ion-molecular reaction:

$$H_2O^+ + H_2O \rightarrow H_3O^+ + OH. \tag{7.47}$$

Taking into account that the ionization potential of the water molecules is relatively low, most of the positive ions initially formed in air have a tendency for the charge exchange as described by Equation (7.46). The discharge energy initially distributed over different air components is therefore somewhat selectively localized on ionization of water and selective production of OH radicals. This explains the relatively low-energy price of production of OH radicals in air plasma: 10–30 eV per radical, although the fraction of water molecules in air is not large (Gutsol, Tak and Fridman, 2005). Representing a hydrocarbon VOC molecule as RH, where R is a relevant organic group, the oxidation of the molecule starts with an elementary reaction of its dehydrogenization:

$$OH + RH \rightarrow R + H_2O. \tag{7.48}$$

The almost immediate attachment of molecular oxygen to the active organic radical R results in the formation of organic peroxide radical:

$$R + O_2 \rightarrow RO_2. \tag{7.49}$$

The peroxide RO_2 radical is then able to react with another saturated hydrocarbon VOC molecule RH, forming saturated organic peroxide RO_2H and propagating the chain mechanism described by Equations (7.49) and (7.50) of RH oxidation in air:

$$RO_2 + RH \rightarrow RO_2H + R. \tag{7.50}$$

Depending on temperature and chemical composition, peroxides RO_2 and RO_2H are further oxidized up to CO_2 and H_2O with or without additional consumption of OH and other plasma-generated oxidizers.

Summarizing the mechanism, energy cost corresponding to the VOC treatment process is of the order 10–30 eV per one molecule of the pollutant RH (Gutsol, Tak and Fridman, 2005). This energy is relatively high, but total energy consumption can be small because of the very low concentration of pollutants. A major conventional approach to the VOC control, the regenerative thermal oxidation (RTO) consumes c. 0.1 eV/mol but this energy is calculated per molecule of air. Energy consumption in plasma is therefore lower than that of conventional RTO, when VOC concentration in air is below 0.3–1% (3000–10 000 ppm).

In the case of plasma treatment of large-volume exhausts, the *radical shower* approach can be effective. It implies that plasma treats only a portion of the total flow, generating active species which then clean the whole gas. While it has been discussed above regarding $deNO_x$ and $deSO_2$ processes (Yan *et al.*, 2001; Chang, 2003; Wu *et al.*, 2005), it can also be useful for VOC removal. Exhaust gases of paper mills (especially the brownstock washer ventilation gases) and wood processing plants (especially strandboard press and dryer ventilation gases) are good examples of the HVLC VOC streams which can be effectively cleaned by non-thermal plasma (Harkness and Fridman, 1999). Compositions, temperature and humidity of the streams are listed in Table 7.2. The same table provides the range of the stream parameters tested in the experiments with the VOC treatment in pulsed corona discharges (Sobacchi *et al.*, 2003; Gutsol, Tak, and Fridman, 2005).

Table 7.2   *Characteristics of the major HVLC VOC exhaust streams of paper mills and wood processing plants and range of parameters tested for plasma treatment using pulse corona discharges.*

Stream composition humidity and temperature	Paper mills, brownstock washer ventilation stream	Wood processing plants, strandboard press ventilation stream	Wood processing plants, strandboard dryer ventilation stream	Range of parameters tested in pulsed corona plasma treatment
Methanol, CH_3OH (ppm)	83	25	8	5–1000
Acetone, CH_3COCH_3 (ppm)	3	1	28	5–1000
α-pinene (ppm)	209	4	16	150–800
Dimethyl sulfide, $S(CH_3)_2$ (ppm)	2	—	—	5–1000
Relative humidity (%)	100	70	90	0–100
Temperature (°C)	43	37	93	25–220

7.4.3   Removal of acetone and methanol from air using pulsed corona discharge

Before discussing plasma cleaning of large-volume VOC exhaust streams of paper mills and wood processing plants using powerful pulsed corona systems (Gutsol, Tak, and Fridman, 2005), we consider laboratory-scale experiments using pulsed corona (wire-in-cylinder coaxial electrode configuration) with power 1–20 W, voltage pulse amplitude 9–12 kV, pulse repetition rate 0.2–2 kHz, pulse duration 100 ns and rise time 10 ns (Sobacchi *et al.*, 2003). Typical atmospheric-pressure gas flow through the plasma system was 2 slm; corresponding residence time is c. 13 s. The pulsed corona reactor was placed inside a tubular furnace to sustain fixed gas temperature between 25 and 220°C. In the experiments with acetone and methanol removal, initial concentration of both CH_3COCH_3 and CH_3OH varied between 5 and 1000 ppm (see Table 7.2), temperature was maintained at a constant 200°C and air flow humidity corresponded to saturation at c. 55°C.

The destruction and removal efficiency (DRE), the mole percentage of VOC removed with respect to the initial amount of acetone and discharge power, are provided in Figure 7.13 as a function of the pulse repetition rate. As can be seen, the discharge power is proportional to the pulse repetition rate. DRE of acetone depends strongly on the initial composition and on the corona power. Higher-power levels result in higher DRE; an increase of the initial acetone concentration requires higher power to reach the same level of DRE. No organic by-products were detected at initial acetone concentrations 5 and 20 ppm. The treatment of air stream with 200 and 1000 ppm of acetone leads to the formation of very small amounts of methanol as a by-product ($\ll$10 ppm). A similar dependence of the DRE of methanol on the discharge power in the pulsed corona system at 200°C is demonstrated in Figure 7.14. No organic by-products and CO were observed during the CH_3OH treatment; the methanol has been entirely converted into CO_2. The methanol DRE dependence on specific energy input (SEI) in the discharge is shown in Figure 7.15. Summarizing the results, the relatively low-energy cost of treatment (c. 20 W h m^{-3} of air) can be sufficient for 98% of DRE, which is acceptable for industrial applications.

7.4.4   Removal of dimethyl sulfide from air using pulsed corona discharge

Dimethyl sulfide (DMS) $S(CH_3)_2$ represents sulfur-containing VOC, which are air pollutants of major concern in different industrial exhausts including the brownstock ventilation gases of paper mills (see Table 7.2). DMS is highly toxic, flammable and emits a foul odor; the DMS odor threshold is of the order 1 ppb. While typical

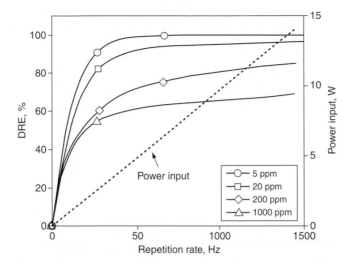

Figure 7.13 *Destruction-removal efficiency (DRE) and electric power input as a function of the pulse repetition rate for treatment of acetone (diluted in air) in pulsed corona discharge. Initial acetone concentrations are: 5, 20, 200 and 1000 ppm.*

DMS concentration in the brownstock ventilation gases is 2 ppm (Table 7.2), it reaches 1700 ppm (0.17 vol. %) in the worst-case-scenario exhausts.

Effective plasma removal of DMS has been demonstrated by the pulsed corona discharges described in the previous section (Sobacchi *et al.*, 2003; Gutsol, Tak, and Fridman, 2005). DRE of DMS is presented in Figure 7.16 as function of the SEI and temperature. Products of the non-thermal plasma DMS removal are CO_2, H_2O and SO_2. At the lower initial DMS concentrations, no DMS has been detected in the after-treatment

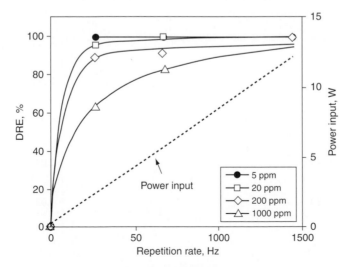

Figure 7.14 *DRE and electric power input as a function of the pulse repetition rate for treatment of methanol (diluted in air) in pulsed corona discharge. Initial methanol concentrations are: 5, 20, 200 and 1000 ppm.*

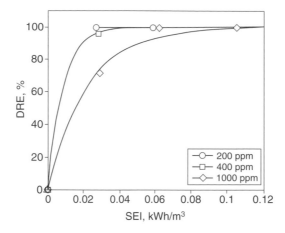

Figure 7.15 *DRE of methanol as a function of SEI in pulsed corona discharge. Initial methanol concentrations are 200, 400 and 1000 ppm.*

gases. For higher initial concentrations, DRE exceeding 98% can be achieved at fairly low levels of SEI. SEI of c. 50 W h m^{-3} is sufficient for processing the highest DMS concentrations (Sobacchi *et al.*, 2003). The only by-product of the DMS treatment is methanol; its concentration in the after-treatment air is shown in Table 7.3 and is dependent on initial DMS concentration and the SEI.

The mechanism of the plasma removal of DMS from air begins with oxidation provided by plasma-generated oxidizers and especially OH (Turnipseed, Barone, and Ravishenkara, 1996a, b; Sun, Sato, and Clements, 1997, 1999):

$$OH + CH_3SCH_3 \rightarrow CH_3S + CH_3OH, \tag{7.51}$$

$$OH + CH_3SCH_3 \rightarrow H_2O + CH_3SCH_2. \tag{7.52}$$

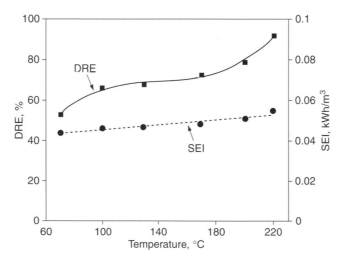

Figure 7.16 *DRE of DMS as a function of SEI in pulsed corona discharge (SEI values are related to temperature).*

Table 7.3 *Concentration of methanol as a by-product of the plasma destruction of dimethyl sulfide.*

Initial DMS concentration (ppm)	SEI (27 W h m^{-3})	SEI (60 W h m^{-3})	SEI (100 W h m^{-3})
5	0	0	0
20	0	0	0
200	5	4	2
400	14	12	8
1000	21	28	37

The CH_3S radical is an intermediate particle in the oxidation of DMS. The CH_3S radical is formed either directly in the elementary reaction (7.51) or through transformation of the CH_3SCH_2 radicals produced in (7.52) into corresponding peroxide radicals by means of attachment of molecular oxygen:

$$CH_3SCH_2 + O_2 + M \rightarrow CH_3SCH_2OO + M. \tag{7.53}$$

The peroxide radical CH_3SCH_2O is then transformed into the radical CH_3SCH_2O by reactions with different active species, but especially by reactions with NO (even when concentration of NO is relatively very low):

$$CH_3SCH_2OO + NO \rightarrow CH_3SCH_2O + NO_2. \tag{7.54}$$

The major intermediate particles of the DMS oxidation – the CH_3S radicals – are then generated by decomposition of the CH_3SCH_2O radicals with additional formation of the formaldehyde molecules as by-products:

$$CH_3SCH_2O + M \rightarrow CH_3S + CH_2O. \tag{7.55}$$

Further oxidation of the major intermediate active particles CH_3S leads to formation of CO_2, H_2O and SO_2 as final products. It should be mentioned that another primary product of the DMS reaction with the plasma-generated OH is the DMS.OH adduct:

$$OH + CH_3SCH_3(+M) \rightarrow OH.CH_3SCH_3(+M). \tag{7.56}$$

The DMS.OH adduct then effectively reacts with molecular oxygen generating dimethyl-sulfoxide $CH_3S(O)CH_3$ and dimethyl-sulfone $CH_3S(O_2)CH_3$:

$$OH.CH_3SCH_3 + O_2 \rightarrow HO_2 + CH_3S(O)CH_3, \tag{7.57}$$

$$OH.CH_3SCH_3 + O_2 \rightarrow OH + CH_3S(O_2)CH_3. \tag{7.58}$$

Both dimethyl-sulfoxide $CH_3S(O)CH_3$ and dimethyl-sulfone $CH_3S(O_2)CH_3$ are known to react rapidly with OH (which leads to effective further oxidation). In the presence of liquid water, they are very likely to be removed by absorption in the liquid phase. The effective absorption of the dimethyl-sulfoxide and dimethyl-sulfone in water is due to their high solubility; the Henri constant is $K_H > 5 \times 10^4$ for both organic sulfoxides. Effective DMS removal from air can therefore be provided not only by its complete oxidation to CO_2, H_2O and SO_2, but also by means of formation of highly soluble compounds with following scrubbing.

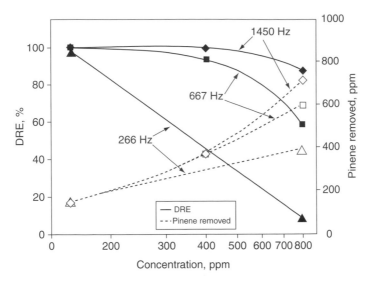

Figure 7.17   *DRE and actual amount of α-pinene removed in pulsed corona discharge as a function of initial concentration of α-pinene.*

7.4.5   Removal of α-pinene from air using pulsed corona discharge

VOC emissions of paper mills and wood processing plants usually contain significant amount of terpenes, especially α-pinene $C_{10}H_{16}$ (see Table 7.2). Effective plasma removal of α-pinene $C_{10}H_{16}$ has been demonstrated by pulsed corona discharges at temperatures of 70–200°C (Sobacchi *et al.*, 2003). Figure 7.17 illustrates removal by different pulse frequencies corresponding to different levels of the discharge power (the SEI is 20, 50 and 100 W h m^{-3} for the frequencies 266, 667 and 1450 Hz). A major initial by-product of the α-pinene plasma removal is acetone, which is shown in Table 7.4 together with its DRE dependence on SEI and gas temperature.

Overall, the α-pinene plasma removal tests demonstrate that 98% DRE for the typical initial VOC concentration of 200 ppm can be reached at low values of SEI to the air of 25 W h m^{-3}. The efficiency of treatment and especially the energy cost of the exhaust treatment can be enhanced at higher humidity (Sobacchi *et al.*, 2003), discussed in the following section. Decomposition of VOC in some specific exhaust mixtures related to the paper mills and wood processing plants is listed in Table 7.5 at two values of the specific energy input: 90 and 190 W h m^{-3} (Sobacchi *et al.*, 2003). As seen from the table, the presence of the large molecules of α-pinene decreases in the exhaust gases with the DRE of other components, which is due to the formation of methanol and acetone as treatment by-products.

Table 7.4   *Characteristics of plasma-assisted removal of 400 ppm of α-pinene from air stream using pulsed corona with repetition rate 667 Hz.*

Plasma VOC treatment, process characteristics	Gas temperature				
	70°C	100°C	130°C	170°C	200°C
SEI (W h m^{-3})	47	48	48.5	49	50
DRE (%)	76.6	81.5	84	89.7	96.4
Treatment energy cost per 1 kg of α-Pinene (kW h kg^{-1})	25.5	24.2	23.7	22.5	21.4
Initial by-product: acetone (ppm)	44	55	59	88	189

Table 7.5 *Decomposition of VOC exhaust mixtures in air related to paper mills and wood processing plants at different values of the pulsed corona power input.*

Input and DRE	Methanol	Acetone	Dimethyl sulfide (DMS)	α-Pinene
Initial fraction, ppm (SEI $= 90$ W h m^{-3})	100	20	20	0
DRE, % (SEI $= 90$ W h m^{-3})	100	91	100	—
Initial fraction, ppm (SEI $= 190$ W h m^{-3})	100	20	20	0
DRE, % (SEI $= 190$ W h m^{-3})	100	99.3	100	—
Initial fraction, ppm (SEI $= 190$ W h m^{-3})	100	20	20	200
DRE, % (SEI $= 190$ W h m^{-3})	86.5	14.5	83.8	96.5

7.4.6 Treatment of paper mill exhaust gases using wet pulsed corona discharge

A significant reduction of the VOC removal energy cost as well as complete elimination of the plasma treatment by-products can be achieved by application of wet or spray pulsed corona discharges, which is a combination of pulsed corona discharge with a scrubber (Gutsol, Tak and Fridman, 2005). First of all, water droplets (or film) effectively absorb soluble VOC (such as methanol) present in the exhaust gases, which obviously simplifies the task for the plasma treatment. Non-soluble organic compounds (in particular, non-polar hydrocarbons) cannot be directly removed by water scrubbing. Their plasma treatment (see Equations (7.48)–(7.50)) however leads to the ormation of soluble peroxides ROOH and peroxide radicals RO_2, which can be effectively removed by water scrubbing. Plasma cleaning of VOC emissions is therefore provided in this plasma approach not by complete oxidation of the organic compounds to CO_2 and H_2O, but by plasma-induced conversion of the non-polar non-soluble compounds into soluble compounds which are then removed by water scrubbing. Complete VOC oxidation to CO_2 and H_2O requires more plasma-generated oxidizers (such as OH) than conversion of the non-polar non-soluble compounds RH to soluble compounds ROOH or ROO, where a single OH radical can be sufficient (Equations (7.48)–(7.50)).

This method results in much lower energy requirements for VOC treatment in pulsed corona discharge, illustrated in Figure 7.18. By-products of the plasma VOC treatment are usually soluble, therefore application

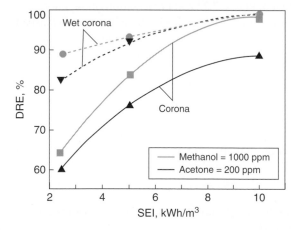

Figure 7.18 *Comparison of the plasma-stimulated VOC removal efficiency using wet pulsed corona and regular (dry) pulsed corona. DRE of methanol (initial concentration 1000 ppm) and acetone (initial concentration 200 ppm) are presented in both types of corona as a function of SEI. Water flow rates in the spray of the wet pulsed corona are: 0.4 mL min^{-1} for methanol removal, and 1 mL min^{-1} for acetone removal.*

Table 7.6   *Major requirements and characteristics of the wet pulsed corona treatment of the brownstock washer ventilation gases (polluted air) from paper mills.*

Total DRE of the VOC mixture	98–100%
Plasma treatment by-products in gas-phase	0% (no by-products)
Energy cost (regular conditions)	10 W h m^{-3}
Energy cost (worst-case-scenario: additional 1700 ppm of DMS)	20 W h m^{-3}
Power requirements for VOC removal from 25 000 scfm brownstock washer ventilation stream	500 kW
Water requirements for VOC removal from 25 000 scfm brownstock washer ventilation stream	25 gpm

of wet pulsed corona discharge removes all of them from the gas flow. In the particular case of plasma cleaning of sulfur-containing exhausts (such as DMS), sulfur oxides and polar by-products of oxidation are also effectively removed from the air flow in the form of relevant acidic solutions. Plasma-stimulated oxidation of organic compounds continues even when they are absorbed in water droplets, which increases the absorbing capacity of the water droplets with respect to the same initial VOC. Water consumption for the VOC exhaust cleaning approach is therefore very low.

The major requirements and characteristics of the wet pulsed corona treatment of the brownstock washer ventilation gases from paper mills (see Table 7.2) are summarized in Table 7.6 (Gutsol, Tak and Fridman, 2005).

Although the VOC removal parameters presented in the table are quite impressive, the organic compounds in this approach are not completely removed but just converted from gas to liquid phase. Such VOC treatment is however very acceptable to the paper industry, where the amount of polluted water from other sources is large and effective water cleaning is necessary anyway.

A mobile plasma laboratory for VOC removal from the exhaust streams typical for paper mills and wood processing plants (see Table 7.2) has been built in Drexel University (Figure 7.19; see Gutsol, Tak and Fridman, 2005) based on 10 kW wet (spray) pulsed corona discharge. Figure 7.20 is a photo of the discharge from one of the six windows; all six windows can be seen in Figure 7.21. The pneumatic and hydraulic scheme of the system is shown in Figure 7.22 and organization of the pilot plant is illustrated in Figure 7.23. Exhaust gas in the pilot plant first goes to a scrubber where it is mostly washed out from soluble VOC. The exhaust gas is then directed to the wet pulsed corona discharge, where non-soluble VOC are converted into soluble compounds to be scrubbed out by water spray in the chamber. Air refining is provided by the mist separator, finalizing the exhaust cleaning process.

The pilot plant is mounted on a trailer (see Figures 7.19 and 7.21), providing opportunities for industrial field experiments. The mobile plasma laboratory for VOC removal includes not only the gas cleaning system (Figures 7.22 and 7.23) but also special exhaust gas characterization capability for gas diagnostics before and after plasma treatment. Note that the mobile plasma pilot plant for VOC removal has been effectively applied not only in combination with the scrubber, but also for 'dry' cleaning (without water shower) of industrial exhaust gases.

7.4.7   Non-thermal plasma decontamination of diluted large-volume emissions of chlorine-containing VOC

Non-thermal atmospheric-pressure plasma systems, including those based on electron beams, are effectively applied in the destruction of different chlorine-containing VOC including: vinyl chloride, trichloroethylene,

Figure 7.19   *Mobile Environmental Laboratory built in Drexel University carrying precise exhaust gas characterization facility as well as 10 kW wet (spray) pulsed corona discharge for treatment of large-volume low-concentration exhaust gases containing VOC (see color plate).*

trichloroethane and carbon tetrachloride. Major specific plasma systems used for the application are electron beams, pulsed corona and DBD (see for example Penetrante and Schultheis, 1993; Penetrante *et al.*, 1996a, b; Hadidi *et al.*, 1996). As an example, consider the destruction of carbon tetrachloride CCl_4 diluted in air. A major contribution to the process kinetics according to Penetrante *et al.* (1996a, b) is provided by plasma-generated OH radicals, O and N atoms and by direct electron impact destruction through dissociative

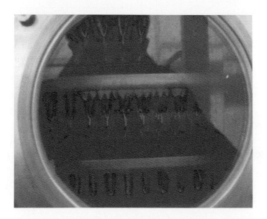

Figure 7.20   *Photo of the 10 kW pulsed corona discharge (generated in the water-spray configuration) through a window of the Mobile Environmental Laboratory (see color plate).*

Figure 7.21 *Photo of the Mobile Environmental Laboratory demonstrating all 6 access windows to the 10 kW pulsed corona discharge system applied for treatment of large volume low concentration exhaust gases containing VOC (see color plate).*

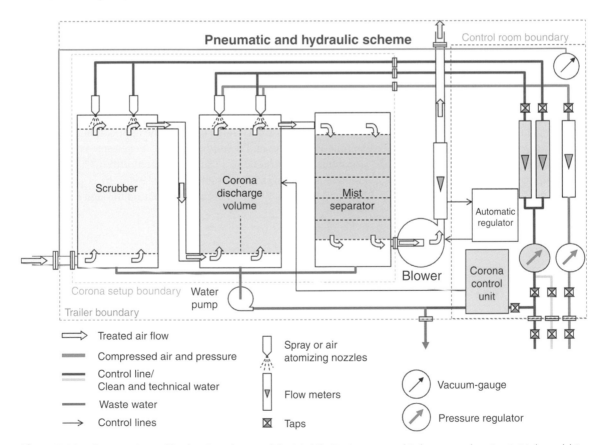

Figure 7.22 *Pneumatic and hydraulic scheme of the Mobile Environmental Laboratory showing initial scrubbing chamber, wet pulsed corona chamber, final mist separation unit and air and water flow channels.*

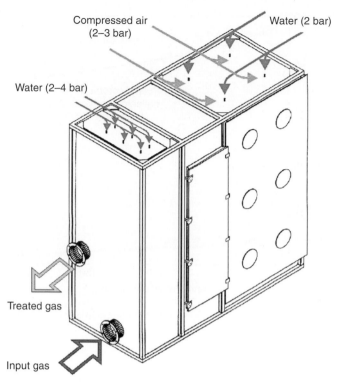

Figure 7.23 *Constructional organization of the exhaust gas treatment unit of the Mobile Environmental Laboratory.*

attachment. The radical-induced decomposition starts with the formation of OH radicals, O and N atoms:

$$e + O_2 \rightarrow O(^3P) + O(^1D) + e, \quad O(^1D) + H_2O \rightarrow OH + OH, \tag{7.59}$$

$$e + N_2 \rightarrow N(^4S) + N(^2D) + e. \tag{7.60}$$

Oxidation of carbon tetrachloride CCl_4 is then initiated by its reactions with O and OH:

$$OH + CCl_4 \rightarrow HOCl + CCl_3, \tag{7.61}$$

$$O(^3P) + CCl_4 \rightarrow ClO + CCl_3. \tag{7.62}$$

N atoms generated in reaction (7.60) promote the reduction of dichloroethylene:

$$N(^4S) + CH_2Cl_2 \rightarrow NH + CHCl_2. \tag{7.63}$$

The major initial reaction of CCl_4 decomposition through dissociative attachment starts with the elementary process (Penetrante *et al.*, 1996a, b):

$$e + CCl_4 \rightarrow CCl_3 + Cl^-. \tag{7.64}$$

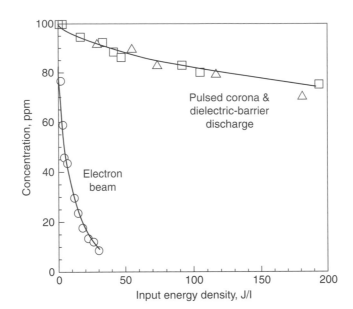

Figure 7.24 *Electron beam, pulsed corona and DBD processing of 100 ppm carbon tetrachloride in dry air at 25°C.*

Figure 7.24 depicts a comparison between electron beam, pulsed corona and DBD processing of 100 ppm CCl_4 in dry air at 25°C (Penetrante *et al.*, 1996a, b). Major products of the plasma processing of carbon tetrachloride CCl_4 in air are Cl_2, $COCl_2$ and HCl. These products can be easily removed from the gas stream; for example, they dissolve and/or dissociate in aqueous solutions and combine with $NaHCO_3$ in a scrubber solution to form NaCl.

Similar approaches can be applied for the plasma cleaning of other diluted chlorine-containing VOC exhausts. The plasma-stimulated decomposition of CCl_4 is strongly dependent on the dissociative attachment (7.64), consuming an electron per decomposition. Energy cost of the CCl_4 decomposition is therefore determined by energy cost of ionization. The ionization energy cost is lower at higher electron energies in a plasma system. The lowest ionization energy cost can be achieved in plasma generated by electron beams (c. 30 eV per anion-electron pair). To compare, the ionization energy cost in pulsed corona and DBD is usually of the order of hundreds of electron-volts. This explains the higher energy efficiency of electron beams in the destruction of carbon tetrachloride (see Figure 7.24).

The advantages of electron beams versus pulsed corona discharges have also been demonstrated (Penetrante *et al.*, 1996a, b) in the plasma treatment of 100 ppm of methylene chloride $C_2H_2Cl_2$ in dry air at 25°C (see Figure 7.25) and in the plasma treatment of 100 ppm of trichloroethylene (TCE, C_2HCl_3) in dry air at 25°C (see Figure 7.26). Reactions with N and O atoms (see above) make a significant contribution in the destruction of these chlorine-containing VOC. However, the higher energies of plasma electrons typical for electron beams allows the energy cost of not only ionization but also generation of the atoms to be reduced, hence the advantage of electron beams in destruction of methylene chloride and TCE (Figures 7.25 and 7.26).

Major by-products of the plasma destruction of TCE diluted in air are: dichloroacetyl chloride (DCAC), phosgene, hydrochloric acid and some amount of CO and CO_2. Plasma destruction of TCE in air requires significantly less energy than that of methylene chloride $C_2H_2Cl_2$ and carbon tetrachloride CCl_4 (compare Figures 7.24–7.26). This can be explained by the plasma-stimulated chain mechanism of TCE destruction,

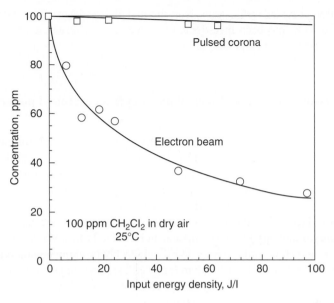

Figure 7.25   *Electron beam and pulsed corona processing of 100 ppm CH₂Cl₂ in dry air at 25°C.*

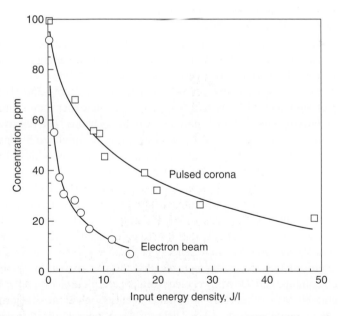

Figure 7.26   *Electron beam and pulsed corona processing of 100 ppm TCE (trichloroethylene) in dry air at 25°C.*

with the chain propagation provided by Cl atoms. Non-thermal plasma destruction of different diluted streams with chlorine-containing VOC has been investigated by several other research groups, including Futamura and Yamamoto (1996) and Oda, Han and Ono (2005).

7.4.8   Non-thermal plasma removal of elemental mercury from coal-fired power plants and other industrial air-based off-gases

Elemental mercury and mercury compounds are extremely toxic, volatile and difficult to remove from off-gases. Hg removal can be effectively performed by using non-thermal plasma, however. Major sources of mercury are combustion facilities (coal-fired power plants), municipal solid waste incinerators, hazardous waste incinerators and medical waste incinerators which, all together, account for 87% of total Hg emissions.

The US Environment Protection Agency (USEPA) established a standard regulating Hg concentration to 40 μg m^{-3} (c. 4.5 ppb). Typical off-gas emissions from incineration or a coal-fired boiler contain 100–1000 μg m^{-3} (0.1–1 ppm) of mercury in both elemental and oxidized form. While oxidized forms of mercury (Hg (II), primarily HgCl$_2$ and HgO) can be removed by wet scrubbers, elemental mercury is chemically inert, insoluble in water solutions and therefore difficult to remove. A significant portion (10–30%) of total mercury load in the coal-fired power plant off-gases stays in the elemental form (Lee *et al.*, 2004).

Typically, a combination of different removal techniques is required today to remove oxidized and elemental mercury. Oxidized mercury is typically removed by wet scrubbing; the basic pH normally used in the scrubbers (to increase efficiency of removal of acidic gases such as SO$_2$ and NO$_2$) promotes reduction of mercury into elemental form followed by re-entering the off-gas. Conventional Hg removal technologies based on adsorption, scrubbing and electrostatic precipitation (ESP) are not effective and also generate secondary wastes. Mercury capture in the conventional control devices is limited to 27% in the cold-side ESP, to 4% in the hot-side ESP and to 58% in the fabric filters.

The application of non-thermal plasma provides a significant improvement in the effectiveness of mercury capture, especially in the case of elemental mercury. Initial investigations of Hg removal from flue gas using a pulsed corona enhanced dry ESP were performed by Masuda *et al.* (1987). The pulsed corona reactor of cylindrical coaxial geometry was powered by a 55 kV peak voltage with 50–250 Hz frequency. The initial Hg concentration in air or simulated flue gas was 500 μg cm^{-3}. The experiments showed high (over 90%) removal efficiency of mercury in pulsed and DC coronas of both polarities. The removal efficiency increased as gas temperature decreased (from c. 350 to 100°C) and was higher in simulated flue gas than in air. Negative polarity discharges performed better. Removal efficiency close to 99% was achieved in the negative pulsed corona regime at energy input level of an ordinary ESP (c. 0.3 kJ m^{-3}).

Experiments on mercury removal by pulsed corona were also performed by Helfritch, Harmon and Feldman (1996). Elemental mercury vapor was introduced into the simulated flue gas at a concentration of 30 μg cm^{-3}. Pulsed corona was initiated with 27 kV pulses at 400 Hz frequency. Hg removal efficiency of 100% was achieved at energy efficiency of 10 W scfm^{-1} (5.9 kWh per 1000 m^3 or 21.2 kJ m^{-3}). For comparison, 500 W scfm^{-1} (1060 kJ m^{-3}) is a typical ratio for energy production by the coal-fired electric power plants.

Powerspan Corp. developed the process named electro-catalytic oxidation (ECO) (Alix, Neister and McLarnon, 1997), which combines DBD with wet electrostatic precipitation (WESP). The process is capable of oxidizing and removing 80–90% of mercury at initial concentrations in the off-gas of c. 26 μg cm^{-3} (out of which c. 6 μg cm^{-3} is in elemental form). The DBD reactor is operated in this case at a power level of 18 W scfm^{-1} (c. 37 kJ m^{-3}) at a flue gas flow rate of 1500 scfm.

MSE Technology Applications, Inc. performed tests on removal of mercury from air and simulated flue gases using a capillary corona reactor (Babko-Malyi *et al.*, 2000). In this reactor, oxygen and water vapor were injected through an electrode into the off-gas (at 0.1% flow rate of the main gas). At initial concentrations of elemental mercury 30–500 µg cm^{-3}, removal efficiencies of up to 97% were achieved at energy input 6 kJ m^{-3}.

7.4.9 Mechanism of non-thermal plasma removal of elemental mercury from air streams

Hg removal from off-gases is quite challenging because mercury is present in both elemental (Hg^0) and oxidized (Hg(II)) forms. These two forms of mercury atoms present in air require different capture approaches. Oxidized mercury (HgO, built from Hg(II)) can be transferred into water solution and separated from the off-gas using conventional scrubber or WESP technologies. Elemental mercury, on the other hand, is very difficult to remove because it is not soluble in water and therefore easily escapes scrubber systems. As an example, wet and dry flue gas desulfurization (FGD) systems can remove 80–90% of oxidized mercury (II), but cannot remove elemental mercury (Hg^0) at all.

Plasma stimulation of the mercury removal is mostly based on the ability of the plasma to oxidize elemental mercury (Hg^0) into mercury (II), which can then be effectively removed using conventional technologies mentioned above. Oxidation of elemental mercury (Hg^0) by molecular oxygen has very high activation barrier (above 3 eV mol^{-1}), and can therefore be completely neglected not only at room temperature but even at elevated temperatures. Oxidation of mercury therefore requires participation of stronger oxidizers. The mechanism of plasma-stimulated oxidation of mercury starts with the generation in air strong oxidizers such as atomic oxygen and electronically excited molecular oxygen, ozone and OH radicals. The mechanism and kinetics of generation of O atoms, electronically excited oxygen and ozone O$_3$ is discussed in Chapter 2. The formation of OH radicals can be due to the interaction of water with atomic oxygen and ozone:

$$O + H_2O \rightarrow OH + OH, \quad O_3 + H_2O \rightarrow 2OH + O_2, \tag{7.65}$$

as well as intensive and highly selective sequence of elementary reactions (Equations (7.46) and (7.47)) based on charge transfer to water molecules. Elemental mercury Hg^0 can be oxidized in plasma into Hg(II) in homogeneous elementary reactions with electronically excited metastable oxygen molecules:

$$Hg^0 + O_2(A^3\Sigma_u^+) \rightarrow HgO + O. \tag{7.66}$$

Another reaction path is oxidation of elemental mercury Hg^0 by ozone. The reaction has been experimentally studied at low concentration of reagents (30 ppbV of ozone) and characterized by relatively slow kinetics:

$$Hg^0 + O_3(A^3\Sigma_u^+) \rightarrow HgO + O_2, \quad k = 3 \times 10^{-20}. \tag{7.67}$$

Oxidation of elemental mercury Hg^0 by peroxide H$_2$O$_2$ also takes place in plasma, but is characterized by relatively slow kinetics similar to Equation (7.67). The presence of water clusters, droplets and microsurfaces (such as fly ash) can significantly accelerate oxidation of elemental mercury Hg^0, in particular due to the surface reaction (Babko-Malyi *et al.*, 2000):

$$Hg^0 + 2OH \rightarrow Hg^{2+} + 2OH^-. \tag{7.68}$$

In liquid or cluster phases, Hg^{2+} is stable (in particular at low pH and may also be transferred into other ionic forms of Hg(II) (Hg$_2^{2+}$, HgCl$^+$, HgCl$_4^{2-}$ etc.; Babko-Malyi *et al.* 2000). Mercury Hg (II) dissolved or captured in the clusters can be then removed from the system using conventional technologies.

The energy cost of mercury oxidation and removal through the above-discussed non-thermal plasma mechanisms is actually determined by the energy cost of plasma generation of strong oxidizers (electronically excited oxygen, OH radicals etc.). Strongly non-equilibrium DBD discharges used in the Powerspan ECO process, for example, requires 10 W h m^{-3} of the combustion of off-gases. The application of non-thermal discharges with special additive injection is able to provide higher selectivity towards removal of mercury and decrease the energy cost of the process (Babko-Malyi *et al.*, 2000).

7.5   Plasma desinfection and sterilization of water

7.5.1   Plasma water disinfection using UV-radiation, ozone and pulsed electric fields

An estimated 1.1 billion people are unable to acquire safe drinking water today, which highlights the need for improved methods of water treatment. Contaminated water can be attributed to a number of factors, including chemical fouling, inadequate treatment and deficient or failing water treatment and distribution systems. An additional important cause of contamination is the presence of untreated bacteria and viruses within the water. As estimated by the Environmental Protection Agency (EPA), nearly 35% of all deaths in developing countries are related directly to contaminated water. The increased presence of *Escherichia coli* along with various other bacteria within some areas of the US has been a cause for national concern. In an effort to inactivate these bacteria, successful experiments and commercial applications of chemical treatments, ultraviolet radiation and ozone injection units have been developed and implemented within potable water delivery systems.

The experimental success and commercialization of these water treatment methods are not, however, without deficiencies. With regard to human consumption, chemical treatments such as chlorination can render potable water toxic. UV radiation and ozone injection have also demonstrated success in bacterial inactivation in water, but the effectiveness of such methods largely depends upon adherence to regimented maintenance schedules. Plasma methods effectively combining the contribution of UV radiation, active chemicals and high electric fields are therefore considered as highly effective approaches to water treatment (Locke *et al.*, 2006; Fridman *et al.*, 2007a–d). Before considering the direct application of plasma to water treatment, we briefly consider the independent application of UV radiation, active chemicals and high electric fields for the deactivation of microorganisms in water.

7.5.1.1   *Pulsed electric fields*

The electric field associated with this technology is not strong enough (membrane potential of more than 1 V can sometimes kill bacteria) to initiate electrical breakdown in water; deactivation of microorganisms is due to electroporation, which is the creation of holes in the cell membranes. This means that plasma-originated electric fields (for example, those in DBD streamers) can be sufficient for electroporation. At nominal conditions, the energy expense for the two-log reduction of number of viable cells is relatively high, c. 30 000 J L^{-1} (Katsuki *et al.*, 2002a, b).

7.5.1.2   *UV radiation*

Plasma-generated UV radiation has proven to be effective in decontamination processes and is gaining popularity, particularly in European countries because chlorination leaves undesirable by-products in water. Most bacteria and viruses require relatively low UV dosages for inactivation, which is usually in the range of 2000–6000 µW s cm^{-2} for 90% kill. For example, *E. Coli* requires a dosage of 3000 µW s cm^{-2} for a 90% reduction. Cryptosporidium, which shows an extreme resistance to chlorine, requires a UV dosage greater than 82 000 µW s cm^{-2}. The criteria for the acceptability of UV disinfecting units include a minimum dosage of 16 000 µW s cm^{-2} and a maximum water penetration depth of c. 7.5 cm. UV radiation in the wavelength

240–280 nm causes irreparable damage to the nucleic acid of microorganisms. The most potent wavelength of UV radiation for DNA damage is c. 260 nm. Total energy cost of UV water treatment is also quite large, similarly to that for pulsed electric fields.

7.5.1.3 Ozonation

Plasma-generated ozone (O_3) is one of the most powerful and popular methods of water treatment. Ozone-gas, generated in plasma aside, is bubbled into a contaminated solution and dissolved. The ozone is chemically active and capable of efficiently inactivating microorganisms at a level comparable to chlorine. Achieving a 4-log reduction at 20°C with an ozone concentration of 0.16 mg L^{-1} requires an exposure time of 0.1 min (Anpilov *et al.*, 2001). At higher temperatures and pH levels, ozone tends to rapidly decay and requires more exposure time.

Plasma discharges, especially DBD, have been used for the production of ozone in the past several decades to kill microorganisms in water. Ozone has a lifetime of approximately 10–60 min which varies depending on pressure, temperature and humidity of surrounding conditions. Because of the relatively long lifetime of O_3, ozone-gas can be produced in air or oxygen plasma, stored in a tank and injected into water.

The bactericide effect of O_3 in water is usually characterized by the Ct factor, defined as the product of ozone concentration C (mg L^{-1}) and the required time t (min) to disinfect a microorganism in water. For example, for *Ditylum brightwelli* (important ballast water species), the Ct value is 50 mg min L^{-1}. In other words, if the ozone concentration is 2 mg L^{-1}, it takes 25 min of contact time to disinfect this organism in ballast water (Dragsund, Andersen and Johannessen, 2001). The energy efficiency of ozonation is limited by O_3 losses during storage and transportation.

Ozone and UV radiation generated in remote plasma sources are therefore effective means of water cleaning and sterilization. If plasma is generated not remotely but directly in water, the effectiveness of the treatment due to plasma-generated UV radiation and active chemicals can be much higher. Generation of plasma within water also leads to additional significant contribution of short-living active species (electronically excited molecules and active radicals such as OH, O etc.), charged particles and plasma-related strong electric fields to cleaning and sterilization (Sun, Sato, and Clements, 1997; Locke *et al.*, 2006; Fridman, Gutsol, and Cho, 2007). Direct water treatment by plasma generated in the water can obviously be much more effective.

7.5.2 Applications of pulsed plasma discharges for water treatment

The application of electrical pulses in the microsecond range to biological cells has been investigated in detail by Schoenbach *et al.* (1997), Abou-Ghazala *et al.* (2002), Joshi *et al.* (2002a), Joshi, Qian and Schoenbach (2002b) and Katsuki *et al.* (2002a, b). They used a point-to-plane geometry to generate pulsed corona discharges for bacterial (*E. coli* or *Bacillus Subtilis*) decontamination of water with a 600 ns, 120 kV square wave pulse. The wire electrode was made of tungsten with 75 μm diameter, 2 cm apart from a plane electrode.

A concentration of *E. coli* could be reduced by 3 orders of magnitude after applying 8 corona pulses to the contaminated water with the corresponding energy expenditure of 10 J cm^{-3} (10 kJ L^{-1}). Plasma pulses cause the accumulation of electrical charges at the cell membrane, shielding the interior of the cell from the external electrical fields. Since typical charging times for the mammalian cell membrane are of the order 1 μs, these microsecond pulses do not penetrate cells. Hence, shorter pulses in the nanosecond range can penetrate the entire cell, nucleus and organelles and affect cell functions, thus disinfecting them. High-voltage pulse generators were used to apply nanosecond pulses as high as 40 kV to small test chambers. Biological cells held in liquid suspension in the cuvettes were placed between two electrodes for pulsing. The power density was up to 10^9 W cm^{-3}, but the energy density was rather low (less than 10 J cm^{-3}, a value that could slightly increase the temperature of the suspension by approximately 2°C).

Heesch *et al.* (2000) also applied pulsed electric fields and pulsed corona discharges to inactivate microorganisms in water. They used four different types of plasma treatment configurations: a perpendicular water flow over two wire electrodes; a parallel water flow along two electrodes; air-bubbling through a hollow needle electrode toward a ring electrode; and a wire cylinder. They used 100 kV pulses (producing a maximum of 70 kV cm^{-1} electric field) with a 10 ns rising time with 150 ns pulse duration at a maximum rate of 1000 pulses per second. The pulse energy varied between 0.5–3 J pulse^{-1} and an average pulse power was 1.5 kW with 80% efficiency. Inactivation of microorganisms was found to be 85 kJ L^{-1} per 1-log reduction for *Pseudomonas flurescens* and 500 kJ L^{-1} per 1-log reduction for spores of *Bacillis sereus*. It was demonstrated that corona directly applied to water was more efficient than pulsed electric fields. With direct corona, Heesch *et al.* (2000) achieved 25 kJ L^{-1} per 1-log reduction for both gram-positive and gram-negative bacteria.

Pulsed plasma discharges in water have been applied for the sterilization and removal of organic compounds such as dyes by Sato, Ohgiyama and Clements (1996), Sun, Sato and Clements (1999), Sugiarto, Ohshima and Sato (2002) and Sugiarto *et al.* (2003). The streamer discharge was produced from a point-to-plane electrode, where a platinum wire of 0.2–1 mm diameter was used for the point electrode, which was positioned 1–5 cm from the ground plane electrode.

Lisitsyn *et al.* (1999) used a dense medium plasma reactor for the disinfection of various waters. The plasma reactor consisted of a rotating upper electrode operating at a range of 500–5000 rpm and a hollow conical cross-sectional end-piece. The advantage of such equipment was that the rotating action spatially homogenized the multiple micro-arcs, activating a larger effective volume of water. In addition, spinning the upper electrode also simultaneously pumped fresh water and vapors into the discharge zone.

Water sterilization in the electrohydraulic (arc) discharge reactor was investigated by Ching *et al.* (2001) and Ching, Colussi, and Hoffmann (2003). Typical operational conditions included a discharge of 135 mF capacitor bank stored energy at 5–10 kV through a 4 mm electrode gap within 40 μs with a peak current of 90 kA. They studied the survival of *E. coli* in aqueous media exposed to the above electrohydraulic discharges. They reported the disinfection of 3 L of a 4×10^7 CFU mL^{-1} *E. Coli* suspension in 0.01 M PBS at pH 7.4 by 50 consecutive electrohydraulic discharges. It was demonstrated that UV radiation emitted from the electrohydraulic discharge was the lethal agent that inactivated *E. coli* colonies rather than the thermal/pressure shocks or the active chemical species.

Detailed specific data regarding efficiency of different thermal and non-thermal discharges for water cleaning and sterilization can be found in reviews by Locke *et al.* (2006) and Fridman, Gutsol and Cho (2007), as well as in the book on this subject by Yang, Cho and Fridman (2012).

7.5.3   Energy-effective water treatment using pulsed spark discharges

The energy cost of water treatment is one the most critical parameters of the plasma process. The highest energy efficiency of water sterilization has been achieved using the pulsed spark discharge (Campbell *et al.*, 2006; Fridman, Gutsol and Cho, 2007), generated in the point-to-plane electrode configuration (see Figure 7.27). Variance in the plasma from corona to spark discharge was observed to be dependent on the gap distance measured from the anode to the grounded cathode.

A stainless steel electrode (0.18 mm) encased in silicon residing in a hollow Teflon tube provided the necessary insulation for the electrodes. The electrode extended c. 1.6 mm beyond the bottom of the glass tube, providing a region for spark discharge initiation. The critical distance between spark discharge and corona discharge was observed to be c. 50 mm between electrodes. Inter-electrode distance greater than 50 mm resulted in the corona discharge, whereas a distance less than 50 mm resulted in the spark discharge.

Figure 7.28 presents voltage, current and power profiles during a typical pulsed spark test. The initial steep rise in the voltage profile indicates the time moment of breakdown in the spark gap, after which the voltage linearly decreased with time over the next 17 μs due to a long delay time while the corona was

Figure 7.27 *Point-to-plane plasma discharge system for pulsed corona discharge and spark in water (see color plate).*

formed and transferred to a spark. The rate of the voltage drop over time depends on the capacitance. The current and power profiles show the corresponding histories which show initially sharp peaks and then very gradual changes over the next 17 μs. The duration of the initial peak was measured to be c. 70 ns. At t ~ 17 μs, there was a sudden drop in the voltage indicating the onset of a spark or the moment of channel appearance, accompanied by sharp changes in both the current and power profiles. The duration of the spark was approximately 2 μs, which was much longer than the duration of the corona.

Plasma sterilization data collected by treatment of *E. coli* in the described pulsed spark discharge (Campbell *et al.*, 2006) are presented in Table 7.7 for two different initial conditions. When the initial cell count was high (1.8×10^8 calls mL^{-1}), the spark discharge could produce a 4-log reduction at 100 pulses and 2-log reduction at c. 65–70 pulses. When the initial cell count was at an intermediate level (2×10^6 cells mL^{-1}), the spark discharge produced a 2-log reduction at 50 pulses. Taking into account energy of one pulse and volume of water, energy per litre of water for 1-log reduction in *E. coli* concentration is as low as 77 J L^{-1}. Note that similar but much more detailed experiments with spark discharge treatment of *E. coli* in water at lower bacterial concentrations (Arjunan *et al.*, 2007) resulted in similar very low values of the plasma sterilization energy cost (c. 100 J L^{-1}).

Table 7.8 compares major plasma discharges applied for water sterilization. The properties of spark discharges in water are quite unique and beneficial with regards to water treatment. The spark discharges in water investigated by Campbell *et al.* (2006) and Arjunan *et al.* (2007) require very low power in comparison to other systems such as arcs, which are too hot and localized, and coronas, which are too weak to generate UV effective for water sterilization. To achieve a 4-log reduction for a typical household flow rate of 6 gallons per minute or gpm (22.7 L min^{-1}), the electrical power requirement is only 120 W. The pulsed spark discharges indicate a potential to accommodate a 1000 gpm (3786 L min^{-1}) water flow rate while retaining the ability to achieve a 4-log reduction in biological contaminant at a power measuring only 20 kW.

7.5.4 Characterization of the pulsed spark discharge system applied for energy-effective water sterilization

7.5.4.1 *Emission spectrum characteristics of the pulsed spark discharge in water*

A TriVista scanning monochromator system (Princeton Instruments) was used to acquire the optical emission spectra from the discharge. A 3-m-long fiber optic cable consisting of a single column of ten 200 μm diameter

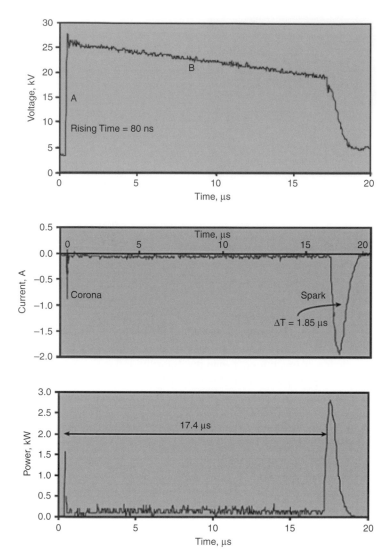

Figure 7.28 *Voltage (upper), current (middle) and power (lower) profiles measured using an oscilloscope during a typical pulsed spark test.*

fibers was pointed at the discharge and connected to the entrance slit of the monochromator in order to collect light from the discharge. The positioning and acceptance angles of the fibers made it such that the spectra acquired correspond to the light emission from the entire discharge. Optical emission spectra of the discharge were acquired in the range 200–900 nm. At the exit of the monochromator, an intensified CCD camera (Princeton Instruments Gen III PI-MAX 2) was mounted, which is triggered by the measured probe voltage to digitally record the spectra.

Figure 7.29 shows the measured spectra for the streamer-corona discharge in the range of 300–900 nm gathered over the whole duration of the discharge. The main features of the spectra indicate combination of a radiation continuum and strong atomic line emission due to the presence of hydrogen and oxygen resulting

Table 7.7 E. coli *bacterial concentrations in water following the pulsed spark discharge treatment of water.*

Experiment 1	Cells mL^{-1}	Cells mL^{-1}	Cells mL^{-1}
Number of pulses	Run A	Run B	Run C
0	1.84×10^8	1.76×10^8	1.60×10^8
20	1.01×10^8	7.40×10^7	—
40	1.54×10^7	1.01×10^7	1.85×10^7
60	1.67×10^6	4.20×10^6	3.20×10^6
80	2.10×10^5	4.80×10^5	3.30×10^5
100	9.30×10^4	2.43×10^4	3.05×10^4
Experiment 2	Cells mL^{-1}	Cells mL^{-1}	Cells mL^{-1}
Number of pulses	Run A	Run B	Run C
0	2.05×10^6	2.04×10^6	1.96×10^6
50	8.70×10^3	1.52×10^4	1.58×10^4

Table 7.8 *Comparison of major discharges used in plasma-based water treatment.*

Water treatment parameters	Pulsed arc discharges	Spark discharges	Pulsed corona discharges
Energy per liter for 1-log reduction in E. coli, J L^{-1}	860	77	30 000–150 000
Power requirement for water consumption at 6 gpm, kW	0.326	0.029	11.4–56.8
Power requirement for water consumption at 1000 gpm, kW	54.3	4.9	1892.7–9463.5
Power available in small power system (10 × 10 × 10 cm), kW	30	10	0.3
Maximum water throughput based on the above power, gpm	553	2058	0.03–0.16

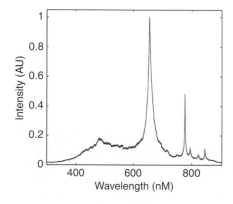

Figure 7.29 *Optical emission spectra for the corona discharge in water.*

from the decomposition of the water. The most intense peak is attributed to the Hydrogen Balmer series alpha line (H_α) at 656 nm. A very broad H_β line is also visible at 484 nm. In the infrared region, the four distinct lines are due to atomic oxygen with the strongest at 777 nm. No features were present in the UVC (200–300 nm) other than the continuing decay of the H_β. Light emission in the VUV due to emission of the hydrogen Lyman series beginning at 121 nm can be expected to occur in parallel with the Balmer lines, but were beyond our measurement capability.

Note that no significant lines due to metal ablation from the electrodes are present in the spectra. The broadening of the H_α can be mainly attributed to Stark broadening and the full width at half maximum of 15 nm measured for the H_α transition in the streamer-corona discharge corresponds to an estimated electron density in the range of 10^{19} cm^{-3}. The Stark effect results in more broadening for the H_β because the transition initiates at a closer to continuum level and is therefore more sensitive to such electronic perturbations. For the spark discharge in water, the H_α line had a width of approximately 35 nm corresponding to an electron density closer to 10^{20} cm^{-3}. Liquid water is largely transparent over this range with an absorption coefficient less than 0.01 cm^{-1} from 200 nm to 800 nm. Below 190 nm, water is strongly absorbing and the IR absorption up to 1000 nm is sensitive to temperature, although typically < 0.3 cm^{-1}.

7.5.4.2 *Preparation of microorganisms*

E. coli is considered as the most reliable measure of public risks in drinking water since its presence is an indicator of fecal pollution and the possible presence of enteric pathogens. A non-pathogenic strain of *E. coli* bacterium, K12, was used in all the experiments. Cell stocks of *E. coli* were prepared by incubating cultures in Luria-Bertani-Miller (LB) broth (10 g of tryptone, 5 g of yeast extract and 10 g of sodium chloride per liter) for 20–22 hours at 37°C. *E. coli*, obtained in stationary phase, were then centrifuged at 3500 rpm for 10 min, washed twice in sterile spring water and finally resuspended to population densities of 10^8 and 10^6 CFU mL^{-1} of water. Water of conductivity varying over the range 100–200 μS cm^{-1} was used in the experiments.

7.5.4.3 *UV absorber*

In order to absorb UV radiation, 2, 2′-dihydroxy-4, 4′- dimethoxybenzophenone-5, 5′-disulfonic acid (benzophenone-9, BASF), a sunscreen agent of the o-hydroxybenzophenone class, was used. It is a non-toxic ingredient commonly found in commercial water-soluble sunscreens which dissipates the absorbed energy without radiation. Various degrees of UV absorption in water were achieved by varying the concentration of benzophenone-9 (BP-9). BP-9 solutions of concentrations 1, 3, 10, 30, 100, 300, 1000 and 3000 mg L^{-1} of water were prepared and absorbance was measured using a Perkin-Elmer Lambda 40 UV-Vis Spectrometer.

7.5.4.4 *Plasma treatment and analysis*

Water contaminated with *E. coli* was treated with a pulsed spark discharge and three samples were taken at regular intervals. Samples for determining the initial population were taken before applying the plasma discharge. To analyze the treatment results, serial dilutions of the samples were prepared using sterile water and enumerated using a spread plate-counting method. 100 μL aliquots of diluted samples were spread on brain heart infusion (BHI) agar plates, incubated at 37°C for 12–18 hours and the colony-forming units (CFU) were counted. The effectiveness of the spark discharge in inactivating *E. coli* has been expressed in terms of D-value, which is identified as the energy required to achieve a 1-log$_{10}$ reduction in bacterial concentration at the specific plasma treatment condition (Moisan *et al*., 2001; Moisan, 2002).

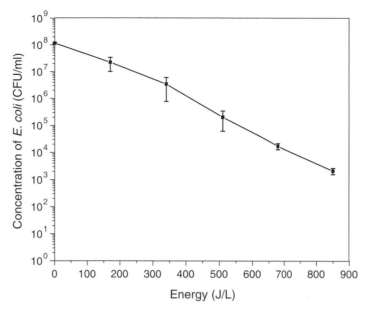

Figure 7.30   *Survival plot obtained for an* E. coli *concentration of 10^8 CFU mL^{-1}.*

7.5.5   Analysis of D-value and role of UV radiation in inactivation of microorganisms in water

7.5.5.1   *Inactivation of E. coli in water*

Inactivation experiments were conducted for *E. coli* concentrations of 10^8 and 10^6 CFU mL^{-1} of water. Two suspensions of different concentrations were chosen to check any dependence of initial bacterial concentration on D-value. Figures 7.30 and 7.31 show the survival curves obtained for these concentrations using spark discharge in water. For an *E. coli* concentration of 10^8 CFU mL^{-1} the D-value was found to be 174 J L^{-1} of water; for a lower concentration of 10^6 CFU mL^{-1}, a low D-value of 14 J L^{-1} was obtained. This indicates some dependence of D-value on initial bacterial concentration. As the initial bacterial concentration increased, the D-value also increased. This loading effect may be due to the inability of UV radiation, produced by spark discharge, to reach *E. coli* through water lacking transparency. At high concentrations, *E. coli* can aggregate to each other partially shielding one another; this prevents the active species from effectively attacking them.

In an attempt to optimize the spark discharge system, the energy per pulse was varied by varying the number of capacitors in the capacitor bank. Figure 7.32 shows the survival plot of *E. coli* for different energies per pulse. For energy per pulse of 1.7 J, the D-value obtained was 187 J L^{-1}. A low D-value of 98 J L^{-1} was obtained for an energy per pulse of 1 J. It may be assumed that only a portion of the energy input into the water contributes to the inactivation of microorganisms, while the rest of the energy is dissipated into the water. For energies per pulse of 0.68 and 0.34 J, D-values were 140 and 366 J L^{-1} respectively. The optimized treatment system corresponds to a minimum D-value of 187 J L^{-1} and energy per pulse of 1 J. Further experiments have been conducted with this optimized system.

7.5.5.2   *Role of ultraviolet radiation in the inactivation of microorganisms*

The transmission spectrum for wavelengths between 200 and 800 nm was obtained for BP-9 solutions of concentrations 3, 30, 300 and 3000 mg L^{-1}. Figure 7.33 shows the relative transmittance for various

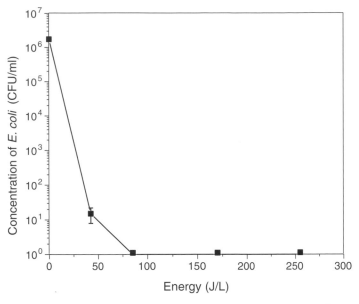

Figure 7.31   *Survival plot obtained for an* E. coli *concentration of 10^6 CFU mL^{-1}.*

concentrations of BP-9 solution. It could be seen that BP-9 solution of concentration 3 mg L^{-1} transmitted a major portion of UV-A, B and C. BP-9 solution of concentration 30 mg L^{-1} absorbed a significant portion of UV-C, UV-B and some UV-A. A concentration of 3000 mg BP-9 L^{-1} completely absorbed UV-A, B and C. The concentrations of 3, 30 and 3000 mg L^{-1} were hence selected for conducting the inactivation

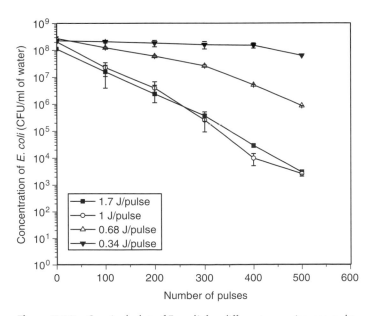

Figure 7.32   *Survival plot of* E. coli *for different energies per pulse.*

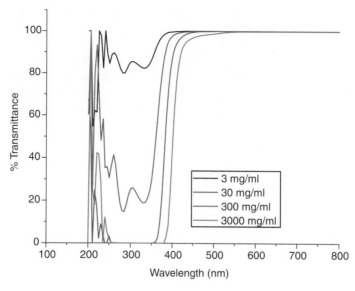

Figure 7.33 *Transmission spectrum for different concentrations of BP-9 solution.*

experiments. As a control, the sensitivity of *E. coli* to BP-9 was tested by exposing 1.44×10^7 CFU mL^{-1} to 7.5 g BP-9 L^{-1} of water for 2 hours. No change in bacterial concentration was observed.

Figure 7.34 shows the survival curves for various concentrations of BP-9. It could be observed that disinfection of *E. coli* in water was almost completely suppressed by adding 30 mg of BP-9 per liter of water. Similar experiments have been conducted to understand the role of UV radiation in the inactivation of

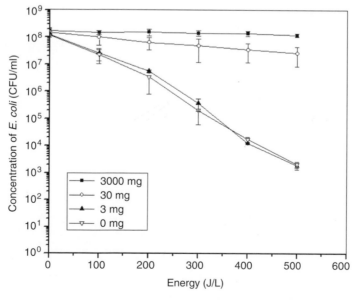

Figure 7.34 *Survival plot of* E. coli *for various concentrations of BP9.*

microorganisms in water by pulsed discharges (Ching, Colussi and Hoffmann, 2003). Ultraviolet radiation is known to kill microorganisms in water (Wolfe, 1990). DNA has an important absorption peak near 254 nm. The UV photons can cause irreparable damage to the bacterial DNA, thus inactivating them (Moisan *et al.*, 2001). These results indicate that UV radiation produced by spark discharge in water plays a major role in inactivating microorganisms. Generally, the spark discharge in water is very effective in sterilizing water. Energy efficiency of the sterilization is high compared to conventional sterilization methods. Of the various factors produced by a spark discharge in water, it was observed that ultraviolet radiation plays a major role in inactivating microorganisms.

Current sterilization techniques are far from being environmentally friendly. In developing countries, about a quarter of the human population has limited access to electricity. This energy-efficient treatment system could be an excellent alternative to the current methods of water sterilization.

8

Plasma Treatment of Blood

8.1 Plasma-assisted blood coagulation

8.1.1 General features of plasma-assisted blood coagulation

Blood coagulation is an important issue in medicine, in particular regarding wound treatment. Thermal plasma has been traditionally used for this application in form of the so-called *cauterization devices* such as argon plasma coagulators (APC) and argon beam coagulators. Widely used in particular in surgery, plasma in these devices is simply a source of local high-temperature heating which cauterizes (actually cooks) the blood. The thermal RF discharge (frequency above 350 kHz) of the APC generates high-temperature (10 000 K) plasma in flowing argon, which leads to rapid cauterization and tissue desiccation. APC devices are widely used today in surgery, and represent a good example of clinical application of plasma.

The recent development of effective non-thermal plasma-medical systems enables effective blood coagulation to be achieved without any of the thermal effects mentioned above. In such systems, the coagulation is achieved through non-thermal plasma stimulation of specific natural mechanisms in blood without any 'cooking' or damage to surrounding tissue (Fridman *et al.*, 2006).

Both coagulating the blood and preventing the coagulation could be needed, depending on the specific application. For example, in wound treatment the surgeon would want to close the wound and sterilize the surface around. Flowing blood, in that case, would prevent wound closure and create the possibility of re-introduction of bacteria into the wound. Where blood coagulation would be detrimental is, for example, in sterilization of stored blood in blood banks. A potential exists for blood to contain or to have somehow acquired bacterial, fungal or viral infection which needs to be removed for this blood to be usable. In this case, the treatment cannot coagulate the blood. An understanding of the mechanisms of blood coagulation by non-thermal plasma is therefore needed (Fridman *et al.*, 2006, 2007a–d). It should be mentioned that Frank Clement and his group (Clement *et al.*, 2009) recently collected interesting plasma-medical results focused on understanding of mechanisms of blood coagulation.

8.1.2 Experiments with non-thermal atmospheric-pressure plasma-assisted in vitro blood coagulation

FE-DBD plasma was experimentally confirmed to significantly hasten blood coagulation in vitro (Fridman *et al.*, 2006, 2007a–d). Visually, a drop of blood drawn from a healthy donor and left on a stainless steel surface

Plasma Medicine, First Edition. Alexander Fridman and Gary Friedman.
© 2013 John Wiley & Sons, Ltd. Published 2013 by John Wiley & Sons, Ltd.

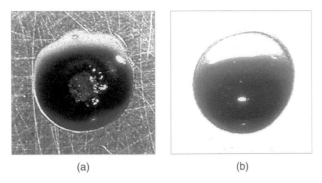

(a)   (b)

Figure 8.1   *Blood drop treated by FE-DBD: (a) after 15 s of FE-DBD and (b) control; photo was taken 1 min after the drops were placed on brushed stainless steel substrate.*

coagulates on its own in c. 15 min; a similar drop treated for 15 s by FE-DBD plasma coagulates in under 1 min (Figure 8.1). FE-DBD treatment of cuts on organs leads to similar results where blood is coagulated without any visible or microscopic tissue damage. Figure 8.2 shows a human spleen treated by FE-DBD for 30 s. Blood is coagulated and tissue surrounding the treatment area looks 'cooked'; the temperature of the cut however remains at room temperature (even after 5 min of FE-DBD treatment) and the wound remains wet, which could potentially decrease healing time.

Additionally, a significant change in blood plasma protein concentration is observed after treatment by plasma of blood plasma samples in healthy patients, patients with hemophilia and blood samples with various anticoagulants. Anticoagulants, like sodium heparin or sodium citrate, are designed to bind various ions or molecules in the coagulation cascade, thus controlling coagulation rate or preventing it. Analysis of changes in

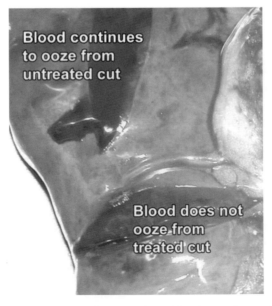

Figure 8.2   *After 30 s of FE-DBD treatment of human spleen, blood coagulates without tissue damage. Top cut: blood continues to ooze from an untreated area; bottom cut: blood coagulates while the wound remains wet.*

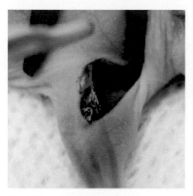

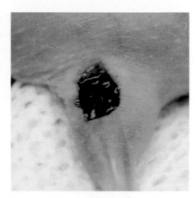

Saphenous vein
is a major blood vessel for a mouse

(a)

If left untreated following a cut
animal will bleed out (control)

(b)

15 seconds at 0.8 Watt/cm² stops the
bleeding completely right after treatment

(c)

Figure 8.3   *Blood coagulation of a live animal. (a) Saphenous vein is a major blood vessel for a mouse; (b) if a cut is left untreated, the animal will bleed out and (c) after 15 s at 0.8 W cm⁻², the bleeding is stopped (see color plate).*

concentration of blood proteins and clotting factors indicates that FE-DBD aids in promoting the advancement of blood coagulation or, in other words, plasma is able to catalyze the biochemical processes taking place during blood coagulation (Fridman *et al.*, 2005a, b, 2006, 2007a–d).

8.1.3   In-vivo blood coagulation using FE-DBD plasma

Plasma stimulation of in vivo blood coagulation has been demonstrated by Fridman *et al.* (2007a–d) in experiments with live hairless SKH1 mice. Fifteen seconds of FE-DBD plasma treatment can coagulate blood at the surface of a cut Saphenous vein (Figure 8.3) and vein of a mouse. Only the ability of direct non-thermal plasma treatment to coagulate blood was tested in these experiments and the animal was not left alive to test improvement in healing times. Full in vivo investigation of ability of plasma to hasten wound healing through sterilization and blood coagulation is discussed in Fridman *et al.* (2007a–d).

8.1.4   Mechanisms of non-thermal plasma-assisted blood coagulation

Detailed biochemical pathways of non-thermal plasma-stimulated blood coagulation remain largely unclear. Several possible mechanisms have however been investigated (Fridman *et al.*, 2006; Kalghatgi *et al.*, 2007a, b). Firstly and most importantly, it was demonstrated that direct non-thermal plasma can trigger natural, rather than thermally induced, coagulation processes. Secondly, it was observed that the release of calcium ions and change of blood pH level, which could be responsible for coagulation, is insignificant. Instead, the evidence points to selective action of direct non-thermal plasma on blood proteins involved in natural coagulation processes. The mechanisms of plasma interaction with blood can be deduced from the facts observed in experiments with FE-DBD plasma:

1. plasma can coagulate both normal and anticoagulated blood, but the rate of coagulation depends on the anticoagulant used;

2. plasma is able to alter the ionic strength of the solution and change its pH, but normal and anticoagulated blood buffers these changes even after long treatment time;
3. plasma changes the natural concentration of clotting factors significantly, thus promoting coagulation;
4. plasma effects are non-thermal and are not related to gas temperature or the temperature at the surface of blood;
5. plasma is able to promote platelet activation and formation of fibrin filaments, even in anticoagulated blood.

Further observations of the blood coagulation effect stimulated by FE-DBD plasma, helpful for understanding the process mechanism, can be summarized as follows.

1. Anticoagulants such as sodium heparin bind thrombin in the coagulation cascade, thus slowing coagulation. Sodium citrate or ethylene diamine tetraacetic acid (EDTA) are however designed to bind calcium, an important factor in the cascade, thereby preventing coagulation altogether. Plasma treatment promotes visible coagulation in blood with all of the above anticoagulants.
2. Initial plasma coagulation hypothesis was focused on increase in concentration of Ca^{2+}, which is an important factor in the coagulation cascade. It was suggested that plasma stimulates generation of Ca^{2+} through the redox mechanism:

$$[Ca^{2+}R^{2-}] + H^+_{H2O} \Leftrightarrow [H^+R^{2-}]_{H2O} + Ca^{2+}_{H2O}$$

    provided by hydrogen ions produced in blood in a sequence of ion-molecular processes induced by plasma ions. The validity of this hypothesis was tested experimentally by measuring Ca^{2+} concentration in the plasma-treated anticoagulated whole blood using a calcium selective microelectrode. Calcium concentration was measured immediately after plasma treatment and remained almost constant for up to 30 s of treatment; it then increased slightly for prolonged treatment times of 60 s and 120 s. Although plasma is capable of coagulating anticoagulated blood within 15 s, no significant change occurs in calcium ion concentration during the typical time of blood coagulation in discharge-treated blood. In vivo, the pH of blood is maintained within a very narrow range of 7.35–7.45 by various physiological processes. The change in pH by plasma treatment (c. 0.1 after 30 s) is less than the natural variation of pH, which indicates that the coagulation is not due to pH change in blood.
3. FE-DBD treatment of whole blood sample changes concentrations of proteins participating in the coagulation cascade. Plasma treatment is shown to 'consume' coagulation factors (proteins and enzymes) and a visible film is formed on the surface of the treated samples. An increase in the sample volume and keeping the surface area fixed decreases the effect, indicating that plasma treatment initiates clot formation at the surface and not in the volume (Figure 8.4). A corresponding kinetic model of the plasma-assisted blood coagulation indicates a two-fold decrease in clot formation time with plasma treatment (Figure 8.5).
4. When the surface of blood is protected by small thin aluminum foil, which prevents contact between blood and FE-DBD plasma but transfers all the heat generated by plasma, no influence of blood is observed. This proves non-thermal mechanism of plasma-stimulated blood coagulation.
5. The final step in the natural biological process of blood coagulation is the production of thrombin which converts fibrinogen into fibrin monomers that polymerize to form fibrin microfilaments. FE-DBD plasma treatment of fibrinogen solution in physiological medium coagulates it, which is confirmed visually through a change in the color of the solution (from clear to milky-white) and through dynamic light scattering. Note that plasma does not influence fibrinogen through a pH or temperature change. FE-DBD treatment, however, is unable to polymerize albumin (directly not participating in coagulation cascade)

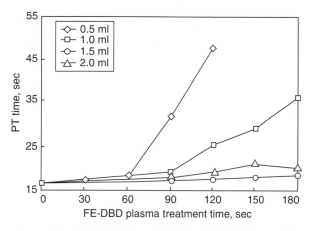

Figure 8.4 *Prothrombin (PT) time for blood samples of different volumes with the same surface area of FE-DBD treatment.*

as no change in its behavior is observed either visually or through dynamic light scattering (DLS). Non-thermal plasma therefore selectively affects proteins (specifically, fibrinogen) participating in the natural coagulation mechanism.

8.1.5 Influence of protein activity

To assess the influence of plasma on protein activity, compared to plasma influence on the protein itself, trypsin (pretreated with L-1-tosylamido-2-phenylethyl chloromethyl ketone or TPCK to inhibit contaminating chymotrypsin activity without affecting trypsin activity) was treated by plasma for up to 2 min and its total protein weight and protein activity analyzed via fluorescence spectroscopy. Total protein weight, or the amount of protein in the treated solution, remains practically intact after up to 90 s of treatment (Figure 8.6) while the enzymatic (catalytic) activity of this protein drops to nearly zero after 10–15 s of treatment.

Similar behavior is also observed for albumin. This proves that the plasma effect on proteins is not just destructive but quite selective and 'natural'. Morphological examination of the clot layer by scanning electron

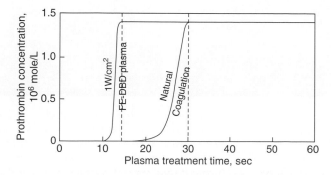

Figure 8.5 *Prothrombin kinetics: twofold decrease in clot formation time with plasma treatment.*

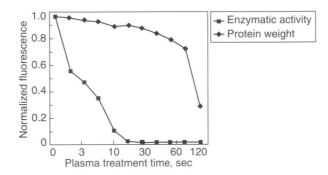

Figure 8.6   *Total protein weight compared to enzymatic activity of Trypsin following plasma treatment.*

microscopy (SEM) further proves that plasma does not 'cook' blood but initiates and enhances the natural sequences of blood coagulation processes. Activation followed by aggregation of platelets is the initial step in the coagulation cascade, and conversion of fibrinogen into fibrin is the final step. Figure 8.7 shows extensive platelet activation, platelet aggregation and fibrin formation following FE-DBD plasma treatment.

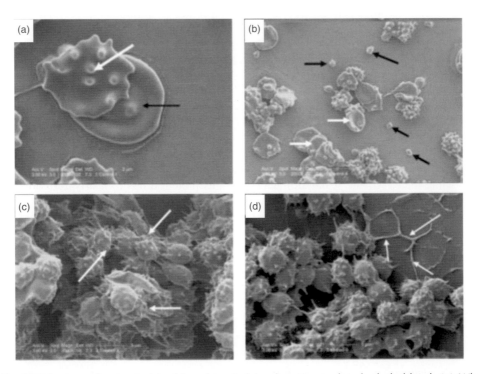

Figure 8.7   *SEM images of untreated (a, b) and treated (c, d) anticoagulated whole blood. (a) Whole blood (control) showing single activated platelet (white arrow) on a red blood cell (black arrow); (b) whole blood (control) showing many non-activated platelets (black arrows) and intact red blood cells (white arrows); (c) whole blood (treated) showing extensive platelet activation (pseudopodia formation) and platelet aggregation (white arrows); and (d) whole blood (treated) showing platelet aggregation and fibrin filament formation (white arrows).*

8.2 Effect of non-thermal plasma on improvement of rheological properties of blood

8.2.1 Control of low-density-lipoprotein (LDL) cholesterol and blood viscosity

Plasma is not only able control blood coagulation, but can also control and regulate multiple physical and biochemical properties of blood. One of these properties is blood viscosity analyzed recently by Jung *et al.* (2012). The viscosity of a fluid represents the frictional resistance between a moving fluid and stationary wall. Blood viscosity is the inherent resistance of blood to flow and represents the thickness and stickiness of blood. Since c. 45% of blood volume is made up of suspended cellular particles, primarily red blood cells, the blood behaves as a non-Newtonian fluid where its viscosity varies with shear rate (i.e. the ratio of flow velocity to lumen diameter). The dynamic range of the whole blood viscosity (WBV) is relatively large, that is, 40–450 millipoise (mP) (or 4–45 centipoise). This highlights the potential utility of this parameter as a biomarker, to the degree that viscosity provides additional incremental prediction of clinical outcomes and is modifiable by therapeutic modalities.

WBV has been strongly associated with cardiovascular disease, stroke and peripheral arterial disease. Mechanical interaction between blood and blood vessels mediated by the increase of WBV has a crucial pathogenic role in the release of endothelium-derived mediators (NO and endothelia), thus causing subsequent vascular remodeling by activation of endothelium, initiation of inflammation, alteration of lipid metabolism and finally progression of atherosclerotic vascular disease.

There are a number of variables for the WBV, which include hematocrit, plasma proteins (i.e. fibrinogen, immunoglobulin and albumin), total cholesterol, low-density lipoprotein (LDL) cholesterol, high-density lipoprotein (HDL) cholesterol and triglyceride. In addition, the aggregation and deformability of erythrocytes critically affect the WBV. Since both the plasma proteins and LDL molecules influence the aggregation of erythrocytes, it is important to keep both levels within the respective normal ranges. For example, lipid-lowering statin drugs have been widely used to keep the LDL cholesterol within the normal range (i.e. 62–130 mg dL^{-1}). Statins are powerful cholesterol-lowering drugs in clinical practice and have a life-saving potential in properly selected patients, particularly those with severe hyperlipidemia and atherosclerotic disease. Results from randomized clinical trials have demonstrated a decrease in congested heart diseases (CHD) and total mortality, reductions in myocardial infarctions, revascularization procedures, stroke and peripheral vascular disease. However, statins are prescribed for less than half of the patients who should receive this therapy, unfortunately, because of the side effect which remains a major impediment to the appropriate use of these drugs: liver and muscle toxicity.

Non-thermal plasma can be a powerful tool to solve the above-described issues. It was demonstrated that selective coagulation of fibrinogen, which is one of the major viscosity determinants in blood, could be induced by the application of dielectric barrier discharges (DBD) in air (Kalghatgi *et al.*, 2007a, b). It is therefore hypothesized that the hemorheological properties of blood may be improved by lowering the levels of fibrinogen and LDL molecules in blood plasma through DBD-assisted coagulation and the subsequent filtration of the coagulated particles from the treated blood plasma. In the following section, we discuss the study of the feasibility of reducing WBV through DBD treatment of blood plasma and subsequent filtration (Jung *et al.*, 2012).

8.2.2 Plasma-medical system for DBD plasma control of blood properties

8.2.2.1 Setup for DBD blood treatment

The DBD system used in the present study was identical to that used by Kalghatgi *et al.* (2007a, b) (Section 8.1), that is, an atmospheric-pressure dielectric barrier discharge (DBD) as illustrated in Figures 8.8 and 8.9. The DBD system was operated with one dielectric-covered powered electrode and the other grounded

Figure 8.8   *Application of DBD to the blood plasma sample (see color plate).*

electrode. Discharge was ignited when the powered electrode approached the surface of the sample to be treated at a distance of less than 3 mm. A pulsed high voltage of 10–35 kV (peak-to-peak) with alternating polarity at 1 kHz frequency was applied between the quartz insulated copper electrode and the surface of the sample (blood plasma) to generate DBD. The power applied to the DBD system was analyzed by measuring the current passing through the discharge gap and the voltage drop in the gap when the discharge was generated. The current and voltage signals were acquired and recorded by a two-channel digital phosphor oscilloscope (DPS). The surface power density which corresponded to the applied power was around 1.5 W cm^{-2}.

The electrode was constructed of a 25 mm diameter copper rod enclosed in polytherimide (Ultem®). A 1-mm-thick fused quartz was used to prevent an arc formation by limiting the current in the DBD discharge. The presence of the insulating layer prevented the buildup of high current and subsequent heating of the gas in the discharge gap, so that biological samples could be treated without thermal damage. The gap between the

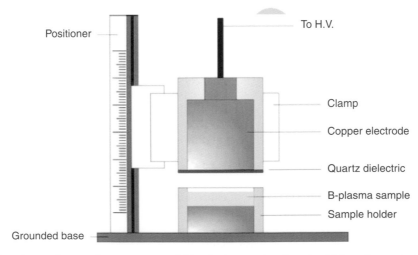

Figure 8.9   *Experimental setup for treatment of blood plasma showing the high voltage electrode, dielectric barrier and the sample holder.*

bottom of the dielectric quartz glass covering the copper electrode and the surface of the blood plasma sample was adjusted to be 2 mm using a precision vertical positioner as shown in Figure 8.9. To treat the sample with DBD, 4 mL of blood plasma were placed in the sample holder (see Figure 8.9) which was composed of a 25-mm-diameter copper rod inserted into polytherimide (Ultem®) shell. The bottom of the copper rod inside the sample holder was placed on a 21-mm-thick polycarbonate plate as the grounded base electrode.

8.2.2.2 Experimental procedure

Human whole blood of 150 mL with 1.5 mL of EDTA as an anticoagulant was obtained from Lampire Biological Laboratories in a glass bottle. Both the WBV and hematocrit for the blood sample were measured to determine the baseline data prior to the DBD treatment. The whole blood in the 150 mL of bottle was then transferred to several 10 mL polyethylene tubes to prepare blood plasma from whole blood using a centrifuge with 1200 G of relative centrifugal force (RCF) for 15 min. The blood plasma viscosity was measured before DBD treatment for the baseline data. Each blood plasma sample separated from whole blood was then treated with DBD for 1, 2, 4, and 8 min to investigate the effect of DBD treatment on the rheological properties of blood plasma and whole blood. Temperature change was monitored before and after the DBD treatment.

The viscosity of the blood plasma after DBD treatment was measured again with viscometers to compare the change in the plasma viscosity before and after DBD treatment. To evaluate the change in the WBV due to the coagulated particles generated by DBD treatment, the treated blood plasma samples were mixed back with the original erythrocytes which had been separated in the previous step. Here, hematocrit concentration was also measured to check whether or not any evaporation of blood plasma had occurred during the DBD treatment, which includes the transfer of blood to containers, the separation of red blood cells from whole blood and mixing treated blood plasma samples with erythrocytes.

After this step, the blood plasma samples treated with DBD were filtered using a syringe filter (25 mm, pore size 0.20 μm), mixed with the original erythrocytes and other cells and the viscosities of the mixed whole blood samples were measured again. Hematocrit was measured again to examine any possible plasma loss at filter medium during DBD treatment and filtration process.

8.2.2.3 Laboratory analysis

WBV was measured at 37°C using Hemathix viscometer over a range of shear rates from 1 to 1000 s^{-1}. The viscosity of the DBD-treated blood plasma was also measured using Brookfield viscometer at two different shear rates of 225 and 450 s^{-1}. The temperature of the test sample at Brookfield viscometer was maintained at a constant temperature of 37°C with a constant temperature water bath and circulator. Since some plasma volume could be lost during the DBD treatment and filtration, the WBV was normalized to a standard hematocrit of 45% using a standard hematocrit-correction method.

8.2.2.4 Statistical analysis

The mean and standard deviations for both whole blood and blood plasma viscosities were calculated. WBV was reported both at the observed (native) hematocrit and normalized hematocrit of 45%.

8.2.3 DBD plasma control of whole blood viscosity (WBV)

The WBVs of blood samples measured using the Hemathix Analyzer at five different shear rates of 1000, 300, 100, 10, 1 s^{-1} before DBD treatment are given in Table 8.1, including the baseline mean and standard deviation values (see 'no-treatment' case) for both uncorrected and normalized hematocrits.

Table 8.1   *The Change of whole blood viscosity (WBV) after DBD treatment.*

DBD plasma treatmenttime	Shear rate (s^{-1})	(Uncorrected) whole blood viscosity (Mean $\pm$ SD; mP)	(Normalized) whole blood viscosity (Mean $\pm$ SD; mP)
No treatment	1000	34.65 $\pm$ 1.63	35.38 $\pm$ 3.28
	300	38.60 $\pm$ 3.39	39.29 $\pm$ 2.48
	100	44.85 $\pm$ 4.74	45.68 $\pm$ 2.82
	10	83.85 $\pm$ 11.95	86.08 $\pm$ 8.13
	1	286.90 $\pm$ 53.60	301.02 $\pm$ 56.17
DBD treated (1 min)	1000	33.79 $\pm$ 3.38	38.99 $\pm$ 0.82
	300	37.25 $\pm$ 3.87	41.92 $\pm$ 0.83
	100	43.21 $\pm$ 6.25	47.89 $\pm$ 0.59
	10	82.94 $\pm$ 11.51	93.23 $\pm$ 5.10
	1	291.25 $\pm$ 49.75	326.12 $\pm$ 44.05
DBD treated (2 min)	1000	43.73 $\pm$ 7.45	41.47 $\pm$ 5.06
	300	46.35 $\pm$ 7.41	43.87 $\pm$ 4.71
	100	52.63 $\pm$ 6.56	49.66 $\pm$ 3.11
	10	101.28 $\pm$ 12.95	93.56 $\pm$ 4.08
	1	370.35 $\pm$ 62.98	327.75 $\pm$ 27.74
DBD treated (4 min)	1000	39.50 $\pm$ 7.91	38.28 $\pm$ 4.48
	300	43.50 $\pm$ 7.77	42.19 $\pm$ 3.90
	100	50.27 $\pm$ 7.86	48.75 $\pm$ 3.04
	10	99.33 $\pm$ 17.95	95.28 $\pm$ 6.14
	1	353.97 $\pm$ 64.75	336.47 $\pm$ 45.00
DBD treated (8 min)	1000	46.27 $\pm$ 9.26	44.50 $\pm$ 3.79
	300	49.30 $\pm$ 9.92	47.36 $\pm$ 3.79
	100	55.23 $\pm$ 9.54	53.23 $\pm$ 2.01
	10	104.90 $\pm$ 15.66	101.01 $\pm$ 5.86
	1	383.83 $\pm$ 60.19	368.10 $\pm$ 65.41

The normalized WBV of untreated blood varied from 35.38 to 301.01 mP when shear rate decreased from 1000 to 1 s^{-1}. When blood plasma was treated with DBD, a white layer formation was found on the top surface of the treated blood plasma samples and at the interfacing edge with a copper surface, whereas the blood plasma without DBD treatment did not exhibit any white particle formation on the sample surface after being exposed to open atmosphere for the same duration (see Figure 8.10). The viscosity of blood plasma treated with DBD measured with Brookfield viscometer at two different shear rates – 225 and 450 s^{-1} – is shown in Figure 8.11. The viscosity of blood plasma showed irregular values in accordance with the DBD treatment time for both shear rates.

The values of WBV for the blood samples mixed with blood plasma treated by DBD are given in Figures 8.11 and 8.12 and in Table 8.1 at five different shear rates at each treatment time (i.e. 1, 2, 4, and 8 min). Over the range of shear rates, the normalized WBV increased with increasing DBD treatment time. WBV measured at a high shear rate of 300 s^{-1} is usually referred to as systolic whole blood viscosity (SBV), whereas WBV measured at a low shear rate of 1 s^{-1} is referred to as diastolic whole blood viscosity (DBV) as WBV varies during a cardiac cycle in the same way as blood pressure. As shown in Figure 8.12 and Table 8.1, DBV values were significantly higher in DBD-treated samples (i.e. 368.10 $\pm$ 65.41 mP for 8 min

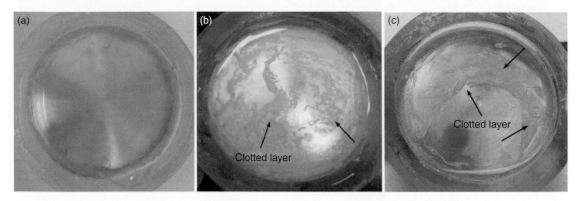

Figure 8.10 *Clotted formation of white layer in blood plasma sample with DBD treatment: (a) before DBD treatment, showing no coagulation; (b) after DBD treatment for 4 min, exhibiting a partially coagulated layer formation; and (c) after DBD treatment for 8 min, showing a white clotted layer on the overall surface of copper (see color plate).*

of DBD treatment) than untreated control samples (301.02 ± 56.17 mP). SBV values were also elevated from 38.60 ± 3.39 mP for control to 49.30 ± 9.92 mP (for 8 min of DBD treatment).

WBVs of blood samples made with DBD-treated blood plasma, which was filtered before being mixed with the original red blood cells, are shown in Figures 8.13 and 8.14 and also in Table 8.1. For SBV values shown in Figure 8.13, the systolic blood viscosity after 8 min of DBD treatment without filtration increased by 5.5% from the baseline value of 35.4 mP. When the DBD-treated plasma was filtered before being mixing with red blood cells, the systolic blood viscosity decreased by 9.1% from the baseline SBV of 35.4 mP. The diastolic blood viscosity of blood treated by DBD but without filtration was elevated with increasing DBD treatment time. For example, the DBV of blood increased by 29.9% from the baseline value of 301.0 mP (see

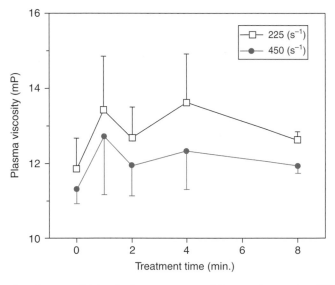

Figure 8.11 *The viscosity change of blood plasma at two different shear rates (225 s⁻¹ and 450 s⁻¹) in accordance with different DBD treatment times.*

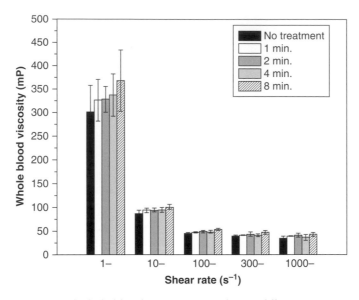

Figure 8.12   *Increase of whole blood viscosity according to different DBD treatment times.*

Figure 8.14). When the DBD-treated plasma was filtered before being mixing with red blood cells, the DBV of blood for 8 min of treatment decreased by 17.7% from the baseline value of 301 mP.

8.2.4   DBD plasma effect on improvement of rheological properties of blood

When blood plasma was treated by DBD (Jung *et al.*, 2012) a white clot layer was formed at the top of blood plasma sample (see Figure 8.10), suggesting that the plasma proteins and lipids in blood plasma might have

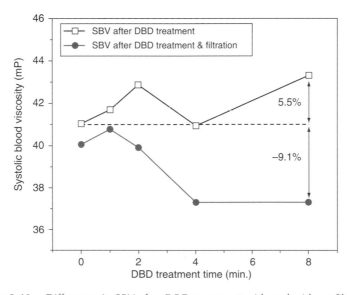

Figure 8.13   *Difference in SBV after DBD treatment with and without filtration.*

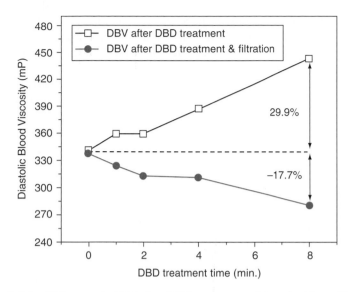

Figure 8.14   *Difference in DBV after DBD treatment with and without filtration.*

precipitated into coagulated particles through DBD treatment. Accordingly, if these coagulated particles are removed, some improvement in the rheological properties of whole blood may be anticipated. Note that the size of fibrinogen (diameter of 5–7 nm and length of 48 nm) and LDL molecules (approximately 22 nm) are too small to be removed by filter. The application of DBD treatment to blood plasma helps the excess fibrinogen and LDL molecules in blood plasma to be removed with a regular filter via DBD-assisted oxidation and coagulation.

Although DBD discharge employed in this study is non-thermal there could be some amount of thermal energy to be transferred to the blood plasma sample, causing the creation of the white layer of coagulated particles. To test this hypothesis, temperature was measured before and after the treatment of the samples with DBD. No significant change in temperature was observed, suggesting that the coagulated particles must have come from the oxidation and subsequent coagulation by active species generated by DBD. In addition, there was no change in hematocrit before and after DBD treatment, indicating no significant evaporation in blood plasma during the DBD treatment procedure.

The blood mixed with the blood plasma treated by DBD without filtration showed a significant increase in WBV. This can be attributed to the fact that plasma proteins and lipids formed large groups of coagulated particles of micron size, increasing frictional resistance to flow over the entire range of shear rates. Since the coagulated particles are visible (i.e. large in size) they could be relatively easily removed with filtration, a process which helped to reduce WBV as the plasma proteins and lipids are the key determinants of WBV.

SBV of blood with DBD treatment and filtration decreased by c. 9.1% from the baseline value, whereas the DBV dropped by 17.7% from the respective baseline value. Both represent significant improvements after the DBD treatment. The reason why the DBD treatment improved the DBV more than the SBV is that the SBV is usually affected by erythrocyte deformability while DBV is affected by the erythrocyte aggregation. Since the plasma proteins (i.e. fibrinogen and immunoglobulins) and LDL molecules are instrumental in the red blood cells (RBC) aggregation, their removal by the DBD treatment and filtration should mitigate the erythrocyte aggregation, subsequently reducing the DBV.

In terms of the mechanism of DBD treatment in the precipitation of plasma proteins and lipids from blood plasma, Kalghatgi *et al.* (2007a, b) reported that DBD treatment might have activated some of the protein coagulation process which resulted in rapid fibrinogen aggregation. It was also mentioned that the selective coagulation of proteins was observed – not of albumin but of fibrinogen only (Kalghatgi *et al.*, 2007a, b). It can be hypothesized that this coagulated fibrinogen and subsequent removal might have affected the WBV of blood with DBD-treated blood plasma.

In addition, oxidized LDL molecules are known to adhere to arterial wall surfaces, playing a key role in the progression of atherosclerosis. In a similar manner, when LDL molecules in blood plasma are oxidized by DBD treatment, the oxidized LDL molecules tend to precipitate and coagulate. Since LDL molecules are one of the important variables in WBV, their removal with DBD treatment and filtration should not only help to reduce WBV but also to reduce the progression of atherosclerosis.

The discussed study of Jung *et al.* (2012) utilized DBD treatment and filtration to reduce blood viscosity. The results presented in this paper indicated that the non-thermal DBD treatment could precipitate and coagulate plasma proteins (i.e. fibrinogen) and LDL molecules in blood plasma. The formation of a white layer on the surface of blood plasma after DBD treatment confirmed the precipitation and coagulation of plasma proteins and lipids in blood plasma. The present study showed that WBV could be significantly reduced by 9.1 and 17.7% for SBV and DBV, respectively, from the respective baseline values when DBD-treated blood plasma was filtered prior to mixing.

Further investigation is necessary to determine the mechanisms of the precipitation and coagulation of plasma proteins and lipids by DBD non-thermal plasma treatment. In general, application of non-thermal plasma has a great potential for effective treatment of blood, including control of blood coagulation as well as effective control of blood composition and critical biophysical and biochemical properties of blood.

9

Plasma-assisted Healing and Treatment of Diseases

Previous chapters have focused on using plasma to sterilize surfaces, tissues and gas (air) and to coagulate blood with minimal collateral damage. In this chapter, we discuss use of plasma for enhanced wound healing, treatment of gastrointestinal disorders, dental applications and treatment of cancer.

9.1 Wound healing and plasma treatment of wounds

9.1.1 Wounds and healing processes

Non-healing (refractory) or slowly healing cutaneous wounds represent a considerable burden for healthcare, markedly reducing the quality of life of patients and increasing cost of care. Although wounds and wound healing pathologies can be relatively complex, to some extent the persistence of refractory wounds is a consequence of inaccurate diagnosis and wound assessment as well as suboptimal management of both wounds and underlying pathologies. For this reason, development of more convenient wound care tools, such as those based on plasma treatment, could be provide substantial value for patients and for the healthcare system as a whole.

9.1.1.1 Skin damage and wound healing

When injured, the skin displays a remarkable ability to repair or regenerate itself through a regulated biochemical cascade. The wound healing process can be categorized into four distinct phases based on a complex array of dynamic mechanisms including cellular migration, adhesion/de-adhesion, proliferation, differentiation and apoptosis.

1. *Hemostatic phase:* immediately following injury, damaged vessels constrict to reduce blood flow and the coagulation cascade is initiated, leading to thrombus formation.
2. *Inflammatory phase*: within 6–8 hours of injury, tissue damage induces release of biochemicals that attract white blood cells, resulting in phagocytosis of invading bacteria and other organisms.

Plasma Medicine, First Edition. Alexander Fridman and Gary Friedman.
© 2013 John Wiley & Sons, Ltd. Published 2013 by John Wiley & Sons, Ltd.

3. *Proliferative (repair) phase*: 2–3 days post-injury, angiogenesis, collagen deposition and granulation tissue development occur together with wound epithelialization and wound contracture.
4. *Maturation phase* (epithelialization and remodeling): this may occur for a considerable period of time following injury and is primarily a process of connective tissue remodeling that results in scar formation and increased wound tensile strength.

Several cell types play a major role in normal wound healing processes including macrophages, fibroblasts and keratinocytes. Within about 24 hours of injury, growth factors released by platelets and other cells attract macrophages to the site of injury (second phase). These immune system cells perform a number of critical functions in normal wound healing, including: direct phagocytosis of bacteria; generation of antimicrobial proteins and reactive oxygen and nitrogen species; secretion of elactase and collagenase (enzymes that break up elastin and collagen fibers) affecting debridement of devitalized tissue; and release of various growth factors, cytokines and enzymes that control cellular proliferation and angiogenesis. In this way, the macrophages not only affect a broad range of actions in the early stages of wound healing, but also push the wound into the next phase of the healing process.

The third phase involves proliferation of fibroblasts that constitute the dominant cell type in the wound by the end of the first week or so. The fibroblasts deposit ground substance into the wound bed and lay down collagen fibers, leading to the formation of a rudimentary granulation tissue and of the extra-cellular matrix (ECM). Protease enzymes released by fibroblasts later aid in the process of ECM remodeling. The fibroblasts also secrete chemo attractants and growth factors, which aid angiogenesis.

The fourth and final phase of healing is dependent on the formation of granulation tissue, since the keratinocytes responsible for re-epithelialization require a viable tissue surface across which to migrate. The keratinocytes detach themselves from the basement membrane and migrate laterally from the wound edges over the ECM, regenerating the basement membrane. Epithelial stem cells in hair follicles in the deep layers of the dermis also proliferate and migrate over the wound bed. Again, keratinocytes (stimulated by nitric oxide) produce a cocktail of growth factors, cytokines and protease enzymes. The growth factors and cytokines promote angiogenesis, while the proteases help dissolve non-viable tissue.

9.1.1.2   Factors affecting wound healing

As a rule, normal healing follows the sequence of events described in the previous section. In many ways, the initial inflammatory response can be regarded as the key process in this sequence responsible not only for the defense against bacterial contaminants, but also for initiating cell proliferation and migration. Prolongation or stasis of the inflammatory stage prevents progression into the proliferative phase and may lead to persistent impairment of healing, destruction of the ECM and development of necrotic tissue. Such prolongation is influenced primarily by (1) wound infection/colonization and (2) hypoxia (low oxygenation and blood circulation).

Hypoxia can be related to a variety of underlying causes such as diabetes, obesity, malignant disease, corticosteroid therapy, chemotherapy treatment or poor nutrition. Wound infection remains a major cause of serious illness and death, however, and is little impacted by antibiotic use. The consequences of wound infections, particularly in non-healing chronic wounds, can persist for years and significantly reduce the quality of life enjoyed by patients. For example, most deaths in severely burn-injured patients are still due to burn wound sepsis or complications due to inhalation injury. Approximately 2–10% of post-operative patients get infected wounds (equivalent to some 3.6 million patients in the USA and some 4 million patients in Europe), requiring up to 10 days extra of in-hospital treatment. Diabetic patients with infected ulcers have a significantly higher overall mortality rate compared to diabetic ulcer patients without infections. Pressure ulcer bacteremia results in more than 50% mortality.

The types of bacteria found colonizing a wound depend on the type of wound and the phase of wound healing. Initial endogenous and exogenous seeding of the wound results in the creation of microenvironments that ironically encourage later colonization by completely different microbial species. As a consequence, chronic wounds will invariably be subject to a polymicrobial colonization in which typically benign commensals, such as *Staphylococcus aureus* and *Pseudomonas aeruginosa*, may become pathogenic. A variety of anaerobes including *Bacteroides spp.*, *Prevotella spp.* and *Peptostreptococcus spp.* are also sometimes isolated from wound samples, although their contribution to delaying/preventing wound healing is probably underestimated due to the difficulty associated in culturing and identifying them.

When combined with changes in the wound milieu, bacterial proliferation also increases the probability of biofilm formation. This is a combination of cellular components and an extracellular polysaccharide matrix secreted by micro- colonies attached to the wound bed. Bacteria found within biofilms tend to be of higher virulence and display an increased resistance to both immunological defense mechanisms and antimicrobial agents. Biofilm structures have been identified in around 60% of biopsies from chronic wounds. Biofilms have also been implicated in prolonging the inflammatory response and delaying wound healing.

9.1.1.3 *Acute wounds*

Classically, wounds are categorized as acute (abrasions, scalds, burns and post-operative incisions) or chronic (long-term wounds such as diabetic ulcers, venous ulcers, arterial ulcers and pressure sores). Acute wounds can develop into a non-healing state and/or become infected, which limits their process through the phases of healing and so can also become chronic in nature. The primary comparative characteristics of these wound types are highlighted in Table 9.1.

Abrasions are largely uncomplicated and easily managed wounds that usually do not require tools beyond those already existing. Surgical wounds and burns are however much more prone to bacterial colonization, which often develops into overt wound infection. Current prevention and treatment of these wound infections relies on the use of topical antimicrobials, which have significant side effects and are of variable clinical efficacy. Significant concerns remain regarding proliferation of antibiotic-resistant microorganisms.

Surgical wounds are often associated with a range of local and systemic changes, not least of which are those that affect the immune system. Surgical trauma, for example, initially causes an early hyper-inflammatory response followed by cell-mediated immunosuppression. Similarly, general anesthetics also produce a short-lived immunosuppression by decreasing T-lymphocyte production, monocyte activity and cytokine secretion and inhibiting a respiratory burst by neutrophils. Factors complicating surgical wound healing include: hypothermia; anesthesia (and therefore immunosuppression); poor aseptic technique; and changes in blood glucose level, oxygenation and perfusion of tissue (affected by many factors such as cardiac output and perioperative fluid management). All of these have an impact on the risk of wound infection and prolongation of wound healing.

Table 9.1 Comparison of acute and chronic wounds.

Wound type	Acute	Chronic (non-healing)
Examples	Abrasions, superficial burns, surgical wounds	Vascular lower limb ulcers, diabetic ulcers, mixed aetiology ulcers, pressure sores, infected wounds
Healing	Rapid and orderly	Prolonged, disorderly
Immunology	Normal	Abnormal
Infection	Some bacteria are present, but colonization is rare	Colonization is common

Burns may be either partial thickness (first and second degree) or full thickness (third degree). Partial-thickness burns involve the epidermis or the epidermis and a portion of the dermis, whereas full-thickness burns involve the entire epidermis and dermis. Full-thickness or extensive partial-thickness burns are associated with significant levels of fluid and protein loss (which may lead to hypovolemia and shock), a hypermetabolic state and a degree of immunosuppression (possibly as a result of T-cell dysfunction with failure of interleukin-2 production). Immediately following injury the surface of a burn is aseptic, but endogenous wound colonization occurs typically within 48 hours. Unfortunately, the necrotic tissue and protein-rich wound exudates found in a burn represent an ideal growth medium for bacteria, and burn wound infection is a common finding.

Burns typically have three zones of involvement, as follows.

1. *Zone of coagulation*: usually occurs in the areas of maximal damage. In this zone there is irreversible tissue loss due to coagulation of the constituent proteins.
2. *Zone of stasis*: typically surrounding the zone of coagulation and characterized by decreased tissue perfusion and sluggish capillary flow. The tissue in this zone is potentially salvageable. The main aim of burns resuscitation is to increase tissue perfusion here and prevent any damage from becoming irreversible. Additional insults, such as prolonged hypotension, infection or edema can convert this zone into an area of complete tissue loss.
3. *Zone of hyperemia*: the outermost zone in which tissue perfusion is increased. The tissue here will invariably recover unless there is severe sepsis or prolonged hypo-perfusion.

9.1.1.4   Chronic wounds

Normal wound healing proceeds in an orderly sequential manner with four discrete stages: *hemostasis, inflammation, proliferation and maturation*. Interruption or prolongation of any of these processes may lead to sub-optimal healing and a failure of the wound to return to functional and anatomical integrity in a timely manner. That said, the majority of chronic wounds are due to persistence of the inflammatory phase with a resulting build-up of reactive oxygen species (ROS), continuing tissue damage and perpetuation of inflammation. Cytokines (particularly interferon gamma) may play a pivotal role in this pro-inflammatory response. The major chronic wound subtypes are venous, arterial, diabetic and pressure ulcers; some examples are shown in Figure 9.1.

It is generally accepted that *venous ulceration* is the result of a functional failure of venous valves in the lower limbs, which leads to increased backflow and venous hypertension. This high pressure is transmitted back to the venules and results in distension and increased permeability of the capillary beds. As a result, fibrinogen leaks into the dermis of the skin and (in the presence of reduced fibrinolysis) forms a fibrin cuff around vessels. This prevents diffusion of oxygen and nutrients, contributes to local tissue hypoxia and ischemia and traps growth factors essential for wound healing in the extravascular space. Venous insufficiency also causes white blood cells (leukocytes) to become trapped in the venules where they release inflammatory factors, proteolytic enzymes and ROS, which together with the physical plugging of small vessels increase local ischemia and capillary pressures. This in turn increases capillary permeability and leads to further extravasation. Substances that extravasate out of the vessel include red blood cells, which release hemoglobin into the extra vascular space. This is digested and the breakdown products are responsible for the characteristic brown staining and thickening of skin seen in chronic venous ulcer patients.

Arterial ulcers are usually caused by progressive atherosclerosis (hardening of the arteries), in which cholesterol plaques gradually narrow and eventually reduce blood flow in large- and medium-sized arteries. Similar changes are found in smaller arteries and arterioles and combine to produce global or regional arterial hypo-perfusion (insufficient perfusion). The accompanying local hypoxia and lack of nutrients result in focal tissue breakdown, particularly in locations most distant from the left side of the heart.

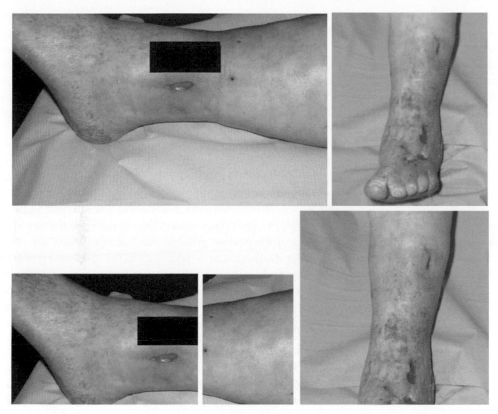

Figure 9.1 *Examples of different chronic wounds: top left is venous leg ulcer, top right is arterial leg ulcer, bottom left is ischemic diabetic foot ulcer, bottom right is grade 3 pressure ulcer (see color plate).*

Foot ulcers are a common complication of *diabetes* and often precede lower-extremity amputation. They have many different causes but, due to a wide number of coexisting metabolic abnormalities, diabetics have a higher incidence of vascular abnormalities including thickening of capillary basement membranes, thickening and calcification of the arterial tunica media (recall that this is the middle cellular layer in arteries) and arteriolar hyalinosis (accumulation of amorphous material in the walls of small arteries and arterioles, a common degenerative change in the elderly). They are also more prone to developing atherosclerotic disease below the knee. This leads to a generalized reduction in arterial inflow, which prevents adequate oxygenation of tissues and predisposes the patient to developing chronic wounds. Also important is the frequent presence of neuropathy (damage to nerves), which inhibits the perception of pain. As a result, patients may not initially notice small wounds to legs and feet and may therefore develop infections or exacerbate the original injury.

Pressure ulcers (bed sores) are caused by ischemia that occurs when external pressure on the tissue leads to capillary occlusion and restriction of blood flow into the area, causing skin and tissue necrosis.

9.1.2 Treatment of wounds using thermal and nitric-oxide-producing plasmas

As long as it is allowed to cool down, thermal plasma can affect living tissues through biochemistry and UV. Several types of active species generated in plasma can be responsible for biological effects including reactive oxygen and nitrogen species. Nitric oxide has received particular attention in this regard. The importance of nitric oxide (NO) as a biological molecule was recognized relatively recently (1980s). Since R.F. Furchgott, L.J. Ignarro and F. Murad received the Nobel Prize for investigating the function of nitric oxide as a signaling molecule in 1998, the idea that a molecule of gas produced by some cells can dissolve in biological fluid, penetrate lipid membranes and transmit regulatory signals to other cells became much better established. Three such molecules seem to play physiological roles in living systems: carbon monoxide (CO), nitric oxide (NO) and hydrogen sulfide (H_2S). All of these molecules can pass through cell membranes, although to a somewhat different extent. All can serve as electron donors to varying extents and therefore behave as antioxidants (serving as scavengers of oxidizers).

In humans, NO appears to serve a multitude of essential biological functions. It offers antimicrobial and antitumor defense and regulates blood vessel tone, blood coagulation, some immune system activity, apoptosis, neural communication, flat bronchial and gastrointestinal muscles and hormonal and sex functions. Nitric oxide also plays an important role in adaptation, stress, tumor growth, immunodeficiency, cardiovascular, liver and gastrointestinal tract diseases. Its role in wounds has been particularly noted. For example, inducible NO-synthase (iNOS) grows substantially in traumatic wounds, burn wound tissues and bone fracture site tissues in the inflammatory and proliferation phases of the healing process. Activation of iNOS was discovered in cultivation of wound fibroblasts. Macrophage activation in a wound, cytokine synthesis, proliferation of fibroblasts, epithelialization and wound healing processes are all linked to the activity levels of iNOS. In animal models, injection of iNOS inhibitors disrupts all of these processes and especially the synthesis of collagen, while NO synthesis promoters accelerate these processes. Animals with iNOS deficiency demonstrate a significant decrease in wound healing rate; however, this can be reversed by injection of iNOS gene. In complicated wound models (e.g. in experimentally induced diabetes) and also in patients with tropic ulcers, lowered activity of iNOS is often found to correlate with slowed healing processes.

Nitric oxide in tissues can be regulated either through drugs that control release of endogenous NO (generated internal to the body by tissues) or administration of exogenous NO (produced externally to the body). Use of exogenous NO in infection and inflammation processes is well studied. It has been linked either to direct antimicrobial effects whereby nitric oxide (partly through interaction with reactive oxygen species) kills bacteria by reacting with various organic molecules, or to stimulation of the immune system responses such as activation of macrophages and T-lymphocytes as well as induction of cytokine and antibody production. The influence of NO on increasing microcirculation also promotes delivery of immune system components to the site of infection. Exogenous NO was demonstrated to be important in traumatic wound processes. Its delivery through NO-donors (nitrogen-containing compounds) to the wound promotes healing processes in animals with complicated wounds and in animals with inhibited iNOS.

The various effects of NO can be complex. In some environments nitric oxide acts as an antioxidant, while in others it acts as an oxidizer. The ultimate effects may depend on the NO concentration. Understanding the effects of different NO concentrations, coupled with theoretical and experimental data on NO generation in air plasmas, can serve as a basis for a series of biomedical experiments focused on the use of the plasma-generated exogenous NO delivered directly to the pathologic site for control of inflammatory processes and increase in the rate of wound healing.

Given the importance of NO, there has been a tendency in some discussions of plasma treatment to focus on nitric oxide as the key molecule generated by plasma. If that were the case, however, we could question the use of plasma as a source of NO in healthcare setting when bottled NO is often available. One argument in favor of plasma is that it requires only electrical energy, which today is widely accessible in various environments. Bottled NO, on the other hand, can be associated with various logistical challenges. Beyond this argument there could be more compelling reasons to employ plasma in various treatments. Plasma can be a source of nitric oxide as well as various other reactive species. In some situations reviewed in the following section, NO may have been the key factor. In other situations however it is not entirely clear what active species play what roles in the treatment. The fact is that, over the last decade or so, different types of plasma treatment have been shown to have substantial beneficial effects. The primary purpose here is to provide an overview of these investigations.

9.1.2.1 *Discharge systems with significant nitric oxide production*

At this time, the most widely investigated plasma device whose therapeutic effects are thought by its inventors to be based primarily on nitric oxide effects is probably the so-called 'Plazon' system based on the jet of hot air plasma rapidly cooled upon exit from the plasma generation region (Shekhter *et al.*, 1998; Pekshev, 2001). This plasma device has been investigated for: sterilization of wound surfaces; destruction and desiccation of dead tissue and pathologic growths; dissection of biological tissues with the plasma jet; and also for stimulation of regenerative processes and wound healing by the gas flow with temperature of 20–40°C.

Plazon generators (Shekhter *et al.*, 1998; Pekshev, 2001) are DC arcs with different configurations of the exit channels corresponding to the different applications (blood coagulation, tissue destruction, therapeutic manipulation/stimulation). The main elements of the system construction are the liquid-cooled cathode, intra-electrode insert and anode. Atmospheric air enters the manipulator through the built-in micro-compressor, passes through the plasma arc, heats up and accelerates and exits through the hole in the anode of the plasma-generating module. Plasma temperature at the anode exit varies for different configurations of the device, corresponding to different medical applications (see Figure 9.2). Temperature drops rapidly away from the anode; at 30–50 mm from the anode, the flow is composed simply of the warm gas and the plasma-generated NO.

Nitric oxide content in the gas flow is mainly determined by the quenching rate. The necessary quenching rate for effective operation of the medical device is c. 10^7–10^8 K s^{-1}. Commonly, the cooling rate of plasma jets is of the order c. 10^6 K s^{-1}. To achieve the cooling rate of c. 10^7–10^8 K s^{-1}, it is necessary to utilize additional cooling of the plasma jet which has been achieved by special construction of the plasma nozzles. The therapeutic manipulator-stimulator configuration of the Plazon discharge system is used solely for therapeutic treatment by exogenous nitric oxide.

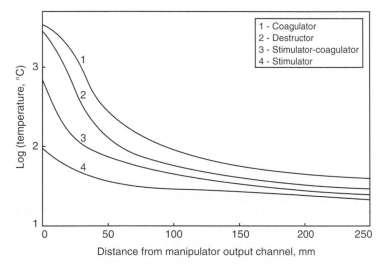

Figure 9.2 *Temperature as a function of distance from the exit of Plazon generator for different configurations of exit channels. Data courtesy of A. Shekhter.*

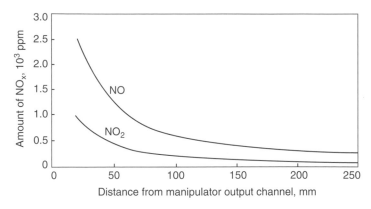

Figure 9.3 *Concentration of nitric oxide and dioxide as a function of distance from the exit of the Plazon generator. Data courtesy of A. Shekhter.*

The principle difference of this manipulator is that the air-plasma jet does not freely exit into the atmosphere. Instead, it exits the anode into a two-step cooling system in which gas channels are created in a maze scheme to force-cool the jet by the liquid circulating from the cooling system. This construction permits NO-containing gas flow (NO-CGF) with sufficiently low temperature, which makes it possible to apply this manipulator for treatment of external body surfaces by using the cooling hose of 150 mm length (temperature of NO-CGF at the exit is c. 36°C). NO content in the gas flow also depends on the distance from the exit channel (Figure 9.3).

For laparoscopic operations, a special manipulator of 350 mm length and 10 mm diameter is utilized. The possible operating regimes of the apparatus are defined by the characteristics of the gas flow exiting from the manipulator, the main parameters of which are its temperature and the nitrogen oxide content. The first group of regimes is that of free-flowing plasma afterglow exiting the manipulator, while the second group of regimes is treatment of biotissues by completely cooled (20°C) NO-CGF. The second set of regimes not only allows the tissues to be directly treated by NO, but also allows the NO to be delivered to a pathologic location through drainage tubes, puncture needles or any endoscopic device (gastroscope, broncoscope, cystoscope, rectascope, etc.).

Another example of nitric-oxide-producing discharge is a special DC spark microplasma called pin-to-hole discharge system (Section 4.5.10). This system was developed for medical applications and investigated in situations where access to the affected site might be limited, such as in corneal infections (Gostev and Dobrynin, 2006). The device allows generation of plasma afterglow with average gas temperature not exceeding 30–40°C. It consists of a tube cathode with a small hole at the end that functions as a nozzle and a pin-like anode positioned coaxially inside the cathode tube. The discharge gap is formed essentially between the tip of pin and the edges of the hole at the end of the cathode tube. The gas is fed through the tube from the back to the discharge gap. The voltage is applied between the pin anode and the tube cathode as a capacitor discharges through the plasma gap. Different gasses, including moist and dry air and Xe have been employed to form the plasma. Typical discharge voltage decays start from 1–3 kV, while the voltage pulse duration is around 50 µs and total power is of the order 1–2 W. The typical diameter of the afterglow sphere outside the nozzle is c. 3 mm.

This microplasma operated in Xe radiates intensively in the UV range; operated in air it generates excited oxygen species, ozone, nitrogen oxides (primarily NO, but also NO_2) and OH radicals (Gostev and Dobrynin, 2006). Both regimes have bactericidal effects and air plasma is also able to aid in tissue regeneration via NO therapy. UV radiation of Xe plasma in this case is: UVA (315–400 nm) 180 µW cm^{-2}, UVB (280–315 nm)

180 μW cm^{-2} and UVC (200–280 nm) 330 μW cm^{-2}. UV-radiation of air plasma is: UVA (315–400 nm) 53 μW cm^{-2}, UVB (280–315 nm) 25 μW cm^{-2} and UVC (200–280 nm) 90 μW cm^{-2}.

The ability of the above microplasma system to sterilize surfaces has been demonstrated by Misyn *et al.* (2000). Staphylococcus culture in liquid media (c. 2×10^6 CFU mL^{-1}) have been treated by the air plasma plume of 3 mm diameter, incubated for 24 hours and counted. A 6-log reduction in viable bacteria is achieved in 25 s of treatment; however, the sterilization efficiency drops off with increasing volume of liquid, which inhibits UV penetration and diffusion of active species generated in plasma. During the investigation of plasma treatment of ulcerous dermatitis of rabbit cornea, two important observations were made: (1) plasma treatment has a pronounced and immediate bactericidal effect and (2) the treatment has an effect on wound pathology and the rate of tissue regeneration and wound healing process.

9.1.2.2   *Nitric oxide effects of Plazon treatment in animal models and cell cultures*

Electron paramagnetic resonance or EPR-spectroscopy, a method commonly employed to detect free radicals, was utilized to investigate the dynamics of levels of endogenic NO in wound tissues and in organs in an animal model (70 rats, Shekhter *et al.*, 2005). Given the relatively short lifetime of many radicals in organic solutions and tissues, special molecules called 'traps' are employed. These traps often react preferentially with a designated radical and can be relatively long lived. In the above-mentioned studies, a nitric oxide trap, diethylthio-carbamate (DETC), was injected into rats with full thickness flat wound of 300 mm^2 area 5 days prior to EPR analysis. Following euthanasia, samples were collected from the animals' blood, granular tissue from the bottom of the wound and from internal organs (heart, liver, kidney and the small intestine). For a portion of the animals, 5 days after initial wound introduction the wound surface was treated by the NO-CGF (500 ppm).

Without the nitric oxide treatment, the results indicated a high content of endogenic NO in wound tissues (10.3 $\pm$ 2.3 μM). The liver of the animals with the wound contained 2.3 $\pm$ 1.4 μM of DETC-iron-mono-nitrosyl complex (IMNC, a complex of the trap, NO and Fe), while the control group (without the wound) had much lower levels of only 0.06 $\pm$ 0.002 μM. Animals without the wound were used to investigate capability of gaseous exogenous NO to penetrate through undamaged tissues of the abdominal wall. Treatment by the NO-CGF was performed for 60 and 180 s. A linear dependence of the amount of DETC-IMNC produced in the liver and blood of the animal on the NO-CGF treatment time was observed. When the animal was euthanized 2 min after the 180 s treatment, a maximum signal was registered in the bowels of the animal which was 2.6 times higher than the control. In the heart, liver and kidney the difference was a factor of 1.7. These results clearly indicate the ability of the exogenous NO molecules to penetrate undamaged tissues and potentially trigger endogenic response.

A more complex relationship was observed in treatment by exogenous NO of the wound tissues. If the animal was euthanized 30–40 min following the 180 s treatment, the NO content in wound tissue and blood was observed to increase by a factor of 9–11 over that observed in the case of the 2 min interval. Several explanations of this effect are possible. One is that an activation of the first cascade of antioxidant defense following exogenous NO treatment leads to a significant decrease in the levels of superoxide. This may have considerably decreased the influence of ROS on DETC-IMNC and the nitrosyl complexes of the hemoproteins (complexes where NO$^-$ ion binds to iron forming Fe–NO). Another possible explanation is activation of iNOS and resulting endogenic NO generation. This partially explains the discovered phenomena of stimulation of wound development processes via the influence of exogenous NO, when there is a deficiency of endogenic NO or excess of free radicals including superoxide.

In experiments on rabbit cornea, mucous membranes of hamster mouths and meninx membrane in rats via lifetime microscopy it was found that the effect of the expansion of the opening of the microvessels under the influence of exogenous NO (500 ppm) lasts with varying intensity up to 10–12 hours, while the lifetime of NO molecules is much shorter (Shekhter *et al.*, 1998, 2005; Vanin, 1998). This provides additional evidence that a single application of exogenous NO initiates a cycle of cascade reactions, including biosynthesis of endogenic NO, which leads to a long-lasting effects and explains the success of the NO therapy.

The action of the exogenous NO on the cellular cultures of the human fibroblasts and rat nervous cells was studied by Stadler *et al.* (1991), Shekhter *et al.* (1998, 2005) and Ghaffari *et al.* (2005). Single treatment by the plasma-generated NO of the cell cultures significantly increases (2.5 times) the cell proliferation rate via the increase of DNA synthesis (tested by inclusion of C^{14} thymidine) and, to a lesser extent (1.5 times), the increase of protein synthesis by the cells (tested by inclusion of C^{14} amino acids). As expected, the stimulating effect is dose-dependent. The action of exogenous NO on the phagocytic activity of the cultured wound macrophages from the washings of the trophic human ulcers, studied by photochemiluminescence (Krotovskii *et al.*, 2002) revealed that a maximum increase in the luminous intensity (a factor of 1.95 greater than control) confirms the activation of the proteolytic enzymes of macrophages under the effect of NO-CGF. A statistically significant increase in fluorescence of macrophages was observed in less than 24 hours following a 30 s treatment.

9.1.2.3   *Clinical experiences of using Plazon for wound treatment*

Application of air-plasma-generated exogenous NO in the treatment of venous and arterial trophic ulcers of lower extremities with an area of 6–200 cm^2 in 318 patients showed high efficiency (Shekhter *et al.*, 1998, 2005). For assessment of the effectiveness of the plasma NO-therapy, clinical and planimetric indices were analyzed in the course of the sanitation and epithelialization of ulcers; a bacteriological study of discharge from the ulcer; cytological study of exudates; a histopathological study of biopsies from the boundary of a trophic ulcer; the indices of microcirculation (according to the data obtained by laser doppler flowmetry or LDF); and transcutaneous partial pressure of oxygen or pO_2). In the main groups of observations trophic ulcers were processed in the regime of NO-therapy (500 and 300 ppm). Prior to beginning the therapy, the ulcer surface was treated in the regime of coagulation until the evaporation of necrotic debris. Following initial treatment, the wound was treated for 10–30 days in the NO-therapy regime. Proteolytic and antimicrobial drugs were used in the control group in the exudation and necrosis phases and wound coatings in the tissue regeneration and epithelialization phases.

Planimetric observation of the dynamics of decrease of the trophic ulcer area showed that, on average, traditional treatment methods applied to the control group led to 0.7% per day decrease, while in the experimental group demonstrated a decrease of 1.7% per day. Cleansing of ulcers from necrosis and exudate and the appearance of granulation and boundary epithelialization were accelerated with NO-therapy by a factor of 2.5. The time to final healing was reduced by a factor of 2.5–4, depending on the initial ulcer size (Figure 9.4). Larger ulcers tended to close faster than smaller ulcers.

LDF investigation of microcirculation in the tissues of trophic ulcers showed that, following the NO-therapy, pathologic changes in the amplitude-frequency characteristics of the microvasculature were normalized and regulatory mechanisms were activated. By 14–18 days, the average index of microcirculation, value of root-mean-square deviation, coefficient of variation and index of fluctuation of microcirculation approached those of the symmetrical sections of healthy skin. In the control group, the disturbances of microcirculation remained. Against the background of treatment, normalization of the level of transcutaneous partial pressure of oxygen (TpO_2) happened at a higher rate in the experimental group than in the control group, especially at the NO concentration of 500 ppm (Figure 9.5).

A bacteriological study of wound discharge from the trophic ulcers showed that for the experimental group, NO-therapy (especially in combination with the preliminary coagulation of ulcerous surface) reduced the

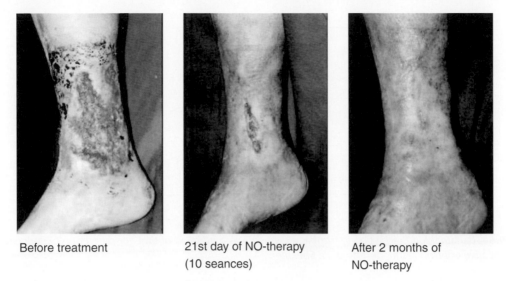

| Before treatment | 21st day of NO-therapy (10 seances) | After 2 months of NO-therapy |

Figure 9.4   *Reduction of ulcer size as a function of time during the Plazon-generator-based NO therapy. Photographs courtesy of A. Shekhter (see color plate).*

degree of bacterial seeding (microbial associations). The level of bacterial seeding fell below the critical level necessary for maintaining the infectious process in the wound by days 7–14 (Figure 9.6).

Use of plasma-generated NO for local treatment of ulcerous and necrotic tissues in patients with diabetes (diabetic foot ulcer) was demonstrated by Shulutko, Antropova and Kryuger (2004). Patients were selected for this study following two months of unsuccessful treatment by state-of-the-art techniques. The improved success was evident from the first few sessions: inflammatory reaction was clearly reduced, patients reported a reduction in pain and cleansing of the ulcer surface was clearly visible. Following 10 sessions, most patients expressed positive healing dynamics and the ulcer size decreased to between one-third and one-quarter of the

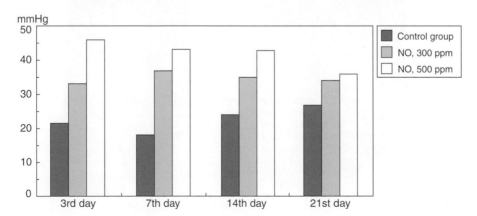

Figure 9.5   *Normalization of the level of transcutaneous partial pressure of oxygen in trophic ulcer tissues for different NO concentrations and for the negative control. Data courtesy of A. Shekhter.*

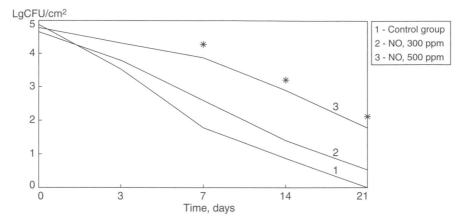

Figure 9.6   *Bacterial load reduction in trophic ulcer tissue as a function of time for different NO concentrations versus negative control. Data courtesy of A. Shekhter.*

original size. LDF markers, pO_2 and bacteriological investigation all showed a positive dynamic. In patients with relatively small-sized ulcers (initial diameter less than 1 cm), full epithelization occurred after 6–8 NO treatment sessions. The period of stationary treatment and full clinical recovery of patients was noticeably shortened (on average by 2.3 times). In the cases of large ulcerating wounds, the necessity for amputation was reduced by a factor of 1.9 (Figure 9.7).

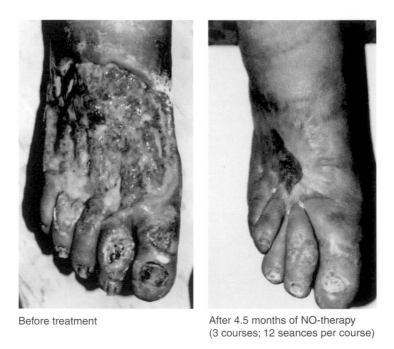

Before treatment   After 4.5 months of NO-therapy
                                    (3 courses; 12 seances per course)

Figure 9.7   *Example of a large ulcerated wound before and after 4.5 month duration of NO treatment. Photographs courtesy of A. Shekhter (see color plate).*

Effectiveness of the exogenic NO and air plasma on healing of pyoinflammatory diseases of soft tissues was demonstrated by studying 520 patients with purulent wounds of different etiology and 104 patients with the phlegmonous-necrotic form of the erysipelatous inflammation (Lipatov *et al.*, 2002). By the fifth day of therapy, wounds on most of the patients in the experimental group (90%) were clear of necrotic tissue (contrary to the control group) and were beginning to be covered by bright spots of granular tissue. Microbial infestation of the wound tissue had lowered from 10^{6-8} colony forming units (CFU) per gram of tissue to 10^{1-2}.

Data from complex analysis of microcirculation (LDF, pO_2) showed significant repair of the microvasculature and blood flow in the wound tissues in most of the patients in the experimental group. The predominant types of cytograms were regenerative and regenerative-inflammatory with a notable increase in fibroblast proliferation (on average, $18.5 \pm 3.1\%$). By day 7–10 of treatment, large suppurated wounds, for example suppurated burn wounds were clear of the pyonecrotic exudate and were beginning to be covered by granular tissue; in other words, these wounds were ready for dermatoplasty.

Effectiveness of plasma NO-therapy is most apparent in the treatment of the pyonecrotic form of erysipelatous inflammation, that is, the most severe cases of purulent surgery departments (Lipatov *et al.*, 2002). The combination of surgical preparation of extensive pyonecrotic centers and local NO-therapy allowed in the majority of the patients with phlegmonous- necrotic erysipelas during 12–14 days of treatment to liquidate heavy pyonecrotic process and to create conditions for completion of reparative procedures.

In maxillofacial surgery, plasma NO-therapy was used to accelerate the healing of post-operative wounds and in preventive maintenance of the formation of hypertrophic and keloid scars, treatment of the formed scars, treatment of pyonecrotic processes (abscesses, phlegmon, etc.). In the latter case, preliminary coagulation of purulent centers is sometimes utilized (Shekhter *et al.*, 2005).

Plasma NO treatment has been also successfully applied to surgical oncology (Kabisov *et al.*, 2000; Reshetov *et al.*, 2000). Inter-operative treatment in the coagulation regime of the Plazon system ensures ablation; it considerably decreases blood plasma and whole blood losses from extensive wound surfaces as a result of thin film formation over the wound surface consisting of coagulative necrotic tissue. As a result of plasma NO-therapy of post-operative wounds, a significant decrease in inflammation is observed along with stimulated proliferation of granular tissue and epithelization. The effect is observed independently of the location of the wound on the body and also of the plastic material used. An additional benefit is the prophylactic treatment of the local relapses of the tumor, which enables this method to be widely applied in oncoplastic surgeries. The effectiveness of NO-therapy in treatment of early and late radiation reactions allows surgeons to carry out a full course of radiation therapy in 88% of the patients. Treatment of radiation tissue fibrosis also yields a statistically significant improvement, confirmed in morphological investigation of these tissues. Plasma NO-therapy is successfully used for the prevention of the formation of post-operative hypertrophic and keloid scars, and for treatment of already formed scars by softening the scar tissue, decreasing fibrosis and preventing their relapse with the surgical removal.

In ophthalmology, treatment by NO-CGF (300 ppm) does not induce a toxic reaction and does not cause intraocular pressure changes or morphological changes in the tissues of eye, but considerably accelerates the healing of wounds and burns of cornea. Such therapy has been used in clinics for the effective treatment of cornea burns, erosions and injuries and burn ischemia of conjunctiva (Chesnokova *et al.*, 2003).

In gynecology, the effectiveness of plasma NO-therapy has been demonstrated for patients with purulent inflammation of appendages of the womb (Davydov *et al.*, 2002, 2004; Kuchukhidze *et al.*, 2004). Where the abdominal cavity was opened in surgery, the purulent wound was treated by air plasma in the coagulation regime. At a later time, plasma NO-therapy was applied remotely through the front abdominal wall and vagina. With operational laparoscopy after dissection and sanitation of the purulent center, the region of surgical incision and the organs of the small basin were treated by NO-CGF delivered locally through the aspiration tube. Plasma NO-therapy was also continued in the post-operation period.

The use of NO-CGF in surgical and therapeutic regimes aided the rapid reduction in the microbial load and in swelling, lowered the risk of post-operative bleeding, encouraged rapid development of reparative processes and decreased the overall time that patients remained in hospital by 6–8 days on average. NO-CGF was also used in organ-saving surgical operations on the womb, the uterine pipes and the ovary.

9.1.3   Experience with other thermal discharges

9.1.3.1   *Pin-to-hole microdischarge applications*

The results of pin-to-hole microdischarge (described in Section 4.5.10) on ulcerous dermatitis of rabbit cornea demonstrated that treatment a pronounced and immediate bactericidal effect and clear benefits for wound pathology, the rate of tissue regeneration and the wound healing process. These results provide strong support for application of the pin-to-hole microplasma system for treatment of human patients with complicated ulcerous eyelid wounds.

A before-and-after pin-to-hole discharge treatment example is shown in Figure 1.15 (Misyn and Gostev, 2000). Necrotic phlegm on the surface of the upper eyelid was treated by an air plasma plume of 3 mm diameter for 5 s once every few days. By the fifth day of treatment (two 5 s plasma treatment sessions) the eyelid edema and inflammation were reduced. By the sixth day (third session), the treated area was free of edema and inflammation and a rose granular tissue appeared. Three more plasma treatments were administered (six in total), and the patient was discharged from the hospital six days after the last treatment.

Microplasma treatment is being further developed for stimulation of reparative processes in various topical wounds, tropic ulcers, chronic inflammatory complications and other diseases of soft tissues and mucous membrane (Misyn and Gostev, 2000).

9.1.3.2   *Microwave-generated thermal argon plasma afterglow*

In 2005 the Max Planck Institute for Extraterrestrial Physics in Garching, Germany developed a plasma device for indirect (jet-based) treatment, called MicroPlaSter (built by ADTEC Plasma Technology Co. Ltd., Hiroshima, Japan and London, UK). This device creates plasma using microwave electromagnetic field. Flow of gas (e.g. argon or air) through the discharge region delivers the cooled afterglow to the skin. The typical torch-skin distance is c. 20 mm. A depiction and operational pictures of this plasma torch are shown in Figure 9.8. The system allows the treatment of relatively large inhomogeneous and topographically uneven areas (c. 5 cm in diameter) below the threshold of thermal damage in contact-free mode. The active agents are carried with inert argon gas from production in the torch to the desired region. Typical flow delivers up to 10^{10} active species per cm^2 per second. A mixture of hydrogen peroxide, nitric oxide and nitrogen dioxide has been reported in argon afterglow of this plasma. Various nitrogen species including nitric oxide and nitrogen dioxide have also been reported when air is employed as the plasma gas.

Clinical trials conducted using the MicroPlaSter focused on lower leg ulcers (Isbary *et al.*, 2010). Chronic ulcers of the lower leg are associated with considerable patient morbidity and account for an estimated 1–2% of the annual healthcare budget in European countries, for example. Bacterial colonization of such wounds is a well-recognized factor contributing to impaired wound healing.

Having carried out Phase I of the trial, focusing on safety and bactericidal dosing, Isbary a group of researchers from the Max Planck Institute for Extraterrestrial Physics and the Department of Dermatology, University of Regensburg, Germany, reported promising results (Isbary *et al.*, 2010) in a Phase II study of 38 chronic infected wound treatments in 36 patients. Patients with chronic infected skin wounds who attended the outpatient and inpatient clinics of the Department of Dermatology, Allergology and Environmental Medicine of Hospital Munich Schwabing in Germany were invited to participate in this trial. Patients could be included

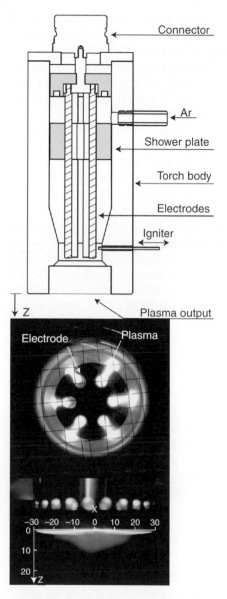

Figure 9.8 *Schematics and photograph of MicroPlaSter operation (courtesy of ADTEC Plasma Technology Co. Ltd., Hiroshima, Japan and London, UK) (see color plate).*

in the trial if they had at least one colonized wound large enough for plasma treatment and a control area of 3 cm^2. In addition to standard wound care, patients received a daily 5 min cold plasma treatment to the randomized wound(s) using the MicroPlaSter plasma torch operating at 2.46 GHz, 86 W and having Ar gas flow of 2.2 slm (standard liters per minute) at a distance of 2 cm. Control wounds remained undressed during plasma treatment. The same standard wound care was given to both plasma-treated and control areas. Because of the uneven surface of most of the wounds, high-pressure water jet or scalpel was used to clean all wounds of debris before the initial plasma therapy.

Bacterial load on the wounds was assessed in several ways. Once each week, two standard bacterial swabs were taken from all control and plasma-treated areas immediately after dressing removal and before re-dressing to detect the types of bacteria present. On the other days of the week, nitrocellulose filters (Sartorius Stedim Biotech GmbH, Aubagne, France) were used to detect changes in bacterial load. These filters were applied to the wounds with gentle pressure before and after the treatment and then placed on Columbia blood agar plates (Oxoid Ltd, Basingstoke, Hampshire, UK) and incubated for 12 hours at 36°C. Semi-quantitative assessment of the plates was carried out by a manual count. An important benefit of the filter technique is that it displays where the bacteria are primarily situated within the wound.

Plasma treatment in this trial resulted in a highly significant (c. 34%) reduction in bacterial count in plasma-treated areas compared to non-treated areas. The corresponding bootstrap test confirmed the high significance level of the results ($n = 500$). This is probably the first clinical evidence of a significant bactericidal effect in patients of plasma treatment when used in addition to standard wound care for chronic infected wounds. Although statistically no clear conclusions regarding faster healing was made in this trial, some instances of faster healing were observed (Isbary *et al.*, 2010).

9.2   Treatment of inflammatory dysfunctions

The possibility of using thermal nitric-oxide-producing plasma for treatment of inflammatory and erosive processes in tissues within the pleural and abdominal cavities, lungs, stomach and bowels and ear-nose-throat (ENT) organs has been noted in several different studies. In some cases, plasma-generated NO-CGF was inhaled or directed through puncture needles, vent lines and endoscopic instruments. In other cases, plasma was placed in direct proximity to the affected tissues using endoscope-like delivery.

9.2.1   Examples of anti-inflammatory treatment by Plazon

The following examples are described in detail by Shekhter *et al.* (1998, 2005).

In *pulmonology*, the strong effect of NO-CGF was demonstrated in the treatment of pleural empyema via insufflation (inhalation) from the Plazon into the cavity of the pleura through the vent lines (Shulutko, Antropova and Kryuger, 2004). Therapy in treatment of 60 patients with pleural empyema showed regulative influence on the development of the wound tissues and stimulated healing. Acceleration of the purification of pleural cavity from the microorganisms and the debris, stimulation of phagocytosis and normalization of microcirculation accelerate the passage of the phase of inflammation during wound regeneration, which leads to a significant decrease in the drainage time for all patient categories in the experimental group (compared to control) and to the reduction in the hospitalization time.

The inhalation application in treatment of patients with complex chronic unspecific inflammatory lung diseases led to the clearly expressed positive dynamics of the endoscopic picture of the tracheobronchial tree: decrease of the degree of inflammatory changes in the mucous membrane of bronchi and the reduction in the quantity and the normalization of the nature of contents of the respiratory tract. Through biopsies of mucosa of bronchi, it was verified that for all cases the liquidation or the considerable decrease of inflammatory changes occurred in addition to a complete or partial restoration of the morphological structure of the bronchi.

Plasma NO-therapy was also employed as an adjunct treatment in patients with infiltrative and fibrous-cavernous pulmonary tuberculosis via NO insufflation through the bronchoscope or cavernostomy for cavernous tuberculosis, through the vent line with tubercular pleurisy or empyema. A significant acceleration of healing of the cavities, tubercular bronchitis and pleurisy was achieved (Seeger, 2005) in 8–10 therapy sessions.

In *dermatology*, NO-therapy was effectively used for the treatment of psoriasis, eczemas, dermatitis, ulcerous injuries with local and systemic angiitises, scleroderma, red flat lishchaya and a number of other skin illnesses (Zaitsev, 2003).

In *gastroenterology*, NO therapy was delivered through endoscopic instruments for the treatment of chronic ulcers, erosions of stomach and duodenum and blowholes of small intestine (Chernekhovskaia *et al.*, 2004). Stomach ulcers healed twice as fast as in the control group. The proliferating activity of the epithelium, according to the data of the immunomorphology of biopsies, was increased by a factor of 8.

In *purulent peritonitis*, caused by diseases of the abdominal cavity organs, a positive effect was achieved by the direct treatment of peritoneum first and in the post-operative period by cooled NO-CGF delivered through the vent lines (Efimenko *et al.*, 2005). It was argued that NO carries bactericidal action, stimulates microcirculation and lympho-drainage, normalizes the indices of cellular and humoral immunity, dilutes inflammatory processes and serves as a factor of the preventive maintenance of adhesions in the abdominal cavity.

9.2.2 Pin-to-hole microdischarge for ulcerative colitis treatment

Inflammatory bowel diseases (IBD) consist of two major chronic, relapsing and debilitative forms of diseases known as ulcerative colitis and Crohn's disease. The etiology (causes) of these diseases remains a mystery although genetic, environmental and immunological factors are found to play a major role in the induction, chronicity and relapses of these diseases. Crohn's disease may appear in any part of the gastrointestinal tract from the mouth to anus and affects the entire thickness of the bowel wall. On the contrary, ulcerative colitis is an inflammatory disorder affecting colonic mucosa and submucosa. There are no known curative therapies for these diseases; however, recent advances in IBD therapeutics have shown that certain biological therapies have been successful in maintaining remission, particularly in Crohn's disease.

A study of the use of pin-to-hole discharge treatment of both healthy colon tissue and experimentally induced ulcerative colitis disease in a live animal model was carried out by a group of researchers from Drexel University (Chakravarthy *et al.*, 2011). The goals of the study were to examine whether this plasma treatment adversely affects the healthy mucosa, and to evaluate if the plasma treatment results in acceleration or worsening of the disease during its induction phase. Additional pilot experiments were conducted to study whether cold plasma discharges provide therapeutic effects, and whether these effects are comparable to a standard therapy or enhance the beneficial effect of a standard therapy.

A modified and further miniaturized pin-to-hole discharge system was employed in this work. It consisted of a central copper needle covered by dielectric material inserted into a grounded stainless steel cylindrical electrode as shown in Figure 9.9. In order to cause minimal mechanical damage to colon tissues, the external electrode was covered by a polyethylene sleeve. The discharge was ignited by applying positive potential to the center electrode from a capacitor that was charged by an external current source. This resulted in the formation of a dense energetic spark that existed for c. 3 μs, with a peak voltage of c. 3.2 kV and an energy per pulse of 0.1 J. Due to the low repetition frequency of c. 7 Hz and short pulse duration, the average gas temperature did not exceed room temperature.

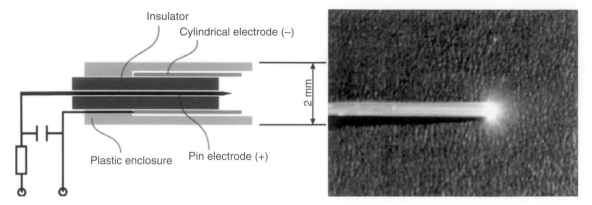

Figure 9.9   *Miniaturized pin-to-hole discharge system employed in IBS treatment demonstration on mice (see color plate).*

Using the Boltzmann plot method and copper emission lines, the temperature of the plasma was estimated to be around 7200 K. This plasma was found to radiate in the UV range at a total power density of c. 5 μW cm^{-2}. The outer tubing was c. 1.5 mm longer than electrode system, covering the discharge from the sides in order to significantly reduce the mucosa UV exposure. The NO production was measured using gas chromatograph Agilent 3000 MicroGC to vary from 900 to 1200 ppm while plasma was applied for 15–60 s.

The Dextran sodium sulfate (DSS) model was used in this study to produce ulcerative colitis in mice, which is representative of human ulcerative colitis. Female Swiss Webster mice 25–30 g in weight aged approximately 6–8 weeks were used. The disease is induced in an animal through a daily oral administration of 2.5% DSS dissolved in drinking water at a concentration of 2.5%. The animal develops an acute form of inflammation beginning on the third day and typically has a full-blown colitis two days later. The primary characteristics of acute inflammation are the increased number of neutrophils in the mucosal layer, shortening of the epithelial crypts and hyalination in the lamina propria, accompanied by severe weight loss, diarrhea and blood in the stool.

This form of the disease provides the opportunity to study efficacy of drugs and compounds. The most striking feature of this model is that it works using a very simple pathway to produce the disease as DSS overcomes the barrier of the epithelium to expose the mucosa to the flora present in the intestine, resulting in an inflammatory response; this in turn leads to activation of macrophages and monocytes. The model also shows close links to the disease in human beings and is simple to induce and reproduce. To quantify the disease induced by the DSS model, a disease activity index (DAI) was used. This index has the scale of 0–4 with 4 being the lethal stage of the disease. DAI is scored on the parameters of weight loss, consistency of the stool and presence of blood in the stool. This index is linearly correlated to the histology score based on changes in the architecture of the crypt.

One of the objectives of the study was to test for any damage caused to the colon tissue or to the animal itself due to the plasma treatment. For this purpose, 12 mice were divided into 4 groups of 3 animals in each group. The first group was a control group which did not receive any treatment, and the other three groups remained as experimental groups. In the experimental groups, a laparotomy was performed and the colon was exposed and kept moist covered with saline gauze. The plasma probe was introduced through the anal verge up to 4 cm into the colon (see Figure 9.10). Plasma treatment was administered for 0, 4, 30 and 60 s in the respective group (for '0' time, the probe was inserted into the colon with no plasma ignited). To check colon tissue damage, the mice were intravenously injected with 30 mg kg^{-1} of Evans Blue (EB); 10 min later the administration colon was washed with 1 mL physiological saline and EB presence

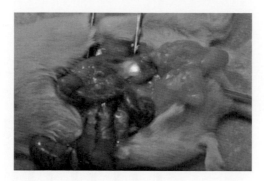

Figure 9.10   *The miniaturized pin-to-hole discharge in operation. The tip was inserted through the anal verge up to 4 cm into the mouse colon. The light due to plasma was observed through the intestinal wall when the abdomen was open (see color plate).*

was analyzed spectrophotometrically. Animals were then euthanized with an overdose of Nembutal®. Colon tissue samples were surgically removed and were preserved in formalin for further histopathological analysis. Spectrophotometrical analysis of the saline fluid collected from the colon showed no traces of EB, indicating that plasma did not affect the tissue integrity. Histology analysis also showed that no macroscopic damage was induced to the colon tissues by plasma treatment or probe manipulation.

To check the response of the disease progression to plasma treatment, 24 animals were divided into 4 groups each receiving 0, 4, 30 or 60 s of plasma treatment every alternate day for 7 days. All animals were fed 2.5% DSS for 7 days in parallel with plasma treatment. DAI was scored every day to see which dosage of plasma was most effective in controlling the progression of the disease. For the plasma probe to be inserted and to go through the colon, the colon has to be cleansed of stool specks. To do so, the animals were fed a polyethylene-glycol-based laxative along with DSS one day before plasma treatment. However, the stool consistency on the next day was compromised as the laxative made the stool consistently loose. Hence, data extracted from the study were bifurcated and analyzed using a three-pronged approach considering:

- weight, stool consistency and presence of blood in the stool;
- weight and presence of blood in the stool (without stool data); and
- mean hemoccult (visible blood in the stool).

To summarize the results of the plasma treatment study, it was found that 30 s of treatment produced the lowest score (the best results) after 7 days, as illustrated in Figure 9.11.

The last stage of the study investigated the effectiveness of plasma treatment as an adjuvant to conventional antioxidant drug (5-amino salicylic acid or 5-ASA) treatment. Based on previous results, where it was shown that 30 s of plasma treatment gives the best results in controlling and reducing the disease progression compared to all other groups, this treatment dose was selected for the next step. In this set of experiments, 24 animals were divided into 4 groups: (1) control group, where no plasma or drug treatment was performed and groups where mice were treated either (2) with plasma alone; (3) 5-ASA alone; or (4) drug and plasma together. All four groups of animals were fed with DSS for 6 days. On days 2, 4 and 6 they received 2.5 % DSS dissolved in water and for days 1, 3 and 5 (one day before plasma treatment) they received 2.5% DSS dissolved in 15% polyethylene-glycol-based laxative to clean the colon. The plasma probe was introduced into the colon 4 cm from anal verge in groups (2) and (4). Group (2) received a 30 s dose of plasma treatment only, while group (4) received the same dose of plasma treatment together with 0.1 mL of 5-ASA treatment.

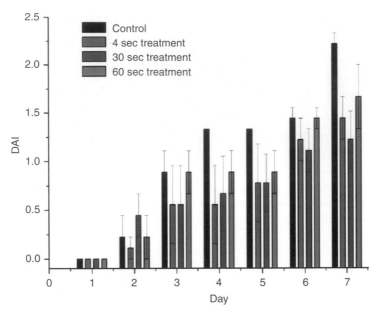

Figure 9.11 *Progression of Crohn's disease activity index after different durations of pin-to-hole microdischarge treatment of mice colons. Plasma treatment significantly slows down disease progression.*

Animals in groups (3) and (4) were treated with 0.1 mL of 5-ASA. The DAI was scored every day during the tenure of the study. On the seventh day of DSS treatment and final plasma treatment, DAI was measured and animals were euthanized with an overdose of Nembutal®.

The disease progression for all four experimental groups is shown in Figure 9.12. The control group showed a steady increase through the course of 7 days with the DAI reaching 2.7 on day 7. The plasma treatment group was administered 30 s of plasma treatment; obtained data are in direct correlation with the data acquired in the previous study and also show a constant increase of disease progression. However, the disease curtailed to near 1.8 on the DAI scale. Group (4) showed very positive results; the treatments in combination proved to be effective in controlling the disease and keeping the DAI at a level of 1.2.

In summary, this study assessed the possibility that pin-to-hole discharge could be used to treat IBD. The plasma treatment did not adversely affect the animals and did not increase the progress of the disease. In fact, it reduced the disease from progressing rapidly as compared to the control. Combination therapy of 5-ASA and cold plasma showed that there is a significant therapeutic relevance when it comes to adding plasma to the drug in controlling colitis in the DSS model during the induction phase. The exact mechanism by which cold plasma induced its beneficial effect remains unknown, although the authors of this study (Chakravarthy 2011) suggest that the observations can be partly attributed to the effects of NO. However, other effects including oxidation of cytokines responsible for activation of the immune system are possible.

9.3   Plasma treatment of cancer

Despite being described by a single menacing word, cancer is no longer viewed as a single disease where some universal therapy has a chance of a cure. Instead, cancer (or a malignant neoplasm, the more formal medical term) is a group of diseases that have some common aspects but also many differences. Cancer is

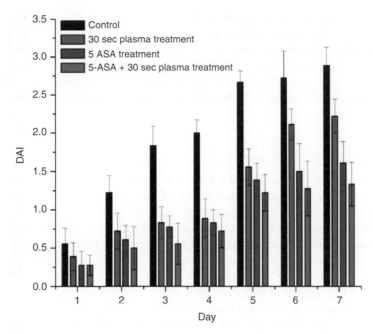

Figure 9.12   *Progression of Crohn's disease activity index in groups where negative control, optimal plasma treatment (30 s), the standard drug treatment (ASA) and combination of plasma and ASA treatment are compared.*

often described as a disease where some cells have increased metabolism and escape the processes of 'normal' regulated growth and cell death for various reasons and, as a result, can accumulate in the body faster than other cells of similar origin. When accumulated in sufficient numbers, these cells can then 'hijack' or divert various processes in the body such as formation of blood vessels for their own benefit leading to various dysfunctions and ultimately death.

Cancers are typically classified by the type of cells that the malignant cells tend to resemble and are, therefore, often presumed to be the origin of the tumor. The different types commonly include the following.

- *Carcinoma*: cancers derived from epithelial cells. This group includes many of the most common cancers, particularly in older individuals, and include nearly all those developing in the breast, prostate, lung, pancreas and colon.
- *Sarcoma*: cancers arising from some connective tissue (i.e. bone, cartilage, fat), each of which develop from cells originating in mesenchymal cells outside the bone marrow.
- *Lymphoma and leukemia*: these two classes of cancer arise from hematopoietic (blood-forming) cells that leave the marrow and tend to mature in the lymph nodes and blood, respectively. These cancer cells do not form solid tumors.
- *Germ cell tumor*: cancers derived from pluripotent cells, most often presenting in the testicle or the ovary.
- *Blastoma*: cancers derived from immature 'precursor' cells or embryonic tissue. These are common in children.

Some cancer cells may acquire the ability to penetrate and infiltrate surrounding normal tissues in the local area, forming a new tumor. The newly formed 'daughter' tumor in the adjacent site within the tissue is called a local metastasis. Other cancer cells acquire the ability to penetrate the walls of lymphatic and/or blood vessels, after which they are able to circulate through the bloodstream (circulating tumor cells) metastasizing to other non-adjacent sites and tissues in the body.

Depending on the type of cancer, treatment options may include surgery often followed by chemotherapy and/or radiation. Photodynamic therapy is a more recent addition to the toolset of cancer treatment options. Interestingly, all of these treatments one way or another involve reactive oxygen species (ROS). It is probably for this reason that plasma treatment can be hypothesized to provide another alternative cancer treatment approach.

9.3.1   Observations of cultured malignant cells

The first reports of the effects of non-thermal plasma specifically on cancer cells (apoptosis in other cells has been considered before) were probably from the Drexel University group (Fridman *et al.*, 2007d), which employed a dielectric barrier discharge (DBD) to treat melanoma cells. Melanoma cancer cell line (ATCC A2058) was propagated using standard cell medium and incubated under well-established processes. After 4–5 days of incubation the cells were detached from the dish surface and transferred to aluminum dishes in dilution ratios of 1:5 to 1:6, where they were incubated for 4–5 more days. The number of cells in each dish at the end of incubation was c. 1.5×10^6 on average. The viability of cultures was determined by Trypan Blue exclusion (a measure of the integrity of the cell membrane) to be 91–97%. Trypen Blue is a dye that tends to enter cells if their membranes have sufficiently large pores, indicating lack of cell viability.

Cell plasma treatment employed in this work is schematically illustrated in Figure 9.13. A plasma gap above the cell medium of 3 mm was used, while peak AC voltage was around 20 kV and power was measured to be around 1 W cm^{-2}. Typical filamentary DBD was observed over the cell medium during the treatment. Cells were treated for times ranging from 5 to 30 s.

The most important finding of this work was that the melanoma cells were not simply destroyed through a process known as necrosis, but that they were induced into apoptotic cell death. Apoptosis is a multi-step, multi-pathway cell-death program that is inherent to most cells in the body. In contrast to necrosis, which is a form of traumatic cell death that results from an acute cellular injury, apoptosis confers certain advantages. For example, unlike necrosis, apoptosis results in cell fragmentation called apoptotic bodies that phagocytic cells are able to engulf and quickly remove before the contents of the cell can spill out onto surrounding cells and cause damage. The removal of apoptotic cell fragments by phagocytes occurs in an orderly manner without eliciting an inflammatory response. Apoptosis can be viewed as a natural part of a cell renewal cycle

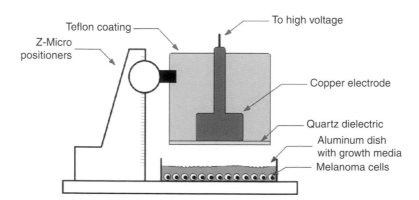

Figure 9.13   *Schematic of the experimental setup where DBD is used to treat cells immersed in some cell growth medium.*

where cells in the body maintain a certain ratio of cell division to apoptosis. In a growing body, this ratio favors cell proliferation. Once the body stops growing, the ratio of cell division to apoptosis is normally stabilized to keep renewing cells, while maintaining the overall size of the body and its organs. In cancer, the apoptosis to cell division ratio is altered, to favor proliferation of cancer cells. It was originally believed that cancer cell accumulation was due to an increase in cellular proliferation, but it is now known that it is also due to a decrease in apoptotic cell death. Many non-surgical cancer treatments such as chemotherapy and irradiation kill malignant cells primarily by inducing apoptosis. The ability of non-thermal plasma treatment to induce apoptosis can therefore be viewed as a promising first step in developing this form of cancer treatment.

The following events are typically observed in apoptotic cells: the phospholipid phosphatidylserine (PS), which is normally hidden within the plasma membrane, is exposed on the surface; cells develop bubble-like blebs on their surface; cells have their mitochondria breakdown with the release of cytochrome c; the chromatin (DNA and protein) in cell nuclei are degraded; cells break into small, membrane-wrapped, fragments; and calls release molecules such as ATP and UTP (uridintriphosphate). These nucleotides bind to receptors on wandering phagocytic cells such as macrophages and dendritic cells, and attract them to the dying cells (a 'find-me' signal). This signal is bound by other receptors on the phagocytes, which then engulf the cell fragments and secrete cytokines that inhibit inflammation (e.g. IL-10 and TGF-β).

Apoptosis may occur through three primary pathways. Two of these pathways, one intrinsic and the other extrinsic, proceed through activation of proteases (enzymes that degrade poypetides) from a family known as caspases. Intrinsic pathways typically proceed through opening of pores in mitochondrial membranes, release of cytochrome c and subsequent activation of caspase 9 followed by other caspases. Extrinsic pathways are typically initiated by the binding of molecules TNF-α, Fas ligand or lymphotoxin to a death receptor followed by activation of caspases beginning with caspase 8. The third caspase-independent pathway proceeds through release of the apoptosis inducing factor (AIF), a molecule found within mitochondrial membranes.

Several biochemical assays are typically employed to detect and follow apoptosis. Cells in their final stages of apoptosis are often recognized by the TUNEL assay which enzymatically labels DNA strands with exposed $3'$-hydroxyl ends exposed in DNA fragmentation. Annexin-V is a calcium-dependent phospholipid-binding protein with high affinity for phosphatidylserine (PS). As mentioned above, PS is a membrane component normally localized to the internal face of the cell membrane. Early in the apoptotic pathways, molecules of PS are translocated to the outer surface of the cell membrane where Annexin-V can readily bind them. Annexin-V is therefore an assay that detects early indicators of apoptosis. Caspase-dependent pathways can be detected through caspase cleavage assays where cleaved caspase are detected from lysed cells.

While the initial work from the Drexel University group detected apoptosis in melanoma cells only by the TUNEL assay (Fridman *et al.*, 2007d), subsequent work by the same group (Sensenig *et al.*, 2011) observed both early apoptosis by Annexin-V assay (see Figure 9.14) and caspase cleavage (caspase 3) in addition to the detection of late apoptotic cells by the TUNEL assay. This was important not only in confirmation of apoptosis, but also in the observation that apoptosis in the DBD-treated melanoma cells occurs via caspase cleavage. In addition, it was observed that scavenging of intracellular reactive oxygen species through pre-incubation of the cells with N-acetyl-cysteine (NAC), a common method of increasing levels of intracellular glutathione that scavenges ROS, reduces or eliminates apoptosis following the plasma treatment. This suggests that the mechanism of apoptosis due to the DBD treatment relies on intracellular generation of ROS.

Current research by the same group goes beyond studying apoptosis in cancer cell line. The key to a successful cancer treatment is not only its ability to induce 'natural' death of cancer cells, but also to reduce the impact on normal cells in the body. Chemotherapy and radiation treatments both tend to have a stronger effect on cancer cells than on the surrounding normal cells, and plasma treatment could be viewed as a viable alternative when it comes to more localized treatment if its effects on transformed (cancer) versus normal cells could be selective.

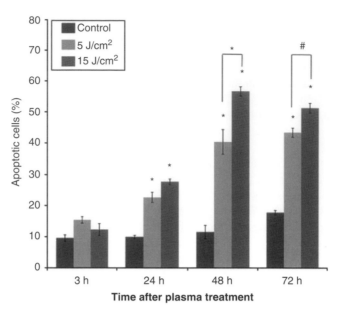

Figure 9.14   *Apoptotic cells as detected by the Annexin-V, an early apoptosis assay at various times following two different doses of DBD plasma treatment and control.*

Initial promising results related to selectivity of the plasma treatment were recently demonstrated using immortalized (in contrast to primary cells taken from the body, immortalized cells typically acquire some mutation to make them capable of dividing indefinitely) but non-invasive and not-tumor-forming mammary epithelial cells MCF10A, and an invasive (tumor-forming) genetic variant of these cells called MCF10A-NUET. These results are shown in Figure 9.15 where, at doses of 7.8 J cm^{-2}, apoptotic effect on the tumor-forming cells is about twice as strong as the apoptotic effect on the non-invasive cells. Importantly, differences in the tumor-forming and non-invasive cell responses could not be attributed to differences in culturing conditions, cell medium or other similar environmental conditions, since they were identical for the two different cell types. Other groups (Zirnheld *et al.*, 2010) recently considered differential effects on some melanoma and keratinocyte cells. It was shown that percent death of melanoma cells was about 6 times as large as keratinocyte cell death following treatment with a He plasma jet powered by a 2.5 kV AC source at a frequency of 80–120 kHz. Treatment of colorectal cancer cells by a He dielectric barrier jet also demonstrated (Kim *et al.*, 2010) loss of invasiveness by the tumor cells as well as apoptosis and phosphorylation of β-catenin, which is a transcription factor that controls colorectal tumorigenesis (its phosphorylation causes its rapid degradation in cells).

9.3.2   Non-thermal plasma treatment of explanted tumors in animal models

The first comprehensive study to evaluate the effect of non-thermal plasma treatment on malignant tumors was performed by a group of researchers from CNRS and GREMI (Orleans University), Orleans in France (Vandamme *et al.*, 2010, 2012). This group employed malignant cells of glioma lineage xenografted (explanted) by subcutaneous injection of the tumor cells suspension (106 cells in 0.1 mL 0.9% NaCl) into the hind legs of Swiss nude female mice about 4 weeks of age. These cells are known to exhibit resistance to chemotherapy and radiotherapy. Floating electrode dielectric barrier discharge (FE-DBD) was used

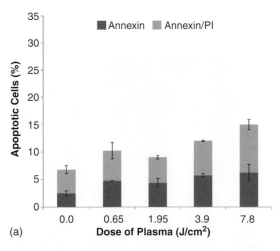
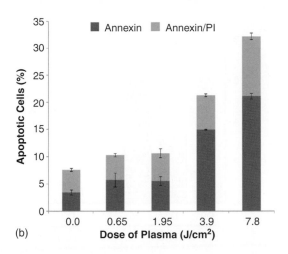

(a)   (b)

Figure 9.15   *Induction of apoptosis in (a) MCF10A immortalized breast epithelial cells and (b) MCF10A-NEUT tumor-forming cells 48 hours after different doses of DBD plasma treatment. The Annexin stains PS that translocate across membranes in early apoptosis, while PI stain is able to enter cells with poor membrane integrity. As apoptosis progresses, cells that initially stain with Annexin will become increasingly stained with PI as well.*

for the non-thermal plasma treatment. The effect of this treatment on tumor volume and the consequences of treatment on cell proliferation, cell cycle and apoptosis induction were investigated in cell culture and in the animal. The cells employed in this work were genetically modified by transfection with firefly lucifcrase gene and, as a result, exhibited bioluminescence that made it possible to observe them even when implanted in the mouse. Monitoring of the growth and activity of the tumor was therefore possible by bioluminescent imaging (BLI).

The key conclusions regarding apoptosis of the employed cells when treated by the plasma in cell medium were confirmed through Annexin and propidium iodide assays. Moreover, the plasma treatment of the cell cultures confirmed the expected DNA damage as well as the fact that intracellular scavenging of the reactive oxygen species using NAC protected the cells from DNA damage and induction of apoptosis. The most interesting result, however, was the reduction of the xenographted tumor activity and size, as illustrated in Figure 9.16. The survival of the plasma-treated mice was extended by c. 60%. These results are remarkable, particularly given the fact that the tumor cells were several millimeters under the surface of the tissue clearly demonstrating that non-thermal plasma treatment can penetrate at least several millimeters. The authors of this study (Vandamme *et al.*, 2012) hypothesized that the possible mechanism of the treatment penetration was similar to the bystander effect observed in radiation treatments, where cells directly exposed to the treatment pass signals to unexposed cells. Another possible mechanism could be related to diffusion of active species created through plasma treatment in tissue.

Another recent study (Keidar *et al.*, 2011) of xenographted cancer cell also reported surprising efficacy and selectivity against several different types of cancer cells. In this study the researchers employed a Helium plasma jet. They compared responses to the plasma treatment of several different cells including lung cancer cells and bronchial epithelial cells as well as murine melanoma (cancer) cells and primary macrophages from mouse bone marrow. The team noted much stronger response to the plasma treatment in increased detachment and reduced survival for the lung cancer compared to the bronchial epithelial cells. For the subcutaneously implanted tumors with bladder cancer cells, the group demonstrated a strong anti-tumor effect of the plasma treatment through mouse skin. In the cases of smaller tumors of c. 5 mm in diameter, the tumor appeared to

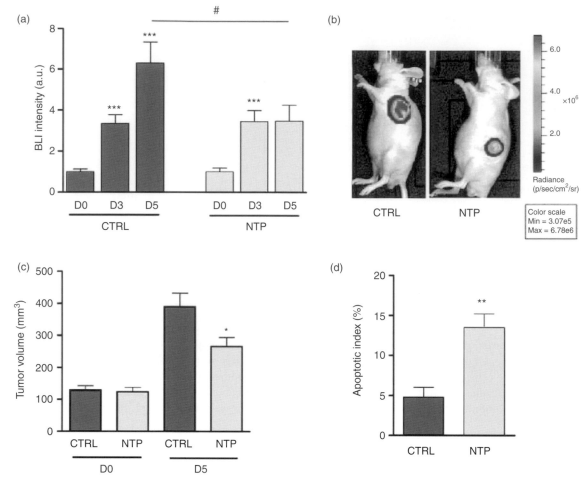

Figure 9.16 *In vivo evaluation of NTP anti-tumor activity. Plasma treatment was delivered each day for five consecutive days (6 min, 200 Hz). (a) BLI imaging performed during treatment course (D3) and 24 hours after the end of treatment protocol (D5) was normalized to total intensity prior to treatment (D0). (b) Representative BLI imaging of CTRL and NTP treated mice at D5. (c) Tumor volume was determined using a caliper 24 hours after the last day of treatment (D5). (d) Apoptosis indexes were determined by immunohistochemical detection of cleaved caspase 3.*

be completely irradiated and did not grow back after a single treatment; this is a remarkable result. Larger tumors shrunk, but did grow back (albeit at a slower rate). Xenographted melanoma tumors also reduced in size during treatment and demonstrated reduced growth after treatment.

9.4   Plasma applications in dentistry

Three primary applications of plasma in dentistry appear to have been discussed in the literature. One application deals with the killing of bacteria on exposed teeth surfaces and inside tooth root canals as well as

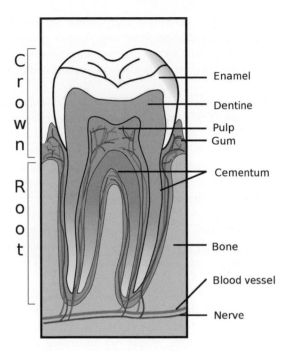

Figure 9.17　*Illustration of tooth structure.*

providing some anti-inflammatory activity. The second is related to improved adhesion and incorporation of materials such as those contained in tooth fillings. The third is whitening of the tooth enamel, which is mostly a cosmetic application. Prior to discussing some of these applications, it may be useful to review some facts about tooth structure.

9.4.1　Brief overview of structure of teeth and dental health

The typical structure of a tooth is illustrated in Figure 9.17, and the various components are described in the following.

Enamel is probably the hardest and most highly mineralized substance of the body. It is normally visible and must be supported by underlying dentin; 96% of enamel consists of hydroxyapatite mineral. The normal color of enamel varies from light yellow to grayish white. At the edges of teeth where there is no dentin underlying the enamel, the color sometimes has a slightly blue tone. Since enamel is semi-translucent, the color of dentin and any restorative dental material underneath the enamel strongly affects the appearance of a tooth. Enamel varies in thickness over the surface of the tooth and is often thickest at the cusp (up to 2.5 mm) and thinnest at its border.

Dentin is a mineralized connective tissue with an organic matrix of collagenous proteins. It has microscopic channels called dentinal tubules which radiate outward through the dentin from the pulp cavity to the exterior cementum or enamel border. The diameter of these tubules ranges from 2.5 μm near the pulp, to 1.2 μm in the middle portion and 900 nm near the dentin/enamel junction. As a result, dentin has a degree of permeability which can increase the sensation of pain and the rate of tooth decay. By weight, 70% of dentin consists of the mineral hydroxyapatite, 20% is organic material and 10% is water. Primary dentin, the most prominent dentin in the tooth, lies between the enamel and the pulp chamber. The outer layer closest to enamel is known as mantle dentin. This layer is unique to the rest of the primary dentin. Mantle dentin is a layer approximately 150 μm thick formed by newly differentiated odontoblast cells. Unlike primary dentin, mantle dentin lacks phosphoryn, has loosely packed collagen fibrils and is less mineralized. Below it lies the circumpulpal dentin, a more mineralized dentin which makes up most of the dentin layer and is secreted by the odontoblasts after the mantle dentin.

Cementum is a specialized bone-like substance covering the root of a tooth. It is approximately 45% inorganic material (mainly hydroxyapatite), 33% organic material (mainly collagen) and 22% water. Cementum is excreted by cementoblast cells within the root of the tooth and is thickest at the root apex. Its coloration is yellowish and it is softer than either dentin or enamel. The principal role of cementum is to serve as a medium by which the periodontal ligaments can attach to the tooth for stability. At the cement–enamel junction, the cementum is acellular (free of cells). This acellular type of cementum covers at least two-thirds of the root. The more permeable form of cementum, cellular cementum, covers about one-third of the root apex.

The *dental pulp* is the central part of the tooth filled with soft connective tissue. This tissue contains blood vessels and nerves that enter the tooth from a hole at the apex of the root. Along the border between the dentin and the pulp are odontoblasts, which initiate the formation of dentin. Other cells in the pulp include fibroblasts, preodontoblasts, macrophages and T-lymphocytes. The pulp is commonly called the 'nerve' of the tooth.

The most common dental problems are probably the result of bacterial activity. One of the reasons bacteria on the teeth can be difficult to fight is formation of bacterial biofilm, known as plaque. If not removed regularly, plaque buildup can lead to dental cavities (caries) and/or periodontal problems such as gingivitis (inflammation of the gum). Given time, plaque can mineralize along the gingiva, forming *tartar*. The microorganisms that form the biofilm are almost entirely bacteria, with the composition varying by location in the mouth. *Streptococcus mutans* is probably the most important bacterium associated with dental caries (cavities).

Certain bacteria in the mouth live off the remains of foods, especially sugars and starches. With insufficient oxygen they produce lactic acid, which dissolves the calcium and phosphorus in the enamel. This process, known as 'demineralization', leads to tooth destruction. Enamel begins to demineralize at a pH of 5.5. Saliva can gradually neutralize the acids raising pH above critical in some locations. However, saliva is often unable to penetrate through plaque to neutralize the acid produced by the bacteria at the foundation of the biofilm.

Tooth restoration typically involves cleaning out infected cavities in teeth and placement of dental filling material in place of removed parts of the teeth. Complete removal of bacteria has been problematic, particularly during root canal procedures. Root canal usually needs to be cleaned out when a tooth's nerve tissue or pulp is damaged and bacteria begin to multiply within the pulp chamber. The bacteria and other decayed debris can cause an infection or abscessed tooth. An abscess is a pus-filled pocket that forms at the end of the roots of the tooth when the infection spreads all the way past the ends of the roots of the tooth. In addition to an abscess, an infection in the root canal of a tooth can cause swelling that may spread to other areas and cause bone loss around the tip of the root. A hole can occur through the side of the tooth with drainage into the gums or through the cheek with drainage into the skin. During a root canal procedure, the nerve and pulp are removed and the inside of the tooth is cleaned and sealed. A tooth's nerve is not considered to be vitally important to a tooth's health and function after the tooth has emerged through the gums. Its only function appears to be sensory, providing the sensation of hot or cold. Removing all bacteria from a root canal is essential because the procedure removes not only pulp and nerves, but also blood vessels which could supply immune cells capable of fighting infection.

Different dental filling materials have been employed in restorative dental procedures. The main difficulties involved is usually poor adhesion and/or material incompatibilities such as shrinkage of the filling material. In the past, dentists created special structures by removing some healthy parts of a tooth to help hold the filling in place. To create a porous surface that the adhesive can infiltrate, current preparation techniques etch and demineralize dentin. The culprit that foils mechanical bonding is a protein layer, the so-called 'smear layer', which is primarily composed of type I collagen that develops at the dentin/adhesive junction. Interactions between demineralized dentin and adhesive gives rise to the smear layer, which actually inhibits adhesive diffusion throughout the prepared dentin surface. This protein layer may be partly responsible for causing premature failure of the composite restoration. It contributes to inadequate bonding that can leave exposed, unprotected collagen at the dentin adhesive interface, allowing bacterial enzymes to enter and further degrade the interface and the tissue.

9.4.2   Recent promising results of plasma applications in dentistry

9.4.2.1   *Gingivitis*

Gingivitis is an inflammation/infection of the gum. Effectiveness of the plasma NO-therapy has been demonstrated on chronic gingivitis using the Plazon plasma generator discussion in the previous sections of this chapter (Section 9.1.2). After the first session of the therapy, gum bleeding ceased. After 1–2 weeks, normalization of tissue and regional blood flow in the tissues of periodontium (Grigorian *et al.*, 2001) occurred. Normalization of cytological signs was observed in 2–3 months, while in the control group normalization was not observed at all. Utilization of NO-CGF from plasma after surgical intervention for periodontal disease showed that normalization of clinical and cytological signs occurs by day 7 in the experimental group, but not until day 14 in the control group. Complications were not observed in the experimental group, but were observed in the control group.

9.4.2.2   *Use of ozone in dentistry*

Recent years have seen a significant rise in interest in the use of ozone for dentistry, although ozone generation has rarely been associated with plasma in the dentistry-related literature. Since inhalation of ozone is considered undesirable, different ozone delivery fittings are available to expose teeth and areas surrounding them, while minimizing inhalation.

The effect of ozonated water on oral microorganisms and plaque has been investigated by several researchers (Nagayoshi *et al.*, 2004; Arita *et al.*, 2005). At 2 and 4 mg L^{-1} of ozone, the water was able to reduce the bacteria and *Candida albicans* dramatically. Other studies pointed out that the effect of ozone gas directly exposed to dentures was even stronger (Oizumi *et al.*, 1998). Irrigation with ozonated water and sonication (Arita *et al.*, 2005; Nagayoshi *et al.*, 2004; Huth *et al.*, 2009) had antimicrobial activity similar to 2.5% NaOCl (sodium hypochloride, a standard root canal rinse used to remove dentin chips and kill bacteria in c. 3 min). Ozonated oil was also shown to be similar in its effect to calcium hydroxide paste (with camphorated paramonochlorophenol and glycerin) in endodontic treatment of teeth (root canal; Silveira *et al.*, 2007).

Key to all these studies is appropriate dozing of ozone. Ozone works best with reduced organic debris present; the use of either ozonated water or ozone gas at the end of the cleaning and shaping process is therefore recommended. Ozone is effective when it is used in sufficient concentration for an adequate time, but is not effective when the dose is insufficient. It was shown, for example, that by increasing the contact time from 10 to 20 s, the bacterial kill rate changed in some cases from ozone being a disinfectant to acquiring sterilizing effect (6-log reduction). This killing was reduced in the presence of saliva, although increasing the ozone application time to 60 s overcame the neutralizing effect of saliva (Johansson *et al.*, 2009) which tends to decompose ozone faster and offers additional organic targets. One study on primary root caries lesion (PRCL) found that ozone application for either 10 or 20 s dramatically reduced most of the microorganisms in PRCLs without any side effects recorded at recall intervals between 3 and 5.5 months (Baysan and Lynch, 2004).

The antimicrobial properties of ozone may not be the only reason for the interest in it. Ozone and ozonated water apparently has strong anti-inflammatory properties. Some researchers (Huth *et al.*, 2007) have examined the effect of ozone on the influence on the host immune response. They considered the NF-kappaB system, a well-known mechanism for inflammation-associated signaling/transcription. Their results demonstrated that NF-kappaB activity in oral cells in periodontal ligament tissue from root surfaces of periodontally damaged teeth was inhibited following incubation with ozonized medium.

9.4.2.3   *Plasma treatment of dental caries and root canals*

One of the first groups to propose applying plasma directly to teeth was probably Eva Stoffels. She and her group at the Eindhoven Institute of Technology developed a RF-powered helium plasma jet called a plasma needle and investigated it as a tool for fighting dental cavities in early 2000 (Sladek *et al.*, 2004; Sladek and Stoffels, 2005). This plasma needle demonstrated the ability to kill relevant bacteria. Substantial work has been performed since that time on development and characterization of various discharges in the form of small jets, many of which have also been referred to as plasma needles. These include discharges operating in various gases and excited by various means using RF sources as well as high-voltage AC sources. Many of these discharges have been shown to be effective against various bacteria in vitro (cultures on dishes).

Some recent studies started looking for antimicrobial activity not only on agar, but also on targets more relevant for dentistry. A microwave-powered non-thermal atmospheric-pressure helium plasma jet was evaluated (Rupf *et al.*, 2010) for its antimicrobial efficacy against adherent oral microorganisms. Agar plates as well as dentin slices were inoculated with 10^6 CFU cm^{-2} of *Lactobacillus casei*, *Streptococcus mutans* and *Candida albicans*, with *Escherichia coli* as a control. Dentin slices were rinsed in liquid media and suspensions were placed on agar plates. The plasma jet treatment reduced the CFU by 3–4 orders of magnitude on the dentin slices in comparison to untreated controls, demonstrating that plasma can have a strong antimicrobial efficacy on relevant dental targets.

Other research groups considered the application of plasma jets to disinfection of root canals. Jiang *et al.* (2009a, b) at the University of California employed a helium/oxygen plasma jet triggered by fast-rising (around 10 ns) high-voltage pulses and demonstrated that bacteria can be killed inside the root canal, although the disinfecting action did not extend through the entire root canal (see Figure 9.18). Another approach where

Figure 9.18   *Disinfection of some parts of the root canal by a helium/oxygen plasma jet triggered by fast-rising (around 10 ns) high-voltage pulses.*

the plasma probe is placed into the tooth (Lu *et al.*, 2009) was probably able to achieve deeper penetration of plasma into the root canal. It was found that this device efficiently kills *Enterococcus faecalis* (an important root canal pathogen) within several minutes of plasma treatment.

9.4.2.4   Plasma treatment for improved filling adhesion

Preliminary data has also shown that plasma treatment increases the bonding strength at the dentin/ composite interface by roughly 60%; that interface-bonding enhancement can then significantly improve composite performance, durability and longevity. Ritts *et al.* (2010) from University of Missouri (Kansas City) investigated the plasma treatment effects on dental composite restoration for improved interface properties. Experimental results showed that atmospheric cold plasma brush (ACPB) treatment can modify the dentin surface and thus increase the dentin/adhesive interfacial bonding. The solution is to introduce bonds that depend on surface chemistry rather than surface porosity.

The effects of plasma treatment on the shear bond strength between fiber-reinforced composite posts and resin composite for core buildup were also studied (Yavirach *et al.*, 2009). It was concluded that plasma treatment appeared to increase the tensile-shear bond strength between post and composite.

9.5   Plasma surgery

Intense heat and even fire have been used in medicine to stop bleeding and prevent infection after injury or surgery for many years. People began experimenting with electrical means to apply heat locally to tissues in the 19th century. It was noticed that relatively high local current density could have intense local heating effects associated with ohmic losses. Using high-frequency currents to avoid neuromuscular stimulation is often attributed to D'Arsonval who performed this work around the end of the 19th century.

At the beginning of the 20th century, researchers including Doyen, Cushing and Bovie experimented with devices that had sufficiently high voltages to generate plasma at the tissue interface. Modern Bovie knifes frequently used in surgery today are a development of this work. The plasma produced at the edge of the Bovie knife and the current passing through plasma into the tissue are sufficiently intense to burn and ablate tissue in a localized fashion, while cauterizing it to avoid bleeding during surgery.

Argon plasma coagulators of various kinds were developed later in the 20th century. They generate plasma remotely within the flow of Ar gas and, subsequently, pass current through the plasma into the tissue to achieve tissue charring, ablative effects and cauterization. In 1990s, thermal plasma was generated in fluids (saline solution) for surgical purposes. In this case, it does not appear to be the current through tissue that performs the desired functions, but rather the plasma in the fluid/gas medium adjacent to the tissue being ablated and cauterized.

Multiple papers and books have reviewed the practice of electrosurgery. For a good overview on the use of plasma in saline solution for surgical applications, the reader is referred to the chapter on electrical discharges in conducting liquids by Ken Stadler and Jean Woloszko in *Plasma Medicine*, recently edited by M. Laroussi, M.G. Kong, G. Morfill and W. Stolz (Laroussi *et al.*, 2012).

10

Plasma Pharmacology

Previous chapters have focused on direct treatment of surfaces and tissues, where plasma is effectively in direct contact with treated material and some charges created in plasma can pass into such tissues. Indirect treatment where a jet or a flow of gas (but not charges) from plasma impacts the surface of the material has also been discussed. The eventual goal of these treatments has varied from killing bacteria or some malignant cells to diminishing inflammatory reactions or enhancing cell proliferation and tissue healing. In many cases it was clear that the organic medium surrounding the cells plays a crucial role in mediating the observed effects of the plasma treatment. This is unsurprising as it is well-known that organic molecules in aqueous phase can be modified (primarily oxidized) by plasma created in the gas phase above or within the fluid. It has also been widely observed that water treated by plasma undergoes a change in acidity.

These facts suggest that perhaps plasma can be used not only for direct or indirect treatment of living systems, but also to modify the medium which could be used separately for treatment of cells and tissues. In this sense, the medium treated by plasma can be viewed as a pharmacological substance. The key issue in this regard is the stability of the plasma-treated medium and its effects on living systems. The ability to store the effects of plasma treatment in the form of plasma-treated material may have advantages in certain medical applications. In some cases it may be difficult, for example, to apply plasma to certain surfaces (e.g. intestinal surfaces). On the other hand, applying plasma-treated fluids to such surfaces can be substantially easier. Moreover, controlling the material that delivers plasma treatment may provide an additional degree of control over the effect of the treatment.

In this chapter some results related to plasma-treated aqueous media, its stability and the stability of its effects on living systems will be reviewed. A large volume of reliable results is not yet available for this topic and a complete understanding of the physical and chemical mechanisms involved is still lacking. However, given the potential importance of this topic in the future, it is well worth some consideration now.

10.1 Non-thermal plasma treatment of water

The antibacterial effects of water, organic molecule oxidation in water and water acidification as a result of treating water surface by different plasma discharges have been widely reported (e.g. Burlica *et al.*, 2004; Moussa *et al.*, 2005; Marouf-Khelifa *et al.*, 2006; Moreau *et al.*, 2007; Satoh *et al.*, 2007; Brissetoussa *et al.*, 2008; Chih Wei, 2008; Doubla *et al.*, 2008; Tang *et al.*, 2008; Ikawa, 2010; Shainsky *et al.*, 2012) including

Plasma Medicine, First Edition. Alexander Fridman and Gary Friedman.
© 2013 John Wiley & Sons, Ltd. Published 2013 by John Wiley & Sons, Ltd.

various jets, gliding arcs and dielectric barrier discharges (DBDs). The correlation between the acidity and antimicrobial effects was also noted (Tang, 2008; Ikawa, 2010). The fact that antimicrobial effect is correlated to lower acidity is not known only in the context of plasma treatment. For example, disinfection of fruit and vegetables with bleach is often performed with addition of acids such as citric to lower acidity of the fluid. In the case of DBD-treated water, it was suggested (Oehmigen *et al*., 2010) that the acid created in water is nitric/nitrous acid and the antimicrobial properties observed are the result of synergetic action between hydrogen peroxide, also generated by plasma and nitric/nitrous acid. Similar suggestions were expressed by other researchers (Tang *et al*., 2008; Ikawa, 2010). The key questions that arise in the context of treating water with plasma are therefore the following.

- What is responsible for acidity where pH has been observed in some cases to be below 2?
- What are the additional organic oxidizing and antimicrobial ingredients?
- How stable is the acidity and other active molecules that are produced by plasma treatment of water?

Most researchers studying plasma treatment of water focus on applications and/or the plasma discharge, while the details related to various water impurities and gases dissolved in water may also play an essential role in the outcomes.

In the following section, we provide a description of a case study which focuses on a particular water treatment experiment where at least some of these issues were carefully considered.

10.2   Deionized water treatment with DBD in different gases: Experimental setup

DBD plasma was generated using the experimental setup depicted in Figure 10.1 where an electrode with a 0.5 mm polished clear fused quartz dielectric barrier was positioned c. 1.5 mm above the surface of deionized water. The discharge was generated by applying an alternating polarity of 1.7 kHz pulsed voltage of c. 17 kV, pulse width 1.5 μs and rise time 5 V ns^{-1} between the insulated high-voltage electrode and a 10 mL water sample. The water served as a second electrode in the case of direct treatment. A grounded mesh was positioned at a distance 1.5 mm away from the insulated electrode and above the water surface to simulate indirect treatment, similarly to that of Dobrynin *et al*. (2009). Recall that in indirect treatment, most of the charged species generated in plasma do not enter the material being treated.

The water sample was held in a pyrex-glass Petri dish with an inner diameter of 37 mm and outer diameter of 40 mm, while the electrode diameter was 33 mm with a 25 mm copper core. The Petri dish was placed on a grounded metal plate. The treatment chamber with the electrode and the Petri dish in it was sealed from the environment with a rubber seal that allowed a gas to flow into and out of the chamber and liquid to be poured into the chamber without opening it.

Deionized water with an electrical resistivity of 18 MΩ was employed for the experiments described in the following sections and a microelectrode was used to measure pH. A UV/VIS Spectrophotometer (Perkin-Elmer Model: Lambda 2) was used for detection of nitric/nitrous acid in the treated water, while hydrogen peroxide concentration was detected using Peroxide test strips.

10.2.1   Changing the working gas and sample degassing

In experiments performed with different gases, the gas inside the treatment chamber was pumped out and the new working gas (e.g. oxygen, argon) was added and circulated in the chamber for 3 hours. The water sample was injected into the glass enclosed treatment chamber with a syringe through the rubber seal after

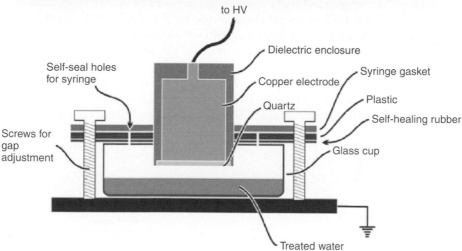

Figure 10.1 *(Lower) Schematic of the experimental setup; DBD electrode is placed in an enclosed glass chamber. The distance between the electrode and the liquid is 1.5 mm. (Upper) Photograph of the experimental setup with plasma (see color plate).*

the 3 hour period. Gases dissolved in the water sample were removed using a relatively standard degassing procedure using the following setup.

A tall glass test tube (16 × 100 mm) was sealed on the top with a rubber stopper. Two needles of different length were inserted through the seal. The longer needle was immersed in the water sample and employed to introduce the new working (plasma treatment) gas into the liquid; the shorter needle was positioned with its tip above the water sample to allow the remaining gas to exit into the environment. The degassing procedure lasted 12 hours, minimizing the concentration of nitrogen and oxygen naturally dissolved in the deionized water. The water was degassed every time with the same gas that was used during the plasma treatment of the water. For example, if we used oxygen gas during the treatment, water degassing was performed with oxygen gas; when plasma treatment was performed with argon, water was degassed with argon gas.

10.3   Deionized water treatment with DBD in different gases: Results and discussion

Figure 10.2 shows the pH reduction of deionized water (black squares) after application of non-thermal atmospheric-pressure DBD, ignited in air, to the surface of the water. As can be seen from this figure, after 15 min of treatment the water becomes strongly acidic having pH around 2. Reduction in the pH depends on the treatment time when the power and liquid volume are constant. The different pH values previously reported can be explained by the different experimental conditions such as treatment method (direct treatment or indirect treatment), treatment duration, treatment dose (J mL^{-1}), properties of the liquid (e.g. deionized water, water containing organic material), treatment gas (e.g. air, oxygen) and discharge type. Additionally, Figure 10.2 demonstrates acidity variation of deionized water for samples treated in pure oxygen with similar time dependence as air-treated samples.

Water degassing prior to the DBD treatment affects the experimental results significantly as shown in Figure 10.3, which compares samples treated in oxygen and argon that were degassed prior the application of DBD with samples of air, oxygen and argon that were not degassed before the application of plasma. The degassed argon sample did not result in any pH change, while the pH of the non-degassed argon sample was closer in pH to the air sample after the application of DBD discharge for 15 min. This suggests that gases dissolved in water can contribute significantly to the observed pH, probably through their presence in the plasma.

No significant pH difference immediately after the treatment was observed between degassed and non-degassed oxygen samples as well as air-treated samples. Since the major difference between the degassed and non-degassed samples is the presence/absence of dissolved nitrogen and oxygen gases in the water and

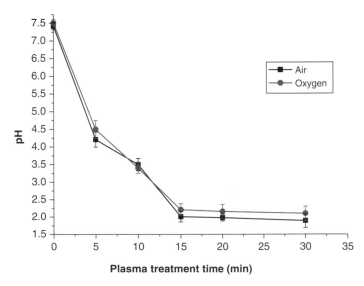

Figure 10.2   *Variations of pH of degassed deionized water after plasma treatment in three different gases: air, oxygen and argon. No pH variation after treatment in Ar was observed.*

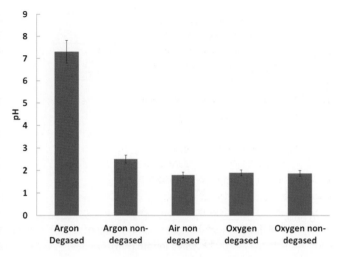

Figure 10.3 *Acidity (pH) of deionized water after plasma degassed and non-degassed treatment for 15 min in argon, oxygen and air samples.*

in plasma, this result might support the hypothesis that the primary effect on pH was due to the presence of oxygen. However, any such definitive conclusion may be a little premature. This experiment certainly demonstrates that details of sample preparation on the acid composition and acidity are important.

Charged species also play a significant role in determining water acidity and composition, as can be observed from Figure 10.4. Direct treatment results in a significantly lower pH of deionized water than the indirect treatment following 15 min of DBD exposure. This suggests that ions play a significant role in the acidification process, possibly through the electron exchange mechanism described by:

$$M^+ + H_2O \rightarrow M + H^+ + OH^* \tag{10.1}$$

where M^+ is a positive ion from plasma.

Researchers using indirect DBD treatment (Oehmigen *et al.*, 2010) previously suggested that acidity of plasma-treated water might be the result of nitric and nitrous acid formation. The presence of nitric/nitrous

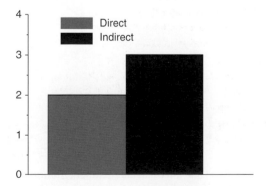

Figure 10.4 *Comparison of pH for deionized water after direct treatment (charges enter water) and indirect treatment (charges do not enter water) after 15 min of DBD treatment.*

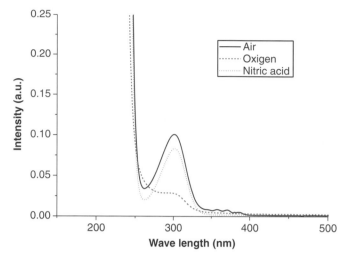

Figure 10.5   *Absorption spectra measured by UV-VIS spectrometer of nitric acid (pH 2.0) and water samples treated with plasma in air (pH 2.07) and in oxygen (pH 2.01).*

acid resulting from direct DBD treatment as described above was examined using UV-VIS spectroscopy. The results are summarized in Figure 10.5. Comparison to nitric acid UV-VIS absorption also shown in the figure clearly demonstrates that air-plasma-treated water has the characteristic nitric acid absorption peak at 300 nm, while oxygen-plasma-treated water may have nitric acid only at the level of impurity. Such impurity-level concentration cannot be responsible for the observed pH level of c. 2.

Nitric/nitrous acid therefore seems to partly explain the low acidity of the air-plasma-treated water. Oxygen plasma treatment results however suggest that some other conjugate base may also be partly responsible for the low level of acidity. This conjugate base may not be as stable as the nitric acid conjugate base (NO_3^-). Indeed, Figure 10.6 demonstrates relative stability of the acid generated through air and oxygen plasma treatments.

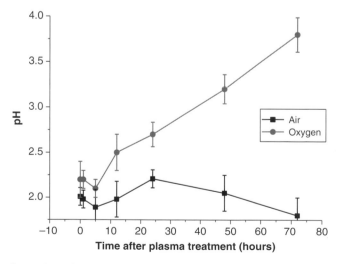

Figure 10.6   *Variation of pH of air-plasma-treated deionized water as a function of time after the treatment. Time 0 indicates measurement immediately after the plasma treatment.*

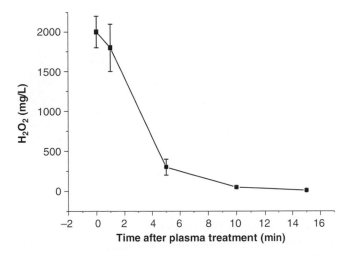

Figure 10.7 *Hydrogen peroxide concentration dependence on time after the air plasma treatment. The highest concentration of 2000 mg L^{-1} corresponds to the molar concentration of c. 59 mM. The initial concentration could be considered quite high.*

The oxygen treated sample shows an increase of pH from c. 2 to c. 4 72 hours after the treatment, while the air-treated sample remains stable longer than 72 hours. In fact, further testing indicates that the air-treated samples retain stable low pH levels for several months. Remarkably, when refrigerated the air-plasma-treated samples also maintain their antimicrobial activity for months.

Despite long-term stable aspects of its behavior, even the air-plasma-treated water is metastable in several other respects. For example, Figure 10.7 illustrates that hydrogen peroxide found in water immediately after the air-plasma treatment at relatively high concentrations of c. 2000 mg L^{-1} decomposes quickly in ambient light and temperature. Sometime between 10 and 15 min after the treatment, the concentration of hydrogen peroxide is reduced below the detection limit. Dilution experiments with the air-plasma-treated water also indicate metastable behavior. Near chemical (and thermodynamic) equilibrium dilution of acid by 10 times the amount of deionized water is expected to raise the pH level by 1 unit. However, as Figure 10.8 indicates, dilution of the air-plasma-treated water by 10 times the amount of deionized water raises the pH by c. 2.5 units, indicating that the conjugate base of the plasma-generated acid in water is much more stable at higher concentrations. This behavior is not unusual for many metastable substances, including substances such as hydrogen peroxide (which cannot explain original low pH values).

Taken together the results are remarkable. Although low pH can be partially explained in air plasma treatment by formation of nitric acid, existence of some other conjugate base that contributes significantly to the low pH is also clear. The fact that oxygen plasma treatment of water generates pH of around 2 is particularly striking, since no relatively stable (stability of several tens of hours) acid consisting of only hydrogen and oxygen has been reported in the literature. This other conjugate base is also likely to provide an oxidizer that is responsible for the observed antimicrobial activity of the air-plasma-treated water.

It has been hypothesized that a stabilized form of the superoxide anion (O_2^-), which is generated directly in plasma or created in water through reactions with the plasma electrons, plays the role of the conjugate base. However, no characteristic UV-VIS absorption has been found (possibly because stabilization of the superoxide somehow changes its absorption spectrum). On the other hand, addition of superoxide dismutase (SOD), which is designed to convert superoxide into hydrogen peroxide, following the plasma treatment has been reported to disproportionally raise the pH and increase concentration of hydrogen peroxide (Shainsky

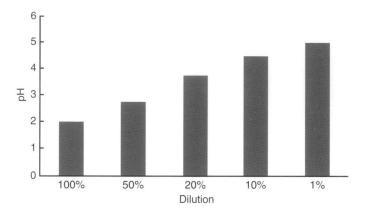

Figure 10.8   *Dependence of pH of air-plasma-generated acid on dilution by deionized water. 100% corresponds to undiluted plasma-treated water, 50% corresponds to 1:1 dilution of the plasma-treated water with deionized water, 20% corresponds to 1:5 dilution, etc.*

et al., 2010; Jae Koo Lee *et al.*, 2011). Questions regarding such experiments however remain, since SOD typically works well at higher pH. In air-plasma-treated water, it has been suggested that the antimicrobial agent may be peroxynitrite, which is generated through the reaction of superoxide and nitric oxide. At the time of writing (2012), these remain poorly tested hypotheses and the exact chemical composition of the plasma-treated water remains a mystery.

10.4   Enhanced antimicrobial effect due to organic components dissolved in water

It has been widely reported that various types of plasma treatment of water can result in oxidation of organic molecules in water. In plasma pharmacology, the goal may be to find organic molecules that can enhance the effect of plasma-treated water alone or can allow longer-term storage of the plasma treatment effects. We can consider various types of organic molecules in the search for enhancement of desired effects. In the case study described in the following section, treated deionized water is compared with phosphate-buffered saline (PBS) and with PBS plus N-acetyl-cystein (NAC) solution in antimicrobial effects. This case study is limited to the use of air plasma treatment.

10.4.1   Setup and sample preparation

Plasma setup and DBD plasma parameters similar to those described in Section 10.1.1 above were employed in this case study. Briefly, the bottom of the insulated electrode is positioned 1.5 mm above the surface of liquid sample and the discharge was generated by applying alternating polarity 1.7 kHz pulsed voltage of c. 17 kV, pulse width 1.5 μs and rise time 5 V ns^{-1}. Only direct plasma exposure was employed in this case study. The volume of the treated liquid is always maintained at 1 mL and the sample holder has the area such that the liquid is always c. 0.6 mm in thickness.

Deionized water (DIW) (MP Biomedicals Inc., Solon, OH), PBS or NAC (Sigma Chemical Co., St. Louis, MO) were treated separately at different time points. NAC (100 mM) stock solution was prepared in 1X sterile PBS, filter sterilized and aliquots stored at -20°C until used. A freshly prepared working solution of 5 mM NAC in PBS was used for subsequent experiments.

Escherichia coli (ATCC25922), *Staphylococcus aureus* (ATCC25923), *Acinetobacter baumannii* (ATCC19606) and *Staphylococcus epidermidis* (ATCC12228) strains were purchased from American Type Culture Collection (ATCC; Manassas, VA). All strains were maintained and used as overnight cultures in trypticase soy broth (TSB) for primary inoculations as per the ATCC guidelines. Reference strains of *Candida albicans* and *Candida glabrata* were obtained from T. Edlind (Drexel University College of Medicine) which had been grown in yeast extract Peptone Dextrose (TPD) medium. Hydrogen peroxide (Sigma) or 70% ethyl alcohol was used as known biocide agents and either TSB alone or PBS alone as negative controls, as appropriate.

Overnight, a culture of a given pathogen was inoculated (10 μL) into TSB medium (10 mL) and incubated at 37°C for 4 hours on orbital shaker incubator. The optical density at 600 nm (OD_{600}) was adjusted to 0.2 before use. The culture dilution thus prepared (1:100) was mixed with plasma-treated liquids (50 μL:50 μL) and held together at room temperature for 0–15 min intervals. After holding, the culture was diluted appropriately with sterile 1X PBS and spread onto TSA plates to be incubated at 37°C for 24 hours. After incubation, the colony-forming units (CFU) were counted to quantify surviving pathogen cells. Some of the experiments were carried out using 10^6–10^9 CFU mL^{-1} (as initial cell numbers) to determine cell-density-dependent rates of inactivation.

Plates which did not show any growth were incubated further up to 72 hours, and observed every 24 hours for possible growth. Similarly, the cultures of 0.2 OD_{600} were diluted (1:100) and exposed to plasma-treated fluid (50 μL:50 μL), mixed and held for variable times. The sample was centrifuged at 8000 rpm for 10 min and the supernatant removed to collect cell pellet. The pellet was re-suspended in XTT reagent to conduct an XTT assay (bacteria metabolism assay), as described by Joshi *et al.* (2010). Some of the experiments were conducted using plasma-treated liquid to treat pathogen biofilms as described by Joshi (2010). The untreated or treated biofilms were processed by XTT assay to determine the inactivation of bacteria embedded in biofilms. Ethanol (70%) was used as a positive control for the biofilm experiment.

The temperature of the fluid during plasma treatment is maintained close to room temperature, as detected by an ultra-sensitive K-type thermocouple. A pH ultra-sensitive probe attached to a digital pH meter (Thermo Orion Research Digital pH, Model 611, Beverly, MA) was used to detect plasma-treatment-associated pH changes in fluid over the treatment time.

'Holding time' is defined as the time that plasma-treated liquid came into contact with bacterial suspension (also known as contact time). For the evaluation of the effect of holding time, plasma-treated fluid was exposed to bacteria for variable periods of time. For 0 min holding time, plasma-treated fluid was exposed to bacteria and then immediately mixed thoroughly by micropipeting. A standard colony-counting assay was performed to evaluate bacterial growth. For longer holding times, the plasma-treated fluid and bacteria were mixed and held together for the desired time in the same tube at room temperature; after the desired time, a standard colony-counting assay was performed.

The delay time was defined as the time immediately after plasma treatment until exposure of bacteria to the plasma-treated liquid. In order to evaluate the effect of delay time, different time points were selected (between 0 min and 3 months). For 0 min delay time, plasma-treated liquid was exposed to bacteria immediately after plasma treatment and a standard colony-counting assay was performed. For a prolonged delay time, plasma-treated liquid was stored either at +4°C (in the refrigerator) or at room temperature in micro-tube, which was sealed with parafilm. Once opened, the sample-containing tube was never reused to avoid contamination of the fluid.

For aging experiments, the plasma-treated NAC solution was kept in a thermostatically controlled incubator which was set to elevated temperature (55°C) and incubated over time. The protocol of the Department of Food and Administration (FDA) of aging pharmaceutical compounds was used in this study (Hemmerich, 1998). The plasma-treated 1 mL DIW was immediately transferred into 3 mL glass vials, screw cap replaced and sealed with parafilm and set upright in the racks kept above the incubator. At the indicated time, one vial

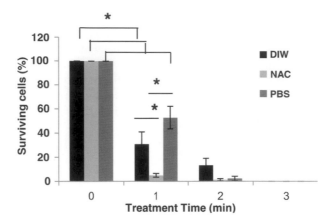

Figure 10.9   *Effect of plasma-treated solutions on planktonic bacteria suspended in solution. The 3 min treatment corresponded to 14.5 J cm⁻² plasma energy input and 5 mM NAC solution was employed.*

was taken out and tested for antimicrobial property of the liquid using the colony-counting assay as described above.

10.4.2   Comparison of DBD treated water, PBS and NAC solutions

Figure 10.9 demonstrates that all three plasma-treated liquids (NAC solution, PBS and deionized water) have strong antimicrobial properties. Less than 3 min of the plasma treatment generated sufficient reactive species in the liquids for complete inactivation of planktonic *E. coli*. Even by the end of 2 min, there was a significant inactivation of either by NAC solution ($p < 0.05$), PBS ($p < 0.05$) or de-ionized water ($p < 0.05$) when compared to respective controls (0 min or no treatment). Plasma-treated NAC solution was the most powerful of all three in carrying antimicrobial effect. After the treatment for 1 min (14.5 J cm⁻²), it inactivated a highly significant amount of *E. coli* ($p < 0.05$) compared to the same time points for DIW or PBS alone. During exposure to plasma-treated liquid, it was found that bacteria (10^7 CFU mL⁻¹) were inactivated in their free-floating planktonic form. Experiments were therefore carried out where *E .coli* cell suspensions of various CFUs were exposed to the plasma-treated liquids.

A range of NAC concentrations (1–20 mM) with or without plasma treatments were tested in order to determine whether NAC solution upon plasma treatment inactivates common pathogens such as *S. aureus*, *S. epidermidis*, *A. baumannii*, *C. albicans* and *C. glabrata* in addition to *E. coli*. The observations demonstrated that 5 mM is sufficient for inactivation and demonstrated a linear relationship during inactivation studies in colony assays. Higher concentrations of NAC did not show significantly higher antimicrobial efficacy. Planktonic form studies showed that 3 min of plasma treatment makes the NAC solution completely inactivate all the pathogens tested (see Figure 10.10). Only *C. glabrata* required 3.2 min (195 seconds) of treated liquid for 100% inactivation (sterilization) of this fungal pathogen (data not shown). Most of the pathogens in their biofilm form were almost equally sensitive to the biocidal effect of treated NAC solution, as illustrated in Figure 10.10.

The contact time of antimicrobial agent with pathogen is generally critical and often the biocidal effect is proportional to the contact time, at least at the initial stages of antibacterial activity. Figure 10.11 demonstrates that a significant inactivation of *E. coli* occurs from 2 min of holding time onwards (*p* values for 2, 3 and 5 min against 0 min are all ≤0.05). After around 5 min, c. 98% of bacterial inactivation is seen. There was complete inactivation (sterilization) after c. 15 min of exposure to plasma-treated NAC solution.

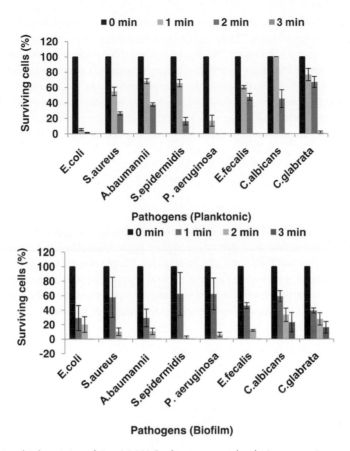

Figure 10.10 *Antimicrobial activity of 5 mM NAC plasma-treated solution at various activation times against planktonic and biofilm forms of pathogens.*

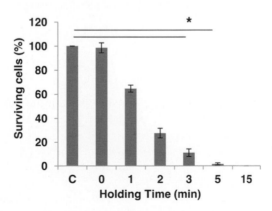

Figure 10.11 *Effect of holding time (time of contact with bacteria) of the plasma-treated (3 min or 14.5 J cm^{-2} exposure) 5 mM NAC solution on bacterial inactivation.*

Similarly, delay time experiments were performed to determine how long plasma-treated fluids retain their antimicrobial effect at room temperature. Here a post-treatment delay was introduced for the plasma-treated NAC solution before mixing with pathogen. No figure is shown here since the treated fluid was apparently holding the same antibacterial effect ($p = <0.05$ against control C) for at least up to 90 days, showing complete inactivation of *E. coli*. Experiments were also performed on the accelerated aging carried out at 37°C and 50°C over time to determine the shelf life of the plasma-generated antimicrobial properties of the 5 mM NAC solution. The accelerated aging was equivalent to 2 years of delay time at room temperature and demonstrated complete inactivation (sterilization) when *E. coli* at 107 CFU mL^{-1} were exposed to the plasma-treated solution.

10.5   Summary

The above experiments revealed the existence of a plasma pharmacological effect of 5 mM and higher concentrations of NAC. This amino acid derivative significantly intensified the antimicrobial effect of the plasma-treated deionized water and PBS solutions. In addition, the plasma-treated NAC solution demonstrated that its effect does not diminish with storage time of several months. Much more careful analysis is required to understand how plasma treatment modifies such organic molecules and how the modified molecules exert their biological effects. Plasma treatment clearly represents a new tool by which pharmacologically relevant compounds could be produced.

11

Plasma-assisted Tissue Engineering and Plasma Processing of Polymers

11.1 Regulation of biological properties of medical polymer materials

11.1.1 Tissue engineering and plasma control of biological properties of medical polymers

Plasma methods are widely applied today in different aspects of tissue engineering. We first consider plasma control of the biological properties of medical polymers and biocompatibility. The main requirement imposed on all polymer biomaterials applied in medicine is a combination of their desired physicochemical and physicomechanical characteristics with biocompatibility. Depending on particular applications, the biocompatibility of polymers can include various requirements which can sometimes be contradictory to each other. In the case of artificial vessels, drainages, intraocular lenses, biosensors or catheters, the interaction of the polymer with a biological medium should be minimized for the reliable operation of the corresponding device after implantation. In contrast, in the majority of orthopedic applications, the active interaction and fusion of an implant with a tissue is required. General requirements imposed on all medical polymers consist in non-toxicity and stability.

The body response to a polymer implant mainly depends on its surface properties: chemical composition, structure and morphology. Physical techniques are required for regulating the biological properties of polymer materials. These techniques should vary the physicochemical, structural and functional properties of surfaces over a wide range without affecting bulk characteristics such as strength, elasticity, transmission factor, refractive index and electrophysical parameters.

Treatment in a low-temperature gas-discharge plasma, which is widely used for modifying the surface characteristics of polymers (Chu *et al.*, 2002; Poncin-Epaillard and Legeay, 2003; Detomazo *et al.*, 2005; Lopez *et al.*, 2005; Section 11.2), is such a multipurpose and multifunctional technique. The cross-linking effect under the action of inert gas plasmas can be used for the immobilization of biocompatible low-molecular-weight compounds and various functional groups on polymer surfaces. For example, the treatment in argon plasma was used for grafting the Pluronic^{TM} 120 copolymer, which is highly biocompatible, onto the surface of polytetrafluoroethylene (PTFE) (Vasilets *et al.*, 2002). The Pluronic^{TM} 120 triblock copolymer (polyethyleneoxide/polypropylene/polyethyleneoxide or PEO/PP/PEO) was initially supported on the polymer surface by physisorption from solution, and then grafted by treatment in low-pressure

Plasma Medicine, First Edition. Alexander Fridman and Gary Friedman.
© 2013 John Wiley & Sons, Ltd. Published 2013 by John Wiley & Sons, Ltd.

argon plasma. A similar procedure was used for the covalent immobilization of sulfo-groups on the surface of polyethylene (PE) (Terlinger, Feijen, and Hoffman, 1993). Plasma etching changes the polymer surface morphology, which plays an important role in biocompatibility. For example, surface smoothing due to plasma etching upon treatment in oxygen- and fluorine-containing plasmas positively affects hemocompatibility because it decreases the probability of thrombosis at surface irregularities in a blood stream. The surface smoothing of a polymethylmethacrylate (PMMA) lens without changes in its bulk optical characteristics is a positive consequence of plasma treatment in a CF_4 discharge (Eloy *et al.*, 1993).

11.1.2   Wettability or hydrophilicity of medical polymer surfaces for biocompatibility

Since most biological fluids are aqueous solutions or contain water, the wettability or hydrophilicity of polymer surfaces is of paramount importance for biocompatibility. According to the current hypothesis of complementarity (Sevastianov, 1991), in order to improve hemocompatibility it is necessary to minimize not only the average interfacial surface energy at the material–blood interface but also at every point in the surface; that is, the character of free energy distribution at the interface of the biomaterial with an adsorbed protein layer should be taken into consideration. In other words, the hydrophilic and hydrophobic regions of a surface that contacts with blood and analogous regions of an adsorbed protein molecule should be complementary in order for the product to be highly hemocompatible.

The hydrophilicity and surface energy of biomaterials can be varied over wide ranges by generating various surface polar groups with the use of treatment of medical polymers in oxygen plasma (Sevastianov, 1991), oxygen-containing CO_2 (Baidarovtsev, Vasilets, and Ponomarev, 1985), H_2O (Simon *et al.*, 1996), SO_2 (Klee and Hocker, 1999) plasmas or nitrogen-containing (N_2, NH_3) plasmas (Ramires *et al.*, 2000). The processes of protein adsorption and cell adhesion can thereby be regulated. Plasma treatment is therefore effective in regulating adhesion of medical polymers (making their surfaces adhesive or repulsive to certain biological agents), as well as controlling their biocompatibility. More details can be found in Vasilets, Kuznetsov, and Sevastianov (2006) and Poncin-Epaillard and Legeay (2003).

11.2   Plasma-assisted cell attachment and proliferation on polymer scaffolds

11.2.1   Attachment and proliferation of bone cells on polymer scaffolds

A significant effect of non-thermal plasma treatment of polymer scaffolds on both attachment of bone cells and also on their further biological activity, specifically on their proliferation, has been demonstrated by Yildirim *et al.* (2007). The experiment in the atmospheric-pressure DBD plasma addressed attachment and proliferation of osteoblasts cultured over poly-ε-caprolactone (PCL) scaffolds. Traditional bone grafting procedures due to pathological conditions, trauma or congenital deformities have disadvantages such as graft rejection, donor site morbidity and disease transmission. Tissue engineering of bone is increasingly becoming the treatment of choice among surgeons to alleviate these problems. Favorable cell-substrate interaction during the early stage of cell seeding is one of the most desirable features of tissue engineering. The ability of bone cells to produce an osteoid matrix on the scaffold can be affected by the quality of the cell–scaffold interaction. Plasma treatment of the scaffolds significantly improves this interaction. PCL is an aliphatic polyester with a promising potential in tissue engineering because of its biocompatibility and mechanical properties. However, the surface hydrophobicity works against PCL when it comes to cell attachment. Plasma makes the PCL surface hydrophilic, which leads to successful formation of tissue constructs and cell proliferation, differentiation and new tissue ingrowth.

The attachment phase of the anchorage-dependent cells starts with formation of a cell adhesive protein layer from serum-containing media on the substrate at the cell–material interface. The cells become attached to these absorbed proteins via cell-surface adhesion receptors such as integrins and extra-cellular matrix (ECM) proteins including fibronectin, vitronectin, fibrinogen and collagen, which all have a cell-binding domain containing arginine-glycine-aspartate (RGD) sequence. Interactions between the proteins and the surface determines the residence time of the initial attachment, thereby influencing the cell proliferation and differentiation capacity on contact biomaterial. Attachment and proliferation of osteoblasts which are anchorage-dependent cells can be effectively controlled by oxygen plasma treatment which increases the cell affinity of the material by binding the cell adhesion proteins (Anselme, 2000; Marxer, 2002).

The experiments of Yildirim *et al.* (2007) investigated the influence of physicochemical properties of atmospheric-pressure oxygen-pulsed DBD plasma-treated PCL on early-time points osteoblastic cell adhesion and proliferation. For this purpose, PCL scaffold were treated with oxygen-based microsecond-pulsed DBD plasma with different exposure time, then mouse osteoblast cells (7F2) were cultured on treated PCL scaffolds for 15 hours. The plasma-treated PCL surfaces were characterized in terms of the surface energy and topography via Owens-Went method and atomic force microscopy (AFM). The mouse osteoblast cell adhesion and proliferation were characterized by live/dead cell viability assay and the AlamarBlue™ assay. Results of the experiments are discussed in the following section.

11.2.2 DBD plasma effect on attachment and proliferation of osteoblasts cultured over PCL scaffolds

The experiments of Yildirim *et al.* (2007) first consider the effect of atmospheric-pressure oxygen DBD plasma on surface hydrophilicity and surface energy of the PCL scaffold by contact angle measurements. The contact angles on untreated and treated PCL scaffolds were measured by sessile drop technique using ultrapure water, diiodomethane and glycerol as probe liquids. This allows the variation of DBD treatment time (at power c. 0.1 W cm^{-2}) with total solid surface energy of PCL (see Figure 11.1) to be calculated, as well as dispersive and polar components of the solid surface energy (see Figure 11.2). The total surface energy of PCL increased with prolonged oxygen plasma treatment time from 95 mN m^{-1} for untreated to 115 mN m^{-1} for 5 min oxygen-plasma-treated scaffold. While the dispersive component was relatively unchanged, the polar component increased significantly. This demonstrates that the major contribution of total solid surface

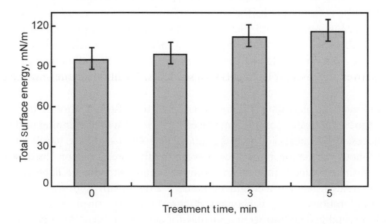

Figure 11.1 *Variation in total solid surface energy of PCL with oxygen plasma treatment time.*

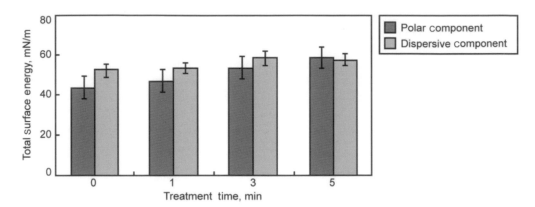

Figure 11.2   *Variation in dispersive and polar components of the total solid surface energy of PCL with oxygen plasma treatment time.*

energy to the increment is due to the formation of polar (mostly peroxide) groups on the polymer surface during its treatment in atmospheric-pressure non-thermal oxygen plasma.

The effect of oxygen plasma treatment on PCL microstructure was characterized in the Yildirim *et al.* (2007) experiments using AFM. The average roughness of untreated film was 13 nm and the surface pattern was relatively smooth. As a result of 1 min and 3 min DBD plasma treatment, the roughness was increased to 20 nm and 26 nm, respectively. For the 5-min treatment time, the change in surface morphology became obvious by large peaks in nanoscale with the roughness increase to 39 nm. The results therefore show that with the prolonged treatment time the mean surface roughness is increased by a factor of almost 3. It is believed that the change in surface morphology is the result of the ion bombardment and selective destruction of the polymer surface layer (Chan, Ko, and Hiraoka, 1996).

The metabolic activity and the morphology of mouse osteoblast cells on plasma treated and untreated PCL scaffold were examined in the Yildirim *et al.* (2007) experiments by the live/dead cell vitality assay and the Alamar Blue™ assay. On the untreated PCL scaffolds, cell proliferation, cell attachment and cell spreading with culturing time were significantly lower as compared to oxygen-plasma-treated PCL scaffolds. The proliferation of mouse osteoblast cells for 15 hour culture period on treated and untreated (control) PCL scaffolds is compared in Figures 11.3 and 11.4. The mouse osteoblasts on untreated PCL scaffolds did not show significant metabolic activity during the incubation time. In contrast, a higher degree of cell attachment and proliferation can be seen on the plasma treated surface. The viable cell count was also increased with prolonged oxygen-plasma treatment. The highest improvement for proliferation rate was observed for the 5-min treated sample.

Normalized fluorescence intensity data (Figure 11.4) shows that after a 15-hour incubation period, the cell proliferation rate on the 5-min treated samples increased by 90% from the beginning of the experiment. From the fluorescence images, it was observed that the mouse osteoblast cells are hardly attached on untreated scaffolds. After a 15-hour incubation period, cells were hardly spread out on the 1-min treated scaffold. With the increased plasma treatment time from 1 min to 5 min, osteoblastic adhesions improved. On the 5-min oxygen-plasma-treated samples, a confluent mouse osteoblast cell layer was observed. A similar trend was found for cell proliferation results, as shown in Figure 11.4. Oxygen-plasma-treated PCL scaffolds are therefore much more favorable than untreated scaffolds for cell attachment, due to the higher hydrophilicity and increased roughness.

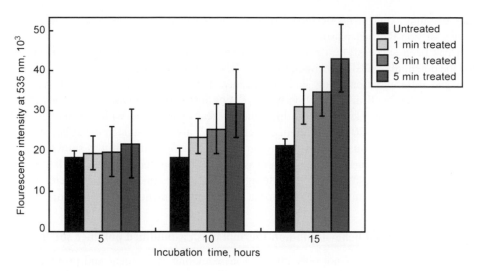

Figure 11.3 *Fluorescence intensity of mouse osteoblast cells cultured on untreated and treated PCL scaffolds for up to 15 hours.*

11.3 Plasma-assisted tissue engineering in control of stem cells and tissue regeneration

11.3.1 About plasma-assisted tissue regeneration

Fracture repair is a dynamic process requiring the mobilization and activation of stem cells to achieve new bone formation and increase resorption of old bone. Research into stem cell use, molecular signaling cascades and application of growth factors is still in the initial phase with minimum impact on patients. At the same time, biophysical applications through either electrical, electromagnetic or ultrasound stimulation have been in use for many years with reasonable success. The application of non-thermal plasma in this regard provides new methods to exploit these phenomena to advance the rapidly growing field of regenerative medicine.

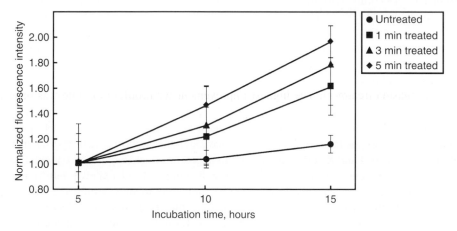

Figure 11.4 *Normalized fluorescence intensity of mouse osteoblast cells according to 5 hour values for different plasma treatment durations.*

From the very first experiments of Terry Freeman and her group from Jefferson University Hospital, a high efficiency of non-thermal plasma on mesenchymal stem cell differentiation and related bone tissue regeneration has been successfully demonstrated in animal models. In particular, Freeman *et al.* (2012) investigated the potential of DBD-generated ROS and RNS to promote mesenchymal cell (MC) proliferation, commitment and differentiation along skeletal lineages. It was shown that DBD-generated ROS can be used to enhance skeletal cell differentiation by increasing intracellular ROS, which leads to the activation of kinases and transcription factors known to influence genes associated with differentiation and skeletal development. Freeman *et al.* (2012) and Shainsky *et al.* (2012) also investigated the effect of plasma-induced intracellular DBD on activation of cell signaling promoting differentiation. It was shown that DBD plasma significantly enhanced both early and late osteoblast differentiation gene expression. It was also observed that DBD plasma alone is not sufficient to initiate significant changes in osteogenic differentiation. However, once differentiation has been initiated, DBD plasma significantly enhances the differentiation.

11.3.2   Control of stem cell behavior on non-thermal plasma-modified polymer surfaces

Plasma treatment of polymers is not only effective in regulating attachment and proliferation of cells as discussed in the previous section, but also in controlling the much more sophisticated behavior of biological systems on treated surfaces. An example of such an application is the selective inhibition of type X collagen expression in human mesenchymal stem cell differentiation on polymer substrates surface modified by low-pressure RF CCP discharge plasma. Such plasma-assisted tissue engineering experiments were carried out by Nelea *et al.* (2005).

Recent evidence indicates that a major drawback of current cartilage- and disc-tissue engineering is that human mesenchymal stem cells (MSCs) rapidly express type X collagen, which is a marker of chondrocyte hypertrophy associated with endochondral ossification. Some studies have attempted to use growth factors to inhibit type X collagen expression, but none to date have addressed the possible effect of the substratum on chondrocyte hypertrophy.

Nelea *et al.* (2005) examine the growth and differentiation potential of human MSCs cultured on two polymer types (polypropylene and nylon-6), both of which have been surface modified by low-pressure RF CCP discharge plasma treatment in ammonia gas.

Plasma treatments were performed in a cylindrical aluminum/steel chamber (Bullett *et al.*, 2004). High-purity ammonia gas was admitted to the chamber at an operating pressure of 300 mTorr (40 Pa). Capacitively coupled radio frequency (13.56 MHz) plasma (RF CCP) was generated at the 10-cm-diameter powered electrode in the center of the chamber, the walls acting as the grounded electrode. The samples were placed on the powered electrode, after which 30 s plasma treatments were performed at 20 W of RF power. The negative DC bias (250 V) developed on the powered electrode led to ion bombardment of the polymer surface during exposure to the plasma.

Cultures in the Nelea *et al.* (2005) experiments were performed for up to 14 days in Dulbecco's modified Eagle medium plus 10% fetal bovine serum. Commercial polystyrene culture dishes were used as control. Reverse transcriptase-polymerase chain reaction was used to assess the expression of types I, II and X collagens and aggrecan using gene-specific primers. Glyceraldehyde-3-phosphate dehydrogenase was used as a housekeeping gene. Types I and X collagens, as well as aggrecan, were found to be constitutively expressed by human MSCs on polystyrene culture dishes. Whereas both untreated and treated nylon-6 partially inhibited type X collagen expression, treated polypropylene (PP) almost completely inhibited its expression. These results indicate that plasma-treated polypropylene or nylon-6 may be a suitable surface for inducing MSCs to a disc-like phenotype for tissue engineering of intervertebral discs in which hypertrophy is suppressed.

The Nelea *et al.* (2005) experiments demonstrate the effectiveness of plasma polymer treatment in controlling not only simple surface properties, but also very specific biological behavior of cells on the treated surfaces.

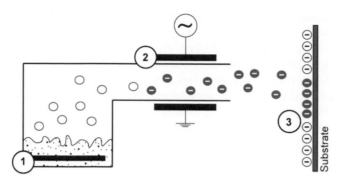

Figure 11.5 *Plasma bio-printer. 1: droplet atomizer; 2: DBD plasma reactor; and 3: charge droplet deposition substrate.*

11.3.3 Plasma-assisted bioactive liquid micro-xerography

Biochemical patterning (bio-printing) which allows for micro-scale resolution on non-planar substrates has been demonstrated by Fridman *et al.* (2003, 2005a, b). Utilizing this method, biomolecules (including DNA, proteins and enzymes) can be delivered to charged locations on surfaces by charged water buffer droplets. Charging of water droplets has been accomplished in using atmospheric DBD stabilized in the presence of a high concentration of micron-size water droplets.

Patterning of biomolecules on surfaces has many applications ranging from biosensors used in genetic discovery and monitoring of dangerous toxins to tissue engineering constructs where surfaces control tissue assembly or adhesion of cells. Most methods of biochemical patterning are only suitable for planar surfaces. In addition, micro- and nanoscale patterning often relies on complex sequences of lithography-based process steps. Plasma provides the biochemical substance patterning or printing allowing micro- and even nanoscale resolution on planar and non-planar substrates. It implies printing droplets of buffer containing biomolecules (including DNA), peptides and cells. The method involves the creation of droplets with relevant molecules or cells in their respective buffer solutions and then DBD charging of the droplets. Finally, the droplets are delivered onto substrate with charge pre-written via conventional xerography (micron-resolution) or via charge stamping (nanometer-resolution); see Figure 11.5.

The key issue of the plasma-assisted bio-printing is protection of the biological agents contained in the buffer droplets from their plasma-based deactivation. Survival of the biological agents in plasma can be achieved by choosing relatively low specific DBD power and a special composition of the protective buffer solutions.

11.4 Plasma-chemical polymerization of hydrocarbons and formation of thin polymer films

11.4.1 Biological and medical applications of plasma polymerization

The formation of high-molecular products (polymers) from initial low-molecular substances (monomers) in non-thermal mostly low-pressure (0.01–10 Torr) discharges is usually referred to as plasma polymerization. Formation of the high-molecular weight products (polymerization) proceeds either on solid surfaces contacting with plasma, which results in growth of polymers films, or in plasma volume which results in the production of polymer powders or some forms of polymer macro-particles. In contrast to conventional polymerization requiring the use of specific monomers, application of plasma allows the polymerization process to start from practically any organic compound.

Plasma-stimulated deposition of polymer films is of significant interest due to their application to surface processing and formation of thin dielectric and protective films, which is especially important in micro-electronics, biology and medicine. Plasma polymers also have unique features for biomedical applications (Fridman, 2008). Numerous research efforts have been focused on investigation of plasma polymerization. The publication of Linder and Davis (1931) describes the plasma synthesis of 57 different polymers. Reviews on plasma polymerization and its biological applications can be found in Yasuda (1985), Biederman and Osada (1992), Inagaki (1996), D'Agostino *et al.* (2003), Oehr (2005), Sardella *et al.* (2005).

Characteristics of the plasma polymerization processes essentially depend on specific power of the non-equilibrium discharges, gas pressure and dilution degree of the initial hydrocarbons in noble gases. At low values of the specific power ($\ll 0.1$ W cm^{-3}), which is typical for discharges in mixtures highly diluted with noble gases, translational gas temperature is usually relatively low $T_0 < 450$ K. In this case, the initial volume dissociation of hydrocarbons is mainly due to non-equilibrium processes stimulated by direct electron impact. At higher levels of specific power (c. 0.1 W cm^{-3}), typical for discharges in non-diluted hydrocarbons, translational gas temperature can exceed 500 K. In this case, thermal decomposition of hydrocarbon and their reactions with atomic hydrogen make a larger contribution to polymerization kinetics. The higher temperatures also lead to higher volume concentration of heavier gas-phase hydrocarbons and to acceleration of plasma polymerization. An increase of pressure above 3–10 Torr in the discharge slows down diffusion of the hydrocarbon radicals to the reactor walls and stimulates their volume reactions and recombination. Nevertheless, interesting experiments with plasma polymerization in non-thermal atmospheric-pressure discharges have been carried out recently.

11.4.2 Mechanisms and kinetics of plasma polymerization

Numerous experiments have focused on the kinetics of plasma polymerization in low-pressure non-equilibrium plasma. For example, experiments with plasma polymerization of benzene, toluene, ethyl-benzene and styrene in glow discharges demonstrate an exponential decrease of the polymer deposition rate with temperature, which indicates the contribution of absorption kinetics to the polymerization process (see kinetic details later in this section). The similar experiments of Tusov *et al.* (1975) show the significant contribution of charged particles in the deposition rate, while the contribution of UV-radiation as well as excited and chemically active neutrals is also essential.

Which plasma components – ions or radicals –dominate polymerization is actually still an open question. Consider as an example the glow discharge plasma polymerization of cyclo-hexane, analyzed by Slovetsky (1981). In this system, the film deposition rate linearly grows with cyclo-hexane concentration in the initial mixture with noble gases, while all plasma parameters including the charged particles flux to the surface remain the same. It shows that gradual cyclo-hexane ion attachment does not contribute to the polymer film deposition. Kinetic analysis results in the following polymerization mechanism in this case. First, cyclo-hexane dissociates in the plasma volume to form a cyclo-hexane radical and atomic hydrogen. The radicals diffuse to the walls and recombine there with open bonds formed on the polymer film surface by plasma particle bombardment. Further bombardment leads to more bond breaking and reorganization of C-C- and C-H- bonds on the polymer surface. Hydrogen is then desorbed from the surface, while free surface bonds recombine between one another and with gas-phase radicals, creating highly cross-linked polymer structure.

Development of the general plasma polymerization model is restricted by the essentially non-linear contribution of different plasma components in the process. Plasma particles are able to simultaneously build and destroy thin polymer films. Nevertheless, some models are able to describe a wide variety of experiments. According to the Osipov-Folmanis model, charged particles activate adsorbed molecules providing their cross-linking with the polymer film. The rate of polymer film deposition from a non-equilibrium low-pressure discharge (number of deposited molecules per unit time) can be calculated in this case as:

$$w_d = \frac{\Phi}{1 + \frac{1}{\sigma j}\left(\frac{1}{\tau} + a\Phi\right)} \tag{11.1}$$

where, Φ is the flux of the polymer creating molecules to the surface; j is the flux of charged particles to the surface; σ is the cross-section for activation of the adsorbed molecules by charged particles; τ is the effective lifetime of the adsorbed molecule; and a is the surface area occupied by an absorbed molecule. When fluxes to the surface of neutral and charged particles are not very high,

$$\sigma j \tau \ll 1, \quad a\Phi\tau \ll 1 \tag{11.2}$$

and the rate of the polymer film deposition, Equation (11.1) can be rewritten taking into account the adsorption time dependence on surface temperature as:

$$w_d \propto \Phi j \sigma \exp\left(\frac{E_a^a + \Delta H_a}{T_s}\right) \tag{11.3}$$

where E_a^a and ΔH_a are activation energy and enthalpy of the adsorption relevant to plasma polymerization and T_s is the surface temperature. Equation (11.3) interprets the plasma polymerization kinetics when the film deposition rate is proportional to fluxes of both neutral and charged particles. It explains the exponential acceleration of the polymer film deposition rate in plasma with reduction of surface temperature. This kinetic effect is due to surface stabilization of intermediate products at lower temperatures, which accelerate the polymerization rate. Polymer film growth rates in non-thermal plasma vary over a large range between 1 nm s^{-1} and 1 μm s^{-1}. These values are time-averaged; instantaneous film growth rates can be higher.

11.4.3 Initiation of polymerization by dissociation of hydrocarbons in plasma volume

We consider the kinetics of the plasma process starting with decomposition of hydrocarbons in low-pressure glow discharges, which initiates polymerization. If the initial concentration of hydrocarbons mixed with an inert gas exceeds 10–20 vol. %, dissociation of the hydrocarbons is mostly due to electronic excitation by direct electron impact. When the concentration of hydrocarbons is less than 3 vol. %, electronically excited atoms of inert gases and hydrogen atoms start to make an essential contribution to the dissociation of hydrocarbons. Ion-molecular processes also can make a contribution to the dissociation at very low hydrocarbon concentrations. If gas temperature in the non-thermal discharges exceeds 500 K, additional decomposition of hydrocarbons is due to their reactions with atomic hydrogen which can double the plasma dissociation rate of hydrocarbons. Vibrational excitation of hydrocarbons stimulates the dissociation if ionization degree is sufficiently high. For low-pressure glow discharges in hydrocarbons, the criterion of significant contribution of vibrational excitation in dissociation can be presented as (Ivanov, 2000):

$$AP_v/p^2 \gg 1 \tag{11.4}$$

where P_v is the specific power in the discharge (W cm^{-3}); p is the gas pressure (Pa) and A is a numeric factor which equals 10^4 for discharges in CH$_4$ and decreases for discharges in higher hydrocarbons. When the pressure is c. 1000 Pa, significant contribution of vibrational excitation requires very high values of specific

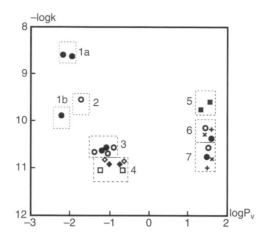

Figure 11.6 *Rate coefficients of dissociation of hydrocarbons by direct electron impact in glow discharges: effective experimental values (k_{diss}) and simulated partial values (k_{ed}). Pressures: 13 Pa (∇), 29 Pa (white squares), 59 Pa (black rhomb), 133 Pa (half white/half black circle), 266 Pa ($\circ$), 400 Pa (X), 530 Pa (half black/half white circle), 660 Pa (+), 2660 Pa (xxx). Effective coefficients of dissociation (k_{diss}): 1a: cis-C_6H_{12} (cis-C_5H_{10} and C_6H_6 are 30–40% lower, cis-$C_6H_{11}CH_3$ and n-C_6H_{14} are lower still), DC glow discharge in Ar + 1% hydrocarbons; 3: C_2H_6, RF glow discharge in ethane; 4: CH_4, RF glow discharge in methane; 5: CH_4 (C_2H_4 and C_2H_6 are 10% higher, n-C_5H_{12} are more than 10% lower), RF glow discharge in hydrocarbons; 6: C_6H_6, microwave discharge in $C_6H_6+H_2$. Simulated partial values of the dissociation rate coefficient (k_{ed}): 1b: cis-C_6H_{12}, DC glow discharge in Ar + 1% hydrocarbons; 2: CH_4, DC glow discharge in Ar + 1% hydrocarbons. Rate coefficients are given in $cm^3\ s^{-1}$, specific power is presented in W cm^{-3}.*

power exceeding 100 W cm^{-3}. At lower pressures, power requirements are not so strong but essential losses of vibrational excitation can be related to heterogeneous relaxation.

Effective rate coefficients for dissociation of different hydrocarbons in different low-pressure non-thermal discharges (glow-, RF and microwave discharges) are summarized in Figure 11.6 as function of specific discharge power P_v (Ivanov, 2000). This dissociation process results in the formation of active hydrocarbon radicals, which are responsible for growth of the thin polymer film. An increase of the specific discharge power between 0.1 and 10 W cm^{-3} leads to an increase of the effective dissociation coefficient due to the increase of gas temperature and reactions of atomic hydrogen with hydrocarbons. The figure also illustrates that dilution (99%) with argon leads to higher values of the effective dissociation coefficient (10^{-9} cm^3 s^{-1}), which is due to additional dissociations in collisions with electronically excited atoms of the inert gas.

11.4.4   Heterogeneous mechanisms of plasma-chemical polymerization of C_1/C_2 hydrocarbons

In contrast to conventional polymerization, plasma polymerization enables a polymer to be built from any organic compound. Building a polymer in plasma does not require specific chemical behavior of monomers such as opening of double-bonds, but is based on attachment of chemically active species (radicals, atoms, non-saturated molecules etc.) generated in plasma beforehand. Such a mechanism of plasma polymerization based on preliminary 'plasma activation of monomers' is sometimes referred to as 'stepwise'. Heterogeneous mechanism of polymerization is quite complicated in this case.

We discuss two examples when homogenous mechanism is better investigated: plasma polymerization of C_1/C_2 hydrocarbons (this section) and plasma-initiated chain polymerization (following section).

In the specific case of plasma polymerization of C_1/C_2 hydrocarbons, mechanisms of heterogeneous stages have been well investigated (Ivanov, 2000). The key heterogeneous process initializing the plasma polymerization is the formation of free bonds near the polymer film surface (depth up to several monolayers), which are called centers of polymer growth. Formation of the centers is mostly due to recombination of electrons with ions of hydrocarbons, which proceeds by electron-tunneling through potential barrier of the dielectric polymer film. The center of polymer growth is usually related to a broken C-H bond; the probability of breaking C-C and C=C bonds is much lower.

Major growth of the polymer films is due to attachment of radicals and non-saturated hydrocarbons produced in plasma volume to the centers of polymer growth. The main mechanism of polymer film growth is accompanied by a wide variety of other heterogeneous processes also making a contribution in the plasma polymerization of C_1/C_2 hydrocarbons. Note that the formation of the centers of polymer growth as a result of recombination of electrons with ions of inert gases result in energy transfer to the hydrocarbons and formation of a free bond. On the other hand, it is not only radicals and non-saturated molecular hydrocarbons that can be attached to the centers of polymer growth (and even to the polymer macromolecules), but also hydrocarbon ions and ion-radicals produced in plasma.

UV radiation from non-thermal plasma also leads to breaking bonds in the macromolecules of hydrocarbons. It is however not absorbed in the near-surface region (several monolayers thin), where polymer film growth actually takes place. The UV radiation generated in plasma is absorbed in the much thicker layer of up to several microns, where it leads to cross-linking of the macromolecules (and not to polymer film growth). Detailed kinetics of the plasma-chemical polymerization of C_1/C_2 hydrocarbons in low-pressure non-equilibrium discharges is described by Ivanov (2000).

Analysis of the heterogeneous mechanisms also indicates that plasma polymerization includes two competing processes: plasma-stimulated polymer film growth and plasma-stimulated etching of the polymer film. Generally, a 'more active' plasma interaction with the surface leads to the domination of etching. Sometimes it is possible to have domination of a polymer film growth in one part of a reactor, while etching dominates in another part of the same reactor. Effective growth of polymer films requires the selection of plasma-chemical conditions that suppress the etching. The contribution of etching becomes especially important when the macromolecules include an oxygen group –O– in the major polymer chain and when the macromolecules include any oxygen-containing side groups.

11.4.5   Plasma-initiated chain polymerization: Mechanisms of plasma polymerization of MMA

Non-thermal plasma can not only be applied for the stepwise polymerization discussed in the previous section, but also for effective stimulation of more conventional chain polymerization processes. Plasma-initiated polymerization of methyl-methacrylate (MMA) with production of the practically important polymer poly-methyl-methacrylate (PMMA) is a good example of such processes (Ponomarev, 1996, 2000). MMA is a quite large organic molecule: $CH_2=C(CH_3)-C(=O)-O-CH_3$, creating polymer (PMMA) by opening of the C=C double bond. Conventional chain propagation reactions of MMA-polymerization can be represented by the elementary MMA attachment to a radical $R(\cdot)$:

$$R(\cdot) + CH_2 = C(CH_3) - C(= O) - O - CH_3 \quad \rightarrow$$
$$\rightarrow \quad (RH_2)C - C(\cdot, CH_3) - C(= O) - O - CH_3. \tag{11.5}$$

Plasma initiation of the chain polymerization is due to formation of a primary free radical $R(\cdot)$, starting the traditional scheme (11.5), and by formation of positive or negative ion-radicals which are also capable of initiating the MMA polymerization. The primary free radical $R(\cdot)$ as well as the charged centers of polymer growth are formed from the absorbed monomers by electron/ion bombardment and UV radiation from plasma. Formation of a positive ion-radical from an adsorbed MMA-molecule on the surface under electron/ion bombardment and UV radiation can be schematically shown as the ionization process:

$$CH_2 = C(R_1) - R_2 \quad \rightarrow \quad CH_2(\cdot) - C^+(R_1) - R_2 + e. \tag{11.6}$$

The positive ion-radical then initiates a sequence of attachments of further and further MMA-molecules in the ion-molecular chain propagation reactions:

$$CH_2(\cdot) - C^+(R_1) - R_2 + CH_2 = C(R_1) - R_2 \quad \rightarrow$$
$$\rightarrow \quad CH_2(\cdot) - (R_1)C(R_2) - CH_2 - C^+(R_1) - R_2. \tag{11.7}$$

Formation of a negative ion-radical (which is also a center of polymer growth) from an adsorbed MMA-molecule on the surface is due to direct electron attachment:

$$CH_2 = C(R_1) - R_2 + e \quad \rightarrow \quad CH_2(\cdot) - C^-(R_1) - R_2. \tag{11.8}$$

Similarly to process described by Equation (11.9), the negative ion-radical operating as a center of polymer growth also initiates a sequence of attachment processes of further MMA-molecules in the ion-molecular chain propagation reactions involving the negative ions:

$$CH_2(\cdot) - C^-(R_1) - R_2 + CH_2 = C(R_1) - R_2 \quad \rightarrow$$
$$\rightarrow \quad CH_2(\cdot) - (R_1)C(R_2) - CH_2 - C^-(R_1) - R_2. \tag{11.9}$$

To deposit the PMMA thin film on a substrate of interest, the partial pressure of MMA should be high enough to provide sufficient MMA concentration in the absorbed surface layer. Such a requirement is common for all plasma-initiated chain polymerization processes. Discharge power should not be high in the plasma-initiated chain polymerization to minimize conversion of MMA (or other monomers) in the plasma volume. The role of plasma should be limited in this case to generation on the surface of relatively low concentration of active centers initiating chain polymerization in the layer of absorbed monomers.

11.4.6   Plasma-initiated graft polymerization

Polymer film deposition on polymer substrates or other surfaces can be accomplished by graft polymerization. In this case, chain polymerization of a monomer proceeds on a polymer (or other material) substrate preliminarily treated in non-thermal plasma. A common approach to such graft polymerization is based on the preliminarily polymer substrate treatment in O_2-containing plasma, which forms organic radicals $R(\cdot)$ on the surface and converts them into organic peroxides. Formation of the organic peroxide compounds is the chain process, which starts with the direct attachment of molecular oxygen and formation of an organic peroxide radical:

$$R(\cdot) + O_2 \rightarrow R - O - O(\cdot). \tag{11.10}$$

Further propagation of the plasma-initiated chain leads to production of the organic peroxides and restoration of organic radicals:

$$R - O - O(\cdot) + RH \rightarrow ROOH + R(\cdot), \tag{11.11}$$

$$R - O - O(\cdot) + RH \rightarrow ROOR_1 + R_2(\cdot). \tag{11.12}$$

The polymer substrate activated this way in plasma (by attachment of the surface peroxide groups) is able to initiate the graft polymerization of gas-phase or liquid-phase monomers. In order to initiate the graft polymerization process, the substrate should be first heated up to dissociate the organic peroxide on the surface (11.11, 11.12) and to form active organic radicals:

$$ROOH \rightarrow RO(\cdot) + (\cdot)OH, \quad ROOR_1 \rightarrow RO(\cdot) + R_1O(\cdot). \tag{11.13}$$

The RO-radicals then function as the centers of polymer growth. They initiate the graft polymerization, which in the specific case of MMA polymerization into PMMA is the sequence of attachment reactions of the monomer to the center of polymer growth:

$$\begin{aligned} RO(\cdot) + CH_2 &= C(CH_3) - C(= O) - O - CH_3 \quad \rightarrow \\ &\rightarrow \quad RO - CH_2 - C(\cdot, \ CH_3) - C(= O) - O - CH_3. \end{aligned} \tag{11.14}$$

The grafting of monomers with special functional groups significantly changes the surface properties of initial polymer substrates, and allows the creation of new special compounds. Such an approach has been effectively applied to create new types of immobilized catalysts on polymer base. In particular, polyethylene powder with specific surface area 2 m^2 g^{-1} has been activated in oxygen-containing plasma for further graft polymerization of acrylic and methacrylic acids from the gas phase. Vanadium (V), titanium (Ti) and cobalt were chemically deposited on the powder to produce highly effective immobilized metal catalysts.

Such plasma-activated powders were successfully used as catalysts for ethylene polymerization into polyethylene (PE), as bi-functional catalysts of C_2H_4-dimerization and C_2H_4-C_4H_8 co-polymerization and for other catalytic applications. The non-thermal plasma-initiated graft polymerization of vinyl-monomers on the surface of PTFE, which significantly change the PTFE surface properties, should be mentioned in particular. The PTFE-film first activated in non-thermal plasma of a low-pressure discharge and then treated in a special liquid monomer significantly enhanced its adhesion to different materials, and especially to steel (Ponomarev, 2000).

11.4.7 Formation of polymer macro-particles in non-thermal plasma in hydrocarbons

When the residence time of hydrocarbons in the discharge volume is relatively long, the polymer film growth on substrates, reactor walls and electrode is sometimes accompanied by the formation of polymer powder in the plasma volume. Polymer macro-particles of the powder are deposited on different surfaces, but can also be incorporated into the growing polymer film. The typical size of spheroidal macro-particles is 0.1 μm and above.

Although the mechanisms of growth of the polymer powder in plasma volume are similar to that of polymer films on surfaces, some specific features are related in this case to the formation of precursors of the macro-particles. The precursors of macro-particles are usually sufficiently large hydrocarbons molecules formed in plasma. The minimal size of a macromolecule to be considered as a macro-particle is determined by the ability of the particle to be effectively negatively charged in plasma. In the case of large-sized hydrocarbon molecules, the ability to be charged usually requires the macro-particle sizes to exceed 5 nm. While growth rate of the plasma polymer film on substrate surfaces usually varies from 1 nm s^{-1} to 1 μm s^{-1}, the growth rate of the polymer macro-particles can be significantly faster at 10^{-8}–10^{-4} cm s^{-1}. The formation of a macro-particle larger than 1 μm therefore requires more than 1 s of residence time.

The polymer macro-particles leave the discharge zone through different channels. Relatively large macro-particles (5–10 μm) simply fall down to the bottom of the plasma-chemical reactor due to gravity. While gas flow drags the polymer powder, stagnation zones lead to longer residence time and to an increase of size and concentration of the macro-particles. The polymer macro-particles are usually charged in plasma, which essentially influences their evolution and their dynamics. Factors that prevent significant formation of the polymer macro-particles include high flow rate of hydrocarbons through the discharge active zone, absence of stagnation zones there and limitation of the film growth rate.

11.4.8 Specific properties of plasma-polymerized films

The most specific property of plasma-polymerized films is a high concentration of free radicals in the films and a large number of cross-links between macromolecules. The concentration of free radicals can actually be very high, up to 10^{19}–10^{20} spin g^{-1}. The cross-linkage immobilizes the free radicals, significantly slowing down their recombination and chemical reactions. The slowly developing chemical processes with the free radicals result in 'slow but sure' changing of gas permeability, electric characteristics and other physical and chemical properties of the plasma-polymerized films, which is usually referred to as the aging effect. Plasma-polymerized films are often characterized by high internal stresses, up to 5–7 $\times$ 10^7 N m^{-2}.

In contrast to conventional polymerization, the internal stresses in the plasma-polymerized films are related to extension of the material. The effect is due to intensive insertion of free-radical fragments between already deposited macromolecules with simultaneous cross-linking during the polymer film growth. Solubility of the plasma-polymerized films in water and organic solvents is usually very low because of the strong cross-linkage between macromolecules. It should be mentioned that if the film growth rate is very high and the formed macromolecules are relatively short, the film can be dissolved in organic solvents and even in water. When polymer film growth rates are 0.1 nm s^{-1} and less however, the plasma-polymerized films are generally highly cross-linked and not soluble in either water or organic solvents.

The majority of the plasma-polymerized films are characterized by high thermal stability. As an example, thin polymer film deposited from a non-thermal low-pressure discharge in methane remains stable and does not lose weight after being treated in argon at temperatures up to 1100 K and in air at temperatures up to 800 K.

The wettability of plasma-polymerized films is related to their surface energy and depends on the type of plasma gas used for polymerization. For example, polymer films formed from fluorocarbon and silicon-organic compounds are characterized by low surface energy and low wettability. Plasma-polymerized hydrocarbon films usually have high wettability. It is important that their wettability is higher than that of their counterparts formed without plasma. The effect is due to oxygen-containing groups usually incorporated into the plasma-polymerized films. These groups usually make a significant contribution to an increase of surface energy and wettability.

Thin plasma-polymerized films are also important for their selective permeability for different gases. A large number of cross-links makes this film somewhat like molecular sieves, which is of interest for applications of such films as membranes for gas separation. Ultra-thin plasma-polymerized films with thicknesses of 0.1 μm and below are usually characterized by very strong adhesion to a substrate. Adhesion of relatively thick films is significantly lower. The highest adhesion is usually achieved at lower rates of the polymer film deposition. As an example, ultra-thin plasma-polymerized films with thickness below 0.2 μm have adhesion to aluminum substrate exceeding 1500 N cm^{-2} (Ponomarev, 2000).

11.4.9 Electric properties of plasma-polymerized films

The electric properties of plasma-polymerized films are especially important in connection with their application as dielectrics in microelectronics. A comparison of major dielectric properties of plasma-polymerized films (dielectric permittivity ε and dielectric loss tangent tan δ) with those characteristics of polymer films produced in a conventional manner from the same monomers is shown in Table 11.1.

Table 11.1 *Dielectric properties of plasma-polymerized films in comparison to conventional polymers at temperature 20°C and frequency 1 kHz.*

Polymer	Dielectric permittivity ε		Dielectric loss tangent tan δ	
	Conventional polymer	Plasma polymer	Conventional polymer	Plasma polymer
Polyethylene	2.3	3.57	$(1–2) \times 10^{-4}$	$(5–30) \times 10^{-4}$
Polystyrene	2.55	2.67–3.3	2×10^{-4}	$(1–3) \times 10^{-3}$
Poly-iso-butylene	2.2–2.3	3.0	2×10^{-4}	9×10^{-3}
Poly-tetra-fluoro-ethylene	2.0	-	$(1–2) \times 10^{-4}$	25×10^{-4}
Hexa-methyl-di-siloxane (silicon rubber)	3.2	2.5–5.23	4×10^{-3}	$(1–10) \times 10^{-3}$

It can be seen from the table that the dielectric permittivity ε is slightly higher and the dielectric loss tangent tan δ is significantly higher in the case of the plasma-polymerized films. This effect is due to the fact that the plasma-polymerized films always include some polar groups. The dielectric properties of the plasma-polymerized films are affected by aging and are also sensitive to humidity. The aging effect means in this case that the dielectric properties change with time. This is due to a high concentration of free radicals in the films, and slow reactions of the radicals with oxygen from air leading to formation of new oxygen-containing polar groups.

The dielectric properties can be stabilized by special treatment of the freshly deposited films, partially decreasing the concentration of free radicals. This can be achieved by heating the films in vacuum or by their additional treatment in hydrogen plasma. Resistance of the plasma-polymerized films with respect to electric breakdown can be quite high. This resistance increases at relatively low specific powers of the non-thermal discharges applied for the polymer film deposition. As an example, plasma-polymerized fluoro-cyclo-butane film is characterized by breakdown electric fields: 5×10^6 V cm^{-1} for a film thickness 150 nm, 6×10^6 V cm^{-1} for a film thickness 100 nm, and $7–8 \times 10^6$ V cm^{-1} for a film thickness of 75 nm. To compare, breakdown of the conventional PTFE film with thickness 0.1–0.2 mm only requires $0.4–0.8 \times 10^6$ V cm^{-1}. Electric conductivity of the plasma-polymerized films is low and strongly depends on temperature with typical activation energy c. 1 eV. Table 11.2 presents electric conductivities of some specific plasma-polymerized films measured at temperatures 150°C and 250°C as well as the corresponding activation energies.

Conductivity of the thin films does not generally follow Ohm's law; data in the table are therefore summarized for the specific film thickness around 1 μm and voltage 1.4 V (Ponomarev, 2000). Plasma-polymerized films usually have high photoelectric conductivity, which is due to effective photo-generation of electric charge carriers in the material of the polymer films. Oxygen-containing groups and free radicals in the material of the films are able to accept free electrons and therefore act as centers for photo-generation of electric charges and photoelectric conductivity.

11.5 Interaction of non-thermal plasma with polymer surfaces

11.5.1 Plasma treatment of polymer surfaces

Non-thermal plasma treatment of polymers leads to significant changes of their surface properties, including: surface energy, wettability, adhesion, surface electric resistance, dielectric loss tangent, dielectric permittivity, catalytic activity, tribological parameters, gas absorption and permeability characteristics. Plasma treatment

Table 11.2   *Electric conductivity of thin plasma-polymerized films, with temperature dependence.*

Initial substance used for plasma-polymerization	Electric conductivity at temperature 150°C, σ $(\Omega\ cm)^{-1}$	Electric conductivity at temperature 250°C, σ $(\Omega\ cm)^{-1}$	Activation energy E_a (eV)
Naphthalene	9×10^{-16}	2.7×10^{-13}	1.1
Styrene	6×10^{-16}	9×10^{-14}	1.2
n-Xylol	5×10^{-17}	1.5×10^{-13}	1.8
Cyclo-pentadiene	1×10^{-16}	1.2×10^{-13}	1.5
Hexa-methyl-benzene	7×10^{-17}	7×10^{-14}	1.5
Ethylene oxide	4×10^{-16}	1.6×10^{-13}	1.1
Methoxi-naphthalene	1.1×10^{-16}	7×10^{-14}	1.5
Thio-carbamide	3.3×10^{-16}	4×10^{-13}	1.7
Chloro-benzene	8×10^{-17}	1.9×10^{-14}	1.4
Picoline	2.2×10^{-14}	6×10^{-12}	1.1
N-nitroso-di-phenyl-amine	8×10^{-15}	3×10^{-12}	1.2
n-Toluidine	7×10^{-16}	2.3×10^{-12}	1.5
Aniline	2.8×10^{-16}	1.4×10^{-12}	1.8
n-Nitro-toluene	5×10^{-16}	2.5×10^{-13}	1.2
Di-phenyl-selenide	3×10^{-18}	8×10^{-13}	1.5
Di-phenyl-mercury	2.8×10^{-15}	2.7×10^{-13}	0.85
Ferrocene	2.7×10^{-13}	4.5×10^{-12}	0.55
Benzene-selenol	2.5×10^{-14}	7×10^{-12}	1.1
Hexa-n-butyl-tin	1.5×10^{-15}	7×10^{-13}	1.1
Tetra-cyanoethylene	1.8×10^{-13}	5×10^{-12}	0.6
Malononitrile	3×10^{-14}	1.8×10^{-12}	0.75
Thianthrene	1.5×10^{-14}	1.6×10^{-12}	0.85
Thiophene	6×10^{-14}	3×10^{-12}	0.75
Thioacetamide	8×10^{-14}	9×10^{-12}	0.85

and modification of polymer surfaces are widely used today in numerous applications such as painting of textiles, printing on synthetic wrapping materials, treatment of photographic materials, microelectronic fabrication and new applications in biology and medicine.

Plasma treatment of cells and different living tissues can often be considered simply as plasma interaction with biopolymers. Plasma treatment of polymers can be performed in non-thermal plasma at low pressures as well as at high pressures. Low-pressure non-thermal discharges are usually applied for polymer surface chemical functionalization and for 'specific and accurate' chemical modification, as in the application for functionalization of photographic materials and in microelectronics. At the same time, high-pressure non-thermal discharges are usually sufficient for less-specific treatment of polymers directed to surface cleaning, changing wettability and so on.

More and more applications which previously required low-pressure discharges now use non-thermal atmospheric-pressure discharges (Roth, 1995, 2001). Taking into account significant practical interest, numerous research efforts have been focused recently on investigation of plasma treatment of polymer surfaces, similarly to the case of plasma polymerization. Reviews and interesting material on the subject can be found in Yasuda (1985), Boenig (1988), Kramer, Yeh and Yasuda (1989), D'Agostino (1990), Biederman and Osada (1992), Liston (1993), Garbassi, Morra and Occhiello (1994), Lipin *et al*. (1994), Ebdon (1995), Gilman and Potapov (1995), Ratner (1995), Chan, Ko and Hiraoka (1996), Inagaki (1996), Maximov, Gorberg and Titov

(1997), Arefi-Khonsari *et al.* (2001), Hocker (2002), D'Agostino *et al.* (2003), Wertheimer, Dennler and Guimond (2003), Tatoulian *et al.* (2004), Oehr (2005), Sardella *et al.* (2005).

11.5.2 Major initial chemical products created on polymer surfaces during non-thermal plasma interaction

Chemical processes in thin polymer surface layers are stimulated by all major plasma components, especially by electrons ions, excited particles, atoms, radicals and ultraviolet (UV) radiation. Major primary products of the plasma polymer treatment are free radicals, non-saturated organic compounds, cross-links between polymer macromolecules, products of destruction of the polymer chains and gas-phase products (mostly molecular hydrogen). The processes for formation of radicals on the polymer surface under the plasma treatment, which are due to electron impact and UV radiation, are related to breaking of R-H and C-C bonds in polymer macromolecules:

$$RH \xrightarrow{e,\hbar w} R(\cdot) + H, \quad RH \xrightarrow{e,\hbar w} R_1(\cdot) + R_2(\cdot). \tag{11.15}$$

The direct formation of non-saturated organic compounds with the double-bonds on the surface of the polymer, which are treated by non-thermal plasma, can be illustrated in similar simple cases as:

$$RH \xrightarrow{e,\hbar w} R_1 - CH = CH - R_2. \tag{11.16}$$

Secondary reactions of atomic hydrogen usually lead to formation of molecular hydrogen through different mechanisms, including recombination and hydrogen transfer with the polymer macromolecule:

$$H + H \to H_2, \quad H + RH \to R(\cdot) + H_2. \tag{11.17}$$

The secondary reaction of atomic hydrogen with the organic radical R can result not only in recombination but also in simultaneous formation of molecular hydrogen and a double-bond in the organic macromolecule, which can be illustrated as:

$$H + R(\cdot) \to R_1 - CH = CH - R_2. \tag{11.18}$$

When the plasma gas contains oxygen, the free organic radical R generated by non-thermal plasma treatment on polymer surfaces (Equation (11.15)) very effectively attaches molecular oxygen from the gas phase forming active organic peroxide radicals:

$$R(\cdot) + O_2 \to R - O - O - . \tag{11.19}$$

The RO_2 peroxide radicals formed on the polymer surface by treatment in non-thermal plasma systems are able to initiate different important chemical surface processes. The simplest processes started by the RO_2-radicals are related to formation of hydro-organic peroxide and other peroxide compounds on the surface of polymers:

$$R - O - O - \ + RH \to \ R - O - O - H \ + R(\cdot), \tag{11.20}$$

$$R - O - O - \ + RH \to \ R - O - O - R_1 + R_2(\cdot). \tag{11.21}$$

Reactions (11.20) and (11.21) together with the attachment process (11.19) create the chain reaction for formation of hydro-organic peroxide and other peroxide compounds on the surface of polymers under treatment by oxygen-containing plasma. Formation of the chemical products takes place in the relatively thin surface layers of the polymer because energies of plasma electrons and ions are quite limited and coefficients of extinction of UV radiation in polymers are high.

Electrons make a significant contribution to the modification of polymer materials. This has been demonstrated in the specific cases of plasma treatment of polystyrene films and biological objects (Rusanov and Fridman, 1978). In the specific conditions of the above experiments, non-thermal plasma electrons penetrate the polymers to a depth of about $x_0 \approx 2 \mu m$. Characteristic depth of modification of polymer material and formation of cross-links between macromolecules can be estimated (Rusanov and Fridman, 1978) as:

$$x^2 \approx x_0^2 (\ln \sigma_{io} N - \ln \ln K) \tag{11.22}$$

where σ_{i0} is the effective cross-section of modification of the macromolecules (in the case of bio-macromolecules it can be estimated as $\sigma_{i0} \approx 10^{-9}$ cm^2); N (cm^{-2}) is the total time-integrated flux of electrons to the polymer surface; and $K \gg 1$ is the degree of the polymer modification. Assuming for example $N = 10^{13}$ and $K = e$, the characteristic depth for modification of a polymer material according to Equation (11.22) is c. 6 μm.

11.5.3   Formation kinetics of main chemical products in pulsed RF treatment of PE

Systematic kinetic investigations of chemical modification of polymer surfaces in non-thermal plasma were carried out by Ponomarev (1982), Vasilets, Tikhomirov and Ponomarev (1979), Ponomarev and Vasilets (2000) and Vasilets (2005) using the treatment of PE in low-pressure strongly non-equilibrium RF discharges as an example. The pulsed RF discharge in these experiments is characterized by voltage amplitude 20–22 kV, pulse duration 5 μs, RF frequency in the pulse 0.5 MHz, repetition frequency of the pulses 50 Hz and average specific discharge power 0.007 W cm^{-3}. High-density PE was used in the form of powder with specific surface area 2 m^2 g^{-1} or in the form of the polymer film. Air, helium and hydrogen were used as plasma gases. The gas pressure was varied over a wide range between 10^{-2} and 50 Torr.

Kinetic curves showing the formation of free radicals in PE at different plasma pressures are presented in Figure 11.7 (Vasilets, 2005). All the curves demonstrate linear growth at first and then saturation. The saturation level of concentration of free radicals significantly decreases at higher pressures, when the reduced electric field E/p is relatively low and the effective depth of plasma treatment on the polymer is less than the size of the polymer particles. At very low pressures 10–100 mTorr, the depth of the plasma polymer treatment is $d \sim 20$–30 μm. At relatively high pressures ≥ 5 Torr, the treatment depth is much shorter ($d < 2$ μm).

The kinetics of free radical formation in PE is determined by the relative contribution of plasma electrons and UV radiation from plasma. The saturation level of the free radical concentration in PE films treated by only UV radiation ($\lambda > 160$ nm) from the plasma does not depend on discharge pressure in the range 1–20 Torr and does not depend on the thickness of the polymer film in the range 2 − 110 μm. The saturation value of the free radicals concentration is c. 2.2×10^{17} radicals g^{-1}, which is much below that corresponding to total plasma treatment at very low pressures of 10–100 mTorr.

These results lead to the conclusion that free radical formation in PE under the pulsed RF discharge treatment is dominated by the contribution of UV radiation at higher pressures, while at very low pressures (10–100 mTorr) it is mostly due to the contribution of plasma electrons. Figure 11.8 illustrates the formation of molecular hydrogen, trans-vinylene bonds and free radicals in PE powder under treatment in the pulsed RF discharge in air at very low pressure (10 mTorr) and temperature (77 K) (Vasilets, 2005). Initial formation rates of intermolecular cross-links and double bonds are close and exceed those of free radicals by a factor

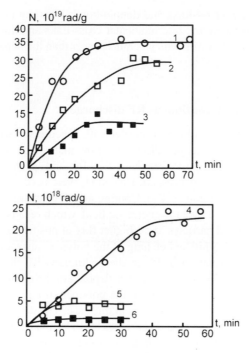

Figure 11.7 *Accumulation of free radicals in polyethylene powder (specific surface 2 m² g⁻¹) in the pulsed RF discharge with pressure. 1: 0.01–0.1 Torr; 2: 1 Torr; 3: 5 Torr; 4: 10 Torr; 5: 20 Torr; 6: 50 Torr.*

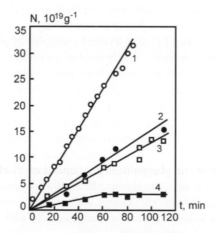

Figure 11.8 *Kinetic curves of formation. 1: molecular hydrogen H2; 2: intermolecular cross-links; 3: transvinylene bondings; and 4: free radicals during treatment of polyethylene powder in pulsed discharge in air, pressure 0.01 Torr, temperature 77 K.*

of 3. This means that formation of cross-links and double-bonds does not proceed through free radicals in this case. The depth of the layer where the intermolecular cross-links and double-bonds are formed at very low pressures (10–100 mTorr) is c. 2–4 μm, which is much lower than the depth of formation of free radicals under these conditions.

11.5.4   Kinetics of PE treatment in continuous RF discharge

Treatment of high-density polyethylene film and powder with specific surface area 2 m^2 g^{-1} has been also performed using low-pressure continuous RF discharge (Ponomarev and Vasilets, 2000; Vasilets, 2005). Voltage amplitude in this case was 0.5–1.2 kV, frequency 1.2 MHz and specific discharge power 0.04–0.35 W cm^{-3}. Similarly to the case of pulsed discharges, air, He and H_2 were used as plasma gases in the continuous RF discharge at pressures between 10^{-2} and 50 Torr. Continuous RF discharge at the same power and pressure as the pulsed discharge is characterized by a lower electric field, which results in a lower fraction of energetic electrons, a higher density of charged particles and a higher flux of positive ions to the polymer surface. The higher flux of positive ions leads to significant etching of the polymer surfaces in this case.

Kinetic curves of free radical formation in PE in the continuous RF discharge are similar to those in the pulsed RF discharge (Figure 11.7), while treatment depths are different. Lower electron energies, less intensive UV radiation and significant ion etching of the polymer surfaces in the continuous RF discharge result in smaller treatment depths of polymer surfaces in this case. The depth of formation of free radicals is 0.27 μm at a specific power of 0.35 W cm^{-3}, and decreases to 0.15 μm and 0.07 μm at lower specific powers of 0.14 W cm^{-3} and 0.04 W cm^{-3}. PE treatment in the continuous RF discharge is characterized by slower formation of trans-vinylene bonds and thinner surface layers where they are formed.

11.5.5   Non-thermal plasma etching of polymer materials

Etching of polymer materials is provided in plasma by two mechanisms: physical sputtering and chemical etching. The physical sputtering is due to simple ion bombardment, while the chemical etching is due to surface reactions and gasification of polymers, provided particularly by atomic oxygen, fluorine, ozone, and electronically excited oxygen molecules $O_2(^1\Delta_g)$. Typical examples of plasma gases applied for polymer etching are CF_4 and CF_4-O_2 mixture, which are widely used in microelectronics for cleaning substrates from organic deposits and for etching of photoresist layers. Treatment of polymers in oxygen-containing plasma initially leads to formation on the surface of the following specific oxygen-containing chemical groups: $-C = O$, $-C(R) - O - H$, $-C(R) - O - O - H$, $-C(R_1) - O - O - C(R_2)-$. Further interaction of the polymer with non-thermal oxygen-containing plasma can result in further oxidation, formation of CO_2 and H_2O and their transition to the gas-phase. Such processes can be interpreted as chemical etching of the polymer oxygen-containing plasma.

Etching rates of different polymers are very different. Fluorocarbon polymers, for example PTFE, are characterized by the slowest etching rates in plasma. Etching of hydrocarbon polymers is much faster. Hetero-atoms and especially oxygen-containing groups are the least resistive to plasma etching. Rates of polymer mass losses during etching in low-pressure oxygen plasma of a RF discharge are illustrated in Figure 11.9 (Ponomarev and Vasilets, 2000; Vasilets, 2005).

Branching of a polymer chain usually leads to higher etching rates, while cross-links between macromolecules usually decrease the etching rate. Etching rates are different for amorphous and crystal regions of a polymer, which result in changes to the polymer surface morphology and can lead to formation of porous surface structure after long plasma treatment. Etching of the amorphous regions is usually faster, which means that the crystal phase can dominate after plasma treatment. This etching effect is applied to increase

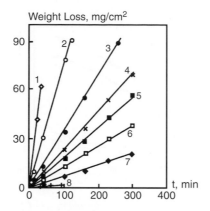

Figure 11.9   *Time evolution of mass losses for different polymers treated by oxygen plasma. 1: polysulfone; 2: polypropylene; 3: low-density polyethylene; 4: poly-ethylene-glycol-terephthalate; 5: polystyrene: 6: poly-tetra-fluorethylene PTFE; and 7: resin based on natural caoutchouc.*

the durability of polymer fibers, in particular. Characteristic plasma etching rates of some specific polymers in helium, nitrogen and oxygen are shown in Table 11.3 (Gilman, 2000).

11.5.6   Contribution of electrons and UV radiation in plasma treatment of polymer materials

The formation of chemical products of plasma polymer treatment (radicals, double-bonds, cross-links, hydrogen, etc.) or, in other words, the chemical effect of the plasma treatment, is due to several plasma components. Usually the primary plasma components active in high-depth polymer treatment are divided into two groups (electrons and UV radiation) which can penetrate relatively far into the surface of polymer material. The contribution of these two deep-penetrating plasma components to polymer treatment is discussed here (Vasilets, 2005; Ponomarev, Vasilets and Talrose, 2002).

The interaction of the chemically active heavy particles with polymers is usually limited to a very thin surface layer (except the effect of ozone and other relatively stable neutrals), and is discussed in the following section.

Although the contribution of 'electrons' is being discussed, the contribution of secondary factors related to the primary plasma electrons such as gas-phase radicals, excited atoms and molecules and so on is sometimes assumed.

Table 11.3   *Etching rates of different polymers in non-thermal low-pressure plasma in helium, nitrogen and oxygen.*

Polymer	Etch rate ($g\ cm^{-2}\ hr^{-1}$)		
	He plasma	N_2 plasma	O_2 plasma
Poly-tetra-fluoroethylene, PTFE	0.2	0.73	5.6
Poly-(tetra-fluoro-ethylene – hexa-fluoro-propylene), PTFE/HFP	0.17	0.45	3.4
Poly-vinyl-difluoride, PVDF	0.83	0.91	10.6
Polyethylene, PE	0.70	0.90	4.2

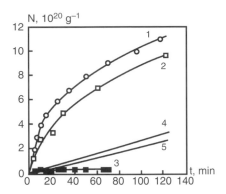

Figure 11.10   *Accumulation of H_2 (1, 4), trans-vinylene double bonds (2, 5) and free radicals (3) in polyethylene powder (specific surface 2 m^2 g^{-1}) due to direct treatment in the pulsed RF discharge in air at pressure 0.01 Torr and due to only VUV radiation of the discharge (4, 5).*

The classification 'UV radiation' includes the contribution of vacuum-ultraviolet (VUV) radiation as well as softer UV radiation. VUV radiation is characterized by short wavelengths (110–180 nm) and very high photon energies sufficient for electronic excitation and direct dissociation of chemical bonds. The high chemical activity of VUV radiation leads, on the other hand, to very short absorption lengths in this case. Effective transfer of this type of radiation across the relatively long distance is possible only in low-pressure gases, which explains the name VUV).

Softer UV radiation is characterized by longer wavelengths exceeding 200 nm and therefore lower photon energies. Softer UV radiation is much less chemically active but able to penetrate much deeper into the polymer film, which is important for some applications. As an example, treatment of polyethylene (PE) by VUV radiation of a low-pressure non-thermal RF discharge results in significant formation of intermolecular cross-links. The VUV radiation is separated from other plasma components able to modify the polymer surface by using a special LiF window. Softer UV radiation from a mercury lamp (typical wavelength 253.7 nm) does not produce the intermolecular cross-links in PE.

Kinetic curves for formation of molecular hydrogen H_2, trans-vinylene bonds and free radicals in PE powder (with specific surface 2 m^2 g^{-1}) by its treatment in the non-thermal pulsed RF discharge in an inert gas are presented in Figure 11.10 (Vasilets, 2005). The figure compares the total chemical effect of the discharge with the partial contribution of the VUV radiation. Initially ($t < 30$ min), the VUV contribution towards formation of molecular hydrogen and trans-vinylene bonds is negligible, but it is significant for longer treatments ($t > 60$ min). The effect of heavy particles is negligible in this case. Analysis of the PE treatment by continuous non-thermal RF discharge shows that formation of H_2 and trans-vinylene bonds to a depth of 1–2 μm after more than 60 min of the plasma-polymer interaction is dominated by VUV in the range 140–160 nm. At the same time, plasma electrons make a significant contribution to the formation of free radicals in the PE powder (Ponomarev and Vasilets, 2000).

It should be mentioned that the effect of plasma-generated active heavy particles can be neglected if depths of only 1–2 μm are being considered. The non-thermal low-pressure plasma treatment of PTFE leads to formation in PTFE of two types of radicals: those formed by detachment of an F-atom ($-CF_2 - CF(\cdot) - CF_2-$) and those formed by breaking a macromolecule ($-CF_2 - CF_2(\cdot)$). During the initial period of the treatment ($t < 10$ min), formation of the radicals is mostly due to contribution of plasma electrons. The partial contribution of VUV (140–155 nm) becomes significant later (Ponomarev and Vasilets, 2000; Ponomarev, Vasilets, and Talrose, 2002). Quantum yields of the free radical formation in this case are 3×10^{-3} for VUV with wavelength

147 nm, and 5×10^{-3} for VUV with wavelength 123 nm. The depth of effective formation of the free radicals in PTFE is c. 0.6 μm.

11.5.7   Interaction of chemically active heavy particles generated in non-thermal plasma with polymer materials

Interaction of the plasma-activated heavy particles with polymers has an important role in plasma-polymer treatment, but is usually localized to the surface only within a few molecular mono-layers. Probably only ozone is an exception here, because of its ability to effectively diffuse and penetrate deeper into the polymer. Molecular and atomic particles provide etching of the polymer surface layers as well as form new functional groups, significantly changing surface characteristics of the polymer. Volume properties of the polymer materials remain the same. The interaction of heavy particles with polymer surfaces varies in different plasma gases; examples of the polymer interaction with some specific gases are therefore discussed here.

11.5.7.1   Plasma-chemical oxidation

We start with the plasma-chemical oxidation of polymer surfaces, which is widely used today in different industrial areas. Interaction with oxygen-containing plasma results in the formation of polar groups on the polymer surfaces, which leads to growth of the polymer surface energy and a significant increase of the polymer wettability and adhesion to metals and different organic compounds. Photo-electronic spectra of plasma-treated PE show that first of all the plasma-chemical oxidation leads to formation on the polymer surfaces of $-C(-)-O-$ bonds. These bonds correspond to specific groups such as peroxides, alcohols, ethers and epoxies. Second in the row are $-C(-)=O$ bonds, which are typical for aldehydes and ketones. The least probable are $-C(-O-)=O$ bonds, corresponding to the carboxyl-acidic groups.

   Plasma-chemical oxidation always includes simultaneous formation of oxygen-containing surface groups and surface etching. As a result, the polymer surface oxidation degree essentially depends on the polymer composition and structure, and can be significantly varied by discharge power, plasma parameters and treatment time. It is possible to avoid the etching of oxygen-containing groups by preliminary treatment of the polymer in plasma of inert gases following contact of the activated surface with oxygen-containing gases outside of the discharge.

11.5.7.2   Plasma-chemical nitrogenation

Interaction with non-thermal plasma of nitrogen and nitrogen compounds (especially NH_3) results, in particular, in the formation of amine-groups $(R-NH_2)$, amide-groups (R_1-NH-R_2) and imine-groups $((R_1, R_2)C=N-H)$ on the polymer surfaces. These groups promote surface metallization or adhesion of different materials to the polymers (Liston, Martinu, and Wertheimer, 1994; Grace *et al.*, 1996; Gerenser *et al.*, 2000).

   The non-thermal plasma nitrogenation approach has been used, for example, for promotion of silver adhesion to polyethylene-terephthalate (PET) (Spahn and Gerenser, 1994). Another practically important example is plasma nitrogenation of polyester webs to promote adhesion of gelatin-containing layers, which are used in the production of photographic film (Grace, 1995). Non-thermal nitrogen plasma is similarly effective in promotion of adhesion on the surface of polyethylene-2,6-naphthalate (PEN) (Gerenser *et al.*, 2000).

   In the above examples, the effective nitrogenation has been achieved in low-frequency RF CCP discharge in N_2 (Conti *et al.*, 2001), and nitrogen has been incorporated in the polymers in the form of amine and amide groups. The types of nitrogen-containing groups incorporated in the polymer surface essentially depend on plasma gas. For example, NH_3 plasma treatment of polystyrene leads mostly to the formation of amine groups NH_2 on the polymer surface while N_2 plasma in similar conditions does not produce the amine groups.

The treatment of polymers in N_2 plasma results in more significant formation of imine groups C=N-H on the surface. Experiments with low-pressure non-equilibrium microwave discharges in molecular nitrogen N_2 indicate that total surface concentration of nitrogen-containing groups can be very high, reaching up to 40 atomic %.

Ammonia (NH_3) plasma in similar conditions results in lower surface concentration of the nitrogen-containing groups (Ponomarev and Vasilets, 2000). Fluorine-containing polymers are the most resistive to nitrogenation and oxidation in non-thermal plasma conditions. For example, treatment of PTFE in N_2 or NH_3 plasma usually leads to the formation of not more than 6 atomic % of the nitrogen-containing groups in a thin 10 nm surface layer.

Plasma nitrogenation of polymer surfaces strongly promotes further oxidation of the surfaces in atmospheric air. For example, nitrogenation of PE in N_2 plasma and following contact of the surface with atmospheric air results in oxygen concentration in PE c. 8 atomic % in the 10 nm layer. A similar procedure in NH_3 plasma leads to oxygen concentration in PE c. 4–6 atomic % in the 10 nm surface layer.

Surface modification of polymers in nitrogen-containing plasmas is widely used to improve biocompatibility of polymer materials, for example, amine groups formed in plasma on polymer surfaces provide effective immobilization of heparin and albumin on the surfaces. Biocompatibility and adhesion of different cells to the polystyrene surface is significantly enhanced by formation of the amine-groups on the surface during treatment in NH_3 plasma. Similarly, the amine-groups created in NH_3-plasma on the surface of PTFE provide adhesion of collagen.

11.5.7.3 Plasma-chemical fluorination

Interaction with non-thermal plasma of fluorine-containing gases leads to a decrease of surface energy for hydrocarbon-based polymers and makes these polymer surfaces hydrophobic; this property is widely used for practical applications.

Interaction of the hydrocarbon polymer materials with fluorine-containing plasmas results in the formation of different surface groups, especially C-F, CF_2, CF_3 and C-CF. Interaction of CF_4 plasma with the hydrocarbon-based polymer materials leads mostly to the formation of C-F and CF_2 groups on the polymer surface, while interaction with CF_3H plasma mostly leads to the formation on the surface of C-CF and CF_3 groups.

Treatment of polymer surfaces with the fluorine-containing plasmas stimulates three groups of processes simultaneously: formation of the fluorine-containing groups (C-F, CF_2, CF_3 and C-CF), polymer etching and plasma polymerization. The relative contribution of these three processes strongly depends on (1) the relative concentration in plasma of CF and CF_2 radicals and (2) the relative concentration of F- atoms. The CF and CF_2 radicals are building blocks for plasma polymerization, while atomic fluorine is responsible for etching and formation of the fluorine-containing surface groups.

The relative concentration in plasma of CF/CF_2 radicals and F-atoms depends on the type of applied plasma gas. Typical fluorine-containing gases applied for polymer treatment are CF_4, C_2F_6, C_2F_4, C_3F_8 and CF_3Cl. Generally, higher C/F ratios in the initial plasma gases leads to an increase of relative concentration of CF and CF_2 radicals in volume with respect to F-atoms. Higher C/F ratios are therefore favorable for plasma polymerization, while lower C/F ratios are favorable for the polymer etching and formation of the fluorine-containing surface groups.

Plasma treatment of PMMA-based contact lenses to minimize their interaction with tissues in the eye is an example of the application of plasma fluorination of polymers. Such treatment can also increase the durability and decrease the friction coefficient of the contact lenses.

11.5.8 Synergetic effect of plasma-generated active particles and UV radiation with polymers

Plasma interaction with polymers and especially biopolymers is a multi-stage and multi-channel process, which is often characterized by strongly non-linear kinetics. The contribution of the different plasma components such as atoms, radicals, active and excited molecules, charged particles and UV radiation to the polymer treatment process is not simply cumulative but essentially synergetic. Obtaining a simple and unambiguous answer to the question of which plasma component dominates the plasma-polymer treatment process is often difficult.

Two or more plasma components can make a synergistic contribution to the total process. An example is the non-thermal plasma etching of PE and poly-vinyl-chloride (PVC) in low-pressure RF discharges in oxygen (Vasilets, 2005). Generally, etching of the polymers in low-pressure oxygen plasma is mostly due to atomic oxygen, electronically excited molecular oxygen and UV radiation. Experiments with these plasma components individually give (after summation) total etching rates of a factor of 3 less than the combined contribution of the components applied together. Ponomarev and Vasilets (2000) claim that PE and PVC etching is a synergetic effect of atomic particles and UV radiation.

A similar situation probably occurs in plasma treatment of biopolymers and in plasma sterilization processes where UV radiation, radicals and chemically active and electronically excited molecules essentially make a non-linear synergistic contribution.

11.5.9 Aging effect in plasma-treated polymers

The composition and space distribution of the products of plasma treatment of polymer materials can keep changing long after the plasma treatment process is finished. This phenomenon is referred to as the aging effect in plasma-treated polymers. The four major mechanisms of the aging process are:

- reorientation and shift of the polar groups formed on the polymer surface inside the polymer material due to thermodynamic relaxation;
- diffusion of the low-molecular mass admixtures and oligomers from volume of the polymer material to the polymer surface;
- diffusion to the polymer surface of the low-molecular mass products formed during the plasma treatment in the relatively thick surface layer; and
- post-plasma treatment reactions of free radicals and other plasma-generated active species and groups between themselves and with the environment.

Aging of the hydrocarbon-based polymer materials treated in oxygen plasma is mostly due to the reorientation and shift of the polar peroxide groups formed on the polymer surface inside of the polymer, which is related to thermodynamic relaxation. The wettability contact angle in H_2O decreases several times immediately after the treatment because of formation of the polar peroxide groups. The wettability contact angle then starts increasing and, because of the aging effect, can return almost to its initial value after several days of storage in atmospheric air.

Aging of the hydrocarbon-based polymers treated in nitrogen is mostly due to the post-processing reactions of the nitrogen-containing surface groups with the environment. As an example, the major effect of plasma nitrogenation of polyethylene in N_2 discharges is related to the formation of imine-groups $((R_1, R_2)C = N - H)$ on the polymer surface. Storage of the plasma-treated polymer in atmospheric air results first of all in hydrolysis of the imine groups:

$$(R_1, R_2)C = N - H + H_2O \rightarrow (R_1, R_2)C = O + NH_3. \tag{11.23}$$

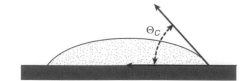

Figure 11.11   *Contact angle θ of a liquid sample.*

Longer storage in atmospheric air results in additional reactions of nitrogen incorporated in polymer with atmospheric water:

$$R_1 - CH = N - R_2 + H_2O \rightarrow R_1 - CH = O + H_2N - R_1. \tag{11.24}$$

Polypropylene is characterized by the strongest aging effect, while PET is affected by aging effect a little less. The most durable with respect to aging are polyethylene and polyimide. Generally, the higher level of crystallinity of polymers leads to their stronger durability with respect to aging after plasma treatment.

11.5.10   Plasma modification of wettability of polymer surfaces

The most important result of plasma treatment of polymers, which are produced on an industrial scale, is the change of their wettability and adhesion characteristics. Plasma treatment can make polymers more hydrophilic as well as more hydrophobic. Both effects are widely used for practical applications. The change of wettability is usually characterized experimentally by the contact angle θ, which is formed on the solid surface along the linear border line of solid-liquid-air (see Figure 11.11). An increase of wettability (making the polymer more hydrophilic) leads to a decrease in the *contact angle*. An example illustrating the improvement of wettability and decrease of the contact angle as a result of plasma treatment of a polymer is shown in Figure 11.12. The change in wettability contact angles with respect to water and glycerin for different polymer materials after treatment with a low-frequency (50 Hz) low-pressure AC discharge in air is listed in Table 11.4 (Gilman, 2000).

(a)   (b)

Figure 11.12   *Increase of wettability and relevant change of the contact angle as a result of the surface treatment by non-thermal atmospheric plasma (a) without plasma treatment and (b) after plasma treatment.*

Table 11.4   *Change of contact angles with respect to water and glycerin due to treatment of different polymers in low-pressure low-frequency AC plasma in air.*

Polymer	Contact angle θ (degrees)			
	Water: before treatment	Water: after treatment	Glycerin: before treatment	Glycerin: after treatment
Poly-propylene, PP	92	46	78	39
Polyimide, PI	76	13	58	6
Poly-(tetra-fluoro-ethylene–hexa-fluoro-propylene), PTFE/HFP	111	85	90	66

The treatment of polymers in oxygen-containing plasma leads to a decrease of contact angle and therefore to an increase of wettability. The wettability increase effect is related to plasma-stimulated formation of polar peroxide groups on the polymer surfaces. Changes in the contact angles for the plasma-modified polymer surfaces depend on the applied plasma gas and conditions of the plasma treatment. The application of discharges in air, oxygen, nitrogen and ammonia transforms initially hydrophobic surfaces into hydrophilic surfaces.

The application of discharges in fluorine-containing gases such as tetra-fluoroethylene (TFE), per-fluoro-propane and octa-fluoro-cyclo-butane significantly enhances the hydrophobic properties of polymer surfaces.

An increase of gas pressure, discharge current and plasma treatment time leads to a reduction of the contact angle θ and to an enhancement of wettability. A major enhancement of hydrophilic properties usually occurs during initial treatment time of 30–120 s (see Figure 11.13).

Plasma-enhanced hydrophilic properties become less strong with time (aging effect). Maintaining the enhanced hydrophilic properties over time is important in practical applications. Plasma treatment of polypropylene allows high wettability (contact angle $\theta < 60°$) to be maintained over 30 days, and plasma treatment of polyimide allows high wettability (contact angle $\theta < 50°$) to be maintained for 12 months. The aging effect is illustrated in Figure 11.14 for the case of plasma-treated polyimide-fluoro-plastic film. As can be seen from the figure, restoration of the wettability contact angle θ takes place mostly during first 10 days; the aging process slows down after this point.

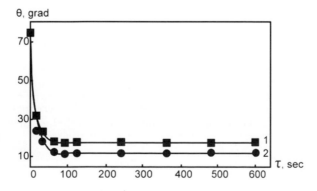

Figure 11.13   *Typical dependence of water/PI contact angle (θ) on plasma treatment time. Air plasma 50 Hz, current 50 mA (1) and 100 mA (2).*

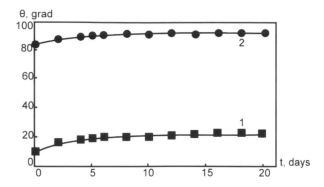

Figure 11.14   *Typical dependence of water contact angle (θ) on storage time in normal air conditions. Plasma-modified film: (1) laminated polyimide fluoroplast film treated in air plasma 50 Hz; and (2) TFE-fluoroplast.*

11.5.11   Plasma enhancement of adhesion of polymer surfaces

Plasma treatment cleans the polymer surface and makes it highly hydrophilic. This results in the enhancement of adhesion properties of the polymer materials, providing a wide variety of relevant applications. It is widely used for gluing polymers in different combinations and the fabrication of composite materials with special mechanical and chemical properties. In particular, plasma-treatment-enhanced adhesion of fibers to binding materials significantly improves the characteristics of composites. Plasma technology for fabrication of high-temperature-resistant electric insulation is an important application. Contact properties of polymer surfaces are often improved in practical applications by plasma deposition of special polymer films, characterized by strong contact properties with respect to the adhesive material as well as to the major polymer.

Plasma enhancement of adhesion of polymer surfaces can be considered as a generalization of the plasma stimulation of wettability. Plasma enhancement of adhesion is widely used in various large-scale industrial applications. The increase of adhesion energy of different polymers with respect to water and glycerin in low-frequency (50 Hz) low-pressure AC discharge in air is illustrated in Table 11.5 (Gilman, 2000). As can be seen from the table, plasma can significantly stimulate the adhesion process. An increase of the adhesion energy of polymer materials by their treatment in oxygen-containing plasma is related to an increase of surface energy for the plasma-treated polymers.

The change of surface energy of different polymers after treatment in the same plasma of low-frequency (50 Hz) low-pressure AC discharge in air is described in Table 11.6 (Gilman, 2000). Data in the table illustrate

Table 11.5   *Change of adhesion energy with respect to water and glycerin due to treatment of different polymers in low-pressure low-frequency AC plasma in air.*

Polymer	Adhesion energy (mJ m^{-2})			
	Water: before treatment	Water: after treatment	Glycerin: before treatment	Glycerin: after treatment
Poly-propylene, PP	70.3	123.4	76.6	112.7
Polyimide, PI	91.6	144.0	100.6	126.4
Poly-(tetra-fluoro-ethylene – hexafluoro-propylene), PTFE/HFP	46.7	79.14	63.4	83.0

Table 11.6 *Change of total surface energy and its components (polar and dispersion) due to treatment of different polymers in low-pressure low-frequency AC plasma in air.*

Polymer	Before/after treatment	Total surface energy (mJ m^{-2})	Dispersion component of surface energy (mJ m^{-2})	Polar component of surface energy (mJ m^{-2})
Poly-propylene, PP	Before	29.3	26.1	3.2
	After	52.5	19.2	33.3
Polyimide, PI	Before	45.73	40.7	5.03
	After	71.55	17.5	54.05
Poly-(tetra-fluoro-ethylene – hexafluoro-propylene), PTFE/HFP	Before	30.19	30.09	0.10
	After	37.83	35.04	2.79

growth of the total surface energy of the polymers as a result of treatment in oxygen-containing non-thermal plasma. The table also shows how the plasma modification changes both the polar and dispersion components of the surface energy.

The polar component of the surface energy characterizes polar interaction between the surface of the polymer material and the working fluid or surface film. This component is determined by the presence of polar groups, electric charges and free radicals on the polymer surface. The polar component of surface energy grows during treatment in oxygen-containing plasma. This effect is due to formation of polar groups, especially peroxide groups, on the polymer surface.

In contrast to that, the dispersion component of surface energy characterizes dispersion interaction between the surface and working liquid or surface film. This component is determined by roughness, unevenness and branching level of the polymer surface. The dispersion component of the surface energy, which is relatively large before plasma treatment, can even be decreased after plasma modification.

Plasma-enhanced adhesion of polymer surfaces enables the following effective metallization of the surfaces using conventional methods such as vacuum thermal and magnetron spraying, deposition by decomposition of metal-organic compounds and so on. A comparison of adhesion of vacuum-thermally-sprayed thin aluminum (Al) films on the surface of different polymer materials treated in plasma and untreated is presented in Table 11.7 together with relevant data on the wettability contact angles with respect to water (Gilman, 2000).

The table shows results of the polymer treatment in non-thermal discharge plasmas generated in different gases: helium He, oxygen O_2 and O_2/CF_4 mixture. Plasma treatment of all considered industrial polymer materials leads to significant improvement of adhesion to the vacuum-thermally-sprayed thin aluminum (Al) films and to a reduction of the wettability contact angles.

11.5.12 Plasma modification of polymer fibers and polymer membranes

Polymers are used in composite materials as a dispersed phase (fibers or powders) as well as a matrix phase. In both cases, adhesion between the phases can be significantly improved by treatment in non-thermal plasma. Some examples of non-thermal plasma modification of polymer fibers (Gilman, 2000), which significantly changes their properties for specific technological applications, are summarized in Table 11.8.

Plasma modification of porous and non-porous polymer substrates allows the production of composite membranes for gas separation, pervaporation (separation of liquids by evaporation through the membrane)

Table 11.7   *Adhesion of plasma-treated industrial polymer films with respect to sprayed aluminum layers and wettability contact angles of the plasma-treated industrial polymer films (adhesion is characterized as A: very strong; B: strong; C: medium strength).*

Polymer	Contact angle θ (adhesion)			
	Before treatment	O_2 plasma	CF_4/O_2 plasma	He plasma
Polycarbonate (PC)	72 (B)	39 (A)	<15 (A)	37 (A)
Polyether-sulfone (PSU)	70 (A)	25 (A)	<15 (A)	26 (A)
Polyethylene-fluoride (PEF)	66 (C)	29 (A)	30 (A)	29 (A)
Polyvinyl-difluoride (PVDF)	71 (C)	40 (A)	70 (A)	57 (A)
Polypropylene (PP)	98 (C)	40 (A)	72 (A)	53 (A)
Polyethylene (PE)	90 (C)	—	20 (A)	50 (A)
Polypropylene (PP)	83 (B)	15 (A)	<15 (A)	26 (A)

Table 11.8   *Modification of properties of polymer fibers after plasma treatment.*

Polymer fiber	Plasma gas	Plasma treatment conditions	Modification of fiber properties
Polyethylene-terephthalate (PET)	Acrylic acid (AA), hexamethyl-di-silazane, air	RF discharge (13.6 MHz), 13.3 Pa, 20–40 W	Reduction of combustibility, hydrophilization: initial contact angle $\theta = 37$, final $\theta = 40$
Polyethylene-terephthalate (PET)	Tetrachloroethylene (TECE), tri-chloro-ethylene (TCE)	RF discharge (13.6 MHz), 13.3 Pa, 20–40 W	Reduction of combustibility, hydrophilization: initial contact angle $\theta = 57$, final $\theta = 46$
Polyethylene (PE), kevlar	Oxygen (O_2)	RF discharge (13.6 MHz), 133 Pa, 20–40 W, 5–60 min	Adhesion enhancement to epoxy-matrix: PE-fiber 118% increase, Kevlar-fiber 45% increase
Polyethylene (PE)	Acrylic acid (AA)	RF discharge (13.6 MHz), 6.6 Pa, 30–70 W, 5–40 min	Adhesion enhancement to epoxy-matrix. Hydrophilization: initial contact angle $\theta = 65$, final contact angle $\theta = 8$
Kevlar	Ar, O_2, NH_3	RF discharge (13.6 MHz), 7-13.3 Pa, 25 W, 2–8 s	20% Adhesion enhancement to polycarbonate (PC) matrix
Cellulose acetate (CA)	Air	RF discharge (13.6 MHz), 13.3 Pa, 100 W	Adhesion enhancement to matrix from polypropylene (PP), polystyrene (PS), chlorinated polyethylene (PE)
Polyethylene	Oxygen (O_2)	RF discharge (13.6 MHz), 17.3 Pa, 67 W, 5 min	Adhesion enhancement to epoxy-matrix

Table 11.9 *Plasma modification of porous polymer membranes.*

Membrane	Plasma gas	Treatment conditions	Properties of treated membranes
Polysulfone (PSU)	Ar	RF-discharge (13.6 MHz), 6.5 Pa, 30 s, sublayer of silicon polymer Sylgard 184	Increase of gas-separation coefficient for O_2/N_2 mixture from 1.2 to 4.6
Polystyrene (PS)	Butyronitrile	RF-discharge (13.6 MHz), 4 Pa, 25–60 W, sublayer	Increase of gas-separation coefficient for H_2/CH_4 mixture from 1 to 15
Polyacrylonitrile (PAN)	He, Ar, O_2	Discharge AC (60 Hz), 13.3 Pa, 350 W, 120 s	Water purification from salts, effectiveness 98%
Polypropylene (PP), Celgard 2400	air	RF-discharge (13.6 MHz), 2.7 Pa, 10 W, 60 s, MAC-grafting in water solution (5%, 70°C, 2 hours)	Pervaporation of ethanol-water mixture, selectivity with respect to water 10%
Polysulfone (PSU)	C_2F_4 (TFE), C_3F_8, C_3F_6	RF-discharge (13.6 MHz), 30–100 W, 230 min	Pervaporation of ethanol-water mixture, separation factor 10, productivity 1 kg m^{-2} hour^{-1}

and water cleaning by reverse osmosis. The plasma modification approaches differ for porous and non-porous substrates. Plasma treatment of porous substrates is focused on the porous size reduction due to: cross-linkage of polymer surface in air, O_2 or inert gas discharges; activation of substrate surface with following grafting; or deposition of thin polymer film (<1 µm) on the porous substrate surface or on the special adhesive sublayer. Some examples of plasma modification of porous membranes (Gilman, 2000) for different applications are summarized in Table 11.9.

Plasma treatment of non-porous membranes can be focused on the functionalization, hydrophilization and cross-linking of the polymer surfaces in plasma of air, O_2, N_2, NH_3 and so on; on plasma deposition of thin polymer films with preliminary surface activation; or on grafting on a membrane surface which has undergone preliminary plasma activation. Some examples of plasma modification of non-porous membranes (Gilman, 2000) for different applications are summarized in Table 11.10.

Generally, the plasma modification of polymer membranes allows excellent results to be achieved in separation of gases, water purification from salts and in concentrating different organic and inorganic compounds. For example, a plasma-modified polyacrylonitrile (PAN) membrane with 98% salt separation efficiency is used in a device cleaning 2 m^3 of water (initial pH of 1–10) per day.

11.5.12.1 *Plasma treatment of wool*

Non-thermal plasma plays an important and multifunctional role in the treatment of natural as well as man-made textile materials (Maximov, 2000; Hocker, 2002). The contribution of plasma technology is not limited to the well-known and widely used positive plasma effect on dyeing and printing of textiles. Plasma is also effectively used for more specific treatment of natural fibers, including enhancement of shrink-resistance of wool and selective oxidation of lignin in cellulose, which transforms the lignin in a water-soluble form for further extraction.

Table 11.10   *Plasma modification of non-porous polymer membranes.*

Membrane	Plasma gas	Treatment conditions	Properties of treated membranes
Polydimethylsiloxane (PDMS)	Ar	RF-discharge (13.6 MHz), 35 W, 6 min	Increase of gas-separation coefficient for O_2/N_2 mixture from 2 to 5.4, for CO_2/CH_4 mixture from 2 to 20
SC(silicon-based membrane)	Methacryl-, benzo-, crotono-nitrile	RF-discharge (13.6 MHz), 133 Pa, 100 W, 60 s	Increase of gas-separation coefficient for H_2/CH_4 mixture from 0.8 to 40
Polyphenylene Oxide (PPO)	Cyanogen Bromide	RF-discharge (13.6 MHz), 67 Pa, 100 W, 30 s	Increase of gas-separation coefficient for H_2/CH_4 mixture from 23 to 297
Nylon 4	Ar, N_2, O_2	RF-discharge (13.6 MHz), 13.3 Pa, 50 W, AC, VP-grafting in solution	Water purification from salts, effectiveness 94%
Poly-vinyl-trimethyl-silane (PVTMS)	air	AC-discharge (50 Hz),40 W, 60 s	Pervaporation of water solution of HNO_3, selectivity coefficient is 0.99 for 3N-solution HNO_3, productivity 2 kg hour^{-1}

To discuss the plasma treatment of wool, it should first be mentioned that the cuticle cells in wool overlap each other to create a directional frictional coefficient and that the very surface is highly hydrophobic. The hydrophobic behavior leads to aggregation of the fibers in aqueous medium, which has the main result of felting and shrinkage of wool. A significant effect of wool treatment in oxygen-containing (in particular, air) plasma is due to oxidation and partial removal of the hydrophobic lipid layer on the very surface of wool. This applies both to the adhering external lipids and to the covalently bound 18-methyl-eicosanoic acid. The plasma-induced reduction of the hydrophobic behavior therefore results in its resistance to shrinkage. The effect can be achieved both from low-pressure and atmospheric-pressure non-thermal plasma discharges.

Another non-thermal plasma effect (which can also be achieved in both low-pressure and atmospheric-pressure discharges) is related to significant reduction of the cross-link density of the exocuticle layer in wool. The exocuticle is the layer located below the fatty acid layer of the very surface, which is called epicuticle. The exocuticle layer is highly cross-linked via disulfide bridges. Treatment of wool in oxygen-containing plasma leads to oxidation and breaking of the disulfide bonds, which results in a significant reduction of the cross-link density and improvement of the wool properties (Hocker, 1995).

Protein loss after even intensive plasma treatment and extraction is very low (c. 0.05%) because of the surface-oriented nature of the plasma treatment. The specific surface area of the wool is significantly increased as a result of plasma treatment from c. 0.1 m^2 g^{-1} to 0.35 m^2 g^{-1}. Due to the surface-directed nature of the plasma treatment, the tenacity of the fibers is only slightly influenced. Plasma modification of the wool surface therefore leads to a reduction of the shrinkage behavior of the wool top. The felting density of the wool top before spinning decreases from more than 0.2 g cm^{-3} to less than 0.1 g cm^{-3}.

Especially strong shrinkage-resistance can be achieved by additional resin coverage of the plasma-treated fiber surface. This combined plasma-resin procedure leads to formation of the smooth surface with reduced scale height and shrinkage of only c. 1% after 50 simulated washing cycles. To compare, the area felting shrinkage of untreated wool is 69% and plasma-treated wool without resin is 21%.

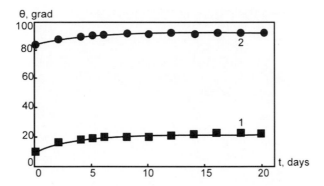

Figure 11.15   *Water drop-induced dye removal from an HMDSO-plasma-treated cotton fabric.*

As the wool is oxidized during treatment in oxygen-containing non-thermal plasma, the hydrophobic behavior of its surface is changed to become increasingly hydrophilic. This results in additional advantages of plasma treatment related to improved dyeing kinetics, enhanced depth of shade and improved bath exhaustion.

11.5.12.2   *Plasma treatment of cotton and synthetic textiles*

Cotton and synthetic textile fibers can be treated in non-thermal plasma under atmospheric pressure as well as under reduced pressure, depending on the required modifications to the materials (Maximov, 2000; Hocker, 2002). We first consider some physical and chemical features of plasma treatment of cotton fibers.

Similar to the wool treatment, non-thermal plasma treatment of cotton in oxygen-containing gases leads to a significant increase of specific surface area of cotton. On the other hand, treatment of cotton using hexa-methyl-di-siloxane (HMDSO) plasma results in a strong hydrophobization effect. It leads to the formation of smooth surfaces with increased contact angle, c. 130° with respect to water. The strong hydrophobization effect of the cotton fiber can be similarly achieved using non-thermal hexa-fluoro-ethane plasma, when fluorine is effectively incorporated into the fiber (Hocker, 2002).

Plasma-induced hydrophobization of cotton fabric in conjunction with increased specific surface area leads to an interesting and practically important effect. Water droplets are able to effectively remove dirt particles from the surface of cotton fabric as illustrated in Figure 11.15 for the case of HMDSO-plasma-treated cotton fabric (Hocker, 2002); this is usually referred to as the *Lotus effect*. Hydrophobic-plasma-treated surfaces of cotton are therefore extremely dust- and dirt-repellant in contact with water. As an important consequence, the plasma-treated surface is also repellant to bacteria and fungi.

The effect is relevant not only to cotton fiber but also to some other materials. Effective treatment of synthetic fibers can be achieved using atmospheric-pressure plasma of DBD or corona. As an example, the treatment of polypropylene (PP) in these discharges significantly increases the hydrophilicity of the surface. In particular, the contact angle with respect to water can be decreased from 90° to 55°; even after 2 weeks the contact angle remains c. 60°. The hydrophilicity increase is due to plasma-induced high oxygen/carbon ratio, which is significant even at the tenth surface layer of the polymer.

The non-equilibrium atmospheric-pressure plasma intensification of the PP surface hydrophilicity can be additionally enhanced by using maleic acid anhydride (MAH) as an assisting reagent. Incorporation of oxygen into the polymer fiber surface is permanent in this case, and a contact angle with respect to water can be decreased to 42°. Polyethylene-terephthalate (PET) fibers are used for enforcing PE matrix in the process of

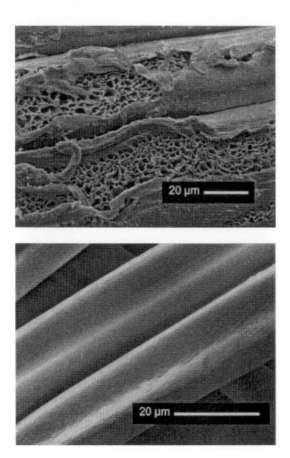

Figure 11.16   *SEM pictures of Nomex-fibers after exposition to diluted H_2SO_4. Top: without plasma treatment and bottom: with preliminary plasma treatment (Hocker, 2002).*

fabrication of polymer composite materials. In this case, non-equilibrium ethylene-plasma treatment of the PET fibers significantly increases their adhesion strength to the PE matrix.

Another example of effective plasma treatment of synthetic textiles is treatment of the polyaramid textile fibers such as Nomex, which are considered to be high-performance fibers but are prone to hydrolysis. Treatment of the fiber in hexa-fluoro-ethane/hydrogen plasma creates a diffusion barrier layer on the surface which is resistant to hydrolysis. The fiber becomes resistant even to 85% H_2SO_4. Contact of the treated fibers with sulfuric acid for 20 hours at room temperature leaves the fibers completely intact, while conventional fluorocarbon finishing under the given conditions is significantly damaging (see Figure 11.16). The damage includes significant shrinkage and loss of properties of the Nomex polyaramid textile fibers (Hocker, 2002).

11.5.12.3   *Specific conditions and results of non-thermal plasma treatment of textiles*

The effect of applying different plasma gases for non-thermal plasma treatment of textiles (details can be found in Maximov, 2000) is summarized in Table 11.11.

The same textile treatment can be achieved using different plasma gases. This effect can be explained by taking into account two factors. Firstly, most of the plasma treatment processes are not chemically specific

Table 11.11 *Application of different plasma gases for non-thermal plasma treatment of different textile polymer materials with major results of treatment.*

Plasma gas	Results of plasma treatment	Treated polymer materials
Inert gases:He, Ne, Ar	1. Enhancement of hydrophilicity 2. Decrease of gas permeability of polymer films 3. Suppressing of losses of plastificator 4. Enhancement of polymer film tearing strength 5. Adhesion properties enhancement 6. Suppressing of surface gloss 7. Stimulation of antistatic properties	1. Wool, cotton, polyamide PA (nylon), polyethylene (PE), polyethylene-terephthalate (PET) 2. Polyethylene (PE), polypropylene (PP), polystyrene (PS) 3. Polyvinylchloride (PVC) 4. Polyethylene (PE), polypropylene (PP), polystyrene (PS) 5. Polyether 6. Polyvinylchloride (PVC) 7. Polyethylene-terephthalate (PET)
Air, oxygen	1. Enhancement of wettability and adhesion properties for gluing fabrics 2. Suppressing of shrinkage and felting, enhancement of stability with respect to fabric wear out 3. Enhancement of wettability and intensification of soaking 4. Enhancement of adhesion for vacuum metallization and gluing polymers with metals 5. Decrease of gas permeability of polymer films 6. Enhancement of polymer film tearing strength 7. Suppressing of surface gloss 8. Stimulation of antistatic properties	1. Cotton, flax, wool, lavsan 2. Wool 3. Rough cotton fabric 4. Polyethylene, polypropylene, poly-oxy-methylene (acetal), polyethylene-terephthalate (PET), poly-tetra-fluoroethylene (PTFE), polyimide (PI), polycarbonate (PC) 5. Polyethylene (PE), polypropylene (PP), polystyrene (PS) 6. Polyethylene (PE), polypropylene (PP), polystyrene (PS) 7. Polyvinylchloride (PVC) 8. Lavsan fabric
Silanes (silicon hydrides)	1. Decrease of gas permeability of polymer films 2. Enhancement of polymer film tearing strength	1. Polyethylene (PE), polypropylene (PP), polystyrene (PS) 2. Polyethylene (PE), polypropylene (PP), polystyrene (PS)
Hydrocarbons: methane, ethane, butane etc.	1. Decrease of gas permeability of polymer films 2. Enhancement of polymer film tearing strength 3. Suppressing of losses of plastificator	1. Polyethylene (PE), polypropylene (PP), polystyrene (PS) 2. Polyethylene (PE), polypropylene (PP), polystyrene (PS) 3. Polyvinylchloride (PVC)
CO_2+Ar	1. Modification of surface morphology	1. Carbon fabrics
CO_2	1. Suppressing of surface gloss 2. Wettability enhancement	1. Polyvinylchloride (PVC) 2. Cotton, wool, lavsan fabric
NO, NO_2, CO	1. Suppressing of surface gloss	1. Polyvinylchloride (PVC)
CO, H_2O	1. Enhancement of adhesion	1. Lavsan films
N_2	1. Enhancement of adhesion 2. Suppressing of losses of plastificator 3. Decrease of gas permeability of polymer films 4. Enhancement of polymer film tearing strength 5. Wettability enhancement 6. Suppressing of surface gloss	1. Polyether 2. Polyvinylchloride (PVC) 3. Polyethylene (PE), polypropylene (PP), polystyrene (PS) 4. Polyethylene (PE), polypropylene (PP), polystyrene (PS) 5. Cotton, wool, lavsan fabric 6. Polyvinylchloride (PVC)

Table 11.11   *Continued*

Plasma gas	Results of plasma treatment	Treated polymer materials
Amines	1. Enhancement of axisymmetric fiber strength 2. Fiber wettability enhancement 3. Enhancement of fiber adhesion 4. Stimulation of fiber's antistatic properties	1. Aromatic polyamides 2. Aromatic polyamides 3. Aromatic polyamides 4. Aromatic polyamides
Alkyl-amines	1. Surface activation of fibers, films and fabrics for further fabrication of epoxy and other composites	1. Aromatic polyamides, poly-tetra-fluoroethylene (PTFE)
NH_3 (ammonia)	1. Wettability enhancement 2. Suppressing of losses of plastificator 3. Enhancement of adhesion	1. Aromatic polyamides, wool, polyether, rough cotton fabric 2. Polyvinylchloride (PVC) 3. Poly-oxy-methylene (acetal), poly-tetra-fluoroethylene (PTFE), polycarbonate (PC), polyacrylate, poly-phenylene-oxide
Hydrogen	1. Wettability enhancement 2. Suppressing of surface gloss 3. Surface flattening	1. Cotton, wool, lavsan fabric 2. Polyvinylchloride (PVC) 3. Polyvinylchloride (PVC)
$H_2 + Ar$	1. Enhancement of adhesion	1. Carbon fibers
Chlorine	1. Suppressing of surface gloss	1. Polyvinylchloride (PVC)
CF_4, $CF_4 + O_2$	1. Enhancement of wettability of plastic surfaces	1. Polyethylene (PE), poly-oxy-methylene (acetal), polystyrene (PS), polypropylene (PP), poly-tetra-fluoroethylene (PTFE), poly-methyl-methacrylate (PMMA), polyimide (PI)
NF_3, BF_3, SiF_4	1. Surface activation for further liquid treatment	1. Silk, wool, cotton
CF_4, C_2F_4, $CHClF_2$, CF_3Br, CF_3Cl, CHF_3	1. Surface activation with changing of wettability	1. Polyethylene (PE), polystyrene (PS), polypropylene (PP)
CF_4, CHF_3	1. Improvement of dyeing properties 2. Suppressing of losses of plastificator	1. Poly-tetra-fluoroethylene (PTFE), polyethylene-terephthalate (PET) 2. Wool, polyvinylchloride (PVC)
Fluoro-carbons / H_2 mixture, CF_4, C_2F_6, SF_6	1. Decrease of surface friction 2. Stimulation of antistatic properties 3. Enhancement of hydrophobicity 4. Enhancement of weather resistance	1. Polyethers 2. Polyethers 3. Polyethers 4. Polyethers

and are related to an integral change of surface energy. Chemical peculiarities of the plasma gas are not so important in this case. Secondly, plasma treatment of polymers usually results in the production of gases with a flow rate close to that of the initial plasma gas. The polymer is therefore treated in plasma which not only includes the initial plasma gas but also the gas generated during the treatment process.

References

Abou-Ghazala, A., Katsuki, S., Schoenbach, K.H., and Moreira, K.R. (2002) "Bacterial decontamination of water by means of pulsed-corona discharges." *IEEE Transactions Plasma Science*, **30**, 1449.

Abraham, R.T. (2004) "PI 3-kinase related kinases: 'Big' players in stress-induced signaling pathways." *DNA Repair*, **3**, 883–887.

Alix, F.R., Neister, S.E., and McLarnon, C.R. (1997) "Barrier discharge conversion of SO_2 and NOX to Acids." US Patent N. 5,871, 703.

Anpilov, A.M., Barkhudarov, E.M, Bark, Y.B., *et al.* (2001) "Electric discharge in water as a source of UV radiation, ozone and hydrogen peroxide." *Journal of Physics, D: Applied Physics*, **34**, 993.

Anselme, K. (2000) "Osteoblast adhesion on biomaterials." *Biomaterials*, **21**, 667.

Arita, M., Nagayoshi, M., Fukuizumi, *et al.* (2005) "Microbicidal efficacy of ozonated water against Candida albicans adhering to acrylic denture plates." *Oral Microbiology and Immunology*, **20**, 206–210.

Arjunan, K.P., Gutsol, A., Vasilets, V., and Fridman, A. (2007) "Water sterilization using a pulsed spark plasma discharge." *18-th International Symposium on Plasma Chemistry, ISPC-18, Kyoto, Japan.*

Ayan, H., Fridman, G., Friedman, G., *et al.* (2007) 18-th Int. Symposium on Plasma Chemistry, ISPC-18, Kyoto, Japan.

Babko-Malyi, S., Battleson, D., Ray, I., *et al.* (2000) "Mercury removal from combustion fuel gases by the plasma-enhanced electrostatic precipitators." Air Quality II: Mercury Trace Elements, and Particular Matter Conference, Proceedings, McLean, VA.

Baidarovtsev, Yu.P., Vasilets, V.N., and Ponomarev, A.N. (1985) "Effect of the nature of the working gas of a low-pressure glow discharge on rate of radical accumulation during plasma treatment of poly(tetrafluoroethylene)." *Soviet Physics, Khimicheskaya Fizika, Chemical Physics*, **4**, 89.

Bailin, L.J., Sibert, M.E., Jonas, L.A., and Bell, A.T. (1975) "Microwave decomposition of toxic vapor simulants." *Environmental Science & Technology*, **9**, 254.

Baranchicov, E.I., Belenky, G.S., Deminsky, M.A., *et al.* (1990) "Plasma-Catalytic SO_2 Oxidation in Air." Kurchatov Institute of Atomic Energy, vol. IAE-5256/12, Moscow.

Baranchicov, E.I., Belenky, G.S., Deminsky, M.A., *et al.* (1992) "Plasma-catalysis SO_2 oxidation in an air stream by a relativistic electron beam and corona discharge." *International Journal of Radiation Applications and Instrumentation. Part C. Radiation Physics and Chemistry*, **40**, 287.

Baydarovtsev, Y.P., Vasilets, V.N., and Ponomarev, A.N. (1985) "The influence of gas nature on the rate of radical accumulation in teflon during low pressure glow discharge treatment." *Russian Journal of Chemical Physics*, **4**(N1), 89–96.

Baysan, A. and Lynch, E. (2004) "Effect of ozone on the oral microbiota and clinical severity of primary root caries." *American Journal of Dentistry*, **17**, 56–60.

Beckman, J.S. and Koppenol, W.H. (1996) "Nitric oxide, superoxide, and peroxynitrite: the good, the bad, and ugly." *American Journal of Physiology - Cell Physiology*, **271**, 1424.

Bettleheiem, F.A. and March, J. (1995) *Introduction to General, Organic and Biochemistry*, 4th edn, Saunders College Pub., Orlando, FL.

Biederman, H. and Osada, Y. (1992) *Plasma Polymerization Process*, Elsevier, Amsterdam.

Boenig, H.V. (1988) *Fundamentals of Plasma Chemistry and Technology*, Technomic Publishing, Lancaster, Pennsylvania.

Bol'shakov, A.A., Cruden, B.A., Mogul, R., *et al.* (2004) "Radio-frequency oxygen plasma as sterilization source." *AIAA Journal*, **42**, 823.

Boucher, R.M. (1980) "Seeded gas plasma sterilization method." US Patent N. 4207286.

Boudam, M.K., Moisan, M., Saoudi, B., Popovici, C., Gherardi, N., and Massines, F. (2006) "Bacterial spore inactivation by atmospheric-pressure plasmas in the presence or absence of UV photons as obtained with the same gas mixture." *Journal of Physics D: Applied Physics*, **39**(16), 3494.

Brissetoussa, J-L., Doubla, A., Hnatiuc, E., *et al.* (2008) "Chemical reactivity of discharges and temporal post-discharges in plasma treatment of aqueous media: examples of gliding discharge treated solutions." *Industrial & Engineering Chemistry Research*, **47**(16), 5761–5781.

Broadwater, W.T. (1973) "Sensitivity of three selected bacterial species to ozone." *Applied and Environmental Microbiology*, **26**, 391.

Bullett, N.A., Bullett, D.P., Truica-Marasecu, F., *et al.* (2004) "Polymer surface micropatterning by plasma and VUV-photochemical modification for controlled cell culture." *Applied Surface Science*, **235**, 395.

Burelson, G.R. (1975) "Inactivation of viruses and bacteria by ozone, with and without sonication." *Applied and Environmental Microbiology*, **29**, 340.

Burlica, R., Finney, W.C., Clark, R.J., and Locke, B.R. (2004) "Organic dye removal from aqueous solution by glidarc discharges." *Journal of Electrostatics*, **62**(4), 309–321.

Campbell, C.A., Snyder, F., Szarko, V., *et al.* (2006) "Water Treatment Using Plasma Technology," SD Report, Drexel University, Philadelphia, PA.

Chae, J.O., Desiaterik, Yu.N., and Amirov, R.H. (1996) Int. Workshop on Plasma Technologies for Pollution Control and Waste Treatment, Ed., MIT, Cambridge, MA.

Chakravarthy, K., Dobrynin, D., Fridman, G., Friedman, G., Murthy, S., and Fridman, A. (2011) "Cold spark discharge plasma treatment of inflammatory bowel disease in an animal model of ulcerative colitis." *Plasma Medicine*, **1**(1), 3–19.

Chan, C.M., Ko, T.M., and Hiraoka, H. (1996) "Polymer surface modification by plasmas and photons." *Surface Science Reports*, **24**, 3.

Chang, B.S. (2003) *Journal of Electrostatics*, **57**, 13.

Chernekhovskaia, N.E., Shishlo, V.K., Svistunov, B.D., and Svistunova, A.S. (2004), "Intra-bronchial NO-therapy in treatment of wound complications in tuberculosis patients." IV-th Army Medical Conference on Intensive Therapy and Profilactic Treatments of Surgical Infections, Moscow, Russia.

Chesnokova, N.V., Gundorova, R.A., Krvasha, O.I., *et al.* (2003) "Experimental investigation of the influence of gas flow containing nitrogen oxide in treatment of corneal wounds." *News of Russian Academy of Medical Sciences*, **5**, 40.

Chih Wei, C. (2008) "Inactivation of aquatic microorganisms by low-frequency AC discharges." *IEEE Transactions on Plasma Science*, **36**(1), 215–219.

Ching, W.K., Colussi, A.J., Sun, H.J., *et al.* (2001) "*Escherichia coli* Disinfection by Electrohydraulic Discharges." *Environmental & Science Technology*, **35**, 4139.

Ching, W.K., Colussi, A.J., and Hoffmann, M.R. (2003) "Soluble sunscreens fully protect *E. coli* from disinfection by electrohydraulic discharges." *Environmental Science & Technology*, **37**, 4901.

Cho, M., Chung, H., Choi, W., and Yoon, J. (2004) "Linear correlation between inactivation of *E. coli* and OH radical concentration in TiO_2 photocatalytic disinfection." *Water Research*, **38**, 1069.

Christophersen, A.G., Jun, H., Jørgensen, K., and Skibsted, L.H. (1991) "Photobleaching of astaxanthin and canthaxanthin." *Zeitschrift für Lebensmittel-Untersuchungund Forschung*, **192**, 433.

Chu, P.K., Chen, J.Y., Wang, L.P., and Huang, N. (2002) "Plasma-surface modification of biomaterials." *Materials Science and Engineering: R Reports*, **36**, 143.

Clement, F., Cambus, J., Panousis, E., Cousty, S., Ricard, A., Held, B. (2009) Study of human plasma coagulation when exposed to afterglows of atmospheric pressure pulsed DBD. Second International Conference on Plasma Medicine, ICPM-2, San Antonio, USA, pg. 64.

Clough, P. and Thrush, B. (1967) "Mechanism of chemiluminescent reaction between nitric oxide and ozone." *Transactions of the Faraday Society*, **63**, 915–925.

Clyne, M., Thrush, B. and Wayne, R. (1964) "Kinetics of the chemiluminescent reaction between nitric oxide and ozone." *Transactions of the Faraday Society*, **60**, 359–370.

Conti, S., Porshnev, P.I., Fridman, A., *et al.* (2001) "Experimental and numerical investigation of a capacitively coupled low-radio frequency nitrogen plasma." *Experimental Thermal and Fluid Science*, **24**, 79.

Cooper, M., Fridman, G., Staack, D., *et al.* (2009) "Decontamination of surfaces from extremophile organisms using nonthermal atmospheric-pressure plasmas." *IEEE Transactions on Plasma Science*, **37**(6), 866.

Cooper, M., Fridman, G., Fridman, A., and Joshi, S.G. (2010) "Biological responses of *Bacillus stratosphericus* to floating electrode-dielectric barrier discharge plasma treatment." *Journal of Applied Microbiology*, **109**, 2039.

Czernichowski, A. (1994) *Pure and Applied Chemistry*, **66**(6), 1301.

D'Agostino, R. (1990) *Plasma Deposition and Etching of Polymers*, Academic Press, Boston, Massachusetts.

D'Agostino, R., Favia, P., Oehr, C., and Wertheimer, M.R. (2003) *Plasma Processes and Polymers*, Wiley-VCH, Berlin.

Davydov, A.I., Strijakov, A.N., Pekshev, A.V., *et al.* (2002) "Possibilities and prospectives in plasma endosurgery with generation of nitrogen monoxide in vaginal surgeries." *Questions in Obstetrics and Gynaecology (OB/GYN)*, **1**(2), 57.

Davydov, A.I., Kuchukhidze, S.T., Shekhter, A.B., *et al.* (2004) "Clinical evaluation of intraoperative application of air-plasma flow enriched by nitrogen monoxide in operations on the uterus and adnexa." *Problems of Gynecology, Obstetrics and Perinatology*, **3**(4), 12.

Deminsky, M.A., Potapkin, B.V., Rusanov, V.D., and Fridman, A. (1990) "Possibility of SO2 Chain Oxidation in Heterogeneous Air Stream by Relativistic Electron Beams." Kurchatov Institute of Atomic Energy, vol. IAE-5260/12, Moscow.

Detomazo, L., Gristina, R., Favia, P., and d'Agostino, R. (2005) 17-th International Symposium on Plasma Chemistry, ISPC-17, Toronto, Canada, p. 1103.

Dobrynin, D., Fridman, G., Mukhin, Y.V., *et al.* (2010) "Cold plasma inactivation of *Bacillus cereus* and *Bacillus anthracis* (anthrax) spores." *IEEE Transactions on Plasma Science*, **38**(8), 1878.

Dobrynin, D., Fridman, G., Friedman, G., and Fridman, A. (2012) *Plasma Chemistry and Plasma Processing, PCPP*, **32**.

Doubla, A., Fotso, M., and Brisset, J.-L. (2008) "Plasmachemical decolourisation of Bromothymol Blue by gliding electric discharge at atmospheric pressure." *Dyes and Pigments*, **77**(1), 118–124.

Dragsund, E., Andersen, A.B., and Johannessen, B.O. (2001) "Ballast Water Treatment by Ozonation." 1st Int. Ballast Water Treatment, R&D Symposium, Global Monograph Series N.5, IMO, London, p. 21.

Efimenko, N.A., Hrupkin, V.I., Marahonich, L.A. *et al.* (2005) "Air-Plasma Flows and NO-therapy-a Novel Technology in a Clinical Setting of the Military Treatment and Profilactic Facilities (Experimental and Clinical Study)." *Journal of Military Medicine (Voenno-Meditsinskii Jurnal)*, **5**, 51.

Eloy, R., Parrat, D., Duc, T.M., *et al.* (1993) "In vitro evaluation of inflammatory cell response after CF4 plasma surface modification of poly(methyl methacrylate) intraocular lenses." *Journal of Cataract and Refractive Surgery*, **19**.

Evans, D., Rosocha, L.A., Anderson, G.K., *et al.* (1993) "Plasma remediation of trichloroethylene in silent discharge plasmas." *Journal of Applied Physics*, **74**, 5378.

Falkenstein, Z. (1997) "The influence of ultraviolet illumination on OH formation in dielectric barrier discharges of Ar/O/HO: The Joshi effect." *Journal of Applied Physics*, **81**, 7158.

Fang, F.C. (2004) "Antimicrobial reactive oxygen and nitrogen species: concepts and controversies." *Nature Reviews Microbiology*, **2**, 820.

Fix, A., Wirth, M., Meister, A., *et al.* (2002) "Tunable ultraviolet optical parametric oscillator for differential absorption lidar measurements of tropospheric ozone." *Applied Physics B: Lasers and Optics, B*, **75**, 153.

Fontijn, A., Sabadell, A.J. and Ronco, R.J. (1970) "Homogeneous chemiluminescent measurement of nitric oxide with ozone. Implications for continuous selective monitoring of gaseous air pollutants." *Analytical Chemistry*, **42**(6), 575–579.

Freeman, T., Steinbeck, M., Fridman, G., Zhang, J., Shainsky, N., Friedman, G., and Fridman, A. (2012) "Nonthermal DBD plasma enhances skeletal cell differentiation and autopod development." Fouth International Conference on Plasma Medicine, ICPM-4, Orleans, France, pg. 17.

Fridman, A. (2003) 16-th Int. Symposium on Plasma Chemistry, ISPC-16, Taormina, Italy, p. 8.

Fridman, A. (2008) *Plasma Chemistry*, Cambridge University Press, Cambridge, N.Y.

Fridman, A. and Kennedy, L.A. (2004) *Plasma Physics and Engineering*, Taylor & Francis, N.Y., London.

Fridman, A. and Kennedy, L.A. (Edition 1, 2004; Edition 2, 2011), *Plasma Physics and Engineering*, Taylor & Francis, N.Y., London.

Fridman, A., Nester, S., Kennedy, L., *et al.* (1998) "Gliding arc gas discharge." *Progress in Energy and Combustion Science*, **25**, 211.

Fridman, A., Gutsol, A., Dolgopolsky, A., and Stessel, E. (2006) "CO$_2$-free energy and hydrogen production from hydrocarbons." *Energy & Fuels*, **20**, 1242.

Fridman, A., Gutsol, A., and Cho, Y. (2007) Transport phenomena in plasma, in *Advances in Heat Transfer*, vol. **40** (eds A. Fridman and Y. Cho), Academic Press.

Fridman, A., Gutsol, A., Gangoli, S., *et al.* (2007a) *J. Propul. Power*, issue ed. S. Macheret, vol. **22**.

Fridman, G., Friedman, G., Gutsol, A., and Fridman, A. (2003) 16-th Int. Symposium on Plasma Chemistry, ISPC-16, Taormina, Italy, p. 703.

Fridman, G., Li, M., Lelkes, P.I., *et al.* (2005a) "Nonthermal plasma bio-active liquid micro and nano-xerography." *IEEE Transactions on Plasma Science*, **33**, 1061.

Fridman, G., Peddinghaus, L., Fridman, A., *et al.* (2005b) 17-th Int. Symposium on Plasma Chemistry (ISPC-17), Toronto, Canada, p. 1066.

Fridman, G., Peddinghaus, L., Ayan, H., *et al.* (2006) "Blood coagulation and living tissue sterilization by floating-electrode dielectric barrier discharge in air." *Plasma Chemistry and Plasma Processing*, **26**, 425.

Fridman, G., Brooks, A., Balasubramanian, M., *et al.* (2007b) "Comparison of direct and indirect effects of non-thermal atmospheric-pressure plasma on bacteria." *Plasma Processing and Polymers*, **4**, 425.

Fridman, G., Friedman, G., Gutsol, A., and Fridman, A. (2007c) 18-th Int. Symposium on Plasma Chemistry, ISPC-18, Kyoto, Japan.

Fridman, G., Shereshevsky, A., Jost, M.M., *et al.* (2007d) "Floating electrode dielectric barrier discharge plasma in air promoting apoptotic behavior in melanoma skin cancer cell lines." *Plasma Chemistry and Plasma Processing*, **27**, 163.

Futamura, S. and Yamamoto, T. (1996) *IEEE Transactions on Industry Applications*, **32**, 1969.

Gadri, R.B. (2000) "Sterilization and plasma processing of room temperature surfaces with a one atmosphere uniform glow discharge plasma (OAUGDP)." *Surface and Coating Technology*, **131**, 528.

Gallagher, M., Friedman, G., Dolgopolsky, A., *et al.* (2005) 17-th International Symposium on Plasma Chemistry, ISPC-17, Toronto, Canada, p. 1056.

Gallagher, M., Gutsol, A., Fridman, A., Friedman, G., and Dolgopolsky, A. (2004) "Non-thermal plasma applications in air sterilization." International Conference on Plasma Science, Baltimore, Maryland.

Gallagher, M., Vaze, N., Gangoli, S., *et al.* (2007) "Rapid inactivation of airborne bacteria using atmospheric pressure dielectric barrier grating discharge." *IEEE Transactions on Plasma Science*, **35**, 1501.

Gallagher, M.J., Vaze, N., Gangoli, S. *et al.* (2007) "Rapid inactivation of airborne bacteria using atmospheric pressure dielectric barrier grating discharge." *IEEE Transactions on Plasma Science*, **35**(5), 1501–1510.

Gangoli, S., Gallagher, M., Dolgopolsky, A., *et al.* (2005) 17-th International Symposium on Plasma Chemistry, ISPC-17, Toronto, Canada, p. 1111.

Garbassi, F., Morra, M., and Occhiello, E. (1994) *Polymer Surfaces, From Physics to Technology*, Wiley & Sons, N.Y.

Gebicki, S. and Gebicki, J.M. (1993) "Formation of peroxides in amino acids and proteins exposed to oxygen free radicals." *Biochemical Journal*, **289**(Pt 3), 743–749.

Gerenser, L.J., Grace, J.M., Apai, G., and Thompon, P.M. (2000) "Surface chemistry of nitrogen plasma-treated poly(ethylene-2,6-naphthalate): XPS, HREELS and static SIMS analysis." *Surface and Interface Analysis*, **29**, 12.

Ghaffari, A., Neil, D.H., Ardakani, A., *et al.* (2005) "A direct nitric oxide gas delivery system for bacterial and mammalian cell cultures." *Nitric Oxide*, **12**, 129.

Gilman, A.B. (2000) Interaction of Chemically Active Plasma with Surfaces of Synthetic Materials, in *Encyclopedia of Low-Temperature Plasma*, vol. **4** (ed. V.E. Fortov), Nauka (Science), Moscow, p. 393.

Gostev, V. and Dobrynin, D. (2006) "Medical microplasmatron." Medical microplasmatron. 3-rd International Workshop on Microplasmas (IWM-2006), Greifswald, Germany.

Gourmelon, M., Cillard, J. and Pommepuy, M. (1994) "Visible light damage to Escherichia coli in seawater: oxidative stress hypothesis." *Journal of Applied Microbiology*, **77**, 105.

Grace, J.M. (1995) "Use of glow discharge of treatment to promote adhesion of aqueous coats to substrate." US Patent N. 5,425,980.

Grace, J.M., Freeman, D.R., Corts, R., and Kozel, W. (1996) "Scaleup of a nitrogen glow-discharge process for silver–poly(ethylene terephthalate) adhesion." *Journal of Vacuum Science and Technology*, **A14**(3), 727.

Gregory, G., Citterio, S., Labra, M., *et al.* (2001) "Resolution of Viable and Membrane-Compromised Bacteria in Freshwater and Marine Waters Based on Analytical Flow Cytometry and Nucleic Acid Double Staining." *Applied and Environmental Microbiology*, **67**, 4662.

Grigorian, A.S., Grudyanov, A.I., Frolova, O.A., Antipova, Z.P., Yerokhin, A.I., Shekhter, A.B., and Pekshev, A.V. (2001) "Application of a new biological factor, exogenous nitric oxide, for the surgical treatment of periodontis." *Stomatology*, **80**, 80–3 (in Russian).

Gutsol, A., Tak, G., and Fridman, A. (2005) 17-th International Symposium on Plasma Chemistry, ISPC-17, Toronto, Canada, p. 1128.

Gutsol, A., Vaze, N., Arjunan, K., Gallagher, M., Yang, Y., Zhu, J., Vasilets, V., and Fridman, A. (2007) "Plasma for air and water sterilization." NATO Advanced Study Institute (ASI), Plasma Assisted Decontamination of Biological and Chemical Agents, Çesme, Turkey.

Hadidi, K., Cohn, D.R., Vitale, S., and Bromberg, L. (1996) Proceedings of the International Workshop on Plasma Technologies for Pollution Control and Waste Treatment, Beijing, China (ed. Y.K. Pu and P.P. Woskov), p. 166.

Harkness, J.B.L. and Fridman, A. (1999) *The Technical and Economic Feasibility of Using Low-Temperature Plasmas to Treat Gaseous Emissions from Pulp Mills and Wood Product Plants*, ed. National Council of the Paper Industry for Air and Stream Improvement, NCASI, Research Triangle Park, N.C., USA.

Heesch, E.J.M., Pemen, A.J.M., Huijbrechts, A.H.J., *et al.* (2000) "A fast pulsed power source applied to treatment of conducting liquids and air." *IEEE Transactions on Plasma Science*, **28**, 137.

Helfritch, D.J., Harmon, G., and Feldman, P. (1996) Emerging Solutions to VOC and Air Toxics Conference, A&WMA, Proceedings, p. 277.

Hemmerich, K. (1998) "General aging theory and simplified protocol for accelerated aging of medical devices." *Metal Plastics and Biomaterials*, **July**, p. 16.

Henle, E.S. and Linn, S. (1997) "Formation, prevention, and repair of dna damage by iron/hydrogen peroxide." *Journal of Biological Chemistry*, **272**(31), 19095–19098.

Herrmann, H.W., Henins, I., Park, J., and Selwyn, G.S. (1999) "Decontamination of chemical and biological warfare (CBW) agents using an atmospheric pressure plasma jet (APPJ)." *Physics of Plasmas*, **6**, 2284.

Hickson, I., Zhao, Y., Richardson, C.J., *et al.* (2004) "Identification and characterization of a novel and specific inhibitor of the Ataxia-Telangiectasia mutated kinase ATM." *Cancer Research*, **64**, 9152–9159.

Hocker, H. (1995) *Int. Text. Bull. Veredlung*, **41**, 18.

Hocker, H. (2002) "Plasma treatment of textile fibers." *Pure and Applied Chemistry*, **74**, 423.

Hong, Y.F., Kang, J.G., Lee, H.Y., *et al.* (2009) "Sterilization effect of atmospheric plasma on *Escherichia coli* and *Bacillus subtilis* endospores." *Letters in Applied Microbiology*, **48**, 33.

Hsiao, M.C., Meritt, B.T., Penetrante, B.M., *et al.* (1995) "Plasma-assisted decomposition of methanol and trichloroethylene in atmospheric pressure air streams by electrical discharge processing." *Journal of Applied Physics*, **78**, 3451.

Huth, K.C., Quirling, M., Maier, S. *et al.* (2009) "Effectiveness of ozone against endodontopathogenic microorganisms in a root canal biofilm model." *International Endodontic Journal*, **42**, 3–13.

Huth, K.C., Saugel, B., Jakob, F.M. *et al.* (2007) "Effect of aqueous ozone on the NF-kappaB system." *Journal of Dentistry Research*, **86**, 451–456.

Ikawa, S. (2010) "Effects of pH on bacterial inactivation in aqueous solutions due to low-temperature atmospheric pressure plasma application." *Plasma Processes and Polymers*, **7**(1), 33–42.

Imlay, J., Chin, S., and Linn, S. (1988) "Toxic DNA damage by hydrogen peroxide through the Fenton reaction in vivo and in vitro." *Science*, **240**(4852), 640–642.

Inagaki, N. (1996) *Plasma Surface Modification and Plasma Polymerization*, CRC Press, FL.

Ivanov, A.V. (2000) Plasma Production of Oxide Materials, in *Encyclopedia of Low-Temperature Plasma*, vol. **4** (ed. V.E. Fortov), Nauka (Science), Moscow, p. 354.

Iza, F. and Hopwood, J. (2003) "Low-power microwave plasma source based on a microstrip split-ring resonator." *IEEE Transactions on Plasma Science*, **31**, 782.

Iza, F., Kim, G.J., Lee, S.M., *et al.* (2008) "Microplasmas: Sources, Particle Kinetics, and Biomedical Applications." *Plasma Processes and Polymers*, **5**, 322.

Jacobs, P.T. and Lin, S.M. (1987) "Hydrogen peroxide plasma sterilization system." US Patent N. 4643876.

Jae Koo Lee, M.S.K., June Ho Byun, Kyong Tai Kim, Gyoo Cheon Kim4, and Gan Young Park (2011) "Biomedical applications of low temperature atmospheric pressure plasmas to cancerous cell treatment and tooth bleaching." *Japanese Journal of Applied Physics*, **50**(08JF0), 1–7.

Jaisinghani, R. (1999) "Bactericidal Properties of Electrically Enhanced HEPA Filtration and a Bioburden Case Study." InterPhex Conference, New York, NY.

Jeong, Y.C. and Swenberg, J.A. (2005) "Formation of M1G-dR from endogenous and exogenous ROS-inducing chemicals." *Free Radical Biology and Medicine*, **39**, 1021–1029.

Jiandong, L., Xianfu, X., Yao, S., and Tianen, T. (1996) *Int. Workshop on Plasma Technologies for Pollution Control and Waste Treatment*, MIT, Cambridge, MA.

Jiang, C., Chen, M.T., Gorur, P.A. *et al.* (2009a) "Nanosecond pulsed plasma dental probe." *Plasma Processes and Polymers*, **6**(8), 479–483.

Jiang, C., Chen, M.T., Gorur, A. *et al.* (2009b) "Atmospheric-pressure cold plasma for endodontic disinfection." *IEEE Transactions on Plasma Science*, **37**(7), 1190–1195.

Johansson, E., Claesson, R., and van Dijken, J.W. (2009) "Antibacterial effect of ozone on cariogenic bacterial species." *Journal of Dentistry*, **37**, 449–453.

Johnstone, R.W., Ruefli, A.A., and Lowe, S.W. (2002) "Apoptosis: A link review between cancer genetics and chemotherapy." *Cell*, **108**, 153.

Joshi, R.P., Hu, Q., Schoenbach, K.H., and Beebe, S.J. (2002a) "Simulations of electroporation dynamics and shape deformations in biological cells subjected to high voltage pulses." *IEEE Transactions on Plasma Science*, **30**, 1536.

Joshi, R.P., Qian, J., and Schoenbach, K.H. (2002b) "Electrical network-based time-dependent model of electrical breakdown in water." *Journal of Applied Physics*, **92**, 6245.

Joshi, S.G., Paff, M., Friedman, G., Fridman, G., Fridman, A., and Brooks, A.D. (2010) "Control of methicillin-resistant *Staphylococcus aureus* (MRSA) in planktonic form and biofilms: a biocidal efficacy study of non-thermal DBD-plasma." *American Journal of Infection Control*, **38**, 293–301.

Jung, J.-M., Yang, Y., Lee, D.H., *et al.* (2012) "Effect of Dielectric Barrier Discharge Treatment of Blood Plasma to Improve Rheological Properties of Blood." *Plasma Chemistry and Plasma Processing*, **32**.

Kabisov, R.K., Sokolov, V.V., Shekhter, A.B., *et al.* (2000) "Experience in application of exogenous NO therapy for treatment of postoperative wounds and radioreactions in oncological patients." *Russian Journal of Oncology*, **1**, 24.

Kalghatgi, S.U., Cooper, M., Fridman, G., *et al.* (2007) "Mechanism of Blood Coagulation by Non-Thermal Atmospheric Pressure Dielectric Barrier Discharge Plasma." 9-th Annual RISC, Drexel University, Philadelphia, PA.

Kalghatgi, S.U., Cooper, M., Fridman, G., *et al.* (2007) "Mechanism of blood coagulation by nonthermal atmospheric pressure dielectric barrier discharge plasma." *IEEE Transactions on Plasma Science*, **35**(5), 1559.

Kalghatgi, S., Kelly, C.M., Cerchar, E., *et al.* (2011) "Effects of non-thermal plasma on mammalian cells." *PLoS One*, **6**(1), e16270.

Karlson, E.L. (1989) "Ozone sterilization." *Journal of Healthcare MaterielManagement*, **7**, 43.

Katsuki, S., Moreira, K., Dobbs, F., *et al.* (2002) "Bacterial decontamination with nanosecond pulsed electric fields." *IEEE Journal* (8), 648.

Katsuki, S., Akiyama, H., Abou-Ghazala, A., and Schoenbach, K.H. (2002) *IEEE Trans. in Dielectrics and Electrical Insulation*, **9**, 498.

Keidar, M., Walk, R., Shashurin, A. *et al.* (2011) "Cold plasma selectivity and the possibility of a paradigm shift in cancer therapy." *British Journal of Cancer*, **105**, 1295–1301.

Kelly-Wintenberg, K. (2000) "An overview of research using the one atmosphere uniform glow discharge plasma (OAUGDP) for sterilization of surfaces and materials." *IEEE Transactions on Plasma Science*, **28**, 2000.

Kelly-Wintenberg, K., Montie, T.C., Brickman, C., *et al.* (1998) "Room temperature sterilization of surfaces and fabrics with a one atmosphere uniform glow discharge plasma." *Journal of Industrial Microbiology and Biotechnology*, **20**, 69.

Khadre, M.A., Yousef, A.E., and Kim, J.G. (2001) "Microbiological aspects of ozone applications in food: a review." *Journal of Food Science*, **66**, 9.

Kieft, I.E., Broers, J.L.V., Caubet-Hilloutou, V., *et al.* (2004) "Electric discharge plasmas influence attachment of cultured CHO K1 cells." *Bioelectromagnetics*, **25**, p. 362.

Kieft, I.E., v.d. Laan, E.P., and Stoffels, E. (2004) "Electrical and optical characterization of the plasma needle." *New Journal of Physics*, **6**, 149.

Kieft, I.E., Darios, D., Roks, A.J.M., and Stoffels, E. (2005) "Plasma treatment of mammalian vascular cells: a quantitative description." *IEEE Transactions on Plasma Science*, **33**, 771.

Kim, C.-H., Jae Hoon Bahn, Seong-Ho Leea, Gye-Yeop Kim, Seung-Ik Jun, Keunho Lee, Seung Joon Baek (2010) "Induction of cell growth arrest by atmospheric non-thermal plasma in colorectal cancer cells." *Journal of Biotechnology*, **150**, 530–538.

Klee, D. and Hocker, H. (1999) "Polymers for biomedical applications: improvement of the interface compatibility." *Advances in Polymer Blends*, **149**, 1.

Kogelschatz, U. (2003a) "Dielectric-barrier discharges: their history, discharge physics, and industrial applications." *Plasma Chemistry and Plasma Processing*, **23**(1), 1.

Kogelschatz, U. (2003b) *Plasma Chemistry and Plasma Processing*, **23**, 191.

Komanapalli, I.R. and B. Lau, H.S., (1996) "Ozone-induced damage of *Escherichia coli K-12*." *Applied Microbiology and Biotechnology*, **46**, p. 610.

Kowalski, W., Bahnfleth, W., and Whittam, T. (1998) "Bactericidal effects of high airborne ozone concentrations on *Escherichia coli* and *Staphylococcus aureus*." *Ozone: Science & Engineering*, **20**, 205.

Kramer, P.W., Yeh, Y.-S., and Yasuda, H. (1989) "Low temperature plasma for the preparation of separation membranes." *Journal of Membrane Science*, **46**, 1.

Krotovskii, G.S., Pekshev, A.V., Zudin, A.M., *et al.* (2002) *Cardio-Vascular Surgery (Grudnaia i serdechno-sosudistaia hirurgia)*, **1**, 37.

Kuchukhidze, S.T., Klihdukhov, I.A., Bakhtiarov, K.R., and Pankratov, V.V. (2004) *Problems of Obstetrics and Gynaecology (OB/GYN)*, **3**(2), 76–82.

Laroussi, M. (2005) "Low temperature plasma-based sterilization: overview and state-of-the-art." *Plasma Processes and Polymers*, **2**, 391.

Laroussi, M. and Leipold, F. (2004) "Evaluation of the roles of reactive species, heat, and UV radiation in the inactivation of bacterial cells by air plasmas at atmospheric pressure." *International Journal of Mass Spectrometry*, **233**, 81.

Laroussi, M., Sayler, G., Galscock, B., *et al.* (1999) "Images of biological samples undergoing sterilization by a glow discharge at atmospheric pressure." *IEEE Trans. Plasma Science*, **27**, 34.

Laroussi, M., Alexeff, I., and Kang, W. (2000) "Biological decontamination by nonthermal plasmas." *IEEE Trans. Plasma Science*, **28**, 184.

Laroussi, M., Richardson, J.P., and Dobbs, F.C. (2002) "Effects of nonequilibrium atmospheric pressure plasmas on the heterotrophic pathways of bacteria and on their cell morphology." *Applied Physics Letters*, **81**, 772.

Laroussi, M., Mendis, D.A., and Rosenberg, M. (2003) "Plasma interaction with microbes." *New Journal of Physics*, **5**, 41.

Lee, S.J., Seo, Y.C., Jurng, J., *et al.* (2004) "Mercury emissions from selected stationary combustion sources in Korea." *Science of the Total Environment*, **325**, 155.

Lerouge, S., Wertheimer, M.R., and Yahia, L.H. (2001) "Plasma Sterilization: A Review of Parameters, Mechanisms, and Limitations." *Plasmas and Polymers*, **6**, 201.

Levitsky, A.A., Macheret, S.O., and Fridman, A. (1983) Kinetic modeling of plasma-chemical processes stimulated by vibrational excitation, in *Chemical Reactions in Non-Equilibrium Plasma* (ed. L.S. Polak), Nauka (Science), Moscow, p. 3.

Levitsky, A.A., Macheret, S.O., Polak, L.S., and Fridman, A. (1983) *Sov. Phys., High Energy Chemistry, (Khimia Vysokikh Energij)*, **17**, 625.

Levitsky, S.M. (1957) "Potential of Space and Electrode Sputtering under High-Frequency Discharge in the Gas." *Soviet Physics Technical Physics, JTPh*, **27**, **970**, 1001.

Linder, W. and Davis, S. (1931) *J. Phys. Chem*, **35**, 3649.

Lipatov, K.V., Kanorskii, I.D., Shekhter, A.B., and Emelianov, A.Y. (2002) *Annals of Surgery*, **1**, 58.

Lipin, Yu.V., Rogachev, A.V., Sydorsky, S.S., and Kharitonov, V.V. (1994) *Technology of Vacuum Metallization of Polymer Materials*, Belorussian Academy of Engineering and Technology, Gomel, Belarus.

Lisitsyn, I.V., Nomiyama, H., Katsuki, S., and Akiyama, H. (1999) *Review of Sci. Instruments*, **70**, 3457.

Liston, E.M. (1993) *Journal of Adhesion Science and Technology*, **7**, 1091.

Liston, E.M., Martinu, L., and Wertheimer, M.R. (1994) *Plasma Surface Modification of Polymers: Relevance to Adhesion* (ed. M. Strobel, C. Lyons, and K.L. Mittal), VSP, Netherlands.

Locke, B.R., Sato, M., Sunka, P., *et al.* (2006) *Ind. Eng. Chem. Res.*, **45**, 882.

Lopez, L.C., Gristina, R., Favia, P., and d'Agostino, R. (2005) 17-th International Symposium on Plasma Chemistry, ISPC-17, Toronto, Canada, p. 1078.

Lu, X., Yinguang Cao, Ping Yang, Qing Xiong, Zilan Xiong, Yubin Xian, Yuan Pan Coll (2009) "An RC plasma device for sterilization of root canal of teeth." *Plasma Science*, **37**, 668–673.

Lu, X.P. and Laroussi, M. (2006) "Dynamics of an atmospheric pressure plasma plume generated by submicrosecond voltage pulses." *Journal of Applied Physics*, **100**, 302.

Lu, X.P., Leipold, F., and Laroussi, M. (2003) "Optical and electrical diagnostics of a non-equilibrium air plasma." *Journal of Physics D: Applied Physics*, **36**, 2662.

Marnett, L. (1999) "Lipid peroxidation—DNA damage by malondialdehyde." *Mutation Research: Fundamental and Molecular Mechanisms of Mutagenesis*, **424**, 83–95.

Marouf-Khelifa, K., Khelifa, A., Belhadj, M., Addou, A., and Brisset, J. (2006) "Reduction of nitrite by sulfamic acid and sodium azide from aqueous solutions treated by gliding arc discharge." *Separation and Purification Technology*, **50**(3), 373–379.

Marxer, C.G. (2002) "Protein and Cell Absorption: Topographical Dependency and Adlayer Viscoelastic Properties Determined with Oscillation Amplitude of Quartz Resonator." Department of Physics, University of Fribourg, Fribourg, Switzerland.

Masuda, S. (1993) *Non-Thermal Plasma Techniques for Pollution Control: Part B – Electron Beam and Electrical Discharge Processing* (ed. B.M. Penetrante and S.E. Schultheis), Springer-Verlag, Berlin.

Masuda, S., Wu, Y., Urabe, T., and Ono, Y. (1987) 3-rd Int. Conf. on Electrostatic Precipitation, Abano-Padova, Italy.

Mattachini, F., Sani, E., and Trebbi, G. (1996) *Int. Workshop on Plasma Technologies for Pollution Control and Waste Treatment*, MIT, Cambridge, MA.

Matzing, H. (1989) "The kinetics of O atoms in the radiation treatment of waste gases." *International Journal of Radiation Applications and Instrumentation. Part C. Radiation Physics and Chemistry*, **33**, 81.

Maximov, A.I. (2000) *Encyclopedia of Low-Temperature Plasma*, vol. **4** (ed. V.E. Fortov), Nauka (Science), Moscow, p. 399.

Mendis, D.A., Rosenberg, M., and Azam, F. (2000) "A note on the possible electrostatic disruption of bacteria." *IEEE Transactions on Plasma Science*, **28**, 1304.

Minayeva, O. and Laroussi, M. (2004) *Proc. IEEE Int. Conf. on Plasma Science (ICOPS)*, IEEE Press, Baltimore, MD, p. 122.

Misyn, F.A. and Gostev, V.A. (2000) Cold plasma application in eyelid phlegmon curing, in *Diagnostics and Treatment of Infectious Diseases*, ed. Petrozavodsk University, Petrozavodsk, Russia.

Misyn, F.A., Besedin, E.V., Komkova, O.P., and Gostev, V.A. (2000) Experimental investigation of bactericidal influence of cold plasma and its interaction with cornea, in *Diagnostics and Treatment of Infectious Diseases*, ed. Petrozavodsk University, Petrozavodsk, Russia.

Moisan, M., (2002) "Plasma sterilization. methods and mechanisms." *Pure and Applied Chemistry*, **74**(3), 349.

Moisan, M., Barbeau, J., Moreau, S., *et al.* (2001) "Low-temperature sterilization using gas plasmas: a review of the experiments and an analysis of the inactivation mechanisms." *International Journal of Pharmaceutics*, **226**, 1.

Montie, T.C., Kelly-Wintenberg, K., and Roth, J.R. (2000) "An overview of research using the one atmosphere uniform glow discharge plasma (OAUGDP) for sterilization of surfaces and materials." *IEEE Transactions on Plasma Science*, **28**, 41.

Moreau, M., Veron, W., Meylheuc, T., Chevalier, S., Brisset, J.-L. and Orange, N. (2007) "Gliding arc discharge in the potato pathogen erwinia carotovora subsp. atroseptica: mechanism of lethal action and effect on membrane-associated molecules." *Applied Environmental Microbiology*, **73**(18), 5904–5910.

Moreau, S., Moisan, M., Barbeau, J., *et al.* (2000) "Using the flowing afterglow of a plasma to inactivate *Bacillus subtilis spores*: influence of the operating conditions." *Journal of Applied Physics*, **88**, 1166.

Moussa, D., Abdelmalek, F., Benstaali, B., Addou, A., Hnatiuc, E., and Brisset, J.-L. (2005) "Acidity control of the gliding arc treatments of aqueous solutions: application to pollutant abatement and biodecontamination." *European Physical Journal of Applied Physics*, **29**(2), 189–199.

Muranyi, P., Wunderlich, J., and Heise, M. (2008) "Effect of humidity on plasma inactivation of spores." *Journal of Applied Microbiology*, **104**, 1659.

Mutaf-Yardimci, O., Kennedy, L., Saveliev, A., and Fridman, A. (1998) *Plasma Exhaust Aftertreatment*, SAE, SP-1395, p. 1.

Nagayoshi, M., Fukuizumi, T., Kitamura, C., Yano, J., Terashita, M., and Nishihara, T. (2004) "Efficacy of ozone on survival and permeability of oral microorganisms." *Oral Microbiology and Immunology*, **19**, 240–246.

Neely, W.C., Newhouse, E.I., Clothiaux, E.J., and Gross, C.A. (1993) *Non-Thermal Plasma Techniques for Pollution Control: Part B – Electron Beam and Electrical Discharge Processing* (eds B.M. Penetrante and S.E. Schultheis), Springer-Verlag, Berlin.

Nelea, V., Luo, L., Demers, C.N., *et al.* (2005) *J. Biomed. Res.*, **75A**, 216.

Neumann, E., Sowers, A.E., and Jordan, C.A. (ed.) (2001), *Electroporation and Electrofusion in Cell Biology*, Springer-Verlag, Berlin, Heidelberg.

Niedernhofer, L.J. *et al.* (2003) "Malondialdehyde, a product of lipid peroxidation, is mutagenic in human cells." *Journal of Biological Chemistry*, **278**, 31426–31433.

Oda, T., Takahashi, T., Nakano, H., and Masuda, S. (1991) The 1991 IEEE Industrial Application Society Meeting, Dearborn, MI, p. 734.

Oda, T., Han, S.B., and Ono, R. (2005) 17-th Int. Symposium on Plasma Chemistry, ISPC-17, Toronto, Canada, p. 1180.

Oehmigen, K., Brandenburg, R., Wilke, C., Weltmann, K.-D. and von Woedtke, T. (2010) "The role of acidification for antimicrobial activity of atmospheric pressure plasma in liquids." *Plasma Processing and Polymers*, **7**, 250–257.

Oehr, C. (2005) 17-th Int. Symposium on Plasma Chemistry, ISPC-17, Toronto, Canada, p. 8.

Ohmi, T., Isagawa, T., Imaoka, T. and Sugiyama, I. (1992) "Ozone decomposition in ultrapure water and continuous ozone sterilization for a semiconductor ultrapure water system." *Journal of the Electrochemical Society*, **139**, 3336.

Oizumi, M., Suzuki, T., Uchida, M., Furuya, J., and Okamoto, Y. (1998) "In vitro testing of a denture cleaning method using ozone." *Journal of Medical and Dental Sciences*, **45**, 135–139.

Okazaki, S. and Kogoma, M. (1993) "Development OP atmospheric pressure glow discharge plasma and its application on a surface with curvature." *Journal Photopolymer Science Technology*, **6**, 339.

Ono, R. and Oda, T. (2001) "OH radical measurement in a pulsed arc discharge plasma observed by a LIF method." *IEEE Transactions on Industry Applications*, **37**, 3.

Park, J., Henins, I., Herrmann, H.W., *et al.* (2001) "Gas breakdown in an atmospheric pressure radio-frequency capacitive plasma source." *Journal of Applied Physics*, **89**(1).

Paur, H.-R. (1991) European Aerosol Conference, Proceedings, KFK, Karlsruhe, Germany.

Paur, H.-R. (1999) *Non-Thermal Plasma Techniques for Pollution Control* (ed. B.M. Penetrante and S.E. Schultheis), Springer-Verlag, Berlin, p. 77.

Pekshev, A.V. (2001) "Use of a novel biological factor - exogenic nitrogen oxide - in surgical treatment of paradontitis." *Stomatology*, **1**, 80.

Penetrante, B.M. and Schultheis, S.E. (ed.) (1993) *Non-Thermal Plasma Techniques for Pollution Control: Part B – Electron Beam and Electrical Discharge Processing*, ed. by Springer-Verlag, Berlin.

Penetrante, B.M., Hsiao, M.C., Bardsley, J.N., *et al.* (1996a) *Pure and Applied Chemistry*, **68**, 1868.

Penetrante, B.M., Hsiao, M.C., Bardsley, J.N., *et al.* (1996b) Proceedings of the International Workshop on Plasma Technologies for Pollution Control and Waste Treatment, Beijing, China (ed. Y.K. Pu and P.P. Woskov), p. 99.

Penetrante, B.M., Hsiao, M.C., Bardsley, J.N., *et al.* (1997) "Identification of mechanisms for decomposition of air pollutants by non-thermal plasma processing." *Plasma Sources Science and Technology*, **6**, 251.

Philip, N., Saoudi, B., Crevier, M.C. *et al.* (2002) "The respective roles of UV photons and oxygen atoms in plasma sterilization at reduced gas pressure: the case of N/sub 2/-O/sub 2/ mixtures." *IEEE Transactions on Plasma Science*, **30**, 1429.

Poncin-Epaillard, F., and Legeay, G., J. (2003) "Surface engineering of biomaterials with plasma techniques." *Journal of Biomaterials Science, Polymers Edition*, **14**, 1005.

Ponomarev, A.N. (1982) *Heat and Mass Transfer in Plasma-Chemical Processes*, vol. **1**, Nauka (Science), Minsk, Belarus, p. 137.

Ponomarev, A.N. (1996) "Fundamentals of Energy-Saving Technologies Based on Non-Equilibrium Low-Temperature Plasma." *News of Russian Academy of Sciences*, **6**, 78.

Ponomarev, A.N. (2000) *Encyclopedia of Low-Temperature Plasma*, vol. **4** (ed. V.E. Fortov), Nauka (Science), Moscow, p. 386.

Ponomarev, A.N., Maksimov, A.I., Vasilets, V.N. and Menagarishvily, V.M. (1989) "Photo-oxidation of polyethylene and polyvinyl chloride in the process of simultaneous action of ultraviolet and active oxygen." *High Energy Chemistry*, **23**(3), 231–232.

Ponomarev, A.N. and Vasilets, V.N. (2000) *Encyclopedia of Low-Temperature Plasma*, vol. **3** (ed. V.E. Fortov), Nauka (Science), Moscow, p. 374.

Ponomarev, A.N., Vasilets, V.N., and Talrose, R.V. (2002) *Chemical Physics (Russian Journal)*, **21**(4), 96.

Porter, J., Edwards, C., Morgan, J.A.W., and Pickup, R.W. (1993) "Rapid, automated separation of specific bacteria from lake water and sewage by flow cytometry and cell sorting." *Applied Environmental Microbiology*, **59**, 3327.

Potapkin, B.V., Rusanov, V.D., and Fridman, A. (1989) *Sov. Phys., Doklady, (Reports of USSR Academy of Sciences)*, **308**, 897.

Potapkin, B.V., Deminsky, M., Fridman, A., and Rusanov, V.D. (1993) *Non-Thermal Plasma Techniques for Pollution Control*, NATO ASI Series, vol. **G 34**, Part A (ed. B.M. Penentrante and S.E. Schultheis), Springer-Verlag, Berlin.

Potapkin, B.V., Deminsky, M., Fridman, A., and Rusanov, V.D. (1995) "The effect of clusters and heterogeneous reactions on non-equilibrium plasma flue gas cleaning." *Radiatation Physics and Chemistry*, **45**, 1081.

Pu, Y.K. and Woskov, P.P. (1996) Int. Workshop on Plasma Technologies for Pollution Control and Waste Treatment, MIT, Cambridge, MA.

Raiser, J. and Zenker, M. (2006) "Argon plasma coagulation for open surgical and endoscopic applications : state of the art." *Journal of Physics D: Applied Physics*, **39**, 3520.

Ramires, P.A., Mirenghi, L., Romano, A.R., *et al.* (2000) "Plasma-treated PET surfaces improve the biocompatibility of human endothelial cells." *Journal of BiomedicalMaterials Research*, **51**, 535.

Reshetov, I.V., Kabisov, R.K., Shekhter, A.B., *et al.* (2000) *Annals of Plastic, Reconstructive and Aesthetic Surgery*, **4**, 24.

Ritts, A., Lin, J., Li, H., Yu, Q., Xu, C., Yao, X., Hong, L., and Wang, Y. (2010). "Dentin surface treatment using a non-thermal plasma brush for interfacial bonding improvement in composite restoration." *European Journal of Oral Sciences*, **118**, 510–516.

Robert, E., Barbosa, E., Dozias, S., *et al.* (2009) "Experimental Study of a Compact Nanosecond Plasma Gun." *Plasma Processes and Polymers*, **6**, 795.

Rosocha, L.A., Anderson, G.K., Bechtold, L.A., *et al.* (1993) *Non-Thermal Plasma Techniques for Pollution Control: Part B – Electron Beam and Electrical Discharge Processing* (ed. B.M. Penentrante and S.E. Schultheis), Springer-Verlag, Berlin.

Roth, J.R. (1995) *Industrial Plasma Engineering*, vol. **1**, Principles, Institute of Physics Publishing, Bristol, Philadelphia.

Roth, J.R. (2001) *Industrial Plasma Engineering*, vol. **2**, Applications for Non-Thermal Plasma Processing, Institute of Physics Publishing, Bristol, Philadelphia.

Roth, J.R. (2006) "The One Atmosphere Uniform Glow Discharge Plasma (OAUGDP) – a Platform Technology for the 21st Century." Plenary talk at the 33-rd IEEE International Conference on Plasma Science (ICOPS-2006), Traverse City, Michigan, USA.

Rupf, S., Lehmann, A., Hannig, M., Schäfer, B., Schubert, A., Feldmann, U., and Schindler, A. (2010) "Killing of adherent oral microbes by a non-thermal atmospheric plasma jet." *Journal of Medical Microbiology*, **59**, 206–212.

Rusanov, V.D. and Fridman, A. (1978) *Sov. Phys., Journal of Physical Chemistry (JFCh)*, **52**, 92.

Rusanov, V.D. and Fridman, A. (1978) *Sov. Phys., Letters to Journal of Technical Physics (Pis'ma v JTF)*, **4**, 28.

Rusanov, V.D. and Fridman, A. (1984) *Physics of Chemically Active Plasma*, Nauka (Science), Moscow.

Sakiyama, Y. and Graves, D.B. (2006) "Finite element analysis of an atmospheric pressure RF-excited plasma needle." *Journal of Physics D: Applied Physics*, **39**, 3451.

Sardella, E., Gristina, R., Gilliland, D., *et al.* (2005) 17-th Int. Symposium on Plasma Chemistry, ISPC-17, Toronto, Canada, p. 608.

Sarron, V., Robert, E., Dozias, S., *et al.* (2011) "Splitting and mixing of high-velocity ionization-wave-sustained atmospheric-pressure plasmas generated with a plasma gun." *IEEE Transactions on Plasma Science*, **39**(11), 2356.

Sato, M., Ohgiyama, T., and Clements, J.S. (1996) "Formation of chemical species and their effects on microorganisms using a pulsed high-voltage discharge in water." *IEEE Trans. Ind. Appl.*, **32**, 106.

Satoh, K., Anderson, J.G., Woolsey, G.A., and Fouracre, R.A. (2007) "Pulsed-plasma disinfection of water containing Escherichia coli." *Japanese Journal of Applied Physics*, **46**(3A), 1137–1141.

Schoenbach, K.H., El-Habachi, A., Shi, W., and Ciocca, M. (1997) "High-pressure hollow cathode discharges." *Plasma Sources Science and Technology*, **6**, 468.

Schoenbach, K.H., Joshi, R.P., Stark, R.H., *et al.* (2000) "Bacterial decontamination of liquids with pulsed electric fields." *IEEE Transactions on Dielectrics and Electrical Insulation*, **7**, 637.

Seeger, W. (2005) *Deutsche Medizinische Wochenschrift*, **130**(25–26), 1543.

Sensenig, R., Kalghatgi, S., Cercha, E. *et al.* (2011) "Non-thermal plasma induces apoptosis in melanoma cells via production of intracellular reactive oxygen species." *Annals of Biomedical Engineering*, **39**(2), 674–687.

Sevastianov, V.I., (1991), *High-Performance Biomaterials*, ed. Pennsylvania Technomic, PA.

Shainsky, N., Dobrynin, D., Ercan, U., Joshi, S., Fridman, G., Friedman, G., and Fridman, A. (2010) "Effect of liquid modified by non-equilibrium atmospheric pressure plasmas on bacteria inactivation rates." 37th IEEE International Conference on Plasma Science (ICOPS), June 20–24, Norfolk, VA, USA.

Shainsky, N., Dobrynin, D., Ercan, U. *et al.* (2012) "Plasma acid: water treated by dielectric barrier discharge." *Plasma Processes and Polymers*, doi: 10.1002/ppap.201100084.

Shekhter, A.B., Kabisov, R.K., Pekshev, A.V., *et al.* (1998) "Experimental and clinical validation of plasmadynamic therapy of wounds with nitricoOxide." *Bulletin of Experimental Biology and Medicine*, **126**(8), 829.

Shekhter, A.B., Serezhenkov, V.A., Rudenko, T.G., *et al.* (2005) "Beneficial effect of gaseous nitric oxide on the healing of skin wounds." *Nitric Oxide-Biology and Chemistry*, **12**, 210.

Shulutko, A.M., Antropova, N.V., and Kryuger, Y.A. (2004) "NO-therapy in the treatment of purulent and necrotic lesions of lower extremities in diabetic patients." *Surgery*, **12**, 43.

Silveira, A.M., Lopes, H.P., Siqueira, J.F. Jr, Macedo, S.B., Consolaro, A. (2007) "Periradicular repair after two-visit endodontic treatment using two different intracanal medications compared to single-visit endodontic treatment." *Brazilian Dental Journal*, **18**, 299–304.

Silverthorn, D.U., Garrison, C.W., Silverthorn, A.C., and Johnson, B.R. (2004) *Human Physiology, an Integrated Approach*. 3rd ed., Benjamin-Cummings Publishing Company.

Simon, F., Hermel, G., Lunkwitz, D., *et al.* (1996) "Surface modification of expanded poly(tetrafluoroethylene) by means of microwave plasma treatment for improvement of adhesion and growth of human endothelial cells." *Macromolecular Symposia*, **103**, 243.

Sladek, R.E.J., Baede, T.A., and Stoffels, E. (2006) "Plasma-needle treatment of substrates with respect to wettability and growth of *Escherichia coli* and *Streptococcus mutans*." *IEEE Transactions on Plasma Science*, **34**, 1325.

Sladek, R.E. and Stoffels, E. (2005) "Deactivation of Escherichia coli by the plasma needle." *Journal of Physics, D: Applied Physics*, **38**, 1716–1721.

Sladek, R.E., Stoffels, E., Walraven, R., Tielbeek, P., and Koolhoven, A. (2004) "Plasma treatment of dental cavities: A feasibility study." *Plasma Science*, **32**, 1540–1543.

Slovetsky, D.I. (1981) *Plasma Chemistry*, vol. **8** (ed. B.M. Smirnov), Energo-Izdat, Moscow, p. 181.

Sobacchi, M.G., Saveliev, A.V., Fridman, A., *et al.* (2003) "Experimental assessment of pulsed corona discharge for treatment of VOC emissions." *Plasma Chemistry and Plasma Processing*, **23**, 347.

Spahn, R.G. and Gerenser, L.J. (1994) "Ion selective electrode." US Patent N. 5,324,414.

Stadler, J., Biliar, T.R, Curran, R.D., Stuehr, D.J., Ochoa, J.B., and Simmons, R. (1991) "Effect of exogenous and endogenous nitric oxide on mitochondrial respiration of hepatocytes." *American Journal of Physiology*, **260**, C910–C916.

Starikovskiy, A., Yang, Y., Cho, Y., and Fridman, A. (2011) "Non-equilibrium plasma in liquid water: dynamics of generation and quenching." *Plasma Sources Science and Technology*, **20**, 024003.

Stoffels, E. (2003) *Proceedings Gaseous Electron. Conference*, AIP, San Francisco, CA, p. 16.

Stoffels, E. (2006) "Gas plasmas in biology and medicine." *Journal of Physics D: Applied Physics*, **39**, 16.

Sugar, I.P. and Neumann, E., (1984) "Stochastic model for electric field-induced membrane pores electroporation." *Biophysical Chemistry*, **19**, 21.

Sugiarto, A.T., Ohshima, T., and Sato, M. (2002) "Advanced oxidation processes using pulsed streamer corona discharge in water" *Thin Solid Films*, **407**, 174.

Sugiarto, A.T., Ito, S., Ohshima, T., *et al.* (2003) "Oxidative decoloration of dyes by pulsed discharge plasma in water." *Journal of Electrostatics*, **58**, 135.

Sun, B., Sato, M. and Clements, J.S. (1997) "Optical study of active species produced by a pulsed streamer corona discharge in water." *Journal of Electrostatics*, **39**, 189.

Sun, B., Sato, M. and Clements, J.S. (1999) "Use of a pulsed high-voltage discharge for removal of organic compounds in aqueous solution." *Journal of Physics D: Applied Physics*, **32**, 1908.

Tang, Y.Z., Laroussi, M., and Dobbs, F.C. (2008) "Sublethal and killing effects of atmospheric-pressure, nonthermal plasma on eukaryotic microalgae in aqueous media." *Plasma Processes and Polymers*, **5**, 552–558.

Terlinger, J.G.A., Feijen, J., and Hoffman, A.S. (1993) *J. Colloid Interface Sci.*, **155**, 55.

Turnipseed, A.A., Barone, S.B., and Ravishenkara, A.R. (1996a) "Reaction of OH with dimethyl sulfide. 2. products and mechanisms." *Journal of Physical Chemistry*, **100**, 14,694.

Turnipseed, A.A., Barone, S.B., Ravishenkara, A.R. (1996b) "Reaction of OH with dimethyl sulfide. 2. products and mechanisms." *Journal Physical Chemistry*, **100**, 14,703.

Tusov, L.S., Kulikov, V.V., Kolotyrkin, Ya.M., and Tunitsky, N.N. (1975) 2-nd USSR Symposium on Plasma Chemistry, Riga, Latvia, vol. 1, p. 234.

Vandamme, M., Robert, E., Pesnel, S. *et al.* (2010) "Antitumor effect of plasma treatment on U87 glioma xenografts: preliminary results." *Plasma Processes and Polymers*, **7**, 264–273.

Vandamme, M., Robert, E., Lerondel, S. *et al.* (2012) "ROS implication in a new antitumor strategy based on non-thermal plasma." *International Journal of Cancer*, **130**, 2185–2194.

Vanin, A.F. (1998) *Biochemistry*, **63**, 867.

Vasilets, V.N. (2005) "Modification of Physical Chemical and Biological Properties of Polymer Materials Using Gas-Discharge Plasma and Vacuum Ultraviolet Radiation." Dr. Sci. Dissertation, Institute of Energy Problems of Chemical Physics, Russian Academy of Sciences, Moscow.

Vasilets, V.N., Tikhomirov, L.A., and Ponomarev, A.N. (1979) *Sov. Phys., High Energy Chemistry (Khimia Vysokikh Energiy)*, **13**, 475.

Vasilets, V.N., Tikchomirov, L.A., and Ponomarev, A.N. (1981) "The study of continues HF low pressure discharge plasma action on the polyethylene surface." *High Energy Chemistry*, **15**(N1), 77–81.

Vasilets, V.N., Werner, C., Hermel, G., *et al.* (2002).

Vasilets, V.N., Kuznetsov, A.V., and Sevastianov, V.I. (2006) "Regulation of the biological properties of medical polymer materials with the use of a gas-discharge plasma and vacuum ultraviolet radiation." *High Energy Chemistry (Khimia Vysokikh Energiy)*, **40**, 105.

Von Gunten, U. (2003) "Ozonation of drinking water: Part I. Oxidation kinetics and product formation." *Water Research*, **37**, 1469.

Walsh, J.L., Iza, F., Janson, N.B., *et al.* (2010) "Three distinct modes in a cold atmospheric pressure plasma jet." *Journal of Physics, D: Applied Physics*, **43**, 075201.

Ward, I.M., Minn, K., and Chen, J. (2004) "UV-induced Ataxia-telangiectasia-mutated and Rad3-related (ATR) activation requires replication stress." *Journal of Biological Chemistry*, **279**, 9677–9680.

Wolfe, R.L. (1990) "Ultraviolet disinfection of potable water." *Environmental Science Technology*, **24**, 768.

Wu, Z.L., Gao, X., Luo, Z.Y., *et al.* (2005) "NO treatment by DC corona radical shower with different geometric nozzle electrodes." *Energy Fuels*, **19**, 2279.

Yamamoto, T., Ramanathan, K., Lawless, P.A., *et al.* (1992) "Control of volatile organic compounds by an AC energized ferroelectric pellet reactor and a pulsed corona reactor." *IEEE Transactions on Industry Applications*, **28**, 528.

Yamamoto, T., Lawless, P.A., Owen, M.K., *et al.* (1993) *Non-Thermal Plasma Techniques for Pollution Control: Part B – Electron Beam and Electrical Discharge Processing* (ed. B.M. Penetrante and S.E. Schultheis), Springer-Verlag, Berlin.

Yamamoto, M., Nishioka, M., and Sadakata, M. (2001) 15-th International Symposium on Plasma Chemistry, ISPC-15, Orleans, France, vol. 2, p. 743.

Yan, K., Yamamoto, T., Kanazawa, S., *et al.* (2001) "NO removal characteristics of a corona radical shower system under DC and AC/DC superimposed operations." *IEEE Transactions on Industry Applications*, **37**, 1499.

Yang, Y., Cho, Y., and Fridman, A., (2012) *Plasma Discharge in Liquid*, CRC Press.

Yasuda, H. (1985) *Plasma Polymerization*, Academic Press, Orlando, USA.

Yavirach, P., Chaijareenont, P., Boonyawan, D., Pattamapun, K., Tunma, S., Takahashi, H., and Arksornnukit, M. (2009) "Effects of plasma treatment on the shear bond strength between fiber reinforced composite posts and resin composite for core build-up." *Dental Materials Journal*, **28**(6), 686–692.

Yildirim, E.D., Ayan, H., Vasilets, V.N., *et al.* (2007) *Plasma Processing and Polymers*, **4**.

Zaitsev, V.M. (2003) "Comparative analysis of effectiveness of treatment of chronic supported ulcers by NO-therapy with addition of ultrasound implantation of miramistine with treatment by ultrasound implantation alone." *Russian Journal of Otorinolaringology*, **1**, 58.

Zheng, S., Zhang, L., Liu, Y., *et al.* (2009) *Vacuum*, **83**, 238.

Zirnheld, J.L., Zucker, S.N., DiSanto, T.M., Berezney, R., Etemadi, K. (2010) "Nonthermal plasma needle: development and targeting of melanoma cells." *IEEE Transactions on Plasma Science*, **38**(4), 948–952.

Index

α-pinene removal from air 368
$\alpha - \gamma$ transition 225

abnormal glow discharge 184
absorption coefficient in continuum 56
absorption of electromagnetic waves 78
acetone removal from air 364
acidic behavior of non-thermal plasma 45
acidic droplets 356
Acinetobacter baumannii 443
active species in tissue 324
acute wounds 405
adenosine triphosphate 93, 95, 111
adiabatic principle 24
adipocytes 108, 131
agar 337
aging in plasma-polymer treatment 471
airborne bacteria sterilization 343
air cleaning from NOx 353
air cleaning from SO2 353
air cleaning from VOC 361
air sterilization sampling 340
albumin 393
Alfven velocity 73
ambipolar diffusion 54
amino acid synthesis 3
anode dark space 178
anode-directed streamer 170
anode glow 178
anode layer of arc discharge 193
anode layer of glow discharge 177
anticoagulants 392
antigen 134, 139–41, 143–4
anti-inflammatory treatment 418
apoptosis 320, 425
applied plasma medicine 15
arc configurations 199
arc discharge 191
arc jet 202

arc with hot cathode spots 192
array of microdischarges 239
arterial ulcers 406
Aspergilus Niger spores 353
associative electron detachment 28
Aston dark space 178
asymmetric DBD discharge 235
atmospheric pressure glow discharge, APG 186
atmospheric pressure plasma jet, APPJ 224
attachment of osteoblasts 449
auto-displacement voltage 218
avalanche 168
Avogadro number 271

Bacillus anthracis (anthrax) 295, 297
Bacillus atrophaeus 337
Bacillus cereus 297
Bacillus stratosphericus 329
Bacillus subtilis spores 294
Bacillus thuringiensis 297
bacteria deactivation by plasma afterglow 294
bactericide effect 379
basic droplets 357
bed sores 407
Bethe-Bloch formula 23
biocompatibility 447
biological property regulations 447
bioluminescent imaging 427
biomaterials 447
biomolecular processes 15
biopolymers 81
Bjerrum length 271
blastoma 423
Bloch-Bradbury mechanism 27
blood clotting factors 391
blood coagulation 389
blood property control 395
blood proteins 391
blood treatment by plasma 389

Plasma Medicine, First Edition. Alexander Fridman and Gary Friedman.
© 2013 John Wiley & Sons, Ltd. Published 2013 by John Wiley & Sons, Ltd.

blood viscosity control 395
bohm velocity 71
Boltzmann distribution 46
Boltzmann vibrational-rotational distribution 48
bond dipole moment 83
bond energies 82
bone cells on polymer scaffolds 448
boomerang resonance 32
born approximation 33
Bouguer law 78
bovine spongiform encephalopathy 303
Breit-Wigner formula 32
bremsstrahlung 56
burns 406

cancer treatment 422
Candida albicans 431
Candida glabrata 443
capacitively-coupled plasma 209
capillary plasma electrode discharge 240
carbohydrates 84–6, 100, 112–13, 133–4, 142, 147
carcinoma 423
cathode dark space 178
cathode-directed streamer 169
cathode glow 178
cathode layer of arc discharge 193
cathode layer of glow discharge 177
cathode spot of arc discharge 193
cathode spot of glow discharges 184
cauterization devices 389
CCP coupling 213
CCP discharge α-regime 216
CCP discharge γ-regime 216
cell attachment on polymer scaffolds 448
cell elongation 336
cell in VBNC state 334
cell membrane integrity 332
cell proliferation on polymer scaffolds 448
cellular respiration 84, 115
cementum 430
chain oxidation in clusters 358
chain polymerization 457
channel model of arc 198
charged particle effect in sterilization 304
Child law 191
chlorine-containing VOC 370
cholesterol 90, 98–100, 138, 146–7, 155, 395
chronic wounds 405
cleaning by wet pulsed corona 369
Clostridium botulinum 337
coagulation cascade 392

coaxial flow-stabilized arc 201
coaxial-hollow micro-DBD 240
collisionless sheath 70
colony-forming unit, CFU 294
combined oxidation of NOx and SO2 in air 360
completely ionized plasma 5
complete thermodynamic equilibrium, CTE 48
condensation reaction 82
contact angle 472
continuous corona 232
corona 232, 272, 275–7
cotton treatment 479
Creutzfeldt-Jakob diseases 303
critical electron density 78
Crohn's disease 419
cross-section of ionization 21
current-voltage characteristics of DC discharges 180
current-voltage characteristics of the cathode layer 182

dark Townsend discharge 180
DBD array 239
Debye length 271
Debye radius 54, 69
Debye shielding 69
decontamination of air 339
deep tissue penetration 327
deformability of erythrocytes 395
Deinococcus radiodurans 295, 302
dental caries 432
dental pulp 430
dentin 429
dentistry 428
deoxyribonucleic acid (DNA) 93
depth of polymer treatment 464
dermatology 419
destruction and removal efficiency, DRE 365
diabetes 407
diabetic ulcer 407
dielectric-barrier discharge, DBD 234
dielectric barrier grating discharge, DBGD 343
dielectric permittivity of plasma 75
diffusion of electrons 54
diffusion of ions 54
diluted emissions 370
dimethyl sulfide removal from air 364
direct ionization by electron impact 20
direct plasma treatment 313
disaccharide 86–7, 100, 147
discharge controlled by diffusion 176
discharge controlled by electron attachment 175
discharge controlled by electron-ion recombination 175

discharge in liquid 253
disease activity index, DAI 422
disease treatment 403
disinfection of water 378
dissociation equilibrium 46
dissociation of hydrocarbons 455
dissociation of molecules 34
dissociative electron attachment 26
dissociative electron-ion recombination 25
distribution of energy between different channels of excitation 35
Ditylum brightwelli 379
Druyvesteyn distribution 60
D-value 294

E. coli 274–7, 279–80, 336, 339, 381
ECR-discharge 227
efficiency α of excitation energy 40
Einstein relation 54
electric breakdown 165
electric field water disinfection 378
electrocautery 16
electrodynamics of plasma 68
electromagnetic field penetration into plasma 72
electromagnetic waves in plasma 77
electron attachment 26
electron beams 23
electron conductivity 52
electron detachment 26
electron detachment by excited particles 28
electron drift in plasma 52
electronegativity 83, 120
electron emission 20, 29
electron energy distribution function, EEDF 20
electron energy losses 23
electronically stabilized DBD 238
electronic excitation 33
electronic relaxation 38
electron impact detachment 28
electron-ion recombination 24
electrons 19
electron's effect in sterilization 304
electrons in treatment of polymers 467
electron temperature 20
electroporation 272
electrostatic influence 270–71
electrostatic water breakdown 256
electrostriction water breakdown 256
Elenbaas-Heller equation 196
emission induced by metastable atoms 31
enamel 429

Engel-Steenbeck relation 176
Engel-Steenbeck theory of cathode layer 182
epithelialization and remodeling 404
equilibrium constant 50
Escherichia coli 320
etching of polymers 466
etching rate of polymers 467
ethane 81
ethylene diamine tetraacetic acid 392
eukaryotes 97, 101, 103–4, 109, 112–14, 119, 124, 139
excited species 31
explanted tumors 426
extremophile organism deactivation 301

Faraday dark space 178
fast reactions of vibrationally excited molecules 67
fatty acid 88–90, 114, 117
Fenton reaction 119, 305
Fermi energy 30
ferroelectric discharge 235
fibrin filament formation 392
fibrinogen 392
fibroblasts 108, 122, 127, 132
field emission 30, 192
filling adhesion 433
filtration of air streams 339
floating-electrode DBD discharge, FE-DBD 241
floating potential 71
flow cytometry 344
flow-stabilized linear arcs 201
fluorination of polymers 470
Fokker-Planck kinetic equation 58, 61
foot ulcer 407
forward contamination 302
forward-vortex stabilization 207
Foucault currents 209
Fowler-Nordheim formula 30
free-burning linear arcs 200
Fridman-Macheret α-model 40
fused hollow cathode discharge 240

gamma radiolysis 322
gas discharge 9
gastroenterology 419
generalized Ohm's law 73
generation of plasma 19
germ cell tumor 423
G-factors 35
gingivitis 431
gliding arc 204
glow discharge 177

glutathione 123
glycerol 88–9, 113, 147
glycolysis 112–14, 134
graft polymerization 458
grey, Gy 322

Hall effect 74
HDP discharge 220
healing of diseases 403
heat-flux potential 197
helical resonator discharge 223
helicon discharge 229
helium microplasma jets 250
Helmholtz free energy 49
hematoxylin and eosin procedure 322
hemostatic phase 403
heterogeneous ionization 29
heterogeneous polymerization 456
high-density lipoprotein (HDL) cholesterol 395
high-energy electrons 23
high-frequency plasma conductivity 75
high-pressure arc discharges 193
histology 323
histone 101, 281
hollow-cathode APG microplasma 191
hollow-cathode discharge 191
homeostasis 96, 122–3, 130, 132, 151, 155
hormones 87, 90, 122, 126, 128, 132, 135–6, 148–50, 155–7
hot thermionic cathode arcs 192
hydration 82
hydrocarbons 81, 84, 88
hydrogen peroxide 102, 105, 114, 117, 119, 276–7, 279, 283, 285–6
hydrogen peroxide biochemistry 305, 311
hydrolysis 82
hydrophilicity of medical polymers 448
hydrophilic properties 473
hydrophobic properties 473
hydroxyl radical 116, 118, 121, 270, 274, 277, 283, 308
hypoxia 404

ICP coupling 213
ICP microplasma 227
ideally conducting streamer channel 172
ideal plasma 68
immobilization of catalysts 459
immunogens 139
inactivation of spores 297
indirect plasma treatment 313
inductively-coupled plasma, ICP 208

inflammatory bowel diseases, IBD 419
inflammatory dysfunctions 418
inflammatory phase 403
initiation of polymerization 455
internal energy 49
ion–atomic charge transfer processes 44
ion conversion 25
ion drift in plasma 52
ion-ion recombination 29
ionization by collisions of heavy particles 20
ionization equilibrium 48
ionization potential 21
ionization processes 19
ion-molecular polarization collisions 43
ion-molecular reactions 43, 45

Joule heating 52

keratinocytes 131
kinase 95, 111, 122–4, 283–5
kinetics of air sterilization 345
kinetics of electronic excitation 65
kinetics of non-equilibrium plasma 58
kinetics of plasma polymerization 453
kinetics of plasma sterilization 295
Krebs cycle 112–14, 160

Lactobacillus casei 432
Landau-Teller formula 37
Langerhan cells 131–2
Langevin cross-section 44
Langevin polarization capture 43
Langevin scattering 43
Langmuir frequency 72
Langmuir plasma oscillations 72
laparoscopic operations 410
leg ulcer 407
Le Roy formula 39
leukemia 423
Lichtenberg figures 237
Lidsky hollow cathode 191
lipid peroxidation 272, 290–92
lipids 84, 87–90, 111, 117, 120–21, 127, 133, 138, 155
liquid-phase chain oxidation of SO2 355
liquid plasma without bubbles 261
living tissue sterilization 320
load line 180
local thermodynamic equilibrium 48
long-lifetime resonance 32
lotus effect 479
low-density lipoprotein, LDL 395

low-pressure arc discharges 193
low-pressure CCP discharge 217
lymphoma 423

macrophages 130–31, 133, 139, 143–4
magnetically stabilized arcs 204
magnetic field frozen in plasma 74
magnetic pressure 74
magnetic Reynolds number 73
magneto-hydrodynamics of plasma 73
magnetron discharge 219
malignant cells 424
Massey parameter 24
maturation phase 404
Maxwell-Boltzmann distribution function 20, 59
mean electron energy 20
medical polymers 447
Meek criterion 171
melanocytes 131–2
memory effect 173
mercury removal from streams 376
mercury removal mechanism 377
Merkel cells 131
mesenchymal stem cells, MSC 452
metallic cylinder model 209
metastasis 423
methane 81, 83
methanol removal from air 364
microdischarge remnants 173
micro glow-discharge 188
microsecond pulsed FE-DBD 243
microwave discharges 208, 210
microwave microdischarge 213
micro-xerography of bioactive liquid 453
Milliken-White formula 37
mitochondria 84, 97, 100–101, 103–4, 110, 113–15,
    118–20, 134, 142, 144, 270, 272
MMA polymerization 457
mobility of electrons 52
moderate-pressure CCP discharge 215
moderate-pressure microwave discharge 230
modification of wettability 472
modified Meek criterion 188
monomers 81–2, 144
monosaccharide 85–7, 147
multi-phase survival curve 296
multi-quantum Fridman approximation 32
multi-temperature plasma 7

N-acetylcysteine (NAC) 123, 282–3, 285
nanosecond pulsed breakdown of liquids 261

nanosecond pulsed FE-DBD 243
Navier-Stokes equation 73
necrosis 320
negative corona 233
negative glow 178
negative ions 19
negative streamer 170
neutral droplets 357
nicotinamide adenine dinucleotide (NAD) 93, 95, 112
nitrogenation of polymers 469
nitrogen dioxide 119–21, 280
nitric oxide 105, 116, 119, 120–21, 134, 280
non-damaging tissue 322
non-equilibrium plasma 7
non-equilibrium statistics 50
non-equilibrium statistics of vibrational excitation 51
non-ideal plasma 68
non-polar bonds 83
non-resonant charge transfer 45
non-thermal plasma 8
non-transferred arcs 202
normal glow discharge 184
NO therapy 407
nucleic acid 84, 93, 95, 111

OAUGDP discharge 237
obstructed arc 200
obstructed glow 185
Ohm's law 73
optically thin transparent plasma 57
optical thickness of plasma 57
oscillations of plasma 72
oxidation of polymers 469
oxidation reaction 84
oxidation state 84, 120
ozone 269, 274, 276, 279–80, 308
ozone water disinfection 378

packed bed discharge 235
paper mill exhaust 362
Paschen curve 165, 167
pathogen detection and remediation facility, 339
Pauling 83–4, 120
PCR technique 337
Peek formula 233
penetrative plasma sterilization 317
Penning ionization 24
peptide bond 92, 94
peroxyl radicals 310
peroxynitrate 306
peroxynitrite 119–21

peroxy radicals 117, 120
pharmacology 435
phosphorylation (phosphorylate) 95, 111–2, 114, 122–4,
    281, 283–6, 289, 292
photodynamic therapy 424
photoelectron emission 31
photo-ionization 20
pin-to-hole discharge, PHD 247
planar-coil ICP discharge 222
Planck radiation intensity 57
plasma-assisted blood coagulation 389
plasma bullets 250
plasma catalysis 46
plasma chemistry 13
plasma control of blood rheology 400
plasma dentistry 428
plasma-induced stress 336
plasma in liquid 253
plasma interaction with cells 14
plasma interaction with living tissues 14
plasma needle 226
plasma pharmacology 435
plasma polarization 69
plasma-polymer interaction 461
plasma polymerization 453
plasma polymerized films 460
plasma statistics 46
plasma sterilization 293
plasma source 9
plasma surgery 433
plasma thermodynamics 46
plasma torch 202
plasma transfer processes 46
plasma-treated NAC 444
plasma-treated PBS 444
plasma-treated water 444
plasma treatment of polymers 461
platelet activation 134–5, 392
platelets 128, 133–5
Plazon device 409
polar bonds 83
poly-ε-caprolactone (PCL) scaffolds 448
polyethylene plasma treatment 464
polymer adhesion enhancement 474
polymer fibers 475
polymerized film electric properties 460
polymer macro-particles 459
polymer membranes 475
polymer processing 447
polymer scaffolds 448
polypeptides, 84, 92, 100

polysaccharide 87, 100, 148
positive column of arc 196
positive column of glow discharge 177, 186
positive corona 233
positive ions 19
positive streamer 169
potential electron emission 31
presheath 71
pressure ulcer 407
prion proteins 303
prokaryotes 97, 101, 103–4, 109, 114, 123
proliferation of osteoblasts 449
proliferative (repair) phase 404
propagation of electromagnetic waves 77
prothrombin 393
pseudomonas aeuroginosa 296
pulmonology 418
pulsed atmospheric-pressure plasma streams, PAPS 253
pulsed corona 232
purulent peritonitis 419

quantum yield 31
quasi-neutral plasma 19
quasi-repulsion of microdischarges 173
quasi-self-sustained streamers 171
quenching of excited species 36

radiation dose 322
radiation energy transfer 56
radiative electron-ion recombination 26
radical shower 361
radio-frequency discharges 208
rate coefficient of direct ionization 21
rate coefficient of stepwise ionization 22
reactions of excited ions 39
reactions of excited neutrals 39
reactions of vibrationally excited molecules 66
reactive nitrogen species, RNS 119–20, 308
reactive oxygen species, ROS 104, 116, 118, 308
rearrangement of chemical bonds 46
recombination component of continuous radiation 56
redox processes 84, 111
reduced electric field 60
reflection of electromagnetic waves 78
relativistic electron beams 23
relaxation of excited species 36
relaxation of vibrational excitation 36
resistive-barrier discharge 188
resonant charge transfer 44
reverse contamination 302
reverse vortex flow 207

rheological properties of blood 395
ribonucleic acid (RNA) 93
root canals 432
ROS in air sterilization 347
rotational excitation 33
rotational relaxation 38
Rutherford formula 21

saccharides 85
Saha equation 25, 46
Salmonella 337
saphenous vein 391
sarcoma 423
Schottky effect 29, 193
secondary electron emission 30
secondary ion-electron emission 30
sheaths 69
short-lifetime resonance 32
silent discharge 234
single-phase survival curve 295
singlet oxygen 117–20, 308
skin damage 403
skin effect 72
skin flora sterilization 322
sliding discharges 236
sliding surface corona 236
sliding surface spark 236
slow reactions of vibrationally excited molecules 67
sodium citrate 392
sodium heparin 392
Sommerfeld formula 29
SO2 oxidation using continuous corona 354
SO2 oxidation using pulsed corona 354
SO2 oxidation using relativistic electron beams 354
spark breakdown 168
spark discharge 243
spores deactivation 294
spores deactivation in closed envelopes 299
Staphylococcus aureus (*Staph*) 142, 337
Staphylococcus epidermidis 142, 443
statistical distribution 46
steady-state discharge regimes 165, 174
Steenbeck minimum power principle 184
Steenbeck–Raizer 'channel' model 198
Stefan–Boltzmann law 58
stem cells control 451
stepwise ionization 20
sterilization active species 304
sterilization by electric field 307
sterilization by heat 311

sterilization by ion bombardment 307
sterilization by low-pressure plasma 293
sterilization by UV-radiation 312
sterilization of living tissue 293
sterilization of surfaces 293
sterilization of water 378
steroids 88, 90, 155
stochastic heating effect 217
stratum corneum 131
streamer breakdown 168
streamer discharges 232
Streptococci 142
Streptococcus mutans 430
stress induced by plasma 336
subnormal glow discharge 184
superoxide 115–22
superoxide biochemistry 305
superoxide dismutase, SOD 119, 305
superoxide dismutation 305
surface discharge 235
surface ionization 20
surface microorganism inactivation 295
surface plasma sterilization 317
surface-wave discharge 229
surgery 433
surgical instrument decontamination 303
surgical wounds 405
survival curve 294
synergy in plasma-polymer treatment 471
synthetic textiles treatment 479

tertiary structure (of protein) 92
theoretical-informational approach 39
thermal breakdown 254
thermal conductivity in plasma 55
thermal plasma 8
thermionic emission 29, 192
thermodynamic functions 49
thermodynamic probability 47
thin polymer films 453
Thomson model 21
three-body electron attachment 27
three-body electron-ion recombination 25
thrombin 393
tissue deep penetration 325
tissue engineering 447
tissue model 324
tissue regeneration 451
tooth structure 429
tornado flow 207
Townsend breakdown 165

Townsend coefficient α 165
Townsend coefficient β 167
Townsend coefficient γ 166
Townsend formula 166
toxicity analysis 322
transferred arcs 201
transpiration-stabilized arc 201
Treanor distribution 51
Treanor effect in thermal conductivity 55
treatment of wounds 403
Trichel pulses 234
triglyceride 88–9, 395
Trypticase soy agar 337
TUNEL assay 425

ulcerative colitis 419
Unsold-Kramers formula 57
UV-absorber 384
UV in treatment of polymers 467
UV-water disinfection 378

vacuum arc 193
VBNC status 331
venous ulceration 406
viable-but-not-culturable, VBNC 331
viability 302, 331
vibrational distribution functions 64
vibrational excitation 31
vibrational quantum 31
vibronic terms 40
vitamin 90, 111, 123, 132, 149

VOC emissions 361
volatile organic compounds, VOC 361
voltaic arc 192
vortex-stabilized arc 202
VT-flux in energy space 62
VT-relaxation 37
VV-flux in energy space 62
VV-relaxation 37

wall-stabilized linear arcs 200
water breakdown 254
water decontamination 339
water ozonation 379
water treatment 435
water treatment by pulsed plasma 379
water treatment by pulsed spark 380
weakly ionized plasma 5
weak vibrational excitation regime 64
wet pulsed corona 369
wettability of medical polymers 448
whole blood viscosity, WBV 397
wood processing exhausts 362
wool treatment 477
work function 29
wound healing 403

xenographted cancer cell 427
xerography of bioactive liquid 453
XTT-assay 335

zygote 98